"In the Working Guides to Estimating and Forecasting Alan has managed to capture the full spectrum of relevant topics with simple explanations, practical examples and academic rigor, while injecting humour into the narrative."

– *Dale Shermon*, Chairman, Society of Cost Analysis and Forecasting (SCAF)

"If estimating has always baffled you, this innovative well illustrated and user friendly book will prove a revelation to its mysteries. To confidently forecast, minimise risk and reduce uncertainty we need full disclosure into the science and art of estimating. Thankfully, and at long last the "Working Guides to Estimating & Forecasting" are exactly that, full of practical examples giving clarity, understanding and validity to the techniques. These are comprehensive step by step guides in understanding the principles of estimating using experientially based models to analyse the most appropriate, repeatable, transparent and credible outcomes. Each of the five volumes affords a valuable tool for both corporate reference and an outstanding practical resource for the teaching and training of this elusive and complex subject. I wish I had access to such a thorough reference when I started in this discipline over 15 years ago, I am looking forward to adding this to my library and using it with my team."

– *Tracey L Clavell*, Head of Estimating & Pricing, BAE Systems Australia

"At last, a comprehensive compendium on these engineering math subjects, essential to both the new and established "cost engineer"! As expected the subjects are presented with the author's usual wit and humour on complex and daunting "mathematically challenging" subjects. As a professional trainer within the MOD Cost Engineering community trying to embed this into my students, I will be recommending this series of books as essential bedtime reading."

– *Steve Baker*, Senior Cost Engineer, DE&S MOD

"Alan has been a highly regarded member of the Cost Estimating and forecasting profession for several years. He is well known for an ability to reduce difficult topics and cost estimating methods down to something that is easily digested. As a master of this communication he would most often be found providing training across the cost estimating and forecasting tools and at all levels of expertise. With this 5-volume set, *Working Guides to Estimating and Forecasting*, Alan has brought his normal verbal training method into a written form. Within their covers Alan steers away from the usual dry academic script into establishing an almost 1:1 relationship with the reader. For my money a recommendable read for all levels of the Cost Estimating and forecasting profession and those who simply want to understand what is in the 'blackbox' just a bit more."

– *Prof Robert Mills*, Margin Engineering, Birmingham City University.
MACOSTE, SCAF, ICEAA

"Finally, a book to fill the gap in cost estimating and forecasting! Although other publications exist in this field, they tend to be light on detail whilst also failing to cover many of the essential aspects of estimating and forecasting. Jones covers all this and more from both a theoretical and practical point of view, regularly drawing on his considerable experience in the defence industry to provide many practical examples to support his comments. Heavily illustrated throughout, and often presented in a humorous fashion, this is a must read for those who want to understand the importance of cost estimating within the broader field of project management."

— *Dr Paul Blackwell*, Lecturer in Management of Projects,
The University of Manchester, UK

"Alan Jones provides a useful guidebook and navigation aid for those entering the field of estimating as well as an overview for more experienced practitioners. His humorous asides supplement a thorough explanation of techniques to liven up and illuminate an area which has little attention in the literature, yet is the basis of robust project planning and successful delivery. Alan's talent for explaining the complicated science and art of estimating in practical terms is testament to his knowledge of the subject and to his experience in teaching and training."

— *Therese Lawlor-Wright*, Principal Lecturer in Project Management at the
University of Cumbria

"Alan Jones has created an in depth guide to estimating and forecasting that I have not seen historically. Anyone wishing to improve their awareness in this field should read this and learn from the best."

— *Richard Robinson*, Technical Principal for Estimating, Mott MacDonald

"The book series of 'Working Guides to Estimating and Forecasting' is an essential read for students, academics and practitioners who interested in developing a good understanding of cost estimating and forecasting from real-life perspectives."

— *Professor Essam Shehab*, Professor of Digital Manufacturing and
Head of Cost Engineering, Cranfield University, UK

"In creating the *Working Guides to Estimating and Forecasting*, Alan has captured the core approaches and techniques required to deliver robust and reliable estimates in a single series. Some of the concepts can be challenging, however, Alan has delivered them to the reader in a very accessible way that supports lifelong learning. Whether you are an apprentice, academic or a seasoned professional, these working guides will enhance your ability to understand the alternative approaches to generating a well-executed, defensible estimate, increasing your ability to support competitive advantage in your organisation."

— *Professor Andrew Langridge*, Royal Academy of Engineering
Visiting Professor in Whole Life Cost Engineering and
Cost Data Management, University of Bath, UK

"Alan Jones's "*Working Guides to Estimating and Forecasting*" provides an excellent guide for all levels of cost estimators from the new to the highly experienced. Not only does he cover the underpinning good practice for the field, his books will take you on a journey from cost estimating basics through to how estimating should be used in manufacturing the future — reflecting on a whole life cycle approach. He has written a must-read book for anyone starting cost estimating as well as for those who have been doing estimates for years. Read this book and learn from one of the best."

— *Linda Newnes*, Professor of Cost Engineering, University of Bath, UK

Best Fit Lines and Curves, and Some Mathe-Magical Transformations

Best Fit Lines and Curves, and Some Mathe-Magical Transformations (Volume III of the Working Guides to Estimating & Forecasting series) concentrates on techniques for finding the Best Fit Line or Curve to some historical data allowing us to interpolate or extrapolate the implied relationship that will underpin our prediction. A range of simple 'Moving Measures' are suggested to smooth the underlying trend and quantify the degree of noise or scatter around that trend. The advantages and disadvantages are discussed and a simple way to offset the latent disadvantage of most Moving Measure Techniques is provided.

Simple Linear Regression Analysis, a more formal numerical technique that calculates the line of best fit subject to defined 'goodness of fit' criteria. Microsoft Excel is used to demonstrate how to decide whether the line of best fit is a good fit, or just a solution in search of some data. These principles are then extended to cover multiple cost drivers, and how we can use them to quantify 3-Point Estimates.

With a deft sleight of hand, certain commonly occurring families of non-linear relationships can be transformed mathe-magically into linear formats, allowing us to exploit the powers of Regression Analysis to find the Best Fit Curves. The concludes with an exploration of the ups and downs of seasonal data (Time Series Analysis). Supported by a wealth of figures and tables, this is a valuable resource for estimators, engineers, accountants, project risk specialists as well as students of cost engineering.

Alan Jones is Principal Consultant at Estimata Limited, an estimating consultancy service. He is a Certified Cost Estimator/Analyst (US) and Certified Cost Engineer (CCE) (UK). Prior to setting up his own business, he has enjoyed a 40-year career in the UK aerospace and defence industry as an estimator, culminating in the role of Chief Estimator at BAE Systems. Alan is a Fellow of the Association of Cost Engineers and a Member of the International Cost Estimating and Analysis Association. Historically (some four decades ago), Alan was a graduate in Mathematics from Imperial College of Science and Technology in London, and was an MBA Prize-winner at the Henley Management College (. . . that was slightly more recent, being only two decades ago). Oh, how time flies when you are enjoying yourself.

Working Guides to Estimating & Forecasting

Alan R. Jones

As engineering and construction projects get bigger, more ambitious and increasingly complex, the ability of organisations to work with realistic estimates of cost, risk or schedule has become fundamental. Working with estimates requires technical and mathematical skills from the estimator but it also requires an understanding of the processes, the constraints and the context by those making investment and planning decisions. You can only forecast the future with confidence if you understand the limitations of your forecast.

The Working Guides to Estimating & Forecasting introduce, explain and illustrate the variety and breadth of numerical techniques and models that are commonly used to build estimates. Alan Jones defines the formulae that underpin many of the techniques; offers justification and explanations for those whose job it is to interpret the estimates; advice on pitfalls and shortcomings; and worked examples. These are often tabular in form to allow you to reproduce the examples in Microsoft Excel. Graphical or pictorial figures are also frequently used to draw attention to particular points as the author advocates that you should always draw a picture before and after analysing data.

The five volumes in the Series provide expert applied advice for estimators, engineers, accountants, project risk specialists as well as students of cost engineering, based on the author's thirty-something years' experience as an estimator, project planner and controller.

Volume I Principles, Process and Practice of Professional Number Juggling
Alan R. Jones

Volume II Probability, Statistics and Other Frightening Stuff
Alan R. Jones

Volume III Best Fit Lines and Curves, and Some Mathe-Magical Transformations
Alan R. Jones

Volume IV Learning, Unlearning and Re-learning Curves
Alan R. Jones

Volume V Risk, Opportunity, Uncertainty and Other Random Models
Alan R. Jones

Best Fit Lines and Curves, and Some Mathe-Magical Transformations

Alan R. Jones

LONDON AND NEW YORK

First published 2019
by Routledge
2 Park Square, Milton Park, Abingdon, Oxon OX14 4RN

and by Routledge
711 Third Avenue, New York, NY 10017

Routledge is an imprint of the Taylor & Francis Group, an informa business

© 2019 Alan R. Jones

The right of Alan Jones to be identified as authors of this work has been asserted by him in accordance with sections 77 and 78 of the Copyright, Designs and Patents Act 1988.

All rights reserved. No part of this book may be reprinted or reproduced or utilised in any form or by any electronic, mechanical, or other means, now known or hereafter invented, including photocopying and recording, or in any information storage or retrieval system, without permission in writing from the publishers.

Trademark notice: Product or corporate names may be trademarks or registered trademarks, and are used only for identification and explanation without intent to infringe.

British Library Cataloguing-in-Publication Data
A catalogue record for this book is available from the British Library

Library of Congress Cataloging-in-Publication Data
Names: Jones, Alan (Alan R.), 1953- author.
Title: Best fit lines and curves : and some mathe-magical transformations / Alan Jones.
Description: Abingdon, Oxon ; New York, NY : Routledge, 2018. |
 Series: Working guides to estimating & forecasting ; volume 3 |
 Includes bibliographical references and index.
Identifiers: LCCN 2017059102 (print) | LCCN 2018000657 (ebook) |
 ISBN 9781315160085 (eBook) | ISBN 9781138065000 (hardback : alk. paper)
Subjects: LCSH: Industrial engineering—Statistical methods. | Regression analysis. |
 Costs, Industrial—Estimates. | Costs, Industrial—Statistical methods.
Classification: LCC T57.35 (ebook) | LCC T57.35 .J66 2018 (print) |
 DDC 519.5/6—dc23
LC record available at https://lccn.loc.gov/2017059102

ISBN: 9781138065000 (hbk)
ISBN: 9781315160085 (ebk)

Typeset in Bembo
by Apex CoVantage, LLC

To my family:
Lynda, Martin, Gareth and Karl
Thank you for your support and forbearance, and for understanding
why I wanted to do this.

My thanks also to my friends and former colleagues at BAE Systems and the wider
Estimating Community for allowing me the opportunity to learn, develop and practice
my profession, . . . and for suffering my brand of humour over the years.
In particular, a special thanks to Tracey C, Mike C, Mick P and Andy L for your
support, encouragement and wise counsel. (You know who you are!)

Contents

List of Figures	xv
List of Tables	xxi
Foreword	xxxi

1 Introduction and objectives 1

1.1 Why write this book? Who might find it useful? Why five volumes? 1

    1.1.1 Why write this series? Who might find it useful? 1

    1.1.2 Why five volumes? 2

1.2 Features you'll find in this book and others in this series 2

    1.2.1 Chapter context 3

    1.2.2 The lighter side (humour) 3

    1.2.3 Quotations 3

    1.2.4 Definitions 3

    1.2.5 Discussions and explanations with a mathematical
slant for Formula-philes 4

    1.2.6 Discussions and explanations without a mathematical
slant for Formula-phobes 5

    1.2.7 Caveat augur 5

    1.2.8 Worked examples 6

    1.2.9 Useful Microsoft Excel functions and facilities 6

    1.2.10 References to authoritative sources 7

    1.2.11 Chapter reviews 7

1.3 Overview of chapters in this volume 7

1.4 Elsewhere in the 'Working Guide to Estimating & Forecasting' series 8

    1.4.1 *Volume I: Principles, Process and Practice of Professional
Number Juggling* 9

    1.4.2 *Volume II: Probability, Statistics and Other Frightening Stuff* 10

    1.4.3 *Volume III: Best Fit Lines and Curves, and
Some Mathe-Magical Transformations* 11

Contents

1.4.4 *Volume IV: Learning, Unlearning and Re-learning curves*	11
1.4.5 *Volume V: Risk, Opportunity, Uncertainty and Other Random Models*	12
1.5 Final thoughts and musings on this volume and series	13
References	14

2 Linear and nonlinear properties (!) of straight lines — **15**

2.1 Basic linear properties	15
2.1.1 Inter-relation between slope and intercept	18
2.1.2 The difference between two straight lines is a straight line	19
2.2 The Cumulative Value (nonlinear) property of a linear sequence	21
2.2.1 The Cumulative Value of a Discrete Linear Function	21
2.2.2 The Cumulative Value of a Continuous Linear Function	26
2.2.3 Exploiting the Quadratic Cumulative Value of a straight line	34
2.3 Chapter review	43
Reference	44

3 Trendsetting with some Simple Moving Measures — **45**

3.1 Going all trendy: The could and the should	45
3.1.1 When should we consider trend smoothing?	45
3.1.2 When is trend smoothing not appropriate?	47
3.2 Moving Averages	48
3.2.1 Use of Moving Averages	49
3.2.2 When not to use Moving Averages	49
3.2.3 Simple Moving Average	50
3.2.4 Weighted Moving Average	54
3.2.5 Choice of Moving Average Interval: Is there a better way than guessing?	58
3.2.6 Can we take the Moving Average of a Moving Average?	66
3.2.7 A creative use for Moving Averages – A case of forward thinking	68
3.2.8 Dealing with missing data	70
3.2.9 Uncertainty Range around the Moving Average	71
3.3 Moving Medians	81
3.3.1 Choosing the Moving Median Interval	83
3.3.2 Dealing with missing data	84
3.3.3 Uncertainty Range around the Moving Median	84
3.4 Other Moving Measures of Central Tendency	85
3.4.1 Moving Geometric Mean	87
3.4.2 Moving Harmonic Mean	87
3.4.3 Moving Mode	88
3.5 Exponential Smoothing	89
3.5.1 An unfortunate dichotomy	89

Contents | xi

3.5.2 Choice of Smoothing Constant, or Choice of
Damping Factor — 92
3.5.3 Uses for Exponential Smoothing — 94
3.5.4 Double and Triple Exponential Smoothing — 95
3.6 Cumulative Average and Cumulative Smoothing — 96
3.6.1 Use of Cumulative Averages — 97
3.6.2 Dealing with missing data — 101
3.6.3 Cumulative Averages with batch data — 103
3.6.4 Being slightly more creative – Cumulative Average
on a sliding scale — 103
3.6.5 Cumulative Smoothing — 105
3.7 Chapter review — 110
References — 112

4 Simple and Multiple Linear Regression — **113**
4.1 What is Regression Analysis? — 113
4.1.1 Least Squares Best Fit — 115
4.1.2 Two key sum-to-zero properties of Least Squares — 120
4.2 Simple Linear Regression — 122
4.2.1 Simple Linear Regression using basic Excel functions — 123
4.2.2 Simple Linear Regression using the Data Analysis
Add-in Tool Kit in Excel — 125
4.2.3 Simple Linear Regression using advanced Excel functions — 127
4.3 Multiple Linear Regression — 129
4.3.1 Using categorical data in Multiple Linear Regression — 131
4.3.2 Multiple Linear Regression using the Data Analysis
Add-in Tool Kit in Excel — 133
4.3.3 Multiple Linear Regression using advanced Excel functions — 136
4.4 Dealing with Outliers in Regression Analysis? — 138
4.5 How good is our Regression? Six key measures — 140
4.5.1 Coefficient of Determination (R-Square):
A measure of linearity?! — 141
4.5.2 F-Statistic: A measure of chance occurrence — 149
4.5.3 t-Statistics: Measures of Relevance or Significant Contribution — 156
4.5.4 Regression through the origin — 162
4.5.5 Role of common sense as a measure of goodness of fit — 171
4.5.6 Coefficient of Variation as a measure of tightness of fit — 172
4.5.7 White's Test for heteroscedasticity . . . and,
by default, homoscedasticity — 174
4.6 Prediction and Confidence Intervals – Measures of uncertainty — 179
4.6.1 Prediction Intervals and Confidence Intervals:
What's the difference? — 180

xii | Contents

4.6.2 Calculating Prediction Limits and Confidence
    Limits for Simple Linear Regression   182
4.6.3 Calculating Prediction Limits and Confidence
    Limits for Multi-Linear Regression   185
4.7 Stepwise Regression   193
    4.7.1 Backward Elimination   197
    4.7.2 Forward Selection   201
    4.7.3 Backward or Forward Selection – Which should we use?   206
    4.7.4 Choosing the best model when we are spoilt for choice   208
4.8 Chapter review   209
References   210

5 Linear transformation: Making bent lines straight   **211**
5.1 Logarithms   212
    5.1.1 Basic properties of powers   213
    5.1.2 Basic properties of logarithms   216
5.2 Basic linear transformation: Four Standard Function types   222
    5.2.1 Linear functions   223
    5.2.2 Logarithmic Functions   225
    5.2.3 Exponential Functions   230
    5.2.4 Power Functions   233
    5.2.5 Transforming with Microsoft Excel   237
    5.2.6 Is the transformation really better, or just a
    mathematical sleight of hand?   242
5.3 Advanced linear transformation: Generalised Function types   244
    5.3.1 Transforming Generalised Logarithmic Functions   245
    5.3.2 Transforming Generalised Exponential Functions   249
    5.3.3 Transforming Generalised Power Functions   250
    5.3.4 Reciprocal Functions – Special cases of Generalised
    Power Functions   253
    5.3.5 Transformation options   254
5.4 Finding the Best Fit Offset Constant   257
    5.4.1 Transforming Generalised Function Types into
    Standard Functions   259
    5.4.2 Using the Random-Start Bisection Method (Technique)   260
    5.4.3 Using Microsoft Excel's Goal Seek or Solver   263
5.5 Straightening out Earned Value Analysis . . . or EVM Disintegration   271
    5.5.1 EVM terminology   271
    5.5.2 Taking a simpler perspective   274
5.6 Linear transformation based on Cumulative Value Disaggregation   279
5.7 Chapter review   281
References   283

Contents | xiii

6 Transforming Nonlinear Regression — **284**

6.1 Simple Linear Regression of a linear transformation — 284
  6.1.1 Simple Linear Regression with a Logarithmic Function — 288
  6.1.2 Simple Linear Regression with an Exponential Function — 291
  6.1.3 Simple Linear Regression with a Power Function — 298
  6.1.4 Reversing the transformation of Logarithmic, Exponential and Power Functions — 299

6.2 Multiple Linear Regression of a multi-linear transformation — 300
  6.2.1 Multi-linear Regression using linear and linearised Logarithmic Functions — 302
  6.2.2 Multi-Linear Regression using linearised Exponential and Power Functions — 312

6.3 Stepwise Regression and multi-linear transformations — 323
  6.3.1 Stepwise Regression by Backward Elimination with linear transformations — 323
  6.3.2 Stepwise Regression by Forward Selection with linear transformations — 330

6.4 Is the Best Fit really the better fit? — 333

6.5 Regression of Transformed Generalised Nonlinear Functions — 337
  6.5.1 Linear Regression of a Transformed Generalised Logarithmic Function — 342
  6.5.2 Linear Regression of a Transformed Generalised Exponential Function — 348
  6.5.3 Linear Regression of a Transformed Generalised Power Function — 351
  6.5.4 Generalised Function transformations: Avoiding the pitfalls and tripwires — 357

6.6 Pseudo Multi-linear Regression of Polynomial Functions — 359
  6.6.1 Offset Quadratic Regression of the Cumulative of a straight line — 361
  6.6.2 Example of a questionable Cubic Regression of three linear variables — 368

6.7 Chapter review — 378
References — 379

7 Least Squares Nonlinear Curve Fitting without the logs — **380**

7.1 Curve Fitting by Least Squares ... without the logarithms — 381
  7.1.1 Fitting data to Discrete Probability Distributions — 381
  7.1.2 Fitting data to Continuous Probability Distributions — 391
  7.1.3 Revisiting the Gamma Distribution Regression — 399

7.2 Chapter review — 406
Reference — 406

xiv | Contents

8 The ups and downs of Time Series Analysis — **407**

8.1 The bits and bats ... and buts of a Time Series — 408

    8.1.1 Conducting a Time Series Analysis — 411

8.2 Alternative Time Series Models — 411

    8.2.1 Additive/Subtractive Time Series Model — 412

    8.2.2 Multiplicative Time Series Model — 413

8.3 Classical Decomposition: Determining the underlying trend — 415

    8.3.1 See-Saw ... Regression flaw? — 416

    8.3.2 Moving Average Seasonal Smoothing — 420

    8.3.3 Cumulative Average Seasonal Smoothing — 422

    8.3.4 What happens when our world is not perfect?
Do any of these trends work? — 424

    8.3.5 Exponential trends and seasonal funnels — 430

    8.3.6 Meandering trends — 436

8.4 Determining the seasonal variations by
Classical Decomposition — 437

    8.4.1 The Additive/Subtractive Model — 438

    8.4.2 The Multiplicative Model — 440

8.5 Multi-Linear Regression: A holistic approach to
Time Series? — 443

    8.5.1 The Additive/Subtractive Linear Model — 444

    8.5.2 The Additive/Subtractive Exponential Model — 449

    8.5.3 The Multiplicative Linear Model — 452

    8.5.4 The Multiplicative Exponential Model — 456

    8.5.5 Multi-Linear Regression: Reviewing the options to
make an informed decision — 460

8.6 Excel Solver technique for Time Series Analysis — 461

    8.6.1 The Perfect World scenario — 462

    8.6.2 The Real World scenario — 465

    8.6.3 Wider examples of the Solver technique — 468

8.7 Chapter review — 468

Reference — 469

Glossary of estimating and forecasting terms — 470
Legend for Microsoft Excel Worked Example Tables in Greyscale — 489
Index — 491

Figures

1.1	Principal Flow of Prior Topic Knowledge Between Volumes	9
2.1	Examples of Lines of Imperfect Fit	17
2.2	Slope as a Function of the Intercepts	18
2.3	Example – Cumulative Enquiries Received Following Marketing Campaign	24
2.4	Example – Weekly Enquiries Received Following Marketing Campaign	25
2.5	Straight Line Discrete Values Expressed as Trapezial Areas Under the Line	28
2.6	Cumulative of a Discrete Value Straight Line as a Trapezial Area Under the Line	28
2.7	Equating Cumulative Values for Discrete and Continuous Straight Lines	29
2.8	Example – Using the Quadratic Formula to Forecast the Cumulative Value	31
2.9	Continuous Linear Functions – The Reporting Paradox	32
2.10	Continuous Linear Functions – The Reporting Paradox Revisited	33
2.11	Example of Cumulative Forecasting Using a Quadratic Trend	35
2.12	Straight Line or Shallow Curve?	39
2.13	Finding the Cumulative Recovery Point to a Linear Plan (1)	41
2.14	Finding the Cumulative Recovery point to a Linear Plan (2)	42
3.1	Simple Moving Average with an Interval of 3	51
3.2	Simple Moving Averages with Increasing Intervals	52
3.3	Example of Simple Moving Average Lagging the True Trend	53
3.4	Weighted Moving Average cf. Simple Moving Average Before Lag Adjustment	56
3.5	Weighted Moving Average cf. Simple Moving Average After Lag Adjustment	57
3.6	Example of Moving Average Interval Difference (MAID)	59
3.7	MAID Minimal Standard Deviation Technique	62
3.8	Moving Average Interval Options Using the MAID Minimal Standard Deviation Technique	62
3.9	Data with Increasing Trend to be Smoothed	64
3.10	Results of the MAID Minimal Standard Deviation Technique with an Increasing Trend	65

xvi | Figures

3.11 Moving Average Interval Options Using the MAID
Minimal Standard Deviation Technique   66
3.12 Moving Average Plot with Missing Data   70
3.13 Moving Average Uncertainty Using Minima and Maxima   73
3.14 Moving Average Uncertainty Using Standard Deviations   77
3.15 Moving Minima and Maxima with Upward and Downward Trends   79
3.16 Moving Standard Deviation with Upward or Downward Trends   80
3.17 Moving Median Compared with Moving Average (Unadjusted for Lag)   82
3.18 Moving Median can Disregard Important Seasonal Data   82
3.19 Moving Median with Moving 10th and 90th Percentiles as
Confidence Limits   86
3.20 Moving Harmonic Mean of Performance cf. Arithmetic Mean   88
3.21 Exponential Smoothing – Effect of Using Different Parameters   93
3.22 Exponential Smoothing – Non-Time-Based Data   94
3.23 Exponential Smoothing – Long Lags with Small Smoothing Constants   95
3.24 Cumulative Average Hours Over Time   99
3.25 Cumulative Average Hours Over Units Completed   99
3.26 Cumulative Average Response to Change in Trend   100
3.27 Cumulative Average with Incomplete Historical Records   103
3.28 Cumulative Average with Partial Missing Data   104
3.29 Cumulative Average Equivalent Unit Costs   106
3.30 When Moving Averages and Cumulative Averages do not Appear to Help   107
3.31 Cumulative Smoothing where Moving Averages and
Cumulative Averages Fail   109
3.32 Cumulative Smoothing where Moving Averages and
Cumulative Averages Fail – Revisited   111
4.1 Determining the Line of Best Fit by Least Squares   115
4.2 Derivation of Line of Best Fit   118
4.3 Line of Best Fit Defined by Least Squares Error cf. Minimum Absolute Error   119
4.4 Error Distribution   119
4.5 Correlation and Covariance in Relation to the Line of Best Fit   123
4.6 Data Input Using Microsoft Excel's Data Analysis Regression Add-In   126
4.7 Multi-Linear Regression Input Example   134
4.8 Preliminary Regression Analysis to Detect Presence of Outliers   139
4.9 Explained and Unexplained Variation in Linear Regression   143
4.10 Adjusted R-Square Decreases as the Number of Independent
Variables Increases   147
4.11 F-Distribution as a Comparison of Explained to Unexplained Variation   150
4.12 Example F-Distribution CDF   151
4.13 Example of a Questionable Linear Regression   155
4.14 Example of a Questionable Linear Regression   156
4.15 Example of a Supportable Linear Regression   158

Figures | xvii

4.16	Example of an Unsupportable Multi-Linear Regression – Insignificant Parameter	161
4.17	Example of a Linear Regression with an Intercept Close to Zero	164
4.18	Example of a Linear Regression with an Intercept Constrained to Zero	164
4.19	Regression Through the Origin Changes the Basis of Measuring the Goodness of Fit	165
4.20	Example of a Linear Regression with a Natural Intercept of Zero	168
4.21	Linear Regression with a Natural Intercept of Zero	168
4.22	Homoscedastic and Heteroscedastic Error Distribution	175
4.23	Prediction and Confidence Intervals	183
4.24	Calculation of Prediction and Confidence Intervals in Microsoft Excel	184
4.25	Calculation of Prediction and Confidence Intervals in Microsoft Excel Using Array Formulae	191
5.1	Slide Rules Exploit the Power of Logarithms	212
5.2	Impact of Taking Logarithms on Small and Large Numbers	221
5.3	Taking Logarithms of a Linear Function	224
5.4	Examples of Increasing and Decreasing Linear Functions	225
5.5	Taking Logarithms of a Logarithmic Function	226
5.6	Examples of Increasing and Decreasing Logarithmic Function	228
5.7	Taking Logarithms of an Exponential Function	230
5.8	Examples of Increasing and Decreasing Exponential Functions	232
5.9	Taking Logarithms of a Power Function	234
5.10	Examples of Increasing and Decreasing Power Functions	235
5.11	Basic Power Function Shapes Change Depending on Power/Slope Parameter	236
5.12	Examples of Alternative Trendlines Through Linear Data	238
5.13	Examples of Alternative Trendlines Through Logarithmic Data	238
5.14	Examples of Alternative Trendlines Through Exponential Data	239
5.15	Examples of Alternative Trendlines Through Power Data	239
5.16	Summary of Logarithmic-Based Linear Transformations	241
5.17	Generalised Increasing Logarithmic Functions	246
5.18	Generalised Decreasing Logarithmic Functions	246
5.19	Generalised Increasing Logarithmic Functions	247
5.20	Generalised Decreasing Logarithmic Functions	248
5.21	Generalised Increasing Exponential Functions	249
5.22	Generalised Decreasing Exponential Functions	250
5.23	Generalised Increasing Power Functions with Like-Signed Offsets Constants	251
5.24	Generalised Increasing Power Functions with Unlike-Signed Offset Constants	251
5.25	Generalised Decreasing Power Functions with Like-Signed Offset Constants	252
5.26	Generalised Decreasing Power Functions with Unlike-Signed Offset Constants	253
5.27	Reciprocal-x as a Special Case of the Generalised Power Function	254

xviii | Figures

5.28	Reciprocal-y as a Special Case of the Generalised Power Function	255
5.29	Finding the Maximum R2 where the SSE does not Converge to a Minimum	258
5.30	Transforming a Generalised Exponential Function to a Standard Exponential Function	259
5.31	Finding an Appropriate Offset Adjustment Using the Random-Start Bisection Method	261
5.32	Example of Offset Adjustment Approximation Using the Random-Start Bisection Method	261
5.33	Using Microsoft Excel Solver to Find the Best Fit Offset Adjustment for a Generalised Logarithmic Function	265
5.34	Finding an Appropriate Offset Adjustment Using Solver (Result)	265
5.35	3-in-1 Solver Graphical Sensibility Check (Before and After)	268
5.36	Lazy S-Curves Typical of EVM Cost and Schedule Performance	272
5.37	Earned Value Management Terminology	274
5.38	EVM Cost Performance Trend	275
5.39	EVM Schedule Performance Trend	278
5.40	EVM Cost and Schedule Outturn Based on Current Trends	278
5.41	Cumulative Design Effort in Response to Design Queries (Perfect World)	280
5.42	Cumulative Design Effort in Response to Design Queries (Real World)	281
6.1	Linear Regression of a Transformed Logarithmic Function	288
6.2	Prediction Interval Around the Linear Regression of a Transformed Logarithmic Function	290
6.3	Linear Regression of a Transformed Exponential Function	291
6.4	Linear Regression of a Transformed Exponential Function Constrained Through Unity	293
6.5	Prediction Interval Around the Linear Regression of a Transformed Exponential Function	294
6.6	Linear Regression of a Transformed Power Function	298
6.7	Natural Pairings of Function Types for Linear Regression	301
6.8	Multicollinearity Between Two Predicator Variables	304
6.9	Indicative Tanker Costs Using Deadweight Tonnage	308
6.10	Regression Slope Parameter – Student's t-Distribution or Normal Distribution	309
6.11	Using an Exponential Function as a "Dimmer Switch" in Multiple Linear Regression	315
6.12	Linear Cost-Weight Regression Model	335
6.13	Power Log(Cost)-Log(Weight) Regression Model	335
6.14	Derivation of Regression Standard Error in Microsoft Excel	336
6.15	3-in-1 Solver Graphical Sensibility Check (Before)	341
6.16	3-in-1 Solver Graphical Sensibility Check (After)	341
6.17	3-in-1 Solver Graphical Sensibility Check for a Generalised Logarithmic Function	343

6.18	Comparison of Generalised Logarithmic Trendline with a Generalised Power Trendline	345
6.19	Extrapolating Standard and Generalised Logarithmic Trendlines	346
6.20	Diverging Prediction Intervals Around Standard and Generalised Logarithmic Functions	347
6.21	3-in-1 Solver Graphical Sensibility Check for a Generalised Exponential Function	349
6.22	Generalised Increasing Power Functions with Unlike-Signed Offset Constants	354
6.23	Comparison of Forward Projection of Alternative Regression Results for Example 3	357
6.24	Using a Cubic Trendline Instead of a Logarithmic Trendline	360
6.25	Extrapolating a Cubic Trendline in Comparison with a Logarithmic Trendline	360
6.26	Extrapolating a Quartic Trendline in Comparison with a Logarithmic Trendline	361
6.27	Unconstrained Quadratic Trendline Example	369
6.28	Gross Tonnage cf. Product of Length, Beam and Depth	370
6.29	Ship Length cf. Beam	371
6.30	Ship depth cf. Length and Beam	371
6.31	Gross Tonnage as a Cubic Function of Ship's Beam (Pre-Regression)	372
6.32	Gross Tonnage as a Cubic Function of Ship's Length (Pre-Regression)	373
6.33	Gross Tonnage as a Cubic Polynomial Regression of the Ship's Beam with Prediction Interval	377
6.34	Gross Tonnage as Power Function Regression of the Ship's Beam with Prediction Interval	378
7.1	Theoretical and Observed Discrete Uniform Distribution	382
7.2	Theoretical and Observed Cumulative Discrete Uniform Distribution	383
7.3	Best Fit Normal Distribution to Golf Tournament Round Scores (Two Rounds)	389
7.4	Normal Distribution of Golf Scores (Two Rounds – All Competitors)	389
7.5	Best Fit Normal Distribution to Golf Tournament Round Scores (Top Half)	390
7.6	Normal Distribution of Golf Scores (Four Rounds – Top Competitors)	391
7.7	Normal Distribution of Golf Scores (Two Rounds – Eliminated Competitors)	392
7.8	Random Rounds from Normal Golfers	392
7.9	Solver Result for Fitting a Continuous Uniform Distribution to Observed Data	395
7.10	Solver Result for Fitting a Normal Distribution to Observed Data	396
7.11	Solver Result for Fitting a Beta Distribution to Observed Data	397
7.12	Solver Result for Fitting a Triangular Distribution to Observed Data	398
7.13	Solver Result for fitting a Gamma Distribution to Observed Data	400
7.14	Solver Result for Fitting a Gamma Distribution to Observed Data on Queue Lengths	401

xx | Figures

8.1	Example of a Time Series with Seasonal Variation and Beginnings of a Cyclical Impact	408
8.2	Example of a Quarterly Time Series	410
8.3	Example Additive/Subtractive Time Series Model – Domestic Gas Consumption	413
8.4	Example Multiplicative Time Series Model – Domestic Gas Consumption	414
8.5	Time Series with Perfect Seasonal Variation Around a Linear Trend	416
8.6	Linear Regression Through Perfect Time Series Additive/Subtractive Model	418
8.7	Linear Regression Through Perfect Time Series Additive/Subtractive Model – 2	418
8.8	Time Series Linear Regression Trends are Unreliable – 1	419
8.9	Time Series Linear Regression Trends are Unreliable – 2	419
8.10	Offset Moving Average Trend of a Perfect Time Series	421
8.11	Cumulative Average Smoothing of a Perfect Time Series – 1	423
8.12	Cumulative Average Smoothing of a Perfect Time Series – 2	423
8.13	Cumulative Smoothing of a Perfect Time Series	424
8.14	Underlying Trend as the Average of the Individual Seasonal Trends	425
8.15	Underlying Trend Based on an Offset Moving Average	427
8.16	Underlying Trend Based on an Offset Cumulative Average	427
8.17	Seasonal Funnelling Effect Around an Exponential Trend	431
8.18	Determining the Exponential Trend Using an Offset Cumulative Geometric Mean (1)	433
8.19	Determining the Exponential Trend Using an Offset Moving Geometric Mean	434
8.20	Determining the Exponential Trend Using an Offset Cumulative Geometric Mean (2)	435
8.21	Splicing a Decreasing Linear Trend with a Steady State Flat Trend	438
8.22	Completed Linear Time Series Model with Additive/Subtractive Seasonal Variation (1)	440
8.23	Completed Exponential Time Series Model with Multiplicative Seasonality Factors (1)	442
8.24	Completed Linear Time Series Model with Additive/Subtractive Seasonal Variation (2)	449
8.25	Completed Exponential Time Series Model with Additive/Subtractive Seasonal Variation	451
8.26	Completed Linear Time Series Model with Multiplicative Seasonality Factors	458
8.27	Completed Exponential Time Series Model with Multiplicative Seasonality Factors (2)	459
8.28	Solver Time Series Set-Up with Initial Parameter Starting Values	463
8.29	Solver Results for Our Perfect World scenario	464
8.30	Solver Time Series Set-Up with Initial Parameter Starting Values for Gas Consumption	466
8.31	Solver Time Series Output with Optimised Parameter Values for Gas Consumption	467

Tables

2.1	Example – Enquiries Received Following Marketing Campaign	23
2.2	Example – Using the Quadratic Formula to Forecast the Cumulative Value	26
2.3	Example – Using the Quadratic Formula and Equivalent Units to Forecast the Cumulative Value	32
2.4	Example Revisited with Steady State Equivalent Unit Completions	33
2.5	Example of Cumulative Forecasting Using a Quadratic Trend	35
2.6	Linear Trend Recovery Point Using a Cumulative Quadratic Trend	40
3.1	Going all Trendy – The Could and the Should	46
3.2	Simple Moving Average with an Interval of 3	51
3.3	Sliding Scale of Weights for Weighted Moving Average of Interval 5	55
3.4	Moving Average Trend Lags	58
3.5	Choosing a Moving Average Interval Using the MAID Minimal Standard Deviation Technique	61
3.6	Moving Average Interval Options Using the MAID Minimal Standard Deviation Technique	63
3.7	Applying the MAID Minimal Standard Deviation Technique to Data with an Increasing Trend	64
3.8	Moving Average Interval Options Using the MAID Minimal Standard Deviation Technique	65
3.9	Moving Average of a Moving Average	67
3.10	Potential Measures of Moving Average Uncertainty	71
3.11	Moving Average Uncertainty Using Minima and Maxima	74
3.12	Moving Average Uncertainty Using Factored Standard Deviation	78
3.13	Confidence Intervals in Relation to Standard Deviations Around the Mean	79
3.14	Potential Measures of Moving Median Uncertainty	86

xxii | Tables

3.15	Exponential Smoothing – Effect of Using Different Parameters	93
3.16	Cumulative Average Over Time and Over Number of Units Completed	98
3.17	Cumulative Average with Incomplete Historical Records	102
3.18	Cumulative Averages with Partial Missing Data	104
3.19	Cumulative Average Equivalent Unit Costs	105
3.20	When Moving Averages and Cumulative Averages do not Appear to Help	107
3.21	When Moving Averages and Cumulative Averages do not Appear to Help – Revisited	110
3.22	Moving Average Steady State – Applying the Offset Lag	111
4.1	Determining the Line of Best Fit by Least Squares	118
4.2	Two Sum-to-Zero Properties of Least Squares Errors	121
4.3	Determining the Line of Best Fit Using Microsoft Excel's Data Analysis Add-In	127
4.4	Determining the Line of Best Fit Using Microsoft Excel's Advanced Functions	129
4.5	LINEST Output Data for a Simple Linear Regression	129
4.6	Example 1 of Using Binary Switched for Multi-Value Categorical Data	132
4.7	Example 2 of Using Binary Switched for Multi-Value Categorical Data	132
4.8	Regression will Eliminate any Redundant Dummy Variable	133
4.9	Determining the Plane of Best Fit Using Microsoft Excel's Data Analysis Add-in	135
4.10	Regression Residuals	135
4.11	Determining the Plane of Best Fit Using Microsoft Excel's Advanced Functions	137
4.12	LINEST Output Data for a Multiple Linear Regression with Two Independent Variables	137
4.13	Example Regression Outlier Test Based on Regression Residuals	140
4.14	Example of Analysis of Variance Sum of Squares	146
4.15	Example 1 – Impact of an Additional Variable on R-Square and Adjusted R-Square	148
4.16	Example 2 – Impact of an Additional Variable on R-Square and Adjusted R-Square	149
4.17	Two Examples of the Significance of the F-Statistic	153
4.18	Example Excel Output for a Questionable Linear Regression	157
4.19	Example Excel Output for a Supportable Linear Regression	158
4.20	Implications of Null Hypothesis that the Parameter Value may be Zero	160
4.21	Regression of a Secondary Driver Only	162
4.22	Regression Through the Origin Changes the Basis of Measuring the Goodness of Fit	166
4.23	Example of a Linear Regression with a Natural Intercept of Zero but Constrained to Zero	169

4.24	Comparison of a Regression Through the Origin with a Natural Regression Through the Origin	171
4.25	Coefficient of Variation as a Measure of Tightness of Fit	173
4.26	Variables Required for White's Test	176
4.27	Example of White's Test – Auxiliary Regression Input Data	177
4.28	Example of White's Test Output Data and Test Result	178
4.29	Calculation of Prediction and Confidence Intervals in Microsoft Excel	183
4.30	Input Array in Microsoft Excel for Prediction and Confidence Intervals	187
4.31	Simple Linear Regression Prediction and Confidence Intervals Using Array Formulae	191
4.32	Simple Linear Regression Prediction and Confidence Intervals Using Array Formulae	192
4.33	Example Data for Stepwise Regression	194
4.34	Unconstrained Simple Linear Regression Based on Total Mileage	195
4.35	Simple Linear Regression Through the Origin Based on Total Mileage	196
4.36	Correlation Matrix for Example Stepwise Regression	197
4.37	Example Backward Elimination – Step 1	197
4.38	Example Backward Elimination – Step 1 Revisited	198
4.39	Example Backward Elimination – Step 2	199
4.40	Example Backward Elimination – Step 3 (Intercept = 0)	200
4.41	Ranking of Independent Data's Likely Contribution to the Model for Trip Hours – Step 1	202
4.42	Example of Forward Selection – Step 2 (Highest Ranked Variable)	202
4.43	Example of Forward Selection – Step 3 (Two Highest Ranked Variables)	202
4.44	Example of Forward Selection – Step 4 (Constrained Through the Origin)	203
4.45	Example of Forward Selection – Step 4 (3 Highest Ranked Variables with Intercept=0)	204
4.46	Example of Forward Selection – Step 5 (One Step Back One Step Forward)	205
4.47	Example of Forward Selection – Step 6 (Last Potential Variable Added)	205
4.48	Comparison of Stepwise Regression Results by Forward Selection and Backward Elimination	206
4.49	Comparison of Regression Residuals Using Forward Selection and Backward Elimination	207
4.50	Advantages and Disadvantages of Forward Selection and Backward Elimination Approaches	208
5.1	Example of the Additive Property of Logarithms	218
5.2	Example of the Multiplicative Property of Logarithms	219
5.3	Example of the Reciprocal Property of Logarithms	220
5.4	Comparing Linear and Power Trendline Errors in Linear Space (1)	243
5.5	Comparing Linear and Power Trendline Errors in Linear Space (2)	244
5.6	Examples of Alternative Standard Functions that might Approximate Generalised Functions	255

xxiv | Tables

5.7	Non-Harmonisation of Minimum SSE with Maximum R2 Over a Range of Offset Values	258
5.8	Using Solver to Maximise the R-Square of Three Generalised Functions – Set-up	266
5.9	Using Solver to Maximise the R-Square of Three Generalised Functions	268
5.10	Key Component Elements of Earned Value Analysis	275
5.11	Potential Response to EVM Cost Performance Trends	277
5.12	Design Effort in Response to Design Queries (Perfect World)	280
5.13	Design Effort in Response to Design Queries (Real World)	281
5.14	Function Type Transformation Summary	282
6.1	Basic Function Types	285
6.2	Means Through which the Nonlinear Regression Passes	286
6.3	Arithmetic Mean of Logarithmic Values Equals the Geometric Mean of the Linear Values	287
6.4	Linear Regression of a Transformed Logarithmic Function	289
6.5	Prediction Interval Around the Linear Regression of a Transformed Logarithmic Function	289
6.6	Linear Regression of a Transformed Exponential Function	292
6.7	Reversing the Logarithmic Transformation of the Regression Coefficients	292
6.8	Prediction Interval Around the Linear Regression of a Transformed Exponential Function	294
6.9	LOGEST Output Data for a Simple Exponential Regression	297
6.10	A Comparison on LOGEST Output with LINEST Output on the Transformed Data	297
6.11	Linear Regression of a Transformed Power Function	298
6.12	Prediction Interval Around the Linear Regression of a Transformed Power Function	299
6.13	Reversing the Transformation of the Standard Functions	300
6.14	Example of a Multi-Linear Regression Using Two Logarithmic Function Transformations	304
6.15	Multi-linear Regression Output with Two Logarithmic Function Transformations	305
6.16	Multi-Linear Regression Sensitivity Analysis Using Prediction Intervals	306
6.17	Tanker Costs and Deadweights	307
6.18	Combined Tanker Regression Model with Linear and Logarithmic Function Parameters	310
6.19	Combined Tanker Regression Model Constrained Through the Origin	311
6.20	Using a "Dimmer Switch" as a Variable Category	313
6.21	Multi-Linear Regression of Multiple Transformed Exponential Functions (1)	314
6.22	Multi-Linear Regression of Multiple Transformed Exponential Functions (2)	314
6.23	Multi-Linear Regression of Multiple Transformed Power Functions	318
6.24	Multi-Linear Regression of Multiple Transformed Power Functions	318

6.25	90% Prediction Interval for a Regression Based on Multiple Power Functions (1)	319
6.26	90% Prediction Interval for a Regression Based on Multiple Power Functions (2)	320
6.27	Combined Exponential and Power Function Regression Input Data	322
6.28	Combined Exponential and Power Function Regression Output Data	322
6.29	Step 1 – Calculate R-Square for all Candidate Variables and their Logarithms	324
6.30	Step 2 – Rank the R-Squares for all Candidate Variables and their Logarithms	325
6.31	Step 4 – Initial Regression Input Variables	326
6.32	Step 4 – Initial Regression Output Report – Rejecting 'Log(x1 Weight)'	326
6.33	Step 5 – Regression Output Report Iteration 2 Rejecting 'x2 Prior Units'	327
6.34	Step 5 – Regression Output Report Iteration 3 – All Variables Significant	328
6.35	Step 6 – Regression Independent Variable Correlation Matrix	329
6.36	Step 7 – Regression Output Report Iteration 4 – All Variables Significant	330
6.37	Forward Regression Output Report Iteration 1 – Best Single Variable	331
6.38	Forward Regression Output Report Iteration 2	331
6.39	Forward Regression Output Report Iteration 4	332
6.40	Forward Regression Output Report Iteration 5	332
6.41	Statistic to Use as the Basis of Comparing Alternative Best Fit models	334
6.42	Derivation of the Standard Error in a Linear Space for a Power Function Regression	337
6.43	Using Solver to Maximise the R-Square of Three Generalised Functions – Set-Up	340
6.44	Using Solver to Maximise the R-Square of Three Generalised Functions – Result	341
6.45	Using 3-in-1 Solver to Identify a Generalised Logarithmic Function	342
6.46	Using 3-in-1 Solver to Identify a Generalised Logarithmic Function	344
6.47	Using 3-in-1 Solver to Identify a Generalised Logarithmic Function	347
6.48	3-in-1 Solver Set-Up for a Generalised Exponential Function Example	348
6.49	3-in-1 Solver Results for the Generalised Exponential Function Example	349
6.50	Solver Regression Results for the Generalised Exponential Function Example	350
6.51	3-in-1 Solver Set-Up for a Generalised Power Function – Example 1	352
6.52	3-in-1 Solver Result for a Generalised Power Function – Example 1	353
6.53	3-in-1 Solver Regression Result for a Generalised Power Function – Example 1	353
6.54	3-in-1 Solver Set-Up for a Generalised Power Function – Example 2	354

xxvi | Tables

6.55	3-in-1 Solver Result for a Generalised Power Function – Example 3	355
6.56	3-in-1 Solver Regression Result for a Generalised Power Function – Example 3	355
6.57	Regression Result Expected for Offset Generalised Power Function – Example 3	356
6.58	Comparison of Forward Projection of Alternative Regression Results for Example 3	356
6.59	Hot Spots where Generalised Functions can be Misinterpreted as Standard Functions	358
6.60	Solver Setup to Determine the Relative Start Time Offset for the Cumulative of a Straight Line	362
6.61	Solver Solution for the Relative Start Time Offset for the Cumulative of a Straight Line	363
6.62	Solver Solution for the Relative Start Time Offset for the Cumulative of a straight Line	364
6.63	Solver Solution for the Relative Start Time Offset for the Cumulative of a Straight Line	364
6.64	Quadratic Polynomial Function Regression Output Through the Origin	365
6.65	Quadratic Polynomial Function Regression Output Rejecting the Intercept	366
6.66	Forecast Data from the Output of a Quadratic Polynomial Function Regression	366
6.67	Ship Dimensions and Gross Tonnage Based on an Internet Search	369
6.68	Regression Input Data for Cubic Polynomial Function Regression	374
6.69	Regression Output for Gross Tonnage as a Cubic Polynomial Function of a Ship's Beam (1)	375
6.70	Regression Output for Gross Tonnage as a Cubic Polynomial Function of a Ship's Beam (2)	375
6.71	Regression Output for Gross Tonnage as a Cubic Polynomial Function of a Ship's Beam (3)	376
6.72	Regression Output for Gross Tonnage as a Power Function of a Ships' Beam	377
7.1	Casting Doubts? Anomalous Quartile Values of a Die	382
7.2	Resolving Doubts? Quartile Values of a Die Re-Cast	384
7.3	Solver Set-Up for Fitting a Discrete Uniform Distribution to Observed Data	385
7.4	Solver Results for Fitting a Discrete Uniform Distribution to Observed Data	386
7.5	Solver Results for Fitting a Normal Distribution to Discrete Scores at Golf (Two rounds)	388
7.6	Solver results for Fitting a Normal Distribution to Discrete Scores at Golf (Top half)	390

7.7	Solver Set-Up for Fitting a Continuous Uniform Distribution to Observed Data	394
7.8	Solver Result for Fitting a Continuous Uniform Distribution to Observed Data	394
7.9	Solver Result for Fitting a Normal Distribution to Observed Data	396
7.10	Solver Result for Fitting a Beta Distribution to Observed Data	397
7.11	Solver Result for Fitting a Triangular Distribution to Observed Data	398
7.12	Solver Result for Fitting a Gamma Distribution to Observed Data	399
7.13	Solver Results for Fitting a Gamma Distribution to Observed Data on Queue Lengths	401
7.14	Regression Input Data Preparation Highlighting Terms to be Omitted	403
7.15	Regression Output Data for Queue Length Data Modelled as a Gamma Function	403
7.16	Interpretation of Regression Coefficients as Gamma Function Parameters	404
7.17	Solver Results for Fitting a Gamma Distribution Using Restricted Queue Length Data	405
7.18	Revised Regression Output Data for Queue Length Data Modelled as a Gamma Function	405
7.19	Interpretation of the Revised Regression Coefficients as Gamma Function Parameters	406
8.1	Time Series with Perfect Seasonal Variation Around a Linear Trend	417
8.2	Range of Regression Results for Slope and Intercept	420
8.3	Linear Regression Lines of Best Fit for Each Season	420
8.4	Offset Moving Average Trend of a Perfect Time Series	421
8.5	Cumulative Average Trend of a Perfect Time Series	422
8.6	Underlying Linear Trend as the Average of the Individual Seasonal Trends	425
8.7	Example of Domestic Gas Consumption – Underlying Linear Trend	426
8.8	Two-Year Forward Forecast Based on Alternative Underlying Linear Trends	428
8.9	Residual Variation of Alternative Underlying Linear Trends Relative to Actual Values	429
8.10	Time Series with Perfect Seasonality Factors Around an Exponential Trend	431
8.11	Determining the Exponential Trend Using an Offset Cumulative Geometric Mean	432
8.12	Underlying Exponential Trend as a Function of the Individual Seasonal Trends	433
8.13	Gas Consumption Example with an Underlying Exponential Trend	434
8.14	Underlying Exponential Trend of Gas Consumed in Relation to the Individual Seasonal Trends	435
8.15	Two-Year Forward Forecast Based on Alternative Underlying Exponential Trends	436

xxviii | Tables

8.16	Residual Variation of Alternative Underlying Exponential Trends Relative to Actual Values	437
8.17	Calculation of Additive/Subtractive Seasonal Variations Using Classical Decomposition	439
8.18	Completed Time Series Model with Additive/Subtractive Seasonal Variations	439
8.19	Calculation of Multiplicative Seasonal Variations Using Classical Decomposition	441
8.20	Completed Time Series Model with Multiplicative Seasonal Variations	442
8.21	Multi-Linear Regression Input Data – Additive/Subtractive Time Series Model	444
8.22	Multi-Linear Regression Output Report – All Parameters Selected	446
8.23	Multi-Linear (Stepwise) Regression Output Report – Excluding Variable x1	446
8.24	Multi-Linear Regression Output Report – Enforcing an Intercept of Zero	447
8.25	Multi-Linear Regression Output Report Comparison	448
8.26	Modelled Data and Two-Year Forecast of Gas Consumption Using Multi-Linear Regression	448
8.27	Transformed Exponential Regression Input Data – Additive/Subtractive Time Series Model	450
8.28	Transformed Exponential Regression Output Report – Enforcing an Intercept of Zero	451
8.29	Two-Year Forecast of Gas Consumption Using Seasonally Adjusted Exponential Regression	452
8.30	Regression Output for Multiplicative Linear Regression Model for Domestic Gas Consumption	453
8.31	Input Data for Multiplicative Linear Regression Model for Domestic Gas Consumption	454
8.32	Stepwise Decomposition – Regression Input for Seasonality Factors in a Multiplicative Model	455
8.33	Stepwise Decomposition – Regression Output for Seasonality Factors in a Multiplicative Model	457
8.34	Stepwise Decomposition – Two-Year Forward Forecast	457
8.35	Stepwise Decomposition – Regression Input for Seasonality Factors in a Multiplicative Model	458
8.36	Stepwise Decomposition – Regression Output for Seasonality Factors in a Multiplicative Model	459
8.37	Stepwise Decomposition – Two-Year Forward Forecast	460
8.38	Comparison of Stepwise Decomposition Regression Results	461
8.39	Solver Set-Up for Perfect Additive/Subtractive Seasonal Variation Around a Linear Trend	462

8.40	Solver Result for Perfect Additive/Subtractive Seasonal Variation Around a Linear Trend	464
8.41	Solver Set-Up for Domestic Gas Consumption Time Series	465
8.42	Solver Result for Domestic Gas Consumption Time Series	466
8.43	Equivalence Between Solver Technique and Stepwise Decomposition Regression	467
8.44	Times Series Trend Analysis – Combination of Regression and Various Moving Measures	469

Foreword to the *Working Guides to Estimating and Forecasting* series

At long last, a book that will support you throughout your career as an estimator and any other career where you need to manipulate, analyse and, more importantly, make decisions using your results. Do not be concerned that the book consists of five volumes as the book is organised into five distinct sections. Whether you are an absolute beginner or an experienced estimator there will be something for you in these books!

Volume One provides the reader with the core underpinning good-practice required when estimating. Many books miss out the need for auditability of your process, clarity of your approach and the techniques you have used. Here, Alan Jones guides you on presenting the basis of your estimate, ensuring you can justify your decisions, evidence these and most of all ensure you keep the focus and understand and focus on the purpose of the model. By the end of this volume you will know how to use, for example, e.g. factors and ratios to support data normalisation and how to evidence qualitative judgement. The next volume then leads you through the realm of probability and statistics. This will be useful for Undergraduate students through to experienced professional engineers. The purpose of Volume Two is to ensure the reader *understands* the techniques they will be using as well as identifying whether the relationships are statistically significant. By the end of this volume you will be able to analyse data, use the appropriate statistical techniques and be able to determine whether a data point is an outlier or not. Alan then leads us into methods to assist us in presenting non-linear relationships as linear relationships. He presents examples and illustrations for single linear relationships to multi-linear dimensions. Here you do need to have a grasp of the mathematics and the examples and key points highlighted throughout the volumes ensure you can. By the end of this volume you will really grasp best-fit lines and curves.

After Volume Three the focus moves to other influences on your estimates. Volume Four brings out the concept of learning curves – as well as unlearning curves! Throughout this volume you will start with the science behind learning curves but unlike other books, you will get the whole picture. What happens across shared projects and learning, what happens if you have a break in production and have to restart learning. This volume

xxxii | Foreword

covers the breadth of scenarios that may occur and more importantly how to build these into your estimation process. In my view covering the various types of learning and reflecting these back to real life scenarios is the big win. As stated many authors focus on learning curves and assume a certain pattern of behaviour. Alan provides you with options, explains these and guides you on how to use them.

The final volume tackles risk and uncertainty. Naturally Monte-Carlo simulation is introduced and a guide on really understanding what you are doing. One of the real winners here is some clear hints on guidance on good practice and what to avoid doing. To finalise the book, Alan reflects on the future of Manufacturing where this encompasses the whole life cycle. From his background in Aerospace he can demonstrate the need for critical path in design, manufacture and support along with schedule risk. By considering uncertainty in combination with queueing theory, especially in the spares and repairs domain, Alan demonstrates how the build-up of knowledge from the five volumes can be used to estimate and optimise the whole lifecycle costs of a product and combined services.

I have been waiting for this book to be published for a while and I am grateful for all the work Alan has undertaken to provide what I believe to be a seminal piece of work on the mathematical techniques and methods required to become a great cost estimator. My advice would be for every University Library and every cost estimating team (and beyond) to buy this book. It will serve you through your whole career.

Linda Newnes
Professor of Cost Engineering
Department of Mechanical Engineering
University of Bath
BA2 7AY

1 Introduction and objectives

This series of books aspires to be a practical reference guide to a range of numerical techniques and models that an estimator might wish to consider in analysing historical data in order to forecast the future. Many of the examples and techniques discussed relate to cost estimating in some way, as the term estimator is frequently used synonymously to mean Cost estimator. However, many of these numerical or quantitative techniques can be applied in other areas other than cost where estimating is required, such as scheduling, or to determine a forecast of physical, such as weight, length or some other technical parameter.

This is the 'trendy' volume in the set of five. It concentrates of determining and using the underlining trend or pattern within the data available to create an estimate or forecast through interpolation or extrapolation. Our journey will consider both linear and non-linear trends, and how with a bit of mathe-magical transformation we can sometimes covert a curved trend into a straight line or linear trend.

1.1 Why write this book? Who might find it useful? Why five volumes?

1.1.1 Why write this series? Who might find it useful?

The intended audience is quite broad, ranging from the relative 'novice' who is embarking on a career as a professional estimator, to those already seasoned in the science and dark arts of estimating. Somewhere between these two extremes of experience, there will be some who just want to know what tips and techniques they can use, to those who really want to understand the theory of why some things work and other things don't. As a consequence, the style of this book is aimed to attract and provide signposts to both (and all those in between).

This series of books is not just aimed at cost estimators (although there is a natural bias there.) There may be some useful tips and techniques for other number jugglers, in

2 | Introduction and objectives

which we might include other professionals like engineers or accountants who estimate but do not consider themselves to be estimators *per se*. Also, in using the term 'estimator', we should not constrain our thinking to those whose estimate's output currency is cost or hours, but also those who estimate in different 'currencies', such as time and physical dimensions or some other technical characteristics.

Finally, in the process of writing this series of guides, it has been a personal voyage of discovery, cathartic even, reminding me of some of the things I once knew but seem to have forgotten or mislaid somewhere along the way. Also, in researching the content, I have discovered many things that I didn't know and now wish I had known years ago when I started on my career, having fallen into it, rather than having chosen it. (*Does that sound familiar to other estimators?*)

1.1.2 Why five volumes?

There are two reasons:

> Size . . . there was too much material for the single printed volume that was originally planned . . . *and that might have made it too much of a heavy reading so to speak.* That brings out another point, the attempt at humour will remain around that level throughout.
> Cost . . . even if it had been produced as a single volume (printed or electronic), the cost may have proved to be prohibitive without a mortgage, and the project would then have been unviable.

So, a decision was made to offer it as a set of five volumes, such that each volume could be purchased and read independently of the others. There is cross-referencing between the volumes, just in case any of us want to dig a little deeper, but by and large the fives volumes can be read independently of each other. There is a common glossary of terms across the five volumes which covers terminology that is defined and assumed throughout. This was considered to be essential in setting the right context, as there are many different interpretations of some words in common use in estimating circles. Regrettably, there is a lack of common understanding by what these terms mean, so the glossary clarifies what is meant in this series of volumes.

1.2 Features you'll find in this book and others in this series

People's appetites for practical knowledge varies from the 'How do I . . . ?' to the 'Why does that work?' This book will attempt to cater for all tastes.

Many text books are written quite formally, using the third person which can give a feeling of remoteness. In this book, the style used is in first person plural, 'we' and 'us'.

Introduction and objectives | 3

Hopefully this will give the sense that this is a journey on which we are embarking together, and that you, the reader, are not alone, especially when it gets to the tricky bits! On that point, let's look at some of the features in this series of *Working Guides to Estimating and Forecasting*.

1.2.1 Chapter context

Perhaps unsurprisingly, each chapter commences with a very short dialogue about what we are trying to achieve or the purpose of that chapter, and sometimes we might include an outline of a scenario or problem we are trying to address.

1.2.2 The lighter side (humour)

There are some who think that an estimator with a sense of humour is an oxymoron. (*Not true, it's what keeps us sane.*) Experience gleaned from developing and delivering training for Estimators has highlighted that people learn better if they are enjoying themselves. We will discover little 'asides' here and there, sometimes at random but usually in italics, to try and keep the attention levels up. (*You're not falling asleep already, are you?*) In other cases, the humour, sometimes visual, is used as an *aide memoire*. Those of us who were hoping for a high level of razor-sharp wit should prepare themselves for a level of disappointment!

1.2.3 Quotations

Here we take the old adage '*A word to the wise*' and give it a slight twist so that we can draw on the wisdom of those far wiser and more experienced in life than I. We call these little interjections '*A word (or two) from the wise?*' You will spot them easily by the rounded shadow boxes. In this one, Kehlog Albran (*and no, he wasn't a 'cereal' writer*) seems to suggest that the current trend will continue. Although he was probably not looking at a set of data values over time, it does remind us subtly that we might expect trends to continue until there is evidence to the contrary or we know of a driver for change.

> ## A word (or two) from the wise?
>
> '*I have seen the future and it is very much like the present, only longer.*'
> **Kehlog Albran**
> Pseudonym of Martin A. Cohen and Sheldon Shacket
> 'On the Future' in *The Profit*
> (1973)

1.2.4 Definitions

Estimating is not just about numbers but requires the context of an estimate to be expressed in words. There are some words that have very precise meanings; there are

4 | Introduction and objectives

others that mean different things to different people (estimators often fall into this latter group). To avoid confusion, we proffer definitions of key words and phrases so that we have a common understanding within the confines of this series of working guides. Where possible we have highlighted where we think that words may be interpreted differently in some sectors, which regrettably, is all too often. I am under no illusion that back in the safety of the real world we will continue to refer to them as they are understood in those sectors, areas and environments.

By way of example, let's define what we mean by a mathematical transformation, as implied in the title:

Definition 1.1   Mathematical transformation

A mathematical transformation is a numerical process in which the form, nature or appearance of a numerical expression is converted into an equivalent but non-identical numerical expression with a different form, nature or appearance.

I dare say that some of the definitions given may be controversial with some of us. However, the important point is that they are discussed and considered, and understood in the context of this book, so that everyone accessing these books have the same interpretation. We don't have to agree with the ones given here forevermore – what estimator ever did that? The key point here is that we are able to appreciate that not everyone has the same interpretation of these terms. In some cases, we will defer to the Oxford English Dictionary (Stevenson & Waite, 2011) as the arbiter.

1.2.5 Discussions and explanations with a mathematical slant for Formula-philes

These sections are where we define the formulae that underpin many of the techniques in this book. They are boxed off with a header to warn off the faint hearted. We will, within reason, provide justification for the definitions and techniques used. For example:

For the Formula-philes: Difference between two straight lines is a straight line

Consider two straight lines with different slopes and intercepts:
Let the first straight line have slope m_1 and intercept c_1: $\qquad y = m_1 x + c_1 \qquad (1)$

Let the second straight line have slope m_2 and intercept c_2:

$$y = m_2 x + c_2 \qquad (2)$$

Let Δ_y be the difference between these two lines, then subtracting (1) from (2)

$$\Delta_y = (m_2 x + c_2) - (m_1 x + c_1) \qquad (3)$$

Re-arranging (3):

$$\Delta_y = (m_2 - m_1)x - (c_2 + c_1) \qquad (4)$$

... which is a straight line with slope $(m_2 - m_1)$ and intercept $(c_2 - c_1)$

1.2.6 Discussions and explanations without a mathematical slant for Formula-phobes

For those less geeky than me, who don't get a buzz from knowing why a formula works (*yes, it's true, there are some estimators like that*), there are the Formula-phobe sections with a suitable header to give you more of a warm comforting feeling. These are usually wordier with pictorial justifications, and with specific particular examples where it helps the understanding and acceptance.

> ## For the Formula-phobes: One way logic is like a dead lobster
>
> *An analogy I remember coming across reading as a teenage fledgling mathematician, but for which sadly I can no longer recall its creator, relates to the fate of lobsters. It has stuck with me, and I recreate it here with my respects to whoever taught it to me.*
>
> Sad though it may be to talk of the untimely death of crustaceans, the truth is that all boiled lobsters are dead! However, we cannot say that the reverse is true – not all dead lobsters have been boiled!
>
> One-way logic is a response to many-to-one relationship in which there are many circumstances that lead to a single outcome, but from that outcome we cannot stipulate what was the circumstance that led to it.
>
> **Please note that no real lobsters were harmed in the making of this analogy.**

1.2.7 Caveat augur

Based on the fairly well-known warning to shoppers, '*caveat emptor*' (let the buyer beware), these call-out sections provide warnings to all soothsayers (or estimators) who try to predict the future, that in some circumstances we many encounter difficulties in

6 | Introduction and objectives

using some of the techniques. They should not be considered to be foolproof or be a panacea to cure all ills.

Caveat augur

These are warnings to the estimator that there are certain limitations, pitfalls or tripwires in the use or interpretation of some of the techniques. We cannot profess to cover every particular aspect, but where they come to mind these gentle warnings are shared.

1.2.8 Worked examples

There is a proliferation of examples of the numerical techniques in action. These are often tabular in form to allow us to reproduce the examples in Microsoft Excel (*other spreadsheet tools are available.*) Graphical or pictorial figures are also used frequently to draw attention to particular points. The book advocates that we should 'always draw a picture before and after analysing data'. In some cases, we show situations where a particular technique is unsuitable (i.e. it doesn't work) and try to explain why. Sometimes we learn from our mistakes; nothing and no one is infallible in the wondrous world of estimating. The tabular examples follow the spirit and intent of Good Practice Spreadsheet Modelling (albeit limited to black and white in the absences of affordable colour printing), the principles and virtues of which are summarised here in Volume I, Chapter 3.

1.2.9 Useful Microsoft Excel functions and facilities

Embedded in many of the examples are some of the many useful special functions and facilities found within Microsoft Excel (*often, but not always, the estimator's toolset of choice because of its flexibility and accessibility.*) Together we explore how we can exploit these functions and features in using the techniques described in this book.

We will always provide the full syntax as we recommend that we avoid allowing Microsoft Excel to use its default settings for certain parameters when they are not specified. This avoids unexpected and unintended results in modelling and improves transparency, an important concept that we discussed in Volume I, Chapter 3.

Example:

The **SUMIF** (*range, criteria, sum_range*) function will summate the values in the *sum_range* if the *criteria* in *range* is satisfied, and exclude other values from the sum where the condition is not met. Note that *sum_range* is an optional parameter of the function in Excel; if it is not specified then the *range* will be assumed instead.

Introduction and objectives | 7

We recommend that we specify it even if it is the same. This is not because we don't trust Excel, but a person interpreting our model may not be aware that a default has been assumed without our being by their side to explain it.

1.2.10 References to authoritative sources

Every estimate requires a documented Basis of Estimate. In common with that principle, which we discussed in Volume I, Chapter 3, every chapter will provide a reference source for researchers, technical authors, writers, and those of a curious disposition, where an original, more authoritative, or more detailed source of information can be found on particular aspects or topics.

Note that an estimate without a Basis of Estimate becomes a random number in the future. On the same basis, without reference to an authoritative source, prior research or empirical observation becomes little more than a spurious unsubstantiated comment.

1.2.11 Chapter reviews

Perhaps not unexpectedly, each chapter summarises the key topics that we will have discussed on our journey. Where appropriate we may draw a conclusion or two just to bring things together or to draw out a key message that may run throughout the chapter.

1.3 Overview of chapters in this volume

This volume concentrates on fitting the 'Best Fit' line or curve through our data, and creating estimates through interpolation or extrapolation, and expressing the confidence we have in those estimates based on the degree of scatter around the 'Best Fit' line or curve.

Chapter 2 of this volume explores the properties of a straight line and how we can exploit them, and our discussion will not be limited to the somewhat straightforward property of projecting straight forwards . . . or backwards; it includes a perhaps surprising non-linear property.

In Chapter 3 we delve into simple data smoothing techniques using a range of 'Moving Measures' including the strengths and weaknesses of the popular Moving Average. We explore alternative Moving Measures, drawing from our collection of Measures of Central Tendency that we discussed in Volume II (*unless you gave that one a miss, but that's OK.*) We show how we can enhance our understanding of the underlying pattern of behaviour in our data by combining these trend measures with various Moving Measures of Dispersion, linking it back to those we discussed in Volume II.

Chapter 3 also sticks a proverbial toe in the undulating waters of exponential smoothing, which appears to be quite promising but in practice can be something of an anti-climax. Cumulative and Cumulative Average Smoothing is perhaps an underused technique that often provides a clear view of the true underlying trend, often linking us back to the properties of a straight line in Chapter 2.

8 | Introduction and objectives

All these techniques from Chapter 2 and Chapter 3 can help us to judge whether we do in fact have an underlying trend that is either linear (straight line) or non-linear (curved).

We begin our exploration of the true delights of Least Squares Regression in Chapter 4 by considering how and why it works with simple straight line relationships. We extend the concept into multiple dimensions where each independent variable (driver) has a linear relationship with the dependent variable that we are trying to estimate, e.g. cost; this is referred to as Multiple Linear, Multi-Linear or Multivariate Regression. However, the problem with Simple or Multiple Linear Regression *per se* is that it will always calculate the "line" or "lines" of Best Fit, even through a set of random numbers! Consequently, we need to be able to assess whether the Regression result is valid and credible. We discuss how we can measure and judge how good the Regression's Line of Best Fit really is, or whether it is really no more than a force fit. We then demonstrate how we can create Confidence Interval around our Regression Line to give us 3-Point Estimates in preference to a single point deterministic value.

Chapter 4 also introduces us to the concept of Stepwise Regression, which is not a different Regression technique but is a procedure for selecting a valid and credible model selected from a range of potential drivers.

Such is the world of estimating, that many estimating relationships are not linear, but there are three groups of relationships (or functions) that can be converted into linear relationships with a bit of simple mathe-magical transformation using logarithms. In Chapter 5 we explore three groups of such functions: Exponential, Logarithmic and Power Functions; some of use will have seen these as different trendline types in Microsoft Excel.

In Chapter 6, we demonstrate how we can use this mathe-magical transformation technique to convert a non-linear relationship into a linear one to which we can exploit the power of Least Squares Regression that we discussed in Chapter 4. It is possible to combine Exponential and Power Functions, or Linear and Logarithmic Functions into single transformed Multi-Linear Regressions. We discuss how we can do this, and why other combinations such as Linear and Exponential Functions are not possible.

Where we have data that cannot be transformed in to a simple or multi-linear form, we explore the options open to us to find the 'Best Fit' curve in Chapter 7 using first principles.

Last, but not least, we look at Time Series Analysis techniques in Chapter 8 in which we consider a repeating seasonal and/or cyclical variation in our data over time around an underlying trend. We explore two basic models (Additive and Multiplicative) and two basic techniques: Classical Decomposition and Multi-Linear Regression.

1.4 Elsewhere in the 'Working Guide to Estimating & Forecasting' series

Whilst every effort has been made to keep each volume independent of others in the series, this would have been impossible without some major duplication and overlap. Whilst there is quite a lot of cross-referral to other volumes, this is largely for those of us who want to explore particular topics in more depth. There are some more fundamental potential pre-requisites. For example, the Regression techniques required to calibrate a Learning Curve in Volume IV are covered in this volume, but for a thorough understanding of the 'what' and the 'why', it assumes we have had access to Volumes I and II.

Figure 1.1 indicates the principal linkages or flows across the five volumes, not all of them.

1.4.1 Volume I: Principles, Process and Practice of Professional Number juggling

This volume clarifies the differences in what we mean by an estimating approach, method or technique, and how these can be incorporated into a closed-loop estimating process. We discuss the importance of TRACEability and the need for a well-documented Basis of Estimate that differentiates an estimate from what would appear in the future to be little more than a random number. Closely associated with a Basis of Estimate is the concept of an Estimate Maturity Assessment, which in effect gives us a health warning on the robustness of the estimate that has been developed. IRiS is a companion tool that allows us to assess the inherent risk in our estimating spreadsheets and models if we fail to follow good practice principles in designing and compiling those spreadsheets or models.

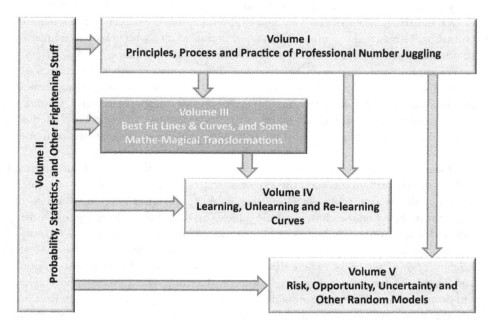

Figure 1.1 Principal Flow of Prior Topic Knowledge Between Volumes

10 | Introduction and objectives

An underlying theme we introduce here is the difference between accuracy and precision within the estimate, and the need to check how sensitive our estimates are to changes in assumptions. We go on to discuss how we can use factors, rates and ratios in support of data normalisation (to allow like-for-like comparisons to be made) and in developing simple estimates using an analogical method.

All estimating basically requires some degree of quantitative analysis, but we will find that there will be times when a more qualitative judgement may be required to arrive at a numerical value. However, in the spirit of TRACEability, we should strive to express or record such subjective judgements in a more quantitative way. To aid this we discuss a few pseudo-quantitative techniques of this nature.

Finally, to round off this volume, we will explore how we might use Benford's Law, normally used in fraud detection, to highlight potential anomalies in third party inputs to our estimating process.

1.4.2 Volume II: Probability, Statistics and Other Frightening Stuff

Volume II is focused on the statistical concepts that are exploited through Volumes III to V (and to a lesser extent in Volume I). It is not always necessary to read the associated detail in this volume if you are happy just to accept and use the various concepts, principles and conclusions. However, a general understanding is always better than blind acceptance, and this volume is geared around making these statistical topics more accessible and understandable to those who wish to adventure into the darker art and science of estimating. There are also some useful 'rules of thumb' that may be helpful to estimators or other number jugglers that are not directly used by other volumes.

We explore the differences between the different statistics that are collectively referred to as 'Measures of Central Tendency' and why they are referred to as such. In this discussion, we consider four different types of Mean (Arithmetic, Geometric, Harmonic and Quadratic) in addition to Modes and the 'one and only' Median, all of which are, or might be, used by estimators; sometimes without our conscious awareness.

However, the Measures of Central Tendency only tell us half the story about our data, and we should really understand the extent of scatter around the Measures of Central Tendency that we use; this gives us valuable insight to the sensitivity and robustness of our estimate based on the chosen 'central value'. This is where the 'Measures of Dispersion and Shape' come into their own. These measures include various ways of quantifying the 'average' deviation around the Arithmetic Mean or Median, as well as how we might recognise 'skewness' (where data is asymmetric or lop-sided in its distribution), and where our data exhibits high levels of Excess Kurtosis, which measures how spikey our data scatter is relative to the absolute range of scatter. The greater the Excess Kurtosis, and the more symmetrical our data, then the greater confidence we should have in the Measures of Central Tendency being representative of the majority of our data. Talking

Introduction and objectives | 11

of 'confidence' this leads us to explore Confidence Intervals and quantiles, which are frequently used to describe the robustness of an estimate in quantitative terms.

Extending this further we also explore several probability distributions that may describe the potential variation in the data underpinning our estimates more completely. We consider a number of key properties of each that we can exploit, often as 'rules of thumb', but that are often accurate enough without being precise.

Estimating in principle is based on the concept of correlation, which expresses the extent to which the value of one 'thing' varies with another, the value of which we know or have assumed. This volume considers how we can measure the degree of correlation, what it means and, importantly, what it does not mean! It also looks at the problem of a system of variables that are partially correlated, and how we might impose that relationship in a multi-variate model.

Estimating is not just about making calculations, it requires judgement, not least of which is whether an estimating relationship is credible and supportable, or 'statistically significant'. We discuss the use of hypothesis testing to support an informed decision when making these judgement calls. This approach leads naturally onto tests for 'outliers'. Knowing when and where we can safely and legitimately exclude what looks like unrepresentative or rogue data from our thoughts is always a tricky dilemma for estimators. We wrap up this volume by exploring several statistical tests that allow us to 'out the outliers'; be warned however, these various outlier tests do not always give us the same advice!

1.4.3 Volume III: Best Fit Lines and Curves, and Some Mathe-Magical Transformations

This is where we are now. This section is included here just to make sure that the paragraph numbering aligns with the volume numbers! (*Estimators like structure; it's engrained, we can't help it.*)

We covered this in more detail in Section 1.3, so we will not repeat or summarise it further here.

1.4.4 Volume IV: Learning, Unlearning and Re-learning Curves

Where we have recurring or repeating activities that exhibit a progressive reduction in cost, time or effort we might want to consider Learning Curves, which have been shown empirically to work in many different sectors.

We start our exploration by considering the basic principles of a learning curve and the alternative models that are available, which are almost always based on Crawford's Unit Learning Curve or the original Wright's Cumulative Average Learning Curve. Later in the volume we will discuss the lesser used time-based Learning Curves and how they differ from unit-based Learning Curves. This is followed by a

12 | Introduction and objectives

healthy debate on the drivers of learning, and how this gave rise to the Segmentation Approach to Unit Learning.

One of the most difficult scenarios to quantify is the negative impact of breaks in continuity, causing what we might term unlearning or forgetting, and subsequent re-learning. We discuss options for how these can be addressed in a number of ways, including the Segmentation Approach and the Anderlohr Technique.

There is perhaps a misconception that unit-based learning means that we can only update our learning curve analysis when each successive unit is completed. This is not so, and we show how we can use Equivalent Units Completed to give us an 'early warning indicator' of changes in the underlying unit-based learning.

We then turn our attention to shared learning across similar products or variants of a base product through multi-variant learning, before extending the principles of the segmentation technique to a more general transfer of learning across between different products using common business processes.

Although it is perhaps a somewhat tenuous link, this is where we explore the issue of collaborative projects in which work is shared between partners, often internationally with workshare being driven by their respective national authority customers based on their investment proportions. This generally adds cost due to duplication of effort and an increase in integration activity. There are a couple of models that may help us to estimate such impacts, one of which bears an uncanny resemblance to a Cumulative Average Learning Curve. (*I said that it was a tenuous link.*)

1.4.5 Volume V: Risk, Opportunity, Uncertainty and Other Random Models

Volume V, the last in the series, begins with a discussion on how we can model research and development, concept demonstration, or design and development tasks when we may only know the objective and not how we are going to achieve it. Possible solutions may be to explore the use of a Norden-Rayleigh Curve, or a Beta, PERT-Beta or even a triangular distribution. These repeating patterns of resource effort have been shown empirically to follow the natural pattern of problem discovery and resolution over the life of such 'solution development' projects.

Based fundamentally on the principles of 3-Point Estimates, we discuss how we can use Monte Carlo Simulation to model and analyse risk, opportunity and uncertainty variation. As Monte Carlo Simulation software is generally proprietary in nature, and is often 'under-understood' by its users, we discuss some of the 'do's and don'ts' in the context of risk, opportunity and uncertainty modelling, not least of which is how and when to apply partial correlation between apparently random events! However, Monte Carlo Simulation is not a technique that is the sole reserve of the risk managers and the like; it can also be used to test other assumptions in a more general modelling and estimating sense.

Introduction and objectives | 13

There are other approaches to risk, opportunity and uncertainty modelling other than Monte Carlo Simulation, and we discuss some of these here. In particular, we discuss the risk factoring technique that is commonly used, and sadly this is often misused to quantify risk contingency budgets.

There is a saying (attributed to Benjamin Franklin) that '*time is money*' and estimators may be tasked with ensuring that their estimates are based on achievable schedules. This links back to Schedule Risk Analysis using Monte Carlo Simulation, but also requires an understanding of the principles (at least) of Critical Path Analysis. We discuss these here and demonstrate that a simple Critical Path can be developed against which we can create a schedule for profiling and to some extent verifying costs.

In the last chapter of this last volume (*ah, sad*) we discuss queueing theory. (*It just had to be last one, didn't it? I just hope that the wait is worth it.*) We show how we might use this in support of achievable solutions where we have random arisings (such as spares or repairs) against which we need to develop a viable estimate.

1.5 Final thoughts and musings on this volume and series

In this chapter, we have outlined the contents of this volume and to some degree the others in this series, and described the key features that have been included to ease our journey through the various techniques and concepts discussed. We have also discussed the broad outline of each chapter of this volume, and reviewed an overview of the other volumes in the series to whet our appetites. We have also highlighted many of the features that are used throughout the five volumes that comprise this series, to guide our journey, and hopefully make it less painful or traumatic.

The main objective of this volume is to review and understand how we might analyse any underlying pattern or trend in our data and use this to create a basic estimate, and also where appropriate to test the sensitivity or uncertainty around that estimate.

The trouble with estimating is that it is rarely right, even when it's not wrong. We could make that observation equally about other professions . . . informed judgement is essential.

However, we must not delude ourselves into thinking that if we follow these techniques slavishly that we won't still get it wrong some of the time, often because assumptions have changed or were misplaced, or we made a judgement call that perhaps we wouldn't have made in hindsight. A recurring theme throughout this volume, and others in the series, is that it is essential that we document what we have done and why; TRACEability is paramount. The techniques in this series are here to help guide our judgement

> ### A word (or two) from the wise?
>
> '*Forecasting is the art of saying what will happen, and then explaining why it didn't!*'
>
> Anonymous
> communicated by
> **Balaji Rajagopalan**

14 | Introduction and objectives

through an informed decision-making process, and to remove the need to resort to 'guess-work' as much as possible.

By documenting our assumptions and following recognised analytical and documented techniques we can learn from our mistakes and continuously improve our estimating capability.

TRACE: Transparent, Repeatable, Appropriate, Credible and Experientially based

References

Albran, K (1974) *The Profit*, Los Angeles, Price Stern Sloan.
Stevenson, A & Waite, M (Eds) (2011), *Concise Oxford English Dictionary, 12th Edition*, Oxford, Oxford University Press.

2 Linear and nonlinear properties (!) of straight lines

Q: When is a straight line not a straight line? **A:** When it has been bent!

Now we might well be thinking '*Why would we want to do that as we have already intimated that straight lines are easier to extrapolate or project than curves?*' However, in the case of sequential linear relationships, we don't have to bend straight lines because they already possess the property that their Cumulative Values always form a Quadratic Function (i.e. a Polynomial of Order 2) – a distinctly curved line. Furthermore, the Quadratic will have coefficients or parameters that are simple functions of the slope and intercept of the straight line.

Before we delve into this perhaps lesser known property, let's remind ourselves of some of the basic properties that are relevant to all straight lines.

2.1 Basic linear properties

The most significant property of a straight line is its constant monotonicity (see Volume II Chapter 5), i.e. its unwavering progression in one direction. Any position on the straight line can be determined by two parameters or constants (the slope and the intercept) and a single variable. Conversely, any two points on a straight line are sufficient to determine the slope and the intercept.

For the Formula-phobes: Two points define a straight line

Consider any two points on a graph. We can only draw one straight line through the two points. The two points uniquely define the slope (or gradient) and the

(Continued)

16 | Linear and nonlinear properties (!) of straight lines

intercept (the value on the vertical axis corresponding with the zero value on the horizontal axis.

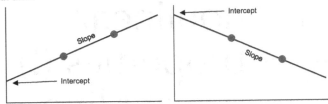

The left-hand example has a positive or increasing slope, whereas the right-hand example has a negative or decreasing slope. Although not shown here, we could re-draw the horizontal axis higher up so that the intercept was negative in both cases.

For the Formula-philes: Two points define a straight line

Consider any two points defined by the paired values (x_1, y_1) and (x_2, y_2)
The equation of a straight line with slope, m, and intercept, c, is:

$$y = mx + c \quad (1)$$

If (x_1, y_1) lies on the line then from (1):

$$y_1 = mx_1 + c \quad (2)$$

If (x_2, y_2) lies on the line then from (1):

$$y_2 = mx_2 + c \quad (3)$$

Eliminating c from (2) and (3):

$$y_1 - mx_1 = y_2 - mx_2 \quad (4)$$

Re-arranging (4), the slope, m, can be defined by the two points:

$$m = \frac{y_2 - y_1}{x_2 - x_1} \quad (5)$$

... which is the ratio of the vertical and horizontal differences between the two points

Re-arranging (2):

$$m = \frac{y_1 - c}{x_1} \quad (6)$$

Re-arranging (3):

$$m = \frac{y_2 - c}{x_2} \quad (7)$$

Eliminating m from (6) by (7):

$$\frac{y_1 - c}{x_1} = \frac{y_2 - c}{x_2} \quad (8)$$

Re-arranging (8):

$$x_2(y_1 - c) = x_1(y_2 - c) \quad (9)$$

Expanding the brackets in (9) and re-arranging:

$$x_2 y_1 - x_1 y_2 = c(x_2 - x_1) \quad (10)$$

Linear and nonlinear properties (!) of straight lines | 17

Re-arranging (10), c can be expressed in terms of the two points:	$c = \dfrac{x_2 y_1 - x_1 y_2}{x_2 - x_1}$	(11)

... which expresses the intercept as a function of any two points on the line

Adding and subtracting $\dfrac{x_1 y_1}{x_2 - x_1}$ in (11), c can be expressed as:	$c = \dfrac{(x_2 - x_1) y_1}{x_2 - x_1} + \dfrac{(y_2 - y_1) x_1}{x_2 - x_1}$	(12)
Substituting (5) and (12) in (1) and simplifying:	$y = \dfrac{y_2 - y_1}{x_2 - x_1}(x - x_1) + y_1$	(13)

... which expresses a straight line as a function of any two points on that line

As we selected the two points at random from the straight line, we can substitute either or both of them with any other point(s) on the line and the property remains intact.

If we have three or more data points formed by two highly (but not perfectly) correlated variables (Volume II Chapter 5) then we can draw a range of different straight lines through the points with varying errors where the line fails to pass through one or more points. However, the range of potential lines will be fairly narrow in terms of the potential values of slope and intercept, as illustrated on the left of Figure 2.1. The range of potential lines will be narrow if we are going to enter into the spirit and intent of making a reasonable effort of passing through the area occupied by the data points.

Where we have three or more such points that are poorly correlated, the range of straight lines we can draw through the points is wider, and the individual errors or scatter around the lines are greater, illustrated on the right of Figure 2.1. *Both these cases are part of the 'less than perfect world' with which the estimator has to contend.*

In Chapter 4 we will be discussing how we can define what we mean by the 'Best Fit straight line' through some data, but in essence when we do this we are implying that the

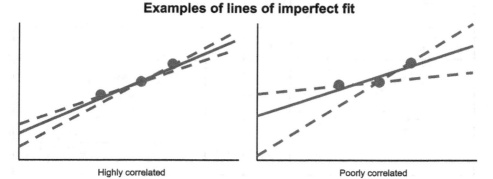

Figure 2.1 Examples of Lines of Imperfect Fit

true underlying relationship approximates to a perfect straight line, and that any disparity between the straight line and the actual data, is an imperfection in the actual. *Well, that is a bit of an overstatement, but we do tend to rely on the line first and then look at the potential scatter around the line as a bit of an afterthought, don't we?*

2.1.1 Inter-relation between slope and intercept

Taking the general equation of a straight line, we can easily show that for both positive and negative slopes or gradients, any straight line can be expressed as a function of its intercepts with the horizontal (x) and vertical (y) axes.

Consider the diagrams in Figure 2.2. Note that with the left-hand pair of graphs that the horizontal intercept d is negative whereas for the right-hand pair it is positive. Also, in the diagonal top-left to bottom-right, the vertical intercept c is positive, whereas on the cross-diagonal it is negative.

This also shows us that for any straight line with a positive slope (i.e. top pair of graphs), then one (but not both) of the intercepts is negative, and the other is positive ... or both are zero. For a straight line which has a negative slope (i.e. the bottom pair of graphs in Figure 2.2), then both intercepts are either positive or negative together, or again, both are zero.

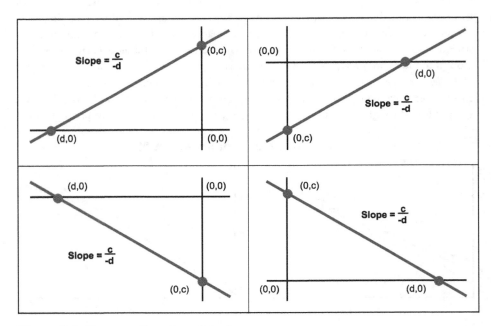

Figure 2.2 Slope as a Function of the Intercepts

Linear and nonlinear properties (!) of straight lines | 19

For the Formula-philes: Straight lines as functions of their intercepts

Let $(d, 0)$ and $(0, c)$ be the intercepts of a straight line with the x and y axes respectively:

Substituting the intercept points in the equation of a straight line, we get:
$$y = \frac{c-0}{0-d}(x-d) + 0 \qquad (1)$$

Simplifying (1):
$$-dy = c(x-d) \qquad (2)$$

Simplifying (2):
$$-dy = cx - cd \qquad (3)$$

Re-arranging (3):
$$cx + dy = cd \qquad (4)$$

Dividing (4) by cd:
$$\frac{x}{d} + \frac{y}{c} = 1 \qquad (5)$$

. . . which expresses a straight line as a ratio function of its two intercepts.

This is an important property when we come to Lines of Best Fit in the Chapter 4 as it illustrates that for any fixed horizontal intercept, the vertical intercept is either perfectly positively or negatively correlated with the slope of the line:

- If the horizontal intercept is negative, then the vertical intercept and the slope are positively correlated (as we increase one, the other increases in direct proportion)
- If the horizontal intercept is positive, then the vertical intercept and the slope are negatively correlated (as we increase one, the other decreases in direct proportion)

2.1.2 The difference between two straight lines is a straight line

It is sometimes a helpful property of straight lines that the difference between any two straight lines is also a straight line. Clearly the two lines should be measuring the things that can be compared such as the planned spend over time and the actual spend over the same time period. However, it might not make sense to subtract the linear trend in the total number of apples being eaten each month from the trend in the total number of oranges being eaten. (*This would clearly be the classic mistake of comparing apples with oranges – unless we're doing a fruit popularity contest, of course!*)

This difference property allows us to compare linear variance trends. Note: Whilst the property is inviolable for perfect straight lines (i.e. it is always true), and can be read across to 'real' data which depict linear trends, there may be some difference in the results between the trend through the differences, and the difference between the two original trendlines. We can get this sometimes due to smoothing or worsening of the data scatter attributable to the difference.

For the Formula-phobes: Difference between two straight lines is a straight line

Draw any two lines on a piece of graph paper (*or the electronic equivalent*). It doesn't matter if they cross.

Pick a point at random on the horizontal x-axis and measure the vertical difference between the two lines for that point. Repeat the exercise for a number of other points. Plot these differences to get another straight line, making sure that you have taken account of the change in sign if the lines have crossed

The reason this will always work is that we are measuring the rate at which the difference changes, which is the difference in the slopes of the two original lines.

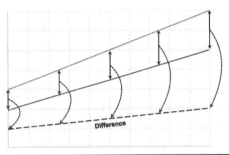

For the Formula-philes: Difference between two straight lines is a straight line

Consider two straight lines:
Let the first straight line have slope m_1
and intercept c_1:

$$y = m_1 x + c_1 \qquad (1)$$

Let the second straight line have slope m_2
and intercept c_2:

$$y = m_2 x + c_2 \qquad (2)$$

Let Δ_y be the difference between these two lines, then subtracting (1) from (2)

$$\Delta_y = (m_2 x + c_2) - (m_1 x + c_1) \qquad (3)$$

Re-arranging (3):

$$\Delta_y = (m_2 - m_1)x - (c_2 + c_1) \qquad (4)$$

... which is a straight line with slope $(m_2 - m_1)$ and intercept $(c_2 - c_1)$

Linear and nonlinear properties (!) of straight lines | 21

2.2 The Cumulative Value (nonlinear) property of a linear sequence

As estimators, we will come across two types of straight line relationships (*and I'm not talking about upward and downward sloping functions.*) We will find that there are discrete functions in which the independent x-variable can only take specific values such as integers, and there are continuous functions which can take any value.

Even then we can sub-divide these relationships into two further types:

- There are those straight lines which represent the relationship between two correlated variables (including all analogical comparisons.) An example might be the cost of component testing in relation to the component's weight.
- Then there are those straight lines which are parametric sequences of data, and for which we might want to consider exploring the cumulative value of the sequence. For example, the cost of successive units produced.

In the case of the former, it would make no sense to aggregate the data. This section is devoted to exploring the second type, sequential linear relationships that can be aggregated across the sequence.

2.2.1 The Cumulative Value of a Discrete Linear Function

Consider the case where the independent variable (depicted normally by the horizontal x-axis) of a linear function can only take discrete integer values, such as a build sequence number, or some time counter like the number of weeks or months from a project start:

- The Cumulative Average of that perfect straight line is another perfect straight line of half the slope.
- A corollary of this is that the Cumulative Value of that straight line forms a perfect quadratic equation (i.e. a polynomial of order 2) that passes through the origin

---

For the Formula-philes: Cumulative Value of a Discrete Linear Function

Consider a straight line with slope m and intercept c, in which the independent variable x represents the sequence of consecutive positive integers x_i from 1 to n:

Expressing y as a function of x:

$$y_i = mx_i + c \qquad (1)$$

The sum of the y_i values in (1) for x_1 to x_n is:

$$\sum_{i=1}^{n} y_i = \sum_{i=1}^{n} mx_i + nc \qquad (2)$$

(Continued)

The values for x_1 to x_n are the integers 1 to n, for which the standard sum is:
$$\sum_{i=1}^{n} x_i = \frac{1}{2}n(n+1) \quad (3)$$

Substituting (3) in (2):
$$\sum_{i=1}^{n} y_i = m\frac{1}{2}n(n+1) + nc \quad (4)$$

Simplifying (4):
$$\sum_{i=1}^{n} y_i = \frac{m}{2}n^2 + \left(\frac{m}{2} + c\right)n \quad (5)$$

...which expresses the cumulative value of a straight line where the independent variable x_i are consecutive positive integers as a standard quadratic equation (polynomial of order two)

If $\bar{y}$ is the Cumulative Average of y_i, dividing (5) by n gives:
$$\bar{y} = \frac{m}{2}n + \left(\frac{m}{2} + c\right) \quad (6)$$

...which is a straight line with a gradient or slope equal to a half of the original data

For the Formula-phobes: The Cumulative Average of a straight line is a straight line

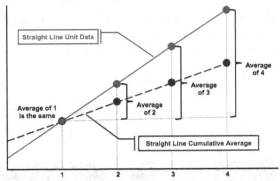

The slope of a line is defined as the vertical movement divided by the horizontal movement (just like on a road). As shown in the diagram, the average of first two points is halfway between them but is plotted at x = 2. The average of the first three is halfway between them, and is plotted at x = 3; and so on, resulting in a line of half the slope of the original.

Both lines have been projected back to the vertical axis to show the impact on the theoretical Cumulative Average Intercept. *(It's theoretical because we can't have a Cumulative Average based on zero quantity!)*

Linear and nonlinear properties (!) of straight lines | 23

Now that's what I call mathe-magic!

However, some of us quite righty, might be asking ourselves, 'Why this is useful? When would we ever use this property?' However, this is not just some academic property, it does have practical applications. For instance:

- The independent variable may be a build sequence number, or a time period, which may be a good indicator of improvement (or degradation) in something that we are trying to estimate, but our detailed records are incomplete for whatever reason (*if only we lived in that perfect world, estimating would be so much easier*).
- Some data we may have been collecting in a natural sequence, may be quite erratic with peaks and troughs, but a Moving Average or Cumulative Average (see Chapter 3) may indicate an underlying linear trend.

Table 2.1 illustrates both these scenarios. In this example we have started to collect the detailed weekly data on the number of sales enquiries received during an ongoing marketing campaign somewhat later than we should have done. (*Yes, this was very remiss of us, and with the benefit of hindsight we should have collected it earlier, but hey, this example is meant to represent the failings of real life!*) However, all is not lost as we do have the cumulative data from the start of the campaign available to us.

The end column in the example, which calculates the Cumulative Average Number of Enquiries per Week, shows that there is an incremental increase week on week.

Figure 2.3 plots the Cumulative Number of Enquiries received and we can fit a Polynomial Trendline (order 2) using Microsoft Excel's Chart utility and project it back

Table 2.1 Example – Enquiries Received Following Marketing Campaign

Weeks from start	Enquiries	Cum enquiries	Comments	Cum average enquiries
0-12			Data not collected by Week prior to Week 14	
13		80		6.15
14	7	87		6.21
15	7	94		6.27
16	6	100		6.25
17	10	110		6.47
18	8	118		6.56
19	9	127		6.68
20			Weekly data not available due to procedural failure	
21		145		6.90
22	9	154		7.00
23	10	164		7.13
24	8	172		7.17
25	9	181		7.24
26	11	192		7.38

24 | Linear and nonlinear properties (!) of straight lines

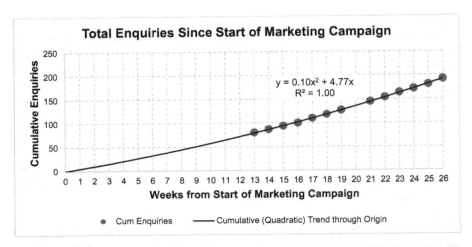

Figure 2.3 Example – Cumulative Enquiries Received Following Marketing Campaign

to the start of the marketing campaign. In this case we appear to have a very good fit, indicated by the Coefficient of Determination R^2 being so close to 1.

Note: Cumulative values are inherently much smoother than the unit data from which they are calculated. In this case it appears to be a near perfect quadratic relationship. However, if we were to expand the Coefficient of Determination to 3 decimal places, we would show that $R^2 \neq 1$.

The underlying linear trend that gives rise to this Cumulative Quadratic Function can be determined from the coefficients of the quadratic equation:

- The slope of this straight line is double the coefficient of the x^2 term in the quadratic; in this case, the slope is 0.20, i.e. 2 times 0.10
- The intercept of the straight line is the difference between the quadratic coefficients of the x and x^2 terms; in this case, 4.77 minus 0.10, or 4.67

Alternatively, we could have simply run a trendline through the weekly data or even the Cumulative Average data, as shown in Figure 2.4. If we use our rule that the Cumulative Average of a straight line is another straight line of half the slope, then based on the trendline through the Cumulative Average, the underlying linear function can be determined as follows:

- The Cumulative Average trendline slope is 0.10, so the underlying data has a slope of 0.2
- The intercept of the underlying straight line is the difference between the intercept and slope of the Cumulative Average. In this case the underlying intercept is 4.79

Linear and nonlinear properties (!) of straight lines | 25

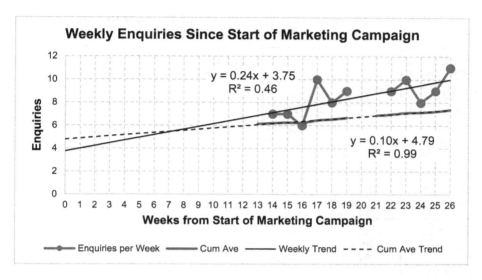

Figure 2.4 Example – Weekly Enquiries Received Following Marketing Campaign

minus 0.1, or 4.69. Note that this is slightly different to the Cumulative Technique used above due to difference in the scatter around each line (e.g. one term is squared.)

These values are very compatible with the results from the Cumulative Quadratic Model (*not surprisingly as they use the same data*), but they are distinctly different to the simple trendline through the raw data, which gives us a steeper slope of 0.24, and smaller intercept of 3.75 in this instance. Now, we may be alarmed by this as they are based on the same data, whereas in reality, we have two more data points for the Cumulative and Cumulative Average Technique (Week 13 and 21 data can be derived by subtracting the weekly data from the cumulative for Weeks 14 and 22). However, notwithstanding that, if we were to remove these additional points (at x = 13 and x = 21) we would still get a different answer because of the degree and manner of the scatter in the raw data.

So, which should we use? We do have a choice, of course, but we may want to take account of the fact that in this case the Coefficient of Determination for the simple trendline through the raw data is not very encouraging (less than 0.5 – see Volume II, Chapter 5.) We should always go back to one of the basic premises of estimating – no single technique is foolproof. So we could use both in order to establish a range estimate.

Now, let's consider other, perhaps more basic uses of the Cumulative Quadratic property. Suppose we wanted to project how many enquiries we would have received after twelve months, and how many a week we could expect to receive if the campaign continued to have the success it appears to be having now?

26 | Linear and nonlinear properties (!) of straight lines

Table 2.2 Example – Using the Quadratic Formula to Forecast the Cumulative Value

Slope, m	0.2	0.1	< Coefficient of x^2 term, m/2
Intercept, c	4.67	4.77	< Coefficient of x term, c+m/2
Weeks from Start	Forecast Weekly Enquiries	Cumulative Enquiries	
39		338	
40	13		
41	13		
42	13		
43	13		
44	13		
45	14		
46	14		
47	14		
48	14		
49	14		
50	15		
51	15		
52	15	518	
Total	180	180	Forecast Enquiries Wks 40-52

A by-product of this Cumulative Quadratic property is that we can very quickly get a cumulative forecast for a future batch or for a period of time:

- We can extrapolate our linear trend to cover the range of values in which we are interested. We can then calculate the value of every valid unit, and add them up (*and yes, it is easy to do that in spreadsheets like Microsoft Excel*)
- Alternatively, we can calculate the cumulative value of our end-point using our quadratic formula and deduct the cumulative value immediately prior to our start-point, again using our quadratic formula

Table 2.2 illustrates both ways using our marketing campaign example.

Note: Due to rounding errors with discrete data, we can sometimes get slightly different answers between these two techniques.

2.2.2 The Cumulative Value of a Continuous Linear Function

Let's turn our attention to straight lines which are Continuous Linear Functions in which the input x-variable can take any value, not just integers. Examples of this might

Linear and nonlinear properties (!) of straight lines | 27

include the equivalent unit build quantity which includes incomplete units e.g. 2.25 units (rather than completed units which would be integer values only.)

If we are taking equivalent units into account rather than just completed units, then we need to ensure than the two are compatible, i.e. they both give the same Cumulative Value in a perfect world. (In reality, they will give comparable results, but we'll come back to that later in this section.)

Let's revisit the Cumulative of the discrete linear relationship and look at it from a different perspective — as the sum of a number of areas.

Each point on the discrete straight line is the mid-point of the two points, half a unit to its left and right. If we multiply this by one, we get the area of the trapezium, T_i. We can repeat this process for each successive integer value representing points on the straight line, giving trapezia T_1 to T_n inclusively, as illustrated in Figure 2.5.

The area of a trapezium is the average length of the two parallel sides multiplied by their distance apart.

Here, we are using the British definition of trapezium with two sides parallel, which is what Americans call a trapezoid, which in British usage is a quadrilateral with no parallel sides. In other words, American definitions are crossed over! What was it that George Bernard Shaw commented?

A word (or two) from the wise?

"England and America are two countries separated by a common language."
George Bernard Shaw
Irish Playwright
(1856–1950)

The sum of all these British trapezia can be added together to get one large trapezium (Figure 2.6) that equals the sum of the consecutive discrete values on a straight line. This trapezium is bounded by the values x = 0.5, x = n+0.5, the x-axis and the straight line. If we slide the trapezium representing the sum of discrete values to the left by half a unit, then we get a trapezium of equal area that is the integral of the Continuous Function (see Figure 2.7). In this case the trapezium is bounded by x = 0 and x = n.

In Volume II Chapter 4 we may have seen that the integral of a function gave the area under the curve and that this was the cumulative probability. We can use the area under this offset graph to represent the cumulative value of a Continuous Linear Function rather than the 'true' Discrete Linear Function. (*Actually, we can use the area under the discrete version but as we have already shown that to get the correct value we have to start at 0.5 and end at n+0.5 rather than 0 through to n. It just doesn't feel right, does it?*)

'And the point of all this?' some of us may be asking. As a consequence, we can use the same quadratic equation to simulate the Cumulative Value of a straight line regardless

Linear and nonlinear properties (!) of straight lines

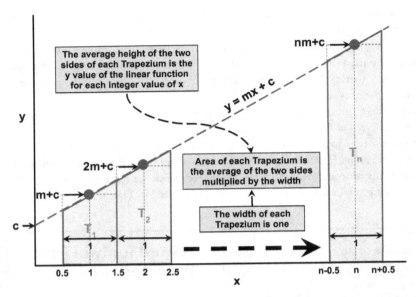

Figure 2.5 Straight Line Discrete Values Expressed as Trapezial Areas Under the Line

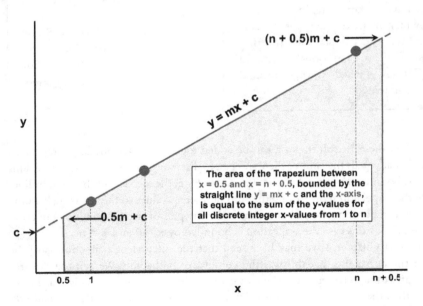

Figure 2.6 Cumulative of a Discrete Value Straight Line as a Trapezial Area Under the Line

of whether it is a discrete or Continuous Linear Function. The Continuous Function allows us to emulate the equivalent number of units built rather than just completed units, i.e. to include work-in-progress.

Linear and nonlinear properties (!) of straight lines | 29

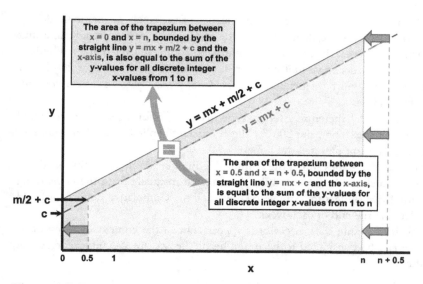

Figure 2.7 Equating Cumulative Values for Discrete and Continuous Straight Lines

Don't you just love it when everything comes together logically like a well laid plan? No? There's just no pleasing some people!

For the Formula-philes: Cumulative Value of a Continuous Linear Function

Consider a straight line with slope m and intercept $\left(c + \dfrac{m}{2}\right)$, in which the independent variable x represents the sequence of consecutive real numbers x_i in the range 0 to n:

Let the straight line express y as a function of x:
$$y_i = mx_i + c + \frac{m}{2} \quad (1)$$

The area under the straight line is the cumulative value of the y_i values. The area under the 'curve' in (1) for 0 to x_n is:
$$\int_0^n y \, dy = \int_0^n \left(mx + c + \frac{m}{2}\right) dx \quad (2)$$

Integrating the right-hand side of (2):
$$\int_0^n y \, dy = \left[\frac{m}{2}x^2 + \left(c + \frac{m}{2}\right)x\right]_0^n \quad (3)$$

Expanding the bracket in (3):
$$\int_0^n y \, dy = \frac{m}{2}n^2 + \left(c + \frac{m}{2}\right)n \quad (4)$$

... which expresses the cumulative value of a continuous straight line as a standard quadratic equation. This Cumulative Model is identical to that we derived in Section 2.2.1 for the discrete value straight line.

30 | Linear and nonlinear properties (!) of straight lines

Figure 2.8 and Table 2.3 illustrate the equivalence of the single quadratic model for both cases. For clarity, we are demonstrating the results for a perfect straight line unit budget which assumes that the man-hours required to complete a recurring task reduce by 10 hours on each successive unit completed, with an intercept value of 210 hours, and that these hours will be expended in a fixed percentage pattern over a constant three-week build cycle.

The quadratic model, therefore, assumes a coefficient of -5 for the square of the build number term (half the slope), and 205 for the coefficient applied to the build number term (i.e. linear intercept plus half the slope):

The lower of the two graphs shows the Cumulative Budget Profile using this quadratic model, against which we have plotted the data created manually from the table, using both the Cumulative Budget for completed units, and the Cumulative Equivalent Build Units completed at the end of each week.

Note: This is not quite the perfect fit it appears, but in the context of an estimate, the difference can be considered to be insignificant (i.e. we are seeking accuracy not precision after all.)

Caveat augur

When we examine continuous sequential data in graphical or tabular form, the interval between successive observations in the sequence may not be regular. When this is the case the value we are measuring or estimating will vary depending on how much of the sequential data has moved on cumulatively. As a consequence, it is unlikely that the data we observe will appear to be linear – even if the underlying relationship is perfectly linear, and the cumulative data is quadratic. This is just a reporting paradox.

If the cumulative sequential data is a fixed interval, then we will see the underlying straight line, or at least a close approximation.

This concept may not be intuitive. However, we can see it in action in the current example. In Figure 2.9 we have plotted the planned budget for each week against the planned equivalent units to be completed – clearly not a linear plot.

However, if we were to re-profile the data (same cumulative hours) so that each week we were to complete the same number of equivalent units, we would get the results in Table 2.4 and Figure 2.10:

We will get the same reporting paradox with discrete sequential linear data also. If we review data ad hoc rather than at fixed sequential intervals, then we may not observe the underlying linear relationship.

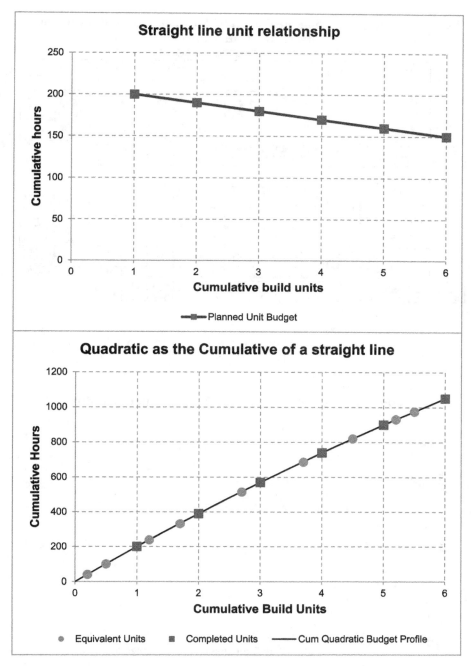

Figure 2.8 Example – Using the Quadratic Formula to Forecast the Cumulative Value

Table 2.3 Example – Using the Quadratic Formula and Equivalent Units to Forecast the Cumulative Value

Build No	Wk 1	Wk 2	Wk 3	Wk 4	Wk 5	Wk 6	Wk 7	Wk 8	Wk 9	Wk 10	Total % per Unit	Cum Build
1	20%	30%	50%								100%	1
2			20%	30%	50%						100%	2
3				20%	30%	50%					100%	3
4					20%	30%	50%				100%	4
5						20%	30%	50%			100%	5
6								20%	30%	50%	100%	6
Units Completed in the Week	20%	30%	70%	50%	100%	100%	80%	70%	30%	50%		
Cumulative Units Completed	0.2	0.5	1.2	1.7	2.7	3.7	4.5	5.2	5.5	6		

Slope: -10
Intercept: 210

Build No	Mhrs	Wk 1	Wk 2	Wk 3	Wk 4	Wk 5	Wk 6	Wk 7	Wk 8	Wk 9	Wk 10	Total Hrs per Unit	Cum Hrs
1	200	40	60	100	0	0	0	0	0	0	0	200	200
2	190	0	0	38	57	95	0	0	0	0	0	190	390
3	180	0	0	0	36	54	90	0	0	0	0	180	570
4	170	0	0	0	0	34	51	85	0	0	0	170	740
5	160	0	0	0	0	0	32	48	80	0	0	160	900
6	150	0	0	0	0	0	0	0	30	45	75	150	1050
Budgeted Hours per Week		40	60	138	93	183	173	133	110	45	75		
Cumulative Hours Budgeted		40	100	238	331	514	687	820	930	975	1050		

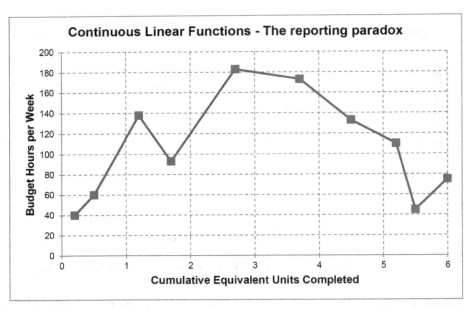

Figure 2.9 Continuous Linear Functions – The Reporting Paradox

Table 2.4 Example Revisited with Steady State Equivalent Unit Completions

Build No	Wk 1	Wk 2	Wk 3	Wk 4	Wk 5	Wk 6	Wk 7	Wk 8	Wk 9	Wk 10	Total % per Unit	Cum Build
1	60%	40%									100%	1
2		20%	60%	20%							100%	2
3				40%	60%						100%	3
4						60%	40%				100%	4
5							20%	60%	20%		100%	5
6									40%	60%	100%	6
Units Completed in the Week	60%	60%	60%	60%	60%	60%	60%	60%	60%	60%		
Cumulative Units Completed	0.6	1.2	1.8	2.4	3	3.6	4.2	4.8	5.4	6		

| Slope | -10 |
| Intercept | 210 |

Build No	Mhrs	Wk 1	Wk 2	Wk 3	Wk 4	Wk 5	Wk 6	Wk 7	Wk 8	Wk 9	Wk 10	Total Hrs per Unit	Cum Hrs
1	200	120	80	0	0	0	0	0	0	0	0	200	200
2	190	0	38	114	38	0	0	0	0	0	0	190	390
3	180	0	0	0	72	108	0	0	0	0	0	180	570
4	170	0	0	0	0	0	102	68	0	0	0	170	740
5	160	0	0	0	0	0	0	32	96	32	0	160	900
6	150	0	0	0	0	0	0	0	0	60	90	150	1050
Budgeted Hours per Week		120	118	114	110	108	102	100	96	92	90		
Cumulative Hours Budgeted		120	238	352	462	570	672	772	868	960	1050		

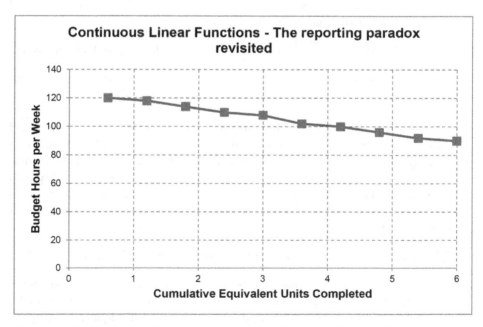

Figure 2.10 Continuous Linear Functions – The Reporting Paradox Revisited

34 | Linear and nonlinear properties (!) of straight lines

2.2.3 Exploiting the Quadratic Cumulative Value of a straight line

There are two obvious ways in which we can exploit this Quadratic Cumulative Value property. If we believe (*or better still have evidence*) that there is an underlying linear trend in some sequential data, then we can answer the following two questions:

1. What is the Cumulative Value of a Linear Trend after a given number of units have been completed (Real or Equivalent), or after a set period of time?
2. At what point in the sequence does the Cumulative of a Linear Trend achieve a particular value?

Let's consider each in turn, abbreviating the question:

Q1. What is the future Cumulative Value of a Linear Trend?

This is simply a shortcut we can use rather than having to calculate each value on the Linear Trend line and then adding them all up. The procedure is simply:

- Identify the slope and intercept of the Best Fit straight line through the data, or the Best Fit Quadratic (polynomial of order 2) through the origin for the Cumulative data
- Create the quadratic equation for the Cumulative Value using the standard result derived in the previous two sections

$$\text{Cumulative} = \frac{m}{2}n^2 + \left(\frac{m}{2} + c\right)n \text{ where } m \text{ is the slope and } c \text{ the intercept}$$

- Calculate the Cumulative Value for the sequence number (Build Number or Time Period) required

Figure 2.11 and Table 2.5 illustrate the technique. The example shows the Monthly and Cumulative Trends with the Best Fit line and curve determined using Microsoft Excel's Chart utility. We can extract the parameter data manually from either or both of these trends in order to express a forecast for the Cumulative Deliveries that can be expected if the current monthly linear trend continues, as shown in Table 2.5. (*We'll be covering how Microsoft Excel does this in Chapter 6 on Nonlinear Regression. I can hardly contain my excitement with the anticipation!*)

Q2. When do we achieve a Given Cumulative Value of a Linear Trend?

For example, we may have detected that Percentage Achievement per Month is increasing linearly, and we want to know when we will attain 100% completion. Here, we can use the general solution of a quadratic equation to help us.

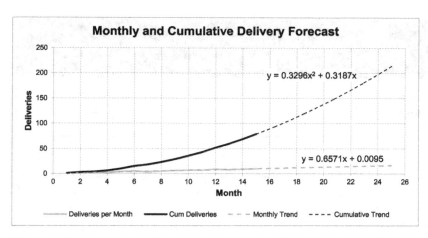

Figure 2.11 Example of Cumulative Forecasting Using a Quadratic Trend

Table 2.5 Example of Cumulative Forecasting Using a Quadratic Trend

Month, x	Actual Deliveries	Cum Deliveries
1	1	1
2	2	3
3	1	4
4	2	6
5	4	10
6	5	15
7	3	18
8	5	23
9	6	29
10	7	36
11	7	43
12	9	52
13	8	60
14	9	69
15	10	79

Best Fit Coefficients from Graph		
	Linear	Quadratic
Slope	0.6571	
Intercept	0.0095	
Coef x^2	0.3286	0.3296
Coef x	0.3381	0.3187

Month	Cum Forecast	
25	214	
25		214

Note that we do not get exactly the same equation if we use the linear trend rather than the Cumulative trend, and this can lead to rounding differences.

This is not a failing in the underlying principle but a recognition that there is a natural scatter around the perfect line and curve. The Cumulative value is inherently smoother than its linear source, i.e. there is less scatter.

Whilst the quadratic coefficients derived from the linear trend are slightly different to those extracted directly from the quadratic trend, the difference is "lost" in this case in the integer rounding:

> 213.8 from Monthly Linear Trend
> 214.0 from Cumulative Quadratic Trend

36 | Linear and nonlinear properties (!) of straight lines

For the Formula-philes: General solution of a quadratic equation

The general solution of any quadratic equation of the form $ax^2 + bx + c = 0$ is:

$$x = \frac{-b \pm \sqrt{b^2 - 4ac}}{2a}$$

Yes, that's right, it's another one of those things we did in school during mathematics classes but we couldn't see its relevance to life as we knew it. Well, that's another outstanding objective to cross off your list!

The procedure is simply:

i. Identify the slope and intercept of the Best Fit straight line through the data, or the Best Fit polynomial (order 2) through the origin for the Cumulative data

ii. Create the Quadratic equation for the Cumulative Value using the standard result derived in the previous two sections

$$\text{Cumulative} = \frac{m}{2}n^2 + \left(\frac{m}{2} + c\right)n \text{ where } m \text{ is the slope and } c \text{ the intercept}$$

iii. Apply the general solution of a quadratic equation to derive the desired solution for the sequence number (Build Number or Time Period) required

We may recall from school that a quadratic equation will have either 0, 1 or 2 solutions (or roots as they were called in class):

0. The sign under the square root is negative and so we cannot solve the equation (*We are not going to delve into the murky waters of imaginary numbers, you'll be pleased to hear.*)

1. An equation of the form $(x - k)^2 = 0$ is a quadratic with two identical roots of $x = k$, so to all intents and purposes it only has one solution

2. This is probably the most common situation we will come across as estimators. However, we should be on our guard to select the right one – see the Formula-phobe discussion.

For the Formula-philes: Finding when a future Cumulative Value occurs

Consider a Quadratic Equation representing the Cumulative Value of a straight line:

Linear and nonlinear properties (!) of straight lines | 37

$y = mx + k$ (constant changed from standard notation used previously to avoid confusion with standard formula cited in Step (1) below)

The general solution of any quadratic equation of the form $ax^2 + bx + c = 0$ is cited in many school textbooks as:

$$x = \frac{-b \pm \sqrt{b^2 - 4ac}}{2a} \qquad (1)$$

From Section 2.2.1 the Cumulative Value of a straight line of the form, $y = mx + k$, has been shown to be:

$$\sum_{i=1}^{x} y_i = \frac{m}{2}x^2 + \left(\frac{m}{2} + k\right)x \qquad (2)$$

Suppose we want to know the value of x for which the Cumulative Value of the straight line equals a target value T:

$$T = \frac{m}{2}x^2 + \left(\frac{m}{2} + k\right)x \qquad (3)$$

Re-arranging (3):

$$\frac{m}{2}x^2 + \left(\frac{m}{2} + k\right)x - T = 0 \qquad (4)$$

Using (1) and substituting $c = -T$, the general solution to the quadratic equation at (4) is:

$$x = \frac{-b \pm \sqrt{b^2 + 4aT}}{2a} \qquad (5)$$

...we want the smallest occurring positive root

We can relate the general solution back to the straight line's parameters by substituting $a = \frac{m}{2}$, $b = \left(\frac{m}{2} + k\right)$ and $c = -T$:

$$x = \frac{-\left(\frac{m}{2} + k\right) \pm \sqrt{\left(\frac{m}{2} + k\right)^2 + 2mT}}{m} \qquad (6)$$

For the Formula-phobes: Quadratic equations generally have two solutions

Quadratic equations are symmetrical about some value and so have two solutions for the same value; we require the lowest positive value root as illustrated.

(Continued)

In the example here for the cumulative of an increasing straight line, there are two values at which Cum y = 100:

$$x = -25 \text{ and } x = 16$$

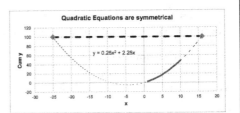

However, in the case of a decreasing straight line, there may be two positive roots, in which case we will want the first, i.e. the lower of the two. In this case, when Cum y = 100:

$$x = 16 \text{ and } x = 25$$

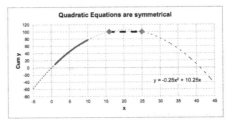

The reason that the Cumulative y value turns back on itself is that the underlying straight line has crossed the x-axis and the y values are all negative.

In this case we should question whether the underlying data line is truly linear, or whether it just appears linear for the data we have (*a shallow curve might be being disguised by the data scatter*).

Caveat augur

It is important that we recognise that any apparent straight-line relationship may only be valid in the context of the data in which we observe it. If the true relationship is in fact a shallow curve through our data, then a straight line might only be a good approximation to that curve in the range local to the data, which may not be the case if we extrapolate too far in either direction.

A word (or two) from the wise?

"The straight line: a respectable optical illusion which ruins many a man."

Victor Hugo
Les Misérables (1862)
French Novelist

Figure 2.12 illustrates the sentiments expressed by Victor Hugo (*albeit in a different context, and not in French.*) Here, the trend is slightly bowed around a straight line. Without the straight line as a reference, it is difficult to see.

Linear and nonlinear properties (!) of straight lines | 39

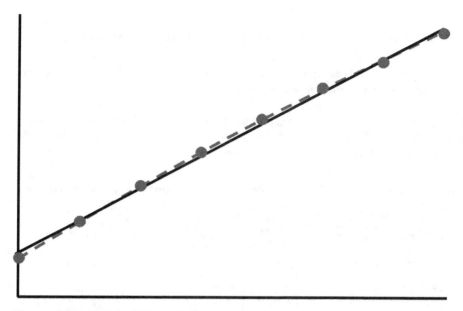

Figure 2.12 Straight Line or Shallow Curve?

A corollary (or natural extension) to this exploits the difference property of two straight lines, which as we showed in Section 2.1.2, is also a straight line. Consequently, the cumulative difference between two straight lines is a quadratic function. This may be really useful to us if we have to estimate a recovery position to some plan. (*And who hasn't?*)

Consider another example where we are behind schedule at the start of a programme, (*does that sound all too familiar?*) but that we aim to recover and are showing positive signs of doing so. Take the scenario in which the requirement or plan is to increase output linearly, but the reality is that we started later than planned, but are managing to increase output at a higher rate than originally planned. The procedure here is simply:

i. Identify the slope and intercept of the planned linear output growth
ii. Identify the slope and intercept of the Best Fit straight line through the actual data
iii. Subtract the difference between the two slopes and the two intercepts (step ii minus step i) to get the Best Fit straight line through the difference. (*Clearly, if the slope from step ii is less than step i, then stop here because we aren't recovering – although we can still predict where we might end up!*)

- Generate the parameters of the Cumulative Quadratic Equation for the Cumulative difference to plan

40 | Linear and nonlinear properties (!) of straight lines

$$\text{Cumulative} = \frac{m}{2}n^2 + \left(\frac{m}{2} + c\right)n \quad \text{where } m \text{ is the slope and } c \text{ the intercept}$$

- Apply the general solution of a quadratic equation to derive the expected recovery position, i.e. when the Cumulative Difference is zero

Table 2.6 illustrates our situation against plan. Both the plan and our actual achievement here are linear, as shown in Figure 2.13 so this allows us to use the 'straight line difference' property to forecast the recovery position.

Table 2.6 Linear Trend Recovery Point Using a Cumulative Quadratic Trend

Time Period	Planned Monthly Deliveries	Actual Monthly Deliveries	Variance to Monthly Planned Deliveries	Cumulative Planned Deliveries	Cumulative Actual Deliveries	Cumulative Delivery Variance
0						
1	1		-1	1		-1
2	1		-1	2		-2
3	1		-1	3		-3
4	2		-2	5		-5
5	2	1	-1	7	1	-6
6	2	1	-1	9	2	-7
7	3	2	-1	12	4	-8
8	3	3	0	15	7	-8
9	3	3	0	18	10	-8
10	4	4	0	22	14	-8
11	4	4	0	26	18	-8
12	5			31		
13	5			36		
14	5			41		
15	6			47		
16	6			53		
17	6			59		
18	7			66		
19	7			73		
20	7			80		

From Monthly Graph Trendlines			Difference
Slope	0.352	0.571	0.219
Intercept	0.305	-2.000	-2.305

From Cumulative Graph Trendlines			Plan	Actual	Difference
Coeff x^2		a	0.178	0.298	0.120
Coeff		b	0.442	-1.869	-2.311
Constant		c	0.000	2.714	2.714

General Solution of a Quadratic Equation = 0 given by:	$x = \dfrac{-b \pm \sqrt{b^2 - 4ac}}{2a}$		Period, x
		First Solution (ref only)	1.256
		Second Solution	18.002

Linear and nonlinear properties (!) of straight lines | 41

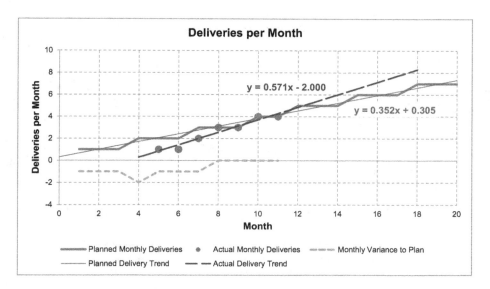

Figure 2.13 Finding the Cumulative Recovery Point to a Linear Plan (1)

From Figure 2.13 we can extract the slope and intercept for both 'Best Fit' trendlines, and calculate the difference between the two slopes and the two intercepts to get an equivalent straight line for the variance recovery. Note that we should resist the temptation to fit a straight line through the variance data as this is a composite curve covering the variance build-up and its recovery. (*It gets worse before it gets better! In other words, it is only a straight line from the point at which recovery commences, in this case Month 4 or arguably Month 5.*)

Figure 2.14 shows the Cumulative position of the linear data, to which we can fit Quadratic Trendlines. Note that we should impose an intercept of zero on the plan because we are assuming that the project starts at Month 0. However, in the case of the recovery we should leave the intercept unconstrained as we are *not* beginning recovery at Month 0 (i.e. Month 4 is the new Month 0).

We can now extract the coefficients for the x^2 and x terms for both these curves, and the constant in the case of the recovery curve. In Table 2.6 we took the difference between the corresponding coefficients in order to get the equivalent Cumulative Variance Recovery Curve. Finally, we can use our rekindled favourite formula from school for the general solution of a quadratic equation to find when the recovery reaches zero. In this case we recover at Month 18.

We can adapt this technique to situations where the plan and the recovery profiles are not linear but the difference is linear.

Linear and nonlinear properties (!) of straight lines

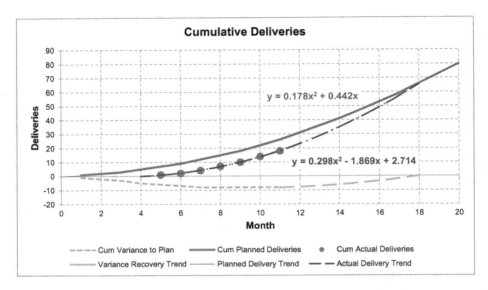

Figure 2.14 Finding the Cumulative Recovery Point to a Linear Plan (2)

OK, *from your facial expression I can see that you want a period of personal recovery. Put the kettle on.*

Example: Using the Quadratic Curve to Estimate a Missing Cumulative Value

Consider the situation in which data has been 'lost' during a migration from a legacy system to a new Enterprise Planning system (*possibly because no-one remembered to ask the estimator what data was important to retain. Is my cynicism beginning to show again?*) Suppose we know how many widgets we have made every month on a project since the new system was introduced, but not how many were made beforehand. (*As usual we want the data now and cannot wait for the archive to be retrieved!*) Suppose also that we know that we appear to be increasing output per week linearly from the data we do have and that the Project commenced ten weeks before the new system was implemented. (*They weren't going to wait just because the IT slipped, were they?*)

If we plot the Cumulative number of Widgets Produced against 'Months from Project Start', we can fit a Quadratic Trendline in Excel but this time we should not constrain the intercept to be zero (that is only valid from the start of the series.) By projecting the Quadratic Trendline back to Month 0 we can get an estimate of the cumulative number of widgets 'lost' in the systems handover using the negative intercept.

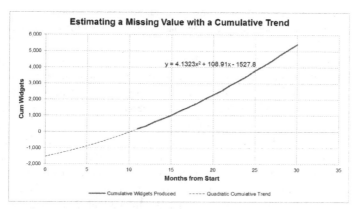

From the example, the intercept of -1528 is an estimate of the number of widgets produced previously, and 'lost' in the systems migration.

2.3 Chapter review

In this chapter, we explored the humble straight line, which can be categorised as being either discrete or continuous linear functions in respect of the input variable (horizontal x-axis), and could be expressed as a function of any two points on the line. These two points allow us to define the slope or gradient of the line and where it intersects the vertical axis (intercept).

Where the straight line represents an estimating relationship in which the data values are independent of each other, then the only property of real note is that the implied best fit relationship is either increasing or decreasing monotonically (incessantly in one direction).

However, where the data points are not independent of each other, but form a natural data sequence, we may want to consider the cumulative value of those relationships. In these cases we found that the Cumulative Value sequence was always a Quadratic Function, whose coefficients were functions of the linear data's slope and intercept. If the Cumulative Value sequence commenced with the first value in the linear sequence, then the Cumulative Value's Quadratic Function passes through the origin. The inherent variability or scatter in the underlying linear trend will be naturally smoothed in the cumulative perspective. Robert Henri (attributed by Cipra, 1998) may have

> ### A word (or two) from the wise?
>
> *"A curve does not exist in its full power until contrasted with a straight line."*
>
> **Robert Henri**
> (1865–1929)
> American Realist Painter and Educator

44 | Linear and nonlinear properties (!) of straight lines

been referring to curves and lines in an artistic sense rather than the mathematical sense employed here, but the same observation can be made. After all is Estimating not part art as well as science?

We discovered that the Cumulative Quadratic property allows us to consider data where we have sequential data, but not necessarily fixed sequential interval reporting periods – which could mask that there is an underlying linear relationship. We may typically report data in fixed time periods which do not necessarily reflect that the profiling of the data (ramp up/ramp down) in question.

We explored the power of this property (*no pun intended, honestly*) and how we might use cumulative data as a smoothing function where we have erratic data, or how in some circumstances we can use it to estimate missing data. We concluded with a foray into our past lives and discovered that there really was a use for the general solution of a quadratic equation that we studied in school . . . and wondered why we had. Well, we know now it was in the syllabus to help prepare us all to be estimators or forecasters!

Reference

Cipra, BA (1998) 'Strength Through Connections at IMA', Philadelphia, *SIAM News*, Vol 31 No 6, July/August.

3 Trendsetting with some Simple Moving Measures

Whenever we are faced with data that changes over time or over some other ordered sequence such as a build number, we usually want to know whether there is any underlying trend: going up, coming down or staying put. There's a variety of Moving Measures that can help us.

3.1 Going all trendy: The could and the should

Very rarely are we provided with the luxury of data with a totally smooth trend. It is more usual for us to be presented with data that is generally smooth but which is also characterised by what appears to be a degree of random behaviour. Failing that, the data might even resemble a random scatter similar to a discharge from a shotgun.

Faced with one of three scenarios, we might respond in different ways (Table 3.1). We might want to consider 'removing' some of the random variations in order to see the underlying pattern or trend in the data using one or more of the basic trend smoothing techniques of Moving Averages, Moving Medians, Exponential Smoothing, Cumulative Averages or Cumulative Smoothing. This chapter will discuss some of the options available.

3.1.1 When should we consider trend smoothing?

If the data can be classified as falling into a natural sequence but the data is characterised by variations, some of which are a consequence of its position in the sequence, but others are due to factors of a more random nature, e.g. where there is a genuine difference in performance between a number of activities, or where the integrity of the booking discipline or data capture is in question.

If the natural data sequence is date based, then it is often referred to as a Time Series, examples of which include escalation indices, exchange rates, delivery rates, arising rates or failure rates, sales, etc., all of which are recorded as values at a regular point in time. (Note: Here time is defined in relation to a calendar date to differentiate it from time as a measure of resource effort expended, e.g. hours of working.)

46 | Trendsetting with some Simple Moving Measures

Table 3.1 Going all Trendy – The Could and the Should

Scenario	We could . . .	We should . . .
Perfectly smooth data trend	Project the line to create an estimate, then sit back with a smug expression	Be suspicious. Check the source and integrity of the data, and question whether there has been a case of creative accounting
Slightly irregular data trend	Draw a line or curve through the middle of the trend, and explain that the estimate might be plus or minus a particular percentage	Try to understand what has caused any of the apparent anomalies, and decide whether it is practical to normalise the data to reduce the effects of the difference before analysing the underlying trend and the degree of uncertainty around it (see normalisation procedures, in Volume I Chapter 6)
Random scatter	Simply throw in the towel, and moan about "nothing spoils a good estimate more than a bagful of bad data"	Check that the data is comparing "apples with apples" and not a whole basket of fruit! Ask whether it is practical to normalise the data to reduce the effects of the difference before analysing the underlying trend and the degree of uncertainty around it[1]

Other natural sequences might be based on the cumulative physical achievement, examples of which include Learning Curves, Number of Design Queries Raised per Production Build Number, etc.

Both of these are discrete sequences where the interval between successive units in the sequence is fixed or constant, and the sequence position might be considered to be an indicator or driver of the value we are trying to predict or understand. There is, of course, another category in which data occurs in a natural sequence but the gap between values is irregular, such as batch costs where the batch size is variable. In such cases an output based Cumulative Average might remain a valid smoothing option, but Moving Average and Moving Median Smoothing would not be appropriate.

Whilst it is recognised that months have different numbers of days, for the purposes of defining a natural date-based sequence, months are usually considered to be of equal duration i.e. constant; the same applies to quarters and years.

For data which occur in a natural sequence, trend smoothing should always be considered when we want to extrapolate to a later (or earlier) value in the sequence, i.e. outside of the range for which data is available. If we want to interpolate a typical value within the range then trend smoothing remains a 'good practice' option but other techniques might also be considered, for example a simple or a weighted average of adjacent points (see Volume II Chapter 2 on Measures of Central Tendency).

3.1.2 When is trend smoothing not appropriate?

It is not appropriate to smooth data which does not fall into a natural sequence or cannot be ordered logically. Nor is it appropriate to smooth sequential data drawn from different populations, e.g. failure rates for a number of different products with no common characteristics. (In other words, trying to compare '*apples with oranges*' or, worse still the complete basket of fruit!).

If there is data missing in the natural sequence (either time or achievement based) then whilst some sequential data smoothing techniques may still present a viable option, we may have to resort to more sophisticated methods for smoothing data such as Curve Fitting. If we were simply to regard it as 'missing data' and assume that it 'would have probably fitted the pattern of the rest of the data anyway', then we are making a potentially dangerous assumption. By ignoring its existence, or rather its absence, we risk introducing statistical bias into the analysis with a potentially flawed conclusion. What if the missing data were extreme, but not to the extent that we might classify it as an 'outlier' (Volume II Chapter 7) if it were known? Where do we draw the line between '*plausible deniability*' and '*unmitigated optimism*'?

Where there is no natural sequence in which the data occurs, but the data can be ordered and plotted in relation to some scale factor which discriminates the data values, e.g. some form of size or complexity measure, then we might still want to consider trend smoothing. Typically the scale factor might represent cost driver values, examples of which might include: number of tests performed, component weight, software lines of code (SLOC), batch size, etc. In this case the scale factor can be a continuous function just as much as a discrete one. However, in this case the trend analysis is more appropriately dealt with through a Curve Fitting technique as opposed to trend smoothing by Moving Averages, Moving Medians, Exponential Smoothing, Cumulative Averages or Cumulative Smoothing. We will consider Curve Fitting in Chapters 5 to 7.

While we are on the subject of batch sizes . . .

You may have spotted that we have cited examples where it is both 'appropriate' and 'not appropriate' to smooth data where the batch size is an indicator of cost.

(Continued)

It is appropriate to use Cumulative Averages or Cumulative Smoothing when the batch size is merely a multiplier of the unit cost which is itself subject to some underlying trend, e.g. a learning curve. In this case the inference is that there are no batch set-up costs.

However, **it is not appropriate** to use any of the basic sequential smoothing techniques in this chapter where there is a distinct difference in the level of unit costs within any batch that may be attributable to the batch size being produced. In other words, where there are batch set-up costs to consider.

Clearly, we can use our judgement here as the estimator's world is rarely so black and white. We need to consider the relative size of the variation in unit cost without set-up with the value of the set-up costs divided by the likely minimum batch size to be produced. If the former is of a similar magnitude to the latter then the batch size influence on the level or driver of unit costs can probably be ignored and a Cumulative Average or Cumulative Smoothing technique used.

Enough chatter, let us look at the basic smoothing techniques for sequential data available to us.

3.2 Moving Averages

> ### Definition 3.1 Moving Average
>
> A Moving Average is a series or sequence of successive averages calculated from a fixed number of consecutive input values that have occurred in a natural sequence. The fixed number of consecutive input terms used to calculate each average term is referred to as the Moving Average Interval or Base.

The Moving Average, sometimes known as Rolling Average or Rolling Mean, is a dynamic sequence in the sense that every time an additional input value in the sequence becomes available (typically the latest actual), a corresponding new Moving Average term can be calculated and added to the end of the Moving Average sequence. The new term is calculated by replacing the earliest value used in the last average term with the latest input value in the natural sequence. Most definitions of Moving Average in the public domain do not include the need for multiple successive average terms from a sequence of values. Without this distinction, a single term from a Moving Average array is nothing more than a simple Average.

3.2.1 Use of Moving Averages

The primary use of Moving Averages is in identifying and calculating an underlying steady 'state value' in a natural sequence. A secondary use is in identifying a change in that steady 'state value' or a change in the underlying trend, or 'momentum' in the case of the stock market. In both cases, if we choose the Moving Average Interval appropriately, the technique can act as a means of normalising data for differences that occur in a repeating cycle, or a seasonal pattern, such as our quarterly gas or electricity bills; we will return to this in Chapter 8.

Moving Averages are widely used in the generation of governmental produced economic indices published for use by businesses and government agencies, typically those that reflect monthly, quarterly or annual economic trends, for example:

*   Retail price index
*   Producer price indices
*   Public service output, inputs and productivity

There are two basic types of Moving Average:

*   Simple Moving Average – a sequence of simple averages or Arithmetic Means
*   Weighted Moving Average – a sequence of simple weighted averages

Note: There are corresponding averages based on the entire range of values in this natural sequence which we will consider in Sections 3.5 and 3.6 on Exponential Smoothing and Cumulative Averages.

3.2.2 When not to use Moving Averages

We should not contemplate applying a Moving Average to data that does not fall into a natural sequence, or to data that is at irregular intervals even though it may fall in a natural sequence. This includes instances where there are multiple missing values scattered at random through a sequence. We can accommodate the occasional missing value in a sequence; this is discussed and demonstrated in Section 3.2.8.

Even though cumulative data occurs by default in a natural sequence, and may well have regular incremental steps such as time periods or units of build, we should resist the temptation to apply moving averages to such data. Whilst we can go through the mechanics of the calculations, it is probably questionable

> As previously noted, although calendar months are not of equal length, and accounting periods may be configured in a pattern of 4 weeks, 4 weeks and 5 weeks, both can be considered to be 'regular' on the pretext of being a month or an accounting period.

50 | Trendsetting with some Simple Moving Measures

from a '*does it make sense*' perspective. We would, in essence, be double counting. Taking the Cumulative Value of a sequence of data points is in itself a trend smoothing technique that removes random variations between successive values. We are probably better using some other form of Curve Fitting routine if the data is not yet smooth enough. Moving Averages are simple, but they are not without 'issues' that we must address.

3.2.3 Simple Moving Average

A Simple Moving Average with Interval N computes the simple average or Arithmetic Mean for the first N terms in a natural sequence, and then repeats the process by deleting the first term used in the calculation and adding the $(N+1)^{th}$ term. The process repeats until the end of the sequence is reached.

For the Formula-phobes

In Microsoft Excel we can use the in-built function **AVERAGE(*range*)** where the range is the interval that represents the input values for a single term in the Moving Average array. Copy the formula across to each successive term in the sequence.

Alternatively, we can use the Data Analysis Toolkit

For the Formula-philes

$$A_i = \sum_{j=i-N+1}^{i} \frac{x_j}{N}$$

where A_i is the Simple Moving Average of the last N terms in the sequence x_j, ending at term x_i

---

In the example below (Table 3.2 and Figure 3.1) a 'steady state' value for 'hours booked per month' can be inferred to be around 160 hours per month. In this case, it has been recognised that there are thirteen weeks in a quarter (52 in a year), and that for accounting purposes these are considered to be a repeating pattern of 4–4–5. For this reason, it should be considered that an appropriate Moving Average Interval would be 3. In this situation, the hours can be considered to have been normalised to reflect an average of 4.33 weeks per month, or 30 to 31 days per month.

Incidentally, if 'trial and error' had been used to determine the Moving Average Interval, then inappropriate interval lengths will be highlighted by Moving Averages that are less smooth than the 'appropriate' one. (*Notice that I resisted saying 'correct'?*) Unfortunately, the simplistic nature of the Simple Moving Average will also give smooth results for any

Table 3.2 Simple Moving Average with an Interval of 3

Hours Booked per Month				
Weeks per Month	Month	Hours Booked	Sum of last 3 months	Moving Average (Interval of 3)
4	1	148		
4	2	145		
5	3	190	483	161
4	4	142	477	159
4	5	145	477	159
5	6	187	474	158
4	7	148	480	160
4	8	145	480	160
5	9	190	483	161

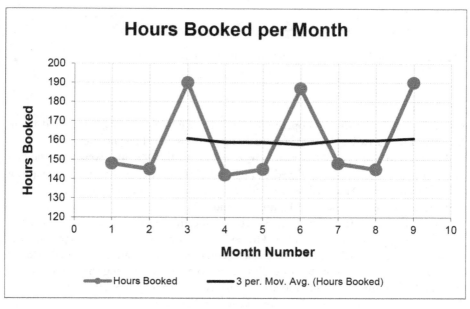

Figure 3.1 Simple Moving Average with an Interval of 3

52 | Trendsetting with some Simple Moving Measures

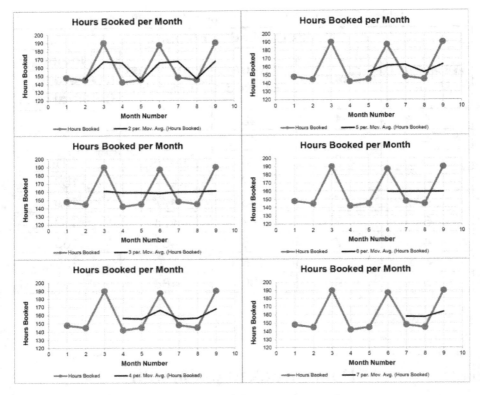

Figure 3.2 Simple Moving Averages with Increasing Intervals

integer multiplier of the 'appropriate' interval. In Figure 3.2, below, this would be an Interval of 6, (i.e. twice 3). So, if you are going to guess, start low and increase!

It is custom and practice in many circles to record the Moving Average at the end of the interval sequence as displayed in Table 3.2. This practice is reflected in Microsoft Excel, in both the Data Analysis Tool for Moving Average, and in the Chart Moving Average Trendline Type (used in Figures 3.1 and 3.2).

If our objective is to identify a 'steady state' value, this practice of recording the average at the end-point is not a problem. However, if our objective is also to understand the point in the series at which that 'steady state' commences, or to ascertain whether there is an underlying upward or downward trend, the practice of plotting the average at the cycle end-point means that the moving average is a misleading indicator as it will always lag the true underlying trend position as shown in Figure 3.3.

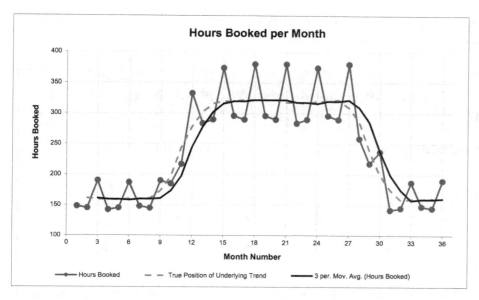

Figure 3.3 Example of Simple Moving Average Lagging the True Trend

Pictorially, this lag is always displaced to the right of the true trend . . . looking like a fugitive leading a line of data in its pursuit. *OK, a bit of a contrived link to the work of William Mauldin, but I liked the quote. (Casaregola, 2009, p. 87.)*

The true underlying trend position will always occur at the mid-point of the Moving Average Interval, and the length of the lag can be calculated as:

> **A word (or two) from the wise?**
>
> '*I feel like a fugitive from the law of averages.*'
> **William H. Mauldin**
> American cartoonist
> 1921–2003

$$\text{Simple Moving Average Lag} = \frac{(\text{Interval} - 1)}{2}$$

Therefore, it is recommended (*but perhaps we should make it our Law of Moving Averages*) that all Simple Moving Averages are plotted at the mid-point of the Interval rather than at the end of the cycle. Whilst this may well mean that we have to introduce an extra calculation step in a spreadsheet (Microsoft Excel will not do it automatically), it may help us in justifying to others that the Moving Average Interval chosen does produce the trend that is representative of the data.

> **For the Formula-phobes: Why is the lag not just half the interval?**
>
> Consider a line of telegraph poles and spaces (or a chain-linked fence):
>
>
>
> The number of spaces between the poles is always one less than the number of poles. For every odd number of poles there are an equal number of spaces on either side of the pole in the middle. which is clearly at the mid-point. In the case of a Moving Average, the number of terms in the Interval is equivalent to the number of telegraph poles, but the mid-point can be described in relation to the spaces.
>
> The same analogy works for an even number of terms or telegraph poles, where the mid-point is defined by the space in the middle (there being an odd number of spaces).

Many quarterly and annual economic indices published by government agencies are Simple Moving Averages. It is important we note that although the indices are published at the end of the quarter or year, they do in fact represent the average for the quarter or year – nominally the mid-point taking account of the lag we have discussed.

3.2.4 Weighted Moving Average

One criticism of the Simple Moving Average is that it gives equal weight to every term in the series, and in the case of non-steady state trends, the Simple Moving Average is slow to respond to changes in the underlying trend. One way to counter this is for us to give more emphasis or weight to the most recent data and less to the oldest or earliest data. One commonly used method is to apply a sliding scale of weights based on integers divided by the sum of the integers in the series; this guarantees that the sum of the weights is always 100%.

For the Formula-phobes

We can use the in-built function **SUMPRODUCT***(range1, range2)* in Microsoft Excel, where *range1* is the array of weightings to be applied, and *range2* is the interval that represents the input values for a single term in the Weighted Moving Average array. Range1 should be anchored with \$ signs, but range 2 should not be anchored. Copy the formula across to each successive term in the sequence.

Note: if the sum of the weightings do not equal 1, then the function **SUMPRODUCT** must be divided by the value **SUM***(range1)*

For the Formula-philes

$$A_i = \sum_{j=i-N+1}^{i} w_j x_j$$

where A_i is the Weighted Moving Average of the last N terms in the sequence x_j at term x_i with weightings w_1 to w_N such that:

$$\sum_{j=1}^{N} w_j = 1$$

For example, in Table 3.3, using an interval of 5, the weights might be calculated as:

The weight of nth term in the Interval, 1 to N, can be calculated as: $\dfrac{2n}{N(N+1)}$

Table 3.3 Sliding Scale of Weights for Weighted Moving Average of Interval 5

Sliding Scale of Weights for Weighted Moving Average of Interval 5			
Term	Arithmetic Weighting	Final Weighting	Equivalent to Fraction
1st	1	0.067	1/15
2nd	2	0.133	2/15
3rd	3	0.200	3/15
4th	4	0.267	4/15
5th	5	0.333	5/15
Total	15	1	15/15

56 | Trendsetting with some Simple Moving Measures

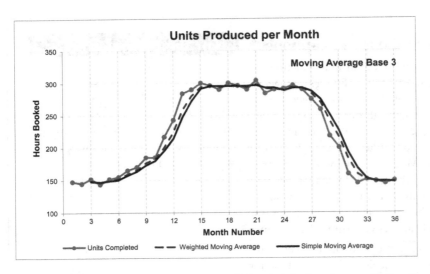

Figure 3.4 Weighted Moving Average cf. Simple Moving Average Before Lag Adjustment

As we can see, in reality, the difference in response to changes in data between Simple and Weighted Moving Averages is marginal, especially for small interval moving averages as shown in Figure 3.4. In order to adjust for the residual lag in response, the estimator can always resort to plotting the Weighted Moving Average at an earlier point in the cycle rather than the end as for the Simple Moving Average. In this case, the true underlying trend position of a Weighted Moving Average Interval using an increasing Arithmetic Progression of weights can be calculated as:

$$\text{Weighted Moving Average Lag} = \frac{(Interval - 1)}{3}$$

However, it is debatable whether this presents any tangible benefit over offsetting the Simple Moving Average, especially for small intervals, as depicted in Figure 3.5 (Spot the Difference if you can!).

For the Formula-philes: Why is the lag not the same as that for a Simple Moving Average?

For a **Simple Moving Average** there are implied equal weights for each term giving a uniform or rectangular distribution, the mean or expected value of which is:

$$\text{Mean} = \frac{\text{Maximum} + \text{Minimum}}{2} \quad (1)$$

The Moving Average Lag is the difference between the Mean position and the Maximum position, hence:	$\text{Lag} = \text{Maximum} - \text{Mean}$	(2)
Substituting (1) in (2) and simplifying:	$\text{Lag} = \dfrac{\text{Maximum} - \text{Minimum}}{2}$	(3)
Substituting, Maximum = Interval, and Minimum =1 in (3), to represent the last and first points in any interval:	$\text{Lag} = \dfrac{\text{Interval} - 1}{2}$	(4)
A **Weighted Moving Average** with an Arithmetic Progression of weights, forms a triangular distribution, the Mean of which is:	$\text{Mean} = \dfrac{\text{Max} + \text{Mode} + \text{Min}}{3}$	(5)
The Arithmetic Progression forms a right-angled triangle in which the Mode = Max, (5) becomes:	$\text{Mean} = \dfrac{2 \times \text{Max} + \text{Min}}{3}$	(6)
From (2) and (6):	$\text{Lag} = \dfrac{\text{Maximum} - \text{Minimum}}{3}$	(7)
Substituting, Maximum = Interval, and Minimum =1 in (7) to represent the last and first points in any interval:	$\text{Lag} = \dfrac{\text{Interval} - 1}{3}$	

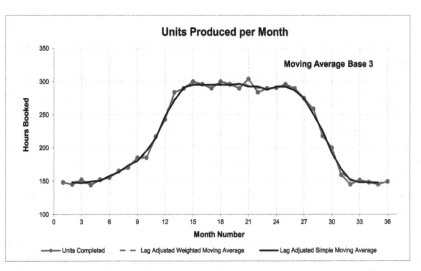

Figure 3.5 Weighted Moving Average cf. Simple Moving Average After Lag Adjustment

58 | Trendsetting with some Simple Moving Measures

Table 3.4 Moving Average Trend Lags

Interval	Simple Moving Average Lag	Weighted Moving Average Lag
2	0.5	0.33
3	1	0.67
4	1.5	1
5	2	1.33
6	2.5	1.67
12	5.5	3.67
N	(N-1)/2	(N-1)/3

Note: The Weighted Average here is based on an Arithmetic Progression of weightings

The lags for both Simple Moving Averages and Weighted Moving Averages can be summarised in Table 3.4 above.

3.2.5 Choice of Moving Average Interval: Is there a better way than guessing?

Basically, we can choose the Moving Average Interval (the number of actual terms used in each average term) in one of three ways:

a) Where possible, it is recommended that the choice of Moving Average Interval is determined by a logical consideration of the data's natural cycle which might explain the cause of any variation, e.g. seasonal fluctuations might be removed by considering four quarters or twelve months.

b) If there is no natural cycle to the data, or it is not clear what it might be, we can determine the Moving Average Interval on a '*trial and error*' basis. In these circumstances, it is recommended that we start with a low interval and increase in increments of one until we are satisfied with the level of smoothing achieved. However, some of us may not feel totally comfortable with this '*trial and error*' approach, tending to regard it as more of a '*hit and miss*' or '*hit and hope*' approach!

c) As an alternative, we can apply a simple statistical test for 'smoothness' that will yield an appropriate Moving Average interval systematically.

In terms of the latter, one such systematic numerical technique would be to calculate the Standard Deviation of the differences between corresponding points using adjacent Moving Average Interval Differences or MAID for short (*but I promise not to milk it*);

Standard Deviation being a Descriptive Statistic that quantifies how widely or narrowly data is scattered or dispersed around its mean (Volume II Chapter 3).

This MAID Minimal Standard Deviation technique requires us firstly to calculate the difference between the start (or end) of each successive Moving Average interval, and then to calculate the Standard Deviation of these Differences. We can also calculate the Difference Range by subtracting the Minimum Difference from the Maximum Difference. If we repeat this procedure for a number of different potential Moving Average Intervals, we can then compare the MAID Standard Deviations; the one with the smallest Standard Deviation is the 'best' or smoothest Moving Average. This is likely to coincide with one of the smaller Difference Ranges, but not necessarily the smallest on every occasion.

The MAID calculation is illustrated in Figure 3.6 for an interval of 3. The difference between each point and the point that occurred three places earlier has been marked with a dotted arrow for the first three data pairs.

The difference between the 4th point and the 1st point is -2
The difference between the 5th point and the 2nd point is +4
The difference between the 6th point and the 3rd point is +2

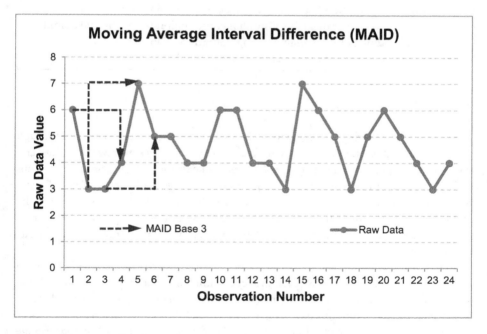

Figure 3.6 Example of Moving Average Interval Difference (MAID)

For the Formula-phobes: Why are we looking at Interval Differences?

The rationale for looking at Interval Differences is fundamentally one of '*swings and roundabouts*'. If we have a pattern in the data that repeats (*roundabouts*) with some values in the cycle higher and others values lower (*swings*), then there will probably be only a small variation in all the highs, and a small variation in all the lows, but also a small variation in all the middle range values.

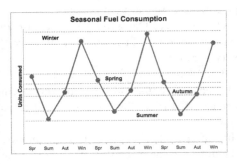

Fuel bills are a typical example of this where winter consumption will vary from year to year, as will that for spring, summer and autumn. However, they are all likely to vary only within the confines of their own basic seasonal level.

If we looked at every third season instead of four, we would get a much wider variation in consumption between the corresponding points in the cycle

If we were to look at fuel costs instead of consumption we might find a similar pattern of banding but with an underlying upward trend in all four seasons (*Come on, can you remember the last time that fuel prices reduced?*)

Table 3.5 summarises the differences for all 24 data observations for intervals of 2 through to 12. An interval of one is simply the difference between successive raw data values and is included as a point of reference only. The standard deviation of each of these columns of differences can then be calculated using **STDEV.S(*range*)** in Microsoft Excel. Finally, we can also calculate the range of the differences for each interval by subtracting the minimum difference from the maximum. The purpose of this is to provide verification that the standard deviation is highlighting the interval with the narrowest range of difference values.

In the example, an interval of 4, 5, 6, 10 and 11 all have a MAID Standard Deviation which is less than that of the raw data differences, implying that they will all smooth the data to a greater or lesser extent. The smallest standard deviation is for a Moving Average of Interval 10, so technically this gives us the MAID Minimum Standard Deviation. However, from a more practical rather than a theoretical viewpoint, a Moving Average Interval of 5 in this case also improves the MAID Standard Deviation substantially in relation to the raw input data differences. It is quite possible that an acceptable result

Trendsetting with some Simple Moving Measures | 61

Table 3.5 Choosing a Moving Average Interval Using the MAID Minimal Standard Deviation Technique

		Moving Average Interval											
		1	2	3	4	5	6	7	8	9	10	11	12
Obs ID	Raw Data	Difference between Raw Data Values at specified Intervals											
1	6	#N/A	#N/A	#N/A	#N/A	#N/A	#N/A	#N/A	#N/A	#N/A	#N/A	#N/A	#N/A
2	3	-3	#N/A	#N/A	#N/A	#N/A	#N/A	#N/A	#N/A	#N/A	#N/A	#N/A	#N/A
3	3	0	-3	#N/A	#N/A	#N/A	#N/A	#N/A	#N/A	#N/A	#N/A	#N/A	#N/A
4	4	1	1	-2	#N/A	#N/A	#N/A	#N/A	#N/A	#N/A	#N/A	#N/A	#N/A
5	7	3	4	4	1	#N/A	#N/A	#N/A	#N/A	#N/A	#N/A	#N/A	#N/A
6	5	-2	1	2	2	-1	#N/A	#N/A	#N/A	#N/A	#N/A	#N/A	#N/A
7	5	0	-2	1	2	2	-1	#N/A	#N/A	#N/A	#N/A	#N/A	#N/A
8	4	-1	-1	-3	0	1	1	-2	#N/A	#N/A	#N/A	#N/A	#N/A
9	4	0	-1	-1	-3	0	1	1	-2	#N/A	#N/A	#N/A	#N/A
10	6	2	2	1	1	-1	2	3	3	0	#N/A	#N/A	#N/A
11	6	0	2	2	1	1	-1	2	3	3	0	#N/A	#N/A
12	4	-2	-2	0	0	-1	-1	-3	0	1	1	-2	#N/A
13	4	0	-2	-2	0	0	-1	-1	-3	0	1	1	-2
14	3	-1	-1	-3	-3	-1	-1	-2	-2	-4	-1	0	0
15	7	4	3	3	1	1	3	3	2	2	0	3	4
16	6	-1	3	2	2	0	0	2	2	1	1	-1	2
17	5	-1	-2	2	1	1	-1	-1	1	1	0	0	-2
18	3	-2	-3	-4	0	-1	-1	-3	-3	-1	-1	-2	-2
19	5	2	0	-1	-2	2	1	1	-1	-1	1	1	0
20	6	1	3	1	0	-1	3	2	2	0	0	2	2
21	5	-1	0	2	0	-1	-2	2	1	1	-1	-1	1
22	4	-1	-2	-1	1	-1	-2	-3	1	0	0	-2	-2
23	3	-1	-2	-3	-2	0	-2	-3	-4	0	-1	-1	-3
24	4	1	0	-1	-2	-1	1	-1	-2	-3	1	0	0
Standard Deviation		1.70	2.16	2.27	1.59	1.08	1.63	2.27	2.31	1.77	0.83	1.57	2.12
Std Dev Rank		6	9	10	4	2	5	11	12	7	1	3	8
Minimum Difference		-3	-3	-4	-3	-1	-2	-3	-4	-4	-1	-2	-3
Maximum Difference		4	4	4	2	2	3	3	3	3	1	3	4
Difference Range		7	7	8	5	3	5	6	7	7	2	5	7
Average Difference (Ref)		-0.09	-0.09	-0.05	0	-0.05	-0.06	-0.18	-0.13	0	0.07	-0.15	-0.17

might be produced using an interval of 5. Perhaps unsurprisingly, the largest two standard deviations fall between these two (at 7 and 8.)

Incidentally, in case you were wondering, searching for an average difference close to zero does not work. For both Intervals 4 and 9 in Table 3.5, the average difference is zero due to the influence of some relatively large differences in comparison to the rest of the set of differences, but these are not the best from a data smoothing perspective.

Figure 3.7 compares the MAID Standard Deviation for the 11 actual intervals (2–12) with the standard deviation for Interval 1 (simple difference between successive raw data values).

Both results for Intervals 5 and 10 are shown in Figure 3.8 and Table 3.6. We can then make the appropriate judgement call – go for a low standard deviation, or simply the lowest? Hence why the technique is called the 'MAID **Minimal** Standard Deviation' as opposed to the 'MAID **Minimum** Standard Deviation'; the intent is to lessen or reduce the significance of the standard deviation, rather than find the absolute minimum *per se*. As a general rule, we are often better opting for the smaller interval if we want to track

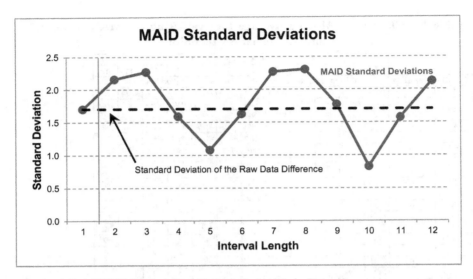

Figure 3.7 MAID Minimal Standard Deviation Technique

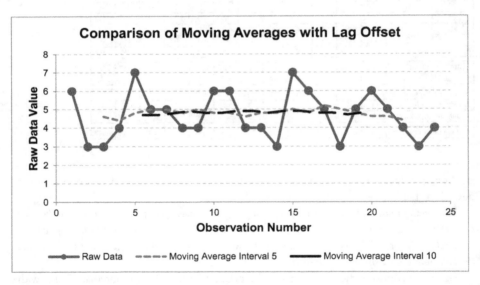

Figure 3.8 Moving Average Interval Options Using the MAID Minimal Standard Deviation Technique

changes in rate earlier – longer intervals are slower to respond to changes in the underlying trend. Unsurprisingly, if a Moving Average Interval of 5 gives an acceptable result, then so will any multiple of 5, e.g. 10, 15, etc.

Table 3.6 Moving Average Interval Options Using the MAID Minimal Standard Deviation Technique

Obs ID	Raw Data	Moving Average Interval 5	Moving Average Interval 10
1	6	#N/A	#N/A
2	3	#N/A	#N/A
3	3	#N/A	#N/A
4	4	#N/A	#N/A
5	7	4.60	#N/A
6	5	4.40	#N/A
7	5	4.80	#N/A
8	4	5.00	#N/A
9	4	5.00	#N/A
10	6	4.80	4.70
11	6	5.00	4.70
12	4	4.80	4.80
13	4	4.80	4.90
14	3	4.60	4.80
15	7	4.80	4.80
16	6	4.80	4.90
17	5	5.00	4.90
18	3	4.80	4.80
19	5	5.20	4.90
20	6	5.00	4.90
21	5	4.80	4.80
22	4	4.60	4.80
23	3	4.60	4.70
24	4	4.40	4.80

The MAID Minimal Standard Deviation Technique can also be made to work (*sorry, bad pun*), where there are underlying increasing or decreasing trends in the data. The following series of figures and tables illustrate the process we have just followed:

Figure 3.9	Data to be smoothed by a Moving Average
Table 3.7	MAID Calculations for Intervals of 2 through to 12
Figure 3.10	Comparison of MAID Standard Deviations highlighting that an Interval of 4 is the best (minimum) option for this dataset
Table 3.8	Moving Average Calculation for an Interval of 4
Figure 3.11	Moving Average Plot of Interval 4 with the inherent lag offset to the Interval mid-point

In this situation, we would be advised going for the shorter interval of 4, which happens to be the one which returns the MAID Minimum Standard Deviation.

However, we should seriously reflect on whether the MAID Minimal Standard Deviation technique is any better than 'trial and error' as it would suggest a level of accuracy

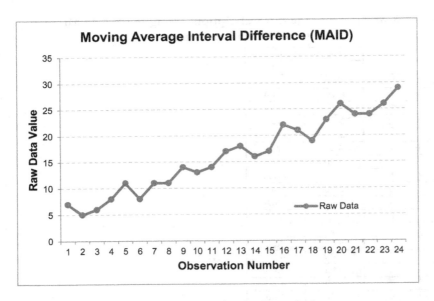

Figure 3.9 Data with Increasing Trend to be Smoothed

Table 3.7 Applying the MAID Minimal Standard Deviation Technique to Data with an Increasing Trend

		Moving Average Interval										
		1	2	3	4	5	6	7	8	9	10	11	12
Obs ID	Raw Data	Difference between Raw Data Values at specified Intervals										
1	7	#N/A	#N/A	#N/A	#N/A	#N/A	#N/A	#N/A	#N/A	#N/A	#N/A	#N/A	#N/A
2	5	-2	#N/A	#N/A	#N/A	#N/A	#N/A	#N/A	#N/A	#N/A	#N/A	#N/A	#N/A
3	6	1	-1	#N/A	#N/A	#N/A	#N/A	#N/A	#N/A	#N/A	#N/A	#N/A	#N/A
4	8	2	3	1	#N/A	#N/A	#N/A	#N/A	#N/A	#N/A	#N/A	#N/A	#N/A
5	11	3	5	6	4	#N/A	#N/A	#N/A	#N/A	#N/A	#N/A	#N/A	#N/A
6	8	-3	0	2	3	1	#N/A	#N/A	#N/A	#N/A	#N/A	#N/A	#N/A
7	11	3	0	3	5	6	4	#N/A	#N/A	#N/A	#N/A	#N/A	#N/A
8	11	0	3	0	3	5	6	4	#N/A	#N/A	#N/A	#N/A	#N/A
9	14	3	3	6	3	6	8	9	7	#N/A	#N/A	#N/A	#N/A
10	13	-1	2	2	5	2	5	7	8	6	#N/A	#N/A	#N/A
11	14	1	0	3	3	6	3	6	8	9	7	#N/A	#N/A
12	17	3	4	3	6	6	9	6	9	11	12	10	#N/A
13	18	1	4	5	4	7	7	10	7	10	12	13	11
14	16	-2	-1	2	3	2	5	5	8	5	8	10	11
15	17	1	-1	0	3	4	3	6	6	9	6	9	11
16	22	5	6	4	5	8	9	8	11	11	14	11	14
17	21	-1	4	5	3	4	7	8	7	10	10	13	10
18	19	-2	-3	2	3	1	2	5	6	5	8	8	11
19	23	4	2	1	6	7	5	6	9	10	9	12	12
20	26	3	7	5	4	9	10	8	9	12	13	12	15
21	24	-2	1	5	3	2	7	8	6	7	10	11	10
22	24	0	-2	1	5	3	2	7	8	6	7	10	11
23	26	2	2	0	3	7	5	4	9	10	8	9	12
24	29	3	5	5	3	6	10	8	7	12	13	11	12
Standard Deviation		2.27	2.72	2.02	1.09	2.43	2.60	1.71	1.38	2.45	2.61	1.55	1.50
Std Dev Rank		7	12	6	1	8	10	5	2	9	11	4	3
Minimum Difference		-3	-3	0	3	1	2	4	6	5	6	8	10
Maximum Difference		5	7	6	6	9	10	10	11	12	14	13	15
Difference Range		8	10	6	3	8	8	6	5	7	8	5	5

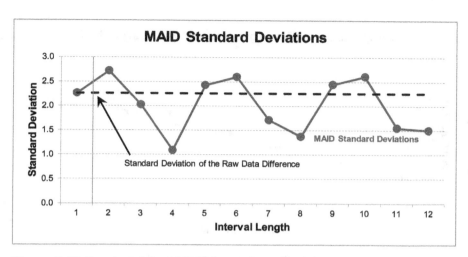

Figure 3.10 Results of the MAID Minimal Standard Deviation Technique with an increasing trend

Table 3.8 Moving Average Interval Options Using the MAID Minimal Standard Deviation Technique

		Moving Average Interval	
Obs ID	Raw Data	4	8
1	6	#N/A	#N/A
2	3	#N/A	#N/A
3	3	#N/A	#N/A
4	4	6.50	#N/A
5	7	7.50	#N/A
6	5	8.25	#N/A
7	5	9.50	#N/A
8	4	10.25	8.38
9	4	11.00	9.25
10	6	12.25	10.25
11	6	13.00	11.25
12	4	14.50	12.38
13	4	15.50	13.25
14	3	16.25	14.25
15	7	17.00	15.00
16	6	18.25	16.38
17	5	19.00	17.25
18	3	19.75	18.00
19	5	21.25	19.13
20	6	22.25	20.25
21	5	23.00	21.00
22	4	24.25	22.00
23	3	25.00	23.13
24	4	25.75	24.00

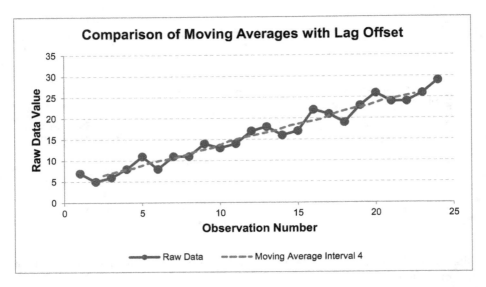

Figure 3.11 Moving Average Interval Options Using the MAID Minimal Standard Deviation Technique

that some might argue is inappropriate for what is after all a simple, even crude, method of trend smoothing. The one distinct advantage that MAID gives is that it fits with our objective of TRACEability (Volume I, Chapter 3.)

3.2.6 Can we take the Moving Average of a Moving Average?

The answer is 'Yes, because it is just a calculation'; the question we really ought to ask is 'Should we?' It is better to consider an example before we lay the proverbial egg on this. (*Sorry, did someone call me a chicken? OK, in short, the answer is 'Yes, but . . . '*)

In Table 3.9 let us consider those economic indices we estimators all love to hate. The quarterly index is a three-month moving average of the monthly index. The annual index is a twelve-month moving average of the monthly index. Does that mean that the annual index is also a four-quarter moving average of the quarterly index?

As we can see from the example, there is a right way (Col I) and a wrong way (Col J).

- Columns B-D is the published Retail Price Index (RPIX) data from the UK Office of National Statistics (ONS, 2017). If we calculate a 3-month moving average (Col F) on the published Monthly indices, we will get the published quarterly indices. As a bonus we will also get the intervening rolling quarterly data (e.g. February–April, etc.).

Table 3.9 Moving Average of a Moving Average

Month	Retail Price Index RPIX-CHAW (Published Data)			Based on Monthly Index		4 Quarter Moving Average of a 3 Month Moving Average	Simple 4 Period Moving Average of a 3 Month Moving Average
(Jan 1987 = 100)	Monthly Index	Quarterly Index	Annual Index	3 Month Simple Moving Average	12 Month Simple Moving Average		
Jan-10	217.9					*Based on every third value from Col(I)*	*Based on Col(I)*
Feb-10	219.2						
Mar-10	220.7	219.3		219.3			**Does not Match Col (G)**
Apr-10	222.8			220.9			
May-10	223.6			222.4			✗
Jun-10	224.1	223.5		223.5			221.5
Jul-10	223.6			223.8			222.6
Aug-10	224.5			224.1			223.4
Sep-10	225.3	224.5		224.5		**Matches Col(G)**	224.0
Oct-10	225.8			225.2			224.4
Nov-10	226.8			226.0		✓	224.9
Dec-10	228.4	227.0	223.6	227.0	223.6	223.6	225.7
Jan-11	229			228.1	224.5	224.5	226.6
Feb-11	231.3			229.6	225.5	225.5	227.7
Mar-11	232.5	230.9		230.9	226.5	226.5	228.9
Apr-11	234.4			232.7	227.4	227.4	230.3
May-11	235.2			234.0	228.4	228.4	231.8
Jun-11	235.2	234.9		234.9	229.3	229.3	233.2
Jul-11	234.7			235.0	230.3	230.3	234.2
Aug-11	236.1			235.3	231.2	231.2	234.8
Sep-11	237.9	236.2		236.2	232.3	232.3	235.4
Oct-11	238			237.3	233.3	233.3	236.0
Nov-11	238.5			238.1	234.3	234.3	236.8
Dec-11	239.4	238.6	235.2	238.6	235.2	235.2	237.6
Jan-12	238			238.6	235.9	235.9	238.2
Feb-12	239.9			239.1	236.7	236.7	238.6
Mar-12	240.8	239.6		239.6	237.3	237.3	239.0
Apr-12	242.5			241.1	238.0	238.0	239.6
May-12	242.4			241.9	238.6	238.6	240.4
Jun-12	241.8	242.2		242.2	239.2	239.2	241.2
Jul-12	242.1			242.1	239.8	239.8	241.8
Aug-12	243			242.3	240.4	240.4	242.1
Sep-12	244.2	243.1		243.1	240.9	240.9	242.4
Oct-12	245.6			244.3	241.5	241.5	242.9
Nov-12	245.6			245.1	242.1	242.1	243.7
Dec-12	246.8	246.0	242.7	246.0	242.7	242.7	244.6

Contains public sector information licensed under the Open Government Licence v3.0

- Similarly, if we calculate a 12-month moving average (Col G) on the published Monthly indices, we will get the published annual indices. Again as a bonus we will also get the intervening rolling annual data (e.g. February–January, etc.).
- In Col I, if we calculate a moving average of every fourth 3-month moving average value then we can replicate the values in Col G. For instance, Cell I14 equals the average of cells F5, F8, F11 and F14.
- However, in Col J, if we blindly take a simple 4-period moving average of the last four 3-month moving average values in Col F, then we will not replicate the values in Col G. For instance, the value in Cell J8 takes the average of Cells F5, F6, F7 and F8. If I can paraphrase the catchphrase from Punch and Judy '*That's not the way to do it!*'

68 | Trendsetting with some Simple Moving Measures

For the Formula-philes: Moving Average of a Moving Average

Consider the monthly data sequence values, $M_1 \ldots M_i \ldots$:

The 12-month Moving Average, A_i for $i \geq 12$:

$$A_i = \sum_{j=i-11}^{i} \frac{M_j}{12}$$

Factoring the Divisor:

$$A_i = \frac{1}{4} \sum_{j=i-11}^{i} \frac{M_j}{3}$$

Subdividing the Sum into four ranges:

$$A_i = \frac{1}{4}\left(\sum_{j=i-11}^{i-9} \frac{M_j}{3} + \sum_{j=i-8}^{i-6} \frac{M_j}{3} + \sum_{j=i-5}^{i-3} \frac{M_j}{3} + \sum_{j=i-2}^{i} \frac{M_j}{3} \right)$$

Simplifying:

$$A_i = \frac{1}{4}\left(Q_{i-9} + Q_{i-9} + Q_{i-9} + Q_{i-9} \right)$$

where Q_i is the Quarterly Moving Average of values M_{j-2} to M_i for months i-2 to i

If in doubt, always go back to the published data and calculate a moving average from scratch.

3.2.7 A creative use for Moving Averages – A case of forward thinking

We need to restrict our thinking on Moving Averages to be from a retrospective standpoint. For instance, we can calculate the level of resource required over a production build line using a Forward Weighted Moving Average, where the weightings are based on the known or expected resource distribution over a standard production build cycle for a unit (see *Example: A 'forward thinking' use for a Weighted Moving Average*).

Example: A 'forward thinking' use for a Weighted Moving Average

Consider a production line of some 20 units, each of four weeks build duration in which it is anticipated resource per unit will be required in the following proportions:

Trendsetting with some Simple Moving Measures | 69

> Week 1 10% or 0.1
> Week 2 25% or 0.25
> Week 3 30% or 0.3
> Week 4 35% or 0.35

Take the weighting factors in reverse order (35%, 30%, 25%, 10%) and calculate a Weighted Forward Moving Average with an interval equal to the build cycle (in this case 4 weeks) of the Planned Number of Completions per Week to give the equivalent number of units produced per week.

Multiply the equivalent units produced by the hours or cost per unit to get the estimated hours or costs required in each week. A similar calculation based on the cumulative number of units completed will give the S-Curve profile of hours or cost over a period of time.

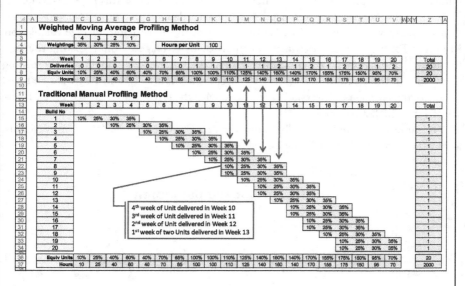

In Microsoft Excel, Cell L8 is **SUMPRODUCT($C$4:$F$4, L7:O7)** where $C$4:$F$4 are the reverse order of the weightings, and L7:O7 are the deliveries per month over the next build cycle commencing at Week 10. This Forward Weighted Moving Average of the deliveries determines the number of equivalent units to be built in that week.

Note: As the weights summate to 100%, there is no need to divide by the Sum of the Weights

The weightings do not have to be those illustrated in the example, we can consider any value of weightings that are pertinent to the data being analysed so long as the weights summate to 100%.

3.2.8 Dealing with missing data

What if we have missing data; what should we do? The worst thing we can do is ignore that it is missing! It is far better that we simply stop the Moving Average calculation at the last point at which data was available; then resume the Moving Average after the next full cycle or interval has elapsed following the missing data.

For example in Figure 3.12, data is missing for months 10 and 11. The moving average, based on an interval of 3, cannot be calculated for months 10 and 11, and consequentially it cannot be calculated for months 12 and 13, as these include values from months 10 and 11.

This will clearly leave a gap in our analysis but at least it is an honest reflection of the data available. The length of the gap will be the number of consecutive missing data points minus one, plus the interval length. In the example above, this is $(2-1)+3$, i.e. 4.

We can always provide an estimate of the missing data (*because that's what we do; we are estimators*), but we should not include that estimate in the Moving Average calculation. Microsoft Excel provides an estimate of the Moving Average by interpolating the gaps for us by changing the Moving Average Interval incrementally by one for every missing data point. In the example in Figure 6, the average at months 10 and 12 would be based on a Moving Average of Interval 2, whereas at Month 11, it would be based on an Interval of 1, i.e. the last data point.

The same advice applies equally to missing data from Weighted Moving Averages just as much as Simple Moving Averages.

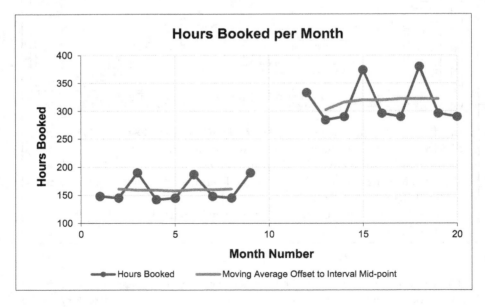

Figure 3.12 Moving Average Plot with Missing Data

In many cases it may not influence your judgement unduly, but why does it always seem to happen when something unusual is happening, like a change in the level? (*Does that sound a little paranoid? In my defence: it can be seen as a positive trait in estimators . . . always suspicious of data!*)

3.2.9 Uncertainty Range around the Moving Average

Whilst Moving Averages are not predictive in themselves, in instances where we are considering them to identify a steady state value, we might choose to use that steady state value as a forward estimate. However, the whole principle of the Moving Average is that we accept that the value will change over time or a number of units of achievement. In short any estimate we infer will have a degree of uncertainty around it. We might want to consider how best to articulate the uncertainty. We have a number of options summarised in Table 3.10.

The absolute measures of minimum and maximum are clearly sensitive to extreme values in the raw data, but are a useful reminder of the fact that things have been 'that good' or 'that bad' before and so might be again. The main benefit of calculating a Moving Average is that it gives us a topical view of the movement in the underlying data. On that basis we might want to consider the minimum or maximum of the Moving Average array rather than the total history. None of these measures need to be adjusted for interval lag.

Table 3.10 Potential Measures of Moving Average Uncertainty

Moving Average Interval	Absolute Measures (All data points)	Moving Measures (based on Interval used)
Small interval (say 2–5)	• Absolute Minimum and Maximum	• Moving Minimum and Moving Maximum
	• Moving Average Minimum and Maximum	
	• Standard Deviation	• Moving Standard Deviation
	• Moving Average Standard Deviation	
Large interval (say >6)	• *All of the above* +	• *All of the above* +
	• Absolute Percentiles (user defined)	• Moving Percentiles (user defined)
	• Moving Average Percentiles (user defined)	

72 | Trendsetting with some Simple Moving Measures

On the other hand in the spirit of 'Moving Statistics', we can always create pseudo prediction limits based on the Moving Minimum and Moving Maximum of the data for which we have calculated the Moving Average.

Definition 3.2   Moving Minimum

A Moving Minimum is a series or sequence of successive minima calculated from a fixed number of consecutive input values that have occurred in a natural sequence. The fixed number of consecutive input terms used to calculate each minimum term is referred to as the Moving Minimum Interval or Base.

Definition 3.3   Moving Maximum

A Moving Maximum is a series or sequence of successive maxima calculated from a fixed number of consecutive input values that have occurred in a natural sequence. The fixed number of consecutive input terms used to calculate each maximum term is referred to as the Moving Maximum Interval or Base.

A variation on this concept is to look at the minimum and maximum of past Moving Averages as 'absolute' lower and upper ranges of uncertainty around the Moving Average. However, the downside of this is that it is really only suitable for a basic 'steady state' system; the bounds will diverge if there is an increasing or decreasing trend. One way around this would be to consider true Moving Minima and Maxima. Figure 3.13 illustrates both of these options for minima and maxima bounds in relation to some reported weekly performance data using a Moving Average Interval of 5. (*Let's hope that we find it to be a 'moving performance' . . . You really didn't expect me to let that one go unsaid, did you?*) Note: the averages, minima and maxima have not been offset by half the interval in this instance to aid clarity. In normal practice, this would be recommended. We will revert to that recommended practice before the end of this section.

Furthermore, as discussed in Volume II Chapter 2 on Measures of Central Tendency, depending on the circumstances, average performance may be better analysed using a Harmonic Mean (in this case, a Moving Harmonic Mean), or a Cumulative Smoothing technique. For illustration purposes, we will consider a simple Moving Average here.

For those who like to follow where the graphs have come from, the data to support this is provided in Table 3.11.

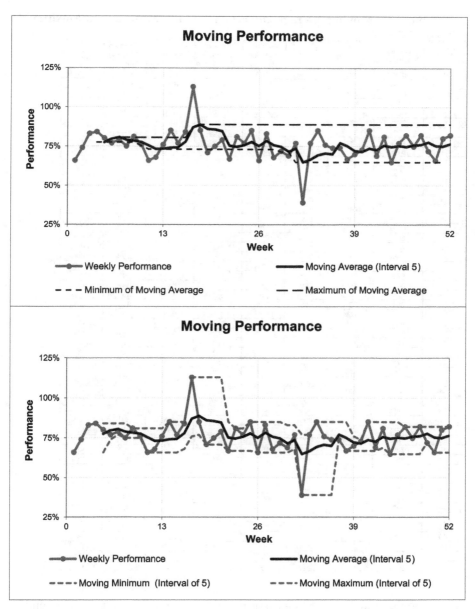

Figure 3.13 Moving Average Uncertainty Using Minima and Maxima

74 | Trendsetting with some Simple Moving Measures

Table 3.11 Moving Average Uncertainty Using Minima and Maxima

Moving Interval (In	5							
Week	Weekly Performance	Moving Average (Interval 5)	Absolute Minimum	Minimum of Moving Average	Moving Minimum (Interval of 5)	Moving Maximum (Interval of 5)	Maximum of Moving Average	Absolute Maximum
1	66%	#N/A	66%	#N/A	#N/A	#N/A	#N/A	66%
2	74%	#N/A	66%	#N/A	#N/A	#N/A	#N/A	74%
3	83%	#N/A	66%	#N/A	#N/A	#N/A	#N/A	83%
4	84%	#N/A	66%	#N/A	#N/A	#N/A	#N/A	84%
5	80%	77%	66%	77%	66%	84%	77%	84%
6	77%	80%	66%	77%	74%	84%	80%	84%
7	79%	81%	66%	77%	77%	84%	81%	84%
8	75%	79%	66%	77%	75%	84%	81%	84%
9	81%	78%	66%	77%	75%	81%	81%	84%
10	76%	78%	66%	77%	75%	81%	81%	84%
11	66%	75%	66%	75%	66%	81%	81%	84%
12	68%	73%	66%	73%	66%	81%	81%	84%
13	76%	73%	66%	73%	66%	81%	81%	84%
14	85%	74%	66%	73%	66%	85%	81%	85%
15	77%	74%	66%	73%	66%	85%	81%	85%
16	84%	78%	66%	73%	68%	85%	81%	85%
17	113%	87%	66%	73%	76%	113%	87%	113%
18	85%	89%	66%	73%	77%	113%	89%	113%
19	71%	86%	66%	73%	71%	113%	89%	113%
20	75%	86%	66%	73%	71%	113%	89%	113%
21	79%	85%	66%	73%	71%	113%	89%	113%
22	67%	75%	66%	73%	67%	85%	89%	113%
23	81%	75%	66%	73%	67%	81%	89%	113%
24	77%	76%	66%	73%	67%	81%	89%	113%
25	85%	78%	66%	73%	67%	85%	89%	113%
26	66%	75%	66%	73%	66%	85%	89%	113%
27	83%	78%	66%	73%	66%	85%	89%	113%
28	68%	76%	66%	73%	66%	85%	89%	113%
29	72%	75%	66%	73%	66%	85%	89%	113%
30	69%	72%	66%	72%	66%	83%	89%	113%
31	77%	74%	66%	72%	68%	83%	89%	113%
32	39%	65%	39%	65%	39%	77%	89%	113%
33	77%	67%	39%	65%	39%	77%	89%	113%
34	85%	69%	39%	65%	39%	85%	89%	113%
35	76%	71%	39%	65%	39%	85%	89%	113%
36	74%	70%	39%	65%	39%	85%	89%	113%
37	74%	77%	39%	65%	74%	85%	89%	113%
38	67%	75%	39%	65%	67%	85%	89%	113%
39	70%	72%	39%	65%	67%	76%	89%	113%
40	73%	72%	39%	65%	67%	74%	89%	113%
41	85%	74%	39%	65%	67%	85%	89%	113%
42	69%	73%	39%	65%	67%	85%	89%	113%
43	81%	76%	39%	65%	69%	85%	89%	113%
44	65%	75%	39%	65%	65%	85%	89%	113%
45	77%	75%	39%	65%	65%	85%	89%	113%
46	82%	75%	39%	65%	65%	82%	89%	113%
47	75%	76%	39%	65%	65%	82%	89%	113%
48	82%	76%	39%	65%	65%	82%	89%	113%
49	72%	78%	39%	65%	72%	82%	89%	113%
50	66%	75%	39%	65%	66%	82%	89%	113%
51	80%	75%	39%	65%	66%	82%	89%	113%
52	82%	76%	39%	65%	66%	82%	89%	113%

A more sophisticated way of looking at the potential spread or dispersion around the Moving Average is to calculate the Standard Deviation (see Volume II Chapter 3). Again we have a choice over whether to use absolute or moving measures. We would typically plot the absolute measure around the absolute average of all data to data (we will revisit this later under Cumulative Averages). Alternatively, we might want to calculate the equivalent Moving Standard Deviation over the same interval that we used to calculate the Moving Average.

Definition 3.4 Moving Standard Deviation

A Moving Standard Deviation is a series or sequence of successive standard deviations calculated from a fixed number of consecutive input values that have occurred in a natural sequence. The fixed number of consecutive input terms used to calculate each standard deviation term is referred to as the Moving Standard Deviation Interval or Base.

If we believe that the raw data is normally distributed around an approximately steady state Moving Average value, then we might want to plot tram-lines 1.96

For the Formula-philes: Standard Deviation in relation to 95% Confidence Intervals

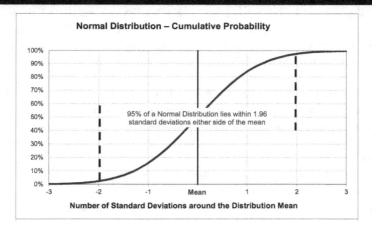

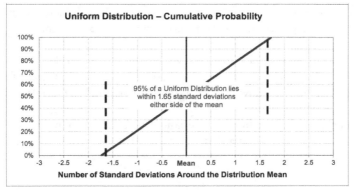

76 | Trendsetting with some Simple Moving Measures

times the standard deviation either side of the average. If we think that raw data is uniformly distributed around the average (i.e. randomly with a band), then we might want to plot tram-lines 1.65 times the standard deviation either side of the average.

Psychic I am not but I can hear many of you thinking, '*Now why on Earth would I choose some fairly random looking numbers like 1.96 or 1.65?*' Well, you are probably right, you just wouldn't; you're more likely to think '*about twice*' and '*about one and two-thirds*', and you would be correct to think that, as they are probably more appropriate levels of imprecision to use with, what is after all, a somewhat simple (some might say 'crude') measure such as a Moving Average. However, these are not random numbers but the values that allow us to plot tram-lines that represent a 95% confidence interval around the Moving Average. We covered this in Volume II Chapters 3 and 4.

Figure 3.14 and Table 3.12 illustrates both the absolute and moving results based on an assumption of a uniform distribution.

The eagle-eyed will have already noted the similarities but also the differences between Figures 3.13 and 3.14. For those who haven't compared them yet, the standard deviation technique largely mirrors the 'worst case' of the minima-maxima technique (in terms of its deviation from the Moving Average). The main difference is that the standard deviation technique then mirrors its deviation either side of the mean.

For your convenience, Table 3.13 provides you with a list of the number of standard deviations either side of the Mean that are broadly equivalent to different confidence intervals for both the normal and uniform distributions.

All the above can be done using standard functions in Microsoft Excel (**MIN**, **MAX** and **STDEV.S** with a sliding range just as we would have for the Moving Average. However, remember that a Moving Average is not a sophisticated measure of trend, so do not over-engineer the analysis of it. ('*Just because we can, does not mean we should.*') Take a step back and ask yourself '*what am I trying to achieve or show here?*'

We can use either the minima-maxima, or the standard deviation technique with both Simple and Weighted Moving Averages for any of the following:

- There is variation around a basic steady state level (discussed above)
- There is a change in the steady state level, or a change in the degree of variation
- There is an upward or downward trend, as illustrated in Figures 3.15 and 3.16.

The latter illustrates the benefit of plotting the data to take account of the lag between the Moving Average calculation and the true underlying trend in the data. Figure 3.15

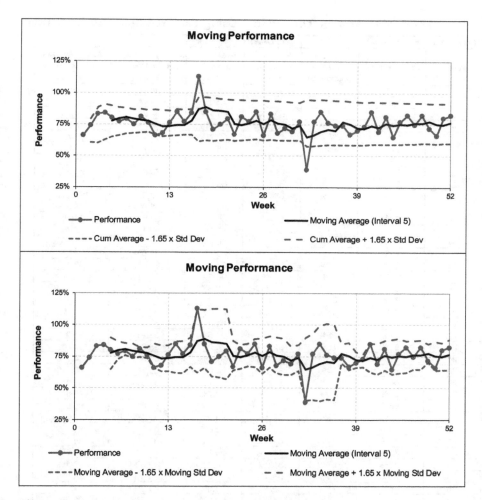

Figure 3.14 Moving Average Uncertainty Using Standard Deviations

uses Moving Minima and Maxima to denote uncertainty ranges, whereas Figure 3.16 uses the factored Moving Standard Deviation approach. The upper graph in each figure is plotted 'as calculated' at the end of the Moving Average interval, whereas the lower graphs are re-plotted to take account of the lag in the various Moving Measures. In both

Table 3.12 Moving Average Uncertainty Using Factored Standard Deviation

Moving Interval (In Weeks)		5	Std Deviation Multiplier		1.65	
Week	Performance	Moving Average (Interval 5)	Average - 1.65 x Std Dev	Average + 1.65 x Std Dev	Moving Average - 1.65 x Moving Std Dev	Moving Average + 1.65 x Moving Std Dev
1	66%	#N/A	#N/A	#N/A	#N/A	#N/A
2	74%	#N/A	60.7%	79.3%	#N/A	#N/A
3	83%	#N/A	60.3%	88.4%	#N/A	#N/A
4	84%	#N/A	62.8%	90.7%	#N/A	#N/A
5	80%	77%	65.1%	89.7%	65.1%	89.7%
6	77%	80%	66.3%	88.4%	72.7%	86.5%
7	79%	81%	67.5%	87.7%	75.8%	85.4%
8	75%	79%	67.8%	86.7%	73.4%	84.6%
9	81%	78%	68.6%	86.8%	74.4%	82.4%
10	76%	78%	68.9%	86.1%	73.6%	81.6%
11	66%	75%	66.5%	86.4%	65.9%	84.9%
12	68%	73%	65.4%	86.1%	63.1%	83.3%
13	76%	73%	65.9%	85.7%	63.1%	83.7%
14	85%	74%	66.1%	86.8%	61.7%	86.7%
15	77%	74%	66.5%	86.4%	61.8%	87.0%
16	84%	78%	66.8%	87.1%	66.6%	89.4%
17	113%	87%	61.6%	96.5%	62.1%	111.9%
18	85%	89%	62.3%	96.5%	65.8%	111.8%
19	71%	86%	62.0%	95.9%	59.4%	112.6%
20	75%	86%	62.2%	95.3%	58.5%	112.7%
21	79%	85%	62.7%	94.9%	57.0%	112.2%
22	67%	75%	62.0%	94.5%	63.9%	86.9%
23	81%	75%	62.4%	94.2%	65.2%	84.0%
24	77%	76%	62.7%	93.8%	66.9%	84.7%
25	85%	78%	63.2%	94.0%	66.7%	88.9%
26	66%	75%	62.5%	93.7%	61.3%	89.1%
27	83%	78%	62.9%	93.7%	66.0%	90.8%
28	68%	76%	62.5%	93.3%	61.6%	90.0%
29	72%	75%	62.4%	93.0%	60.4%	89.2%
30	69%	72%	62.2%	92.6%	60.5%	82.7%
31	77%	74%	62.4%	92.4%	63.5%	84.1%
32	39%	65%	57.7%	94.7%	40.3%	89.7%
33	77%	67%	58.0%	94.4%	40.5%	93.1%
34	85%	69%	58.4%	94.6%	39.8%	99.0%
35	76%	71%	58.6%	94.3%	40.9%	100.7%
36	74%	70%	58.8%	94.0%	40.6%	99.8%
37	74%	77%	59.0%	93.7%	69.7%	84.7%
38	67%	75%	58.8%	93.4%	64.5%	85.9%
39	70%	72%	58.8%	93.1%	66.2%	78.2%
40	73%	72%	58.9%	92.8%	66.6%	76.6%
41	85%	74%	59.2%	93.0%	62.5%	85.1%
42	69%	73%	59.1%	92.7%	61.0%	84.6%
43	81%	76%	59.4%	92.7%	64.0%	87.2%
44	65%	75%	59.1%	92.4%	60.9%	88.3%
45	77%	75%	59.3%	92.3%	61.7%	89.1%
46	82%	75%	59.6%	92.3%	62.4%	87.2%
47	75%	76%	59.7%	92.1%	64.8%	87.2%
48	82%	76%	60.0%	92.1%	64.7%	87.7%
49	72%	78%	60.0%	91.9%	70.4%	84.8%
50	66%	75%	59.8%	91.7%	64.1%	86.7%
51	80%	75%	60.0%	91.7%	64.4%	85.6%
52	82%	76%	60.2%	91.7%	64.6%	88.2%

Table 3.13 Confidence Intervals in Relation to Standard Deviations Around the Mean

Confidence Interval	Approximate number of Standard Distributions either side of the Mean	
	Normal Distribution	Uniform Distribution
99.73%	3.00	1.73
95%	1.96	1.65
90%	1.65	1.56
80%	1.28	1.39
67%	0.97	1.16
50%	0.67	0..87

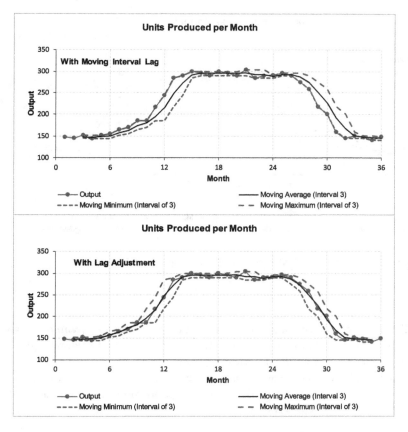

Figure 3.15 Moving Minima and Maxima with Upward and Downward Trends

80 | Trendsetting with some Simple Moving Measures

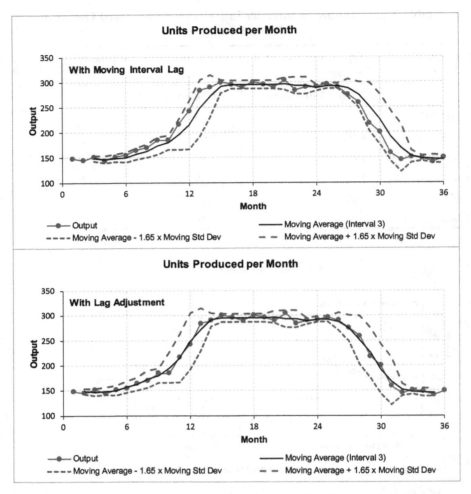

Figure 3.16 Moving Standard Deviation with Upward or Downward Trends

'unadjusted' cases the actual data hugs the upper limit of the uncertainty tramline for an increasing trend, and hugs the lower limit for a decreasing trend. The lag adjusted graph gives a more acceptable perspective of what is happening.

The adjusted graphs highlight that the uncertainty range is wider for increasing and decreasing trends than for 'steady state' values (unless there are more extreme, or atypical, data in the so-called 'steady state' values.)

3.3 Moving Medians

Critics of the Moving Average who argue that it is too sensitive to extreme values, especially for small intervals, might want to consider the alternative but closely related technique of Moving Medians.

Definition 3.5   Moving Median

A Moving Median is a series or sequence of successive medians calculated from a fixed number of consecutive input values that have occurred in a natural sequence. The fixed number of consecutive input terms used to calculate each median term is referred to as the Moving Median Interval or Base.

From the discussion on Measures of Central Tendency in Volume II Chapter 2, a median is one of the other Measures of Central Tendency; the Median puts the spotlight on the middle value in a range. Depending on its intended use, we may find this quite helpful.

If we try this with our earlier weekly performance example, we will notice that it gives comparable results to the Moving Average where the data is 'well behaved', but it is unaffected by the two spikes that drag the Moving Average up or down for the duration of the 'moving' cycle (Figure 3.17). This can be done quite easily in Microsoft Excel using the in-built function **MEDIAN** with a sliding range representing the data points in the Interval.

However, we win some and we lose some, there are different issues with a Moving Median compared with those of a Moving Average. If we use a Moving Median over a Time Series where there is a seasonal pattern, there is a risk that it will completely disregard some key seasonal data. This is especially so where there are only one or two seasonal extremes. For example, our thirteen-week accounting calendar example will be reduced to one that reflects only four-week months. For annual data on labour salaries containing annual bonus payments a Moving Median will treat the bonus payments as if they never existed (*Try telling the employees that their bonuses don't matter!*). Figure 3.18 illustrates the point with Hours Booked per Month.

In essence, we have to decide what we are seeking to achieve:

- Elimination of extreme values that occur at random? Moving Medians work well.
- Smooth data that represents the middle ground through seasonal data? Moving Medians are probably less appropriate than Moving Averages.

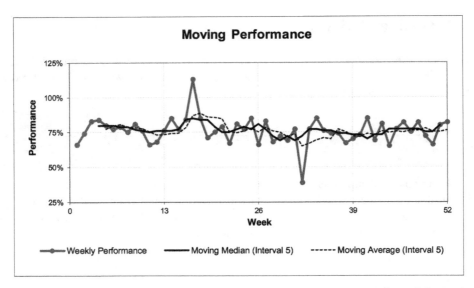

Figure 3.17 Moving Median Compared with Moving Average (Unadjusted for Lag)

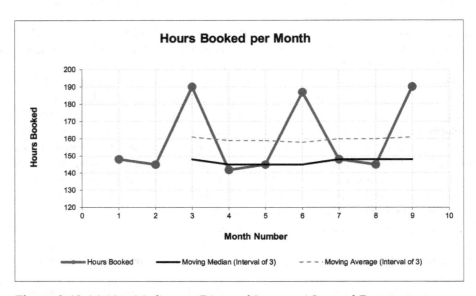

Figure 3.18 Moving Median can Disregard Important Seasonal Data

3.3.1 Choosing the Moving Median Interval

Just as we discussed for Moving Averages, the interval for a Moving Median should be based on the natural cycle of data if there is one. If the natural cycle is not obvious, we can iterate for different interval sizes, starting with a low number until we achieve a degree of smoothing that we are happy to accept. We can use the MAID Minimum Standard Deviation techniques discussed in Section 3.2.1 and apply that interval to the Moving Median.

Note that a Moving Median can only be used as a distinct measure for intervals of 3 or more, as the Moving Median of two points defaults to being the Moving Average of those two points.

For the Formula-phobes: A Moving Median of Interval of 2 defaults to Moving Average?

Recall the analogy of the telegraph poles and spaces from Volume II Chapter 2

If we have an odd number of telegraph poles, then there will always be one that is in the middle, and that would represent the Median.

If we have an even number of telegraph poles, then there will always be a space in the middle; there will never be a pole. The convention in that case is for the Median to equal the average of the two points that flank that middle space.

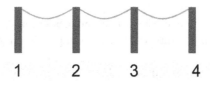

In the case of an interval of two, the median would be simply the average of the two points either side of the middle space making the Median equal the Arithmetic Mean, so a Moving Median of Interval 2 is identical to a Moving Average of Interval 2.

84 | Trendsetting with some Simple Moving Measures

3.3.2 Dealing with missing data

What if we have missing data, what should we do? Well, it is the same advice as for Moving Averages – do not ignore it! (*Sorry am I beginning to sound repetitive?*) It is better practice to truncate the Moving Median calculation at the last point at which data was available; then resume the Moving Median after the next full cycle or interval has elapsed following the missing data.

As estimators we can always provide an estimate of the missing data, but we should not include that estimate in the Moving Median calculation. Microsoft Excel provides an estimate of the Moving Median by interpolating the gaps for us by changing the Moving Median Interval incrementally by one less for every missing data point, similar to the discussion we had on missing data for Moving Averages.

3.3.3 Uncertainty Range around the Moving Median

We can use the Moving Minima and Moving Maxima just as we did with Moving Averages to give a pseudo Confidence Interval around a steady state value. However, we should not use the factored Moving Standard Deviations approach in conjunction with the Median as that principle is predicated on a scatter around an Average or Mean. However, the natural corollary to selecting a Moving Median is to consider Moving Percentiles.

If we want to quote a particular Confidence Interval around the Moving Median, or we want to use asymmetric confidence levels, e.g. a lower limit at the 20% Confidence Level and an upper limit at the 90% Confidence Level, we can always calculate any percentile for a range of data; the Median is by definition the 50th percentile, or 50% if expressed as a Confidence Level.

The procedure for calculating percentiles is quite simple:

1. Arrange the data in the range in ascending size order
2. Assign the 0th percentile (0%) to the smallest value, and the 100th percentile (100%) to the largest
3. The percentile for each successive data point, n, between the first and the last is calculated as being at (n-1)/(Interval-1). For example, for an interval of 5 data points, the percentile of each is 0%, 25%, 50%, 75%, 100%
4. For any percentiles in between these, they can be calculated by linear interpolation between the two points

There are two specific **PERCENTILE** functions in Microsoft Excel that allows us to do this with ease without having to worry about the procedure or the mathematics. An open interval version with the suffix **.EXC** and a closed interval version with suffix **.INC** (see Volume II Chapter 3 for the difference.)

However, for small intervals, there is an argument that citing Specific Confidence Intervals is misleading, giving the impression of more precision than the data actually supports. For larger intervals, the approach can be rationalised quite easily. For an interval of eleven, each constituent value in the range is equivalent to ten percentile points from the previous one. (*Yes, it is our favourite telegraph poles and spaces analogy again.*) However, we should note that the percentiles approach does have a potential issue with granularity. If we have several repeating values in our data then we may get the situation where two distinct percentile calculations can return the same value (*think of them all stacked up vertically*), and all those interpolated between them will do so also. The percentiles approach does not allow for random errors in the observed values.

For the Formula-phobes: How do we calculate percentiles?

Consider a line of five telegraph poles and spaces:

The first telegraph pole on the left is 0%; the last on the right is at 100%. The remaining poles are equally spread at 25% intervals:

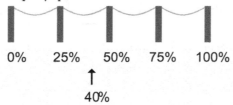

The 40th percentile or 40% is just past halfway between 25% and 50%. More precisely it is 60% or $^3/_5$ of the way between the two.

Figure 3.19 illustrates that whilst the Median disregards any extreme values we might collect, the Moving Percentiles, used as Confidence Levels or Intervals, do take account of their existence if used appropriately. As with Minima, Maxima and factored Standard Deviations used in conjunction with Moving Averages, we can use percentiles at either an absolute level, based on the entire history, or as a rolling or moving measure of confidence. A summary of appropriate measures of uncertainty around a Moving Median are provided in Table 3.14.

3.4 Other Moving Measures of Central Tendency

As we have already intimated, there are occasions where we might want to consider some of the other Moving Measure of Central Tendency, around which we might also want to consider applying our Uncertainty Range as before.) We will briefly discuss three other Moving Measures here.

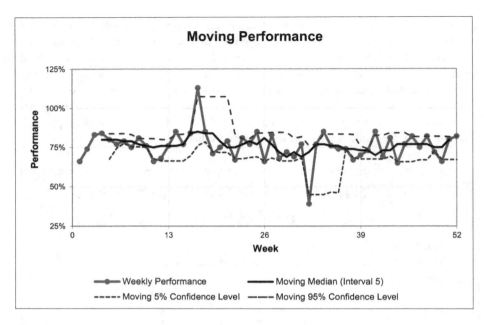

Figure 3.19 Moving Median with Moving 10th and 90th Percentiles as Confidence Limits

Table 3.14 Potential Measures of Moving Median Uncertainty

Moving Median Interval	Absolute Measures (All data points)	Moving Measures (based on Interval used)
Small interval (say 2–5)	• Absolute Minimum and Maximum • Moving Median Minimum and Maximum	• Moving Minimum and Moving Maximum
Large interval (say >6)	• All of the above + • Absolute Percentiles (user defined) • Moving Median Percentiles (user defined)	• All of the above + • Moving Percentiles (user defined)

3.4.1 Moving Geometric Mean

As we will see in Chapter 8 on Time Series Analysis, there are conditions where we should be using a Geometric Mean instead of a simple Arithmetic Mean to calibrate an underlying trend that is accelerating or decelerating in relation to time. This is particularly true in the case of economic indices relating to costs and prices.

Definition 3.6   Moving Geometric Mean

A Moving Geometric Mean is a series or sequence of successive geometric means calculated from a fixed number of consecutive input values that have occurred in a natural sequence. The fixed number of consecutive input terms used to calculate each geometric mean term is referred to as the Moving Geometric Mean Interval or Base.

If we have access to Volume II Chapter 2 we may recall that we can only use a Geometric Mean with positive data, so if there is any chance that the observed data might be zero or negative (such as variation from a target or standard) then we cannot use a Moving Geometric Mean.

3.4.2 Moving Harmonic Mean

If we recall from Volume II Chapter 2 (*unless that was when you decided it was time to put the kettle on and missed that discussion*), a Harmonic Mean is one we should consider using when we are dealing with data that is fundamentally reciprocal in nature i.e. where the 'active' component being used to derive a statistic is in the denominator (bottom line of a fraction or ratio.) Performance is one such statistic, measuring Target Value divided by Actual Value; the Actual Value is the 'active' component.

Definition 3.7   Moving Harmonic Mean

A Moving Harmonic Mean is a series or sequence of successive harmonic means calculated from a fixed number of consecutive input values that have occurred in a natural sequence. The fixed number of consecutive input terms used to calculate each harmonic mean term is referred to as the Moving Harmonic Mean Interval or Base.

Again, we can only use a Harmonic Mean with positive data (Volume II Chapter 2), so if there is any chance that the observed data might be zero or negative then we cannot use a Moving Harmonic Mean.

In many practical cases where the performance is pretty much consistent or only varies to a relatively small degree, then it will make very little difference to any conclusion we draw from using a Moving Harmonic Mean in comparison with one we would make using a Simple Arithmetic Moving Average. Where we have obvious outliers, the Harmonic Mean will tend to favour the lower values.

As we previously indicated, the Harmonic Mean may often be a better trend measure for Performance than a simple Arithmetic Mean. Figure 3.20 compares our earlier Moving Average Performance with a Moving Harmonic Mean Performance.

3.4.3 Moving Mode

Don't do it!

It is highly unlikely that we would find ourselves in circumstances where this would be a reasonable option to consider. Modes really only come into their own with large datasets. Pragmatically thinking, in the context of Moving Measures, the intervals we are likely to be considering will be small, and the chances of getting consistently meaningful modes would be small – remote even.

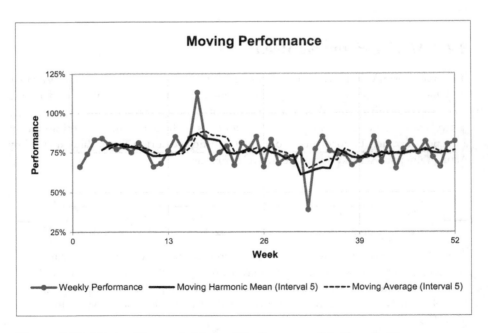

Figure 3.20 Moving Harmonic Mean of Performance cf. Arithmetic Mean

3.5 Exponential Smoothing

Now we wade into the muddy waters of Exponential Smoothing. In essence, Exponential Smoothing is a self-correcting trend smoothing and predictive technique. *Sound too good to be true? Hmm, perhaps you should be the judge of that . . .*

> ### Definition 3.8 Exponential Smoothing
>
> Exponential Smoothing is a 'single-point' predictive technique which generates a forecast for any period based on the forecast made for the prior period, adjusted for the error in that prior period's forecast.

Exponential Smoothing generates an array (i.e. a series or range of ordered values) of successive weighted averages based on the prior forecast and the prior actual reported against the prior forecast. It does this by applying either a percentage Smoothing Constant or a percentage Damping Factor, the magnitude of which determines how strongly forecasts respond to errors in the previous forecast. Exponential Smoothing only predicts the next point in the series, hence the reference to it being a 'single-point predictive technique' in the definition above. However, we may choose to exercise our judgement and use this 'single-point prediction' as an estimate of a steady state constant value for as many future points that we think are appropriate.

The technique was first suggested by Holt (1957), but there is an alternative form of the equation which is usually attributed to Brown (1963).

3.5.1 An unfortunate dichotomy

However, it must be noted before any formulae are presented in terms of what it is and how it works, that statisticians, mathematicians, computer programmers etc have not got their communal acts together! (*No change there, did I hear you say?*) It appears to be generally (but not universally) accepted in many texts on the subject that the terms 'Smoothing Constant' and 'Damping Factor' are complementary terms such that:

Smoothing Constant + Damping Factor = 100%

In terms of the weighted average, the Smoothing Constant is the weighting applied to the Previous Actual, and the Damping Factor is the weighting applied to the Previous Forecast. However, this is not always the case, and some authors appear to use the terms

90 | Trendsetting with some Simple Moving Measures

interchangeably. If we are generating the formula from first principles, then it does not really matter, so long as it is documented properly and clearly for others, but if we are using a pre-defined function in a toolset such as Microsoft Excel (available using the Data Analysis ToolPak Add-In which comes as standard with Excel), then it would be a good idea if we knew which version was being used.

Unfortunately, Microsoft Excel appears to be on the side of '*laissez faire*'. In Excel 2003, 2007 and 2010, Excel quite correctly prompts (in terms of how it works) for the Damping Factor in respect of the weighting to be applied to the Previous Forecast, whereas the Excel Help Facility (normally an excellent resource) refers to the Smoothing Constant when it appears to mean Damping Factor, and this appears to infer that they are interchangeable terms.

Regardless of this unfortunate dichotomy, this gives two different, yet equivalent, standard formulae for expressing the Exponential Smoothing rule algebraically, depending on whether the Smoothing Constant, α, or Damping Factor, d, is the parameter being expressed.

For the Formula-philes: Exponential Smoothing Calculation

Consider a sequence of Time Series actual data, $A_1 \dots A_t$ for which a Forecast, F_i is generated for the next time period, i

Let α be a Smoothing Constant, with
$0 < \alpha < 1$, then the Forecast at Time = t
can be expressed by the weighted Average
of the Last Actual A_{t-1} and the Last
Forecast F_{t-1} such that:

$$F_t = \alpha A_{t-1} + (1 - \alpha) F_{t-1} \qquad (1)$$

Alternatively, by re-arranging (1):

$$F_t = F_{t-1} + \alpha (A_{t-1} - F_{t-1}) \qquad (2)$$

... which expresses the forecast for the next period as the forecast for the last period plus an adjustment for the error in the last forecast

If we assume that the Forecast for Time = 2
is the same as the Actual for Time = 1
(only one data point), then from (1):

$$F_2 = \alpha A_1 + (1 - \alpha) F_1 = A_1 \qquad (3)$$

Simplifying (3):

$$F_2 = F_1 = A_1 \qquad (4)$$

... there is an inherent assumption that the first two forecasts are equal to the first observed data value

Alternatively, we can imply from (1) that for
some prior constants A_0 and F_0 :

$$F_1 = \alpha A_0 + (1 - \alpha) F_0 \qquad (5)$$

For convenience, we can define:

$$F_0 = A_0 = A_1 \qquad (6)$$

We will return to the significance of this shortly.

The equation proposed by Brown (1963) applied the Smoothing Constant to the previous Actual, and it's residual from 100%, as the Damping Factor to the previous Forecast.

In this format, Exponential Smoothing could be said to smooth the previous Forecast by a proportion of the previous Forecast Error.

An alternative form in common use, and also the one adopted by Microsoft Excel (*if that were to influence you one way or another*) applies the Damping Factor to the previous Forecast, and it's residual from 100%, as the Smoothing Constant to the previous Actual:

For the Formula-philes: Exponential Smoothing Calculation

Consider a sequence of Time Series actual data, $A_1...A_t$ for which a Forecast, F_i is generated for the next time period, i

Let δ be a Damping Constant, with
$0 < \delta < 1$, then the Forecast at Time = t
can be expressed by the weighted Average
of the Last Forecast F_{t-1} and the
Last Actual A_{t-1} such that:

$$F_t = \delta F_{t-1} + (1 - \delta)A_{t-1} \qquad (1)$$

Alternatively, by re-arranging (1):

$$F_t = A_{t-1} + \delta(F_{t-1} - A_{t-1}) \qquad (2)$$

... which expresses the forecast for the next period as the actual for the last period plus an adjustment for the error in the last forecast

As with the previous format, there is an inherent assumption that the first two forecasts are equal to the first observed data value

---

In this format Exponential Smoothing could be said to be damping the previous Actual by a proportion of the previous Forecast Error. (Note: the sign of the error term in this version is the reverse of the former!)

The two formats do not give different forecasts, they are just different conventions, and the Damping Factor and Smoothing Constants are complementary values:

$$F_t = \alpha A_{t-1} + \delta F_{t-1}$$

Where:

$$\alpha + \delta = 1$$

Perhaps the logical thing to do to avoid confusion would be for us to write it in terms of '*Smoothing the Actual and Damping the Forecast*': or choosing one of the F-A:D-S (Forecast or Actual implies either Dampen or Smooth in that order).

92 | Trendsetting with some Simple Moving Measures

For the Formula-philes: Why is it called Exponential Smoothing?

Consider Brown's version of the Forecast value, F_t at period t based on the Actual, A_{t-1} and Forecast, F_{t-1} at Period t-1 with a Smoothing Constant of α:

The Forecast, F_t at period t is given by:
$$F_t = \alpha A_{t-1} + (1-\alpha)F_{t-1} \qquad (1)$$

If (1) is true for any t then it is true for t-1:
$$F_{t-1} = \alpha A_{t-2} + (1-\alpha)F_{t-2} \qquad (2)$$

Substituting (2) in (1):
$$F_t = \alpha A_{t-1} + (1-\alpha)\alpha A_{t-2} + (1-\alpha)^2 F_{t-2} \qquad (3)$$

Similarly, by iteration, we can deduce:

$$F_t = \alpha A_{t-1} + (1-\alpha)\alpha A_{t-2} + (1-\alpha)^2 \alpha A_{t-3} + (1-\alpha)^3 F_{t-3} \qquad (4)$$

By iteration, we eventually get:

$$F_t = \alpha\left(A_{t-1} + (1-\alpha)A_{t-2} + (1-\alpha)^2 A_{t-3} + \ldots + (1-\alpha)^{t-1} A_0\right) + (1-\alpha)^t F_0 \qquad (5)$$

Simplifying (5):
$$F_t = \alpha\sum_{i=1}^{t}(1-\alpha)^{i-1} A_{t-i} + (1-\alpha)^t F_0 \qquad (6)$$

...in which the weightings of the previous time-based sequence of actuals decay exponentially to zero. The sum of the geometric progression of the weighting factors converge to $\alpha/(1-(1-\alpha))$ or 1. We previously defined F_0 be the value A_1 for convenience, but as t increases its weighting tends to zero, so its value become immaterial. (*We said we'd get back to it.*)

3.5.2 Choice of Smoothing Constant, or Choice of Damping Factor

You may find it documented in some sources (e.g. Microsoft Excel Help) that Smoothing Constants values between 0.2 and 0.3 are considered reasonable. However, as we have already discussed, Excel appears to use the terms Smoothing Constant and Damping Factor interchangeably; it prompts the user to provide a Damping Factor, and even assigns a default of 0.3 if it is not input by the user, but then in the Help Facility it refers to 0.2 to 0.3 as being reasonable values for the Smoothing Constant. (Clearly they cannot both be values in this range if we accept the position of others that they must summate to 1.) However, all this is an over-generalisation, and we might be better choosing the value of either based on the objective of the analysis in question. If we choose smaller Damping Factors (i.e. larger Smoothing Constants), this will give us a faster response to changes in the underlying trend, but the results will be prone to producing erratic forecasts. On the other hand, if we choose larger Damping Factors (i.e. smaller Smoothing Constants), we will be rewarded with considerably more smoothing, but as a consequence the downside is that we will observe much longer lags before changes in forecast trend values are seen to be relevant. These are illustrated in Table 3.15 and Figure 3.21.

Table 3.15 Exponential Smoothing – Effect of Using Different Parameters

Smoothing Constant α		90%	40%	10%
Damping Factor δ		10%	60%	90%
Period, t	Actual, A_t	Forecast, F_t (90%,10%)	Forecast, F_t (40%,60%)	Forecast, F_t (10%,90%)
1	150	#N/A	#N/A	#N/A
2	155	150	150	150
3	167	154.50	152.00	150.50
4	145	165.75	158.00	152.15
5	148	147.08	152.80	151.44
6	142	147.91	150.88	151.09
7	163	142.59	147.33	150.18
8	158	160.96	153.60	151.46
9	151	158.30	155.36	152.12
10	#N/A	151.73	153.61	152.01
Exponential Smoothing		$F_t = \alpha F_{t-1} + \delta A_{t-1}$		

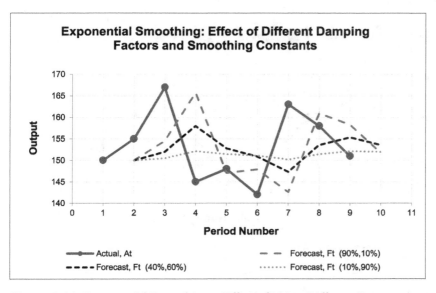

Figure 3.21 Exponential Smoothing – Effect of Using Different Parameters

Rather than pick a Smoothing Constant or Damping Factor at random, or empirically, we could derive a more statistically pleasing result perhaps, (*yes, I know I need to get out more*), by determining the best fit value of α or δ for a given series of data by using a Least Squares Error approach (these principles are discussed and used later commencing with Chapter 4 on Linear Regression – '*Oh, what joy*', did you say?)

3.5.3 Uses for Exponential Smoothing

The most common use of Exponential Smoothing perhaps is with time-based data or Time Series data, but this is not necessarily its only use. As with general Moving Averages, any data series in which there is a natural order, with nominally constant incremental steps (e.g. integer build numbers) might be analysed and predicted using Exponential Smoothing. In these situations, it is aesthetically more appropriate to substitute an alternative letter, such as **n**, in place of the time period, t:

$$F_n = \alpha A_{n-1} + \delta F_{n-1}$$

Figure 3.22 provides a non-time-based example of Exponential Smoothing with a Smoothing Constant of 75% and a Damping Factor of 25%.

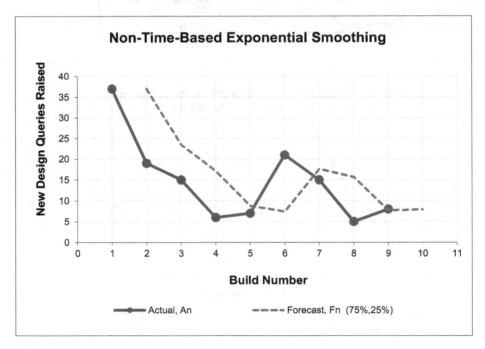

Figure 3.22 Exponential Smoothing – Non-Time-Based Data

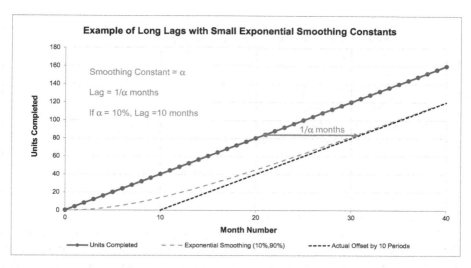

Figure 3.23 Exponential Smoothing – Long Lags with Small Smoothing Constants

Just as with ordinary Moving Averages, wherever there is an upward or downward trend, Exponential Smoothing will always lag the true trend. Furthermore, the smaller the Smoothing Constant (larger Damping Factor) is then the more the calculated trend will appear to diverge from our true trend initially, before settling down to run parallel to it, some $1/\alpha$ periods or units to the right – *so, not really that helpful perhaps as an indicator* (see Figure 3.23). For data that oscillates around a steady state or constant value, then Exponential Smoothing might be the solution you have been looking for. (*Exponential Smoothing is also known by the less snappy name of 'Exponentially Weighted Moving Average'.*)

As you can probably tell, I am not the biggest fan of Exponential Smoothing.

3.5.4 Double and Triple Exponential Smoothing

Just when you thought that it couldn't get any better . . .

Where there is a trend in the data, Double Exponential Smoothing was suggested as a way of dealing with it. One method *(yes, there is more than one)* is the Holt-Winters Method (Winters, 1960) which is an algorithm best performed in a spreadsheet.

> **For the Formula-philes: Double Exponential Smoothing Calculation for Underlying Trends**
>
> Double Exponential Smoothing can be defined using a set of three equations in which:

(Continued)

96 | Trendsetting with some Simple Moving Measures

t = first period at which the forecast is required
F_t = Forecast at period t
A_t = Actual at period t
T_t = Trend Adjustment at period t
α = Data Smoothing Factor, with ($0 < \alpha < 1$)
β = Trend Smoothing Factor, with ($0 < \beta < 1$)
k = number of periods in advance to be forecast
For convenience: $F_1 = A_1$ and $T_1 = A_2 - A_1$

$$F_t = \alpha A_t + (1-\alpha)(F_{t-1} + T_{t-1})$$

$$T_t = \beta(F_t - F_{t-1}) + (1-\beta)T_{t-1}$$

$$F_{t+k} = F_t + kT_t$$

Did you know that your eyes have just glazed over again?

Although it may look quite scary it is easily performed within a spreadsheet such as Microsoft Excel. The main difficulty is in choosing appropriate values for α and β, which we might as well calibrate using a Least Squares Error approach if we are going to all this trouble with double exponential Smoothing. As mentioned previously, this will be discussed in Chapter 7 on Curve Fitting. However, Double Exponential Smoothing does allow an estimator to forecast more than one period or unit in advance, unlike simple Exponential Smoothing.

If that wasn't enough, Holt-Winters' Triple Exponential Smoothing was evolved to take account of both trend and seasonal data patterns (Winters, 1960). However, for Time Series data where there is a distinct Seasonal Pattern, e.g. gas and electric consumption, and a clear linear trend or steady state value, we might be better considering using the alternative techniques covered under Time Series Analysis (see Chapter 8).

So, the good news is that we are not going to discuss Double or Triple Exponential Smoothing further here; they are merely offered up as potential topics for further research if other simpler techniques elsewhere in this book do not work for you – good luck.

3.6 Cumulative Average and Cumulative Smoothing

Definition 3.9 Point Cumulative Average

A Point Cumulative Average is a single term value calculated as the average of the current and all previous consecutive recorded input values that have occurred in a natural sequence.

Trendsetting with some Simple Moving Measures | 97

> **Definition 3.10   Moving Cumulative Average**

A Moving Cumulative Average, sometimes referred to as a Cumulative Moving Average, is an array (a series or range of ordered values) of successive Point Cumulative Average terms calculated from all previous consecutive recorded input values that have occurred in a natural sequence.

In reality, we will probably find that the term 'Cumulative Average' is generally preferred, and is used to refer to either the single instance term or to the entire array of consecutive terms, possibly because the term 'Moving Average' is typically synonymous with a fixed interval average, i.e. the number of input terms used in the calculation. With the Cumulative Average, the fixed element is not the number of input terms, but the start point. In the case of time-based data, this will be a fixed point in time, whereas for achievement based data, it will be from a fixed reference unit; more often than not (but not always) it will be the first unit.

With Moving Averages, the denominator used (i.e. the number we divide by) is always a whole number of time periods or units. In the case of Cumulative Averages, the denominator usually increments by one for every successive term used (month 1, 2, 3 ..., or unit 1, 2, 3 ... etc). (*Yes, you are right, well spotted – 'usually' implies not always – we will come back to this later in this section when we discussed equivalent units completed.*)

3.6.1 Use of Cumulative Averages

Sometimes we are not interested in just the current value or level, but what is happening in the long run. Just as with professional sportsmen and sportswomen, there will be good days and bad days; ultimately, as Jesse Jackson (Younge, 1999) pointed out, we judge their performance by what is happening over a sustained period: if the Cumulative Average is rising or falling then the underlying metric is rising or falling too.

Cumulative Averages are frequently used to iron out all those unpleasant wrinkles that we get with actual data due to the natural variation in performance, integrity of bookings and the like. It is the hoarder's equivalent of a moving average, taking the attitude that all past data is relevant and nothing shown be thrown away.

> ### A word (or two) from the wise?
>
> *'We have to judge politicians by their cumulative score ... In one they score a home run, in another they strike out. But it is their cumulative batting average that we are interested in.'*
> **Jesse Jackson**
> American Baptist minister
> and politician
> b.1941

98 | Trendsetting with some Simple Moving Measures

Table 3.16 Cumulative Average Over Time and Over Number of Units Completed

Productivity per Month					Hours per Unit Completed			
Month	Hours per Month	Units Achieved per Month	Output per 1000 Hours	Cum Average Output per 1000 Hours	Build No	Hours per Unit Built	Cumulative Hours over Units Built	Cum Average Hours per Unit Built
0	-	-						
1	576	0.10	0.17	0.17	1	4,612	4,612	4,612
2	573	0.15	0.26	0.22	2	5,443	10,055	5,028
3	1,279	0.23	0.18	0.21	3	3,948	14,003	4,668
4	1,275	0.27	0.21	0.21	4	3,338	17,341	4,335
5	1,791	0.39	0.22	0.21	5	3,245	20,586	4,117
6	2,741	0.68	0.25	0.22	6	3,157	23,743	3,957
7	3,207	0.82	0.26	0.22	7	3,123	26,866	3,838
8	4,155	1.11	0.27	0.23	8	2,972	29,838	3,730
9	4,212	1.27	0.30	0.24	9	2,538	32,376	3,597
10	5,328	1.69	0.32	0.24	10	2,820	35,196	3,520
11	5,313	1.81	0.34	0.25	11	2,700	37,896	3,445
12	5,070	1.91	0.38	0.26	12	2,367	40,263	3,355
13	4,874	1.90	0.39	0.27	13	2,222	42,485	3,268
14	5,172	2.10	0.41	0.28	14	2,139	44,624	3,187
15	4,449	1.81	0.41	0.29	15	2,266	46,890	3,126
16	3,601	1.52	0.42	0.30	16	2,860	49,750	3,109
17	2,549	1.10	0.43	0.31	17	2,446	52,196	3,070
18	1,914	0.81	0.42	0.31	18	2,185	54,381	3,021
19	723	0.33	0.46	0.32	19	2,146	56,527	2,975
20	-	-			20	2,275	58,802	2,940

Table 3.16, and corresponding Figures 3.24 and 3.25, illustrate a Cumulative Average over time, and over the number of units completed. (Note that the similarity of the number of data points is coincidental and will be dependent on the build cycle and programme rate- they have been chosen for this example purely for the aesthetics of fitting on the page.)

It will generally be the case that Cumulative Averages, over either Time or Units Completed are far smoother than any Simple or Weighted Moving Average for that same data, and that the degree of smoothing increases as the number of terms increases. However, the downsize is that where there is an upward or downward trend lag in the raw data the lag between the raw data and the Cumulative Average will be quite excessive in comparison with Simple or Weighted Moving Averages, and worse still, it will diverge! As Figure 3.26 illustrates, where there is a change in the underlying trend, a Cumulative Average is very slow to respond.

We might be forgiven for thinking, '*Well, a fat lot of use they are to us then*', but perhaps we should not be too hasty to judge ... they do have a special property.

Where there is a continuous underlying linear trend of the raw data, then the underlying trend of the Cumulative Average will also be linear, but with the 'magic' property of being half the gradient or slope of the actual data (*yes, always*), and of course, it will be much smoother than the underlying data because that is what averages do, and therefore it is easier to identify any linear trend in the data plotted. In Figure 3.26 the actual data rises from 100 to 200 in 20 periods before achieving a steady state output of 200 units

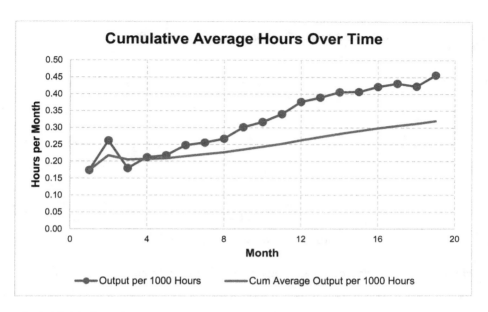

Figure 3.24 Cumulative Average Hours Over Time

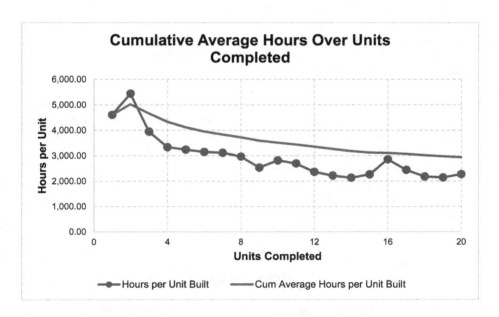

Figure 3.25 Cumulative Average Hours Over Units Completed

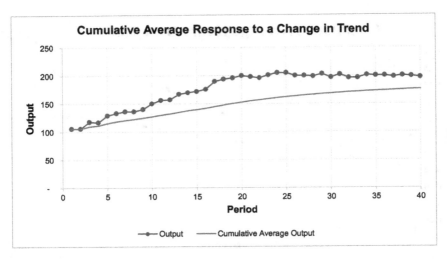

Figure 3.26 Cumulative Average Response to Change in Trend

per period – that is an increasing slope of 5 units per period on average. Meanwhile the Cumulative Average rises from 100 to 150 over the same number of periods, equivalent to an increase of 2.5 units per period, or half the true underlying slope (see Chapter 2). (*Not so useless after all perhaps, we might conclude.*)

This property of Cumulative Averages leads us to a range of potential uses for detecting underlying trends in data; for example:

- If we find that a Cumulative Average plot indicates a nominally steady state constant value, then our underlying raw data trend will also be a steady state constant value, (*even though it might look like the data is in trauma*). Incidentally, if we were to plot the Cumulative value of the raw data, in this scenario, it would be a straight line, the slope of which would be the steady state value.
- If the Cumulative Average of a series of batch cost data values follows a straight line then the unit costs will follow a straight line of twice the slope. (See Section 4.2.2 below on Dealing with Missing Data.)

Caveat augur

If the Cumulative Average cost rate is reducing, then the underlying input cost data will be reducing at twice the rate. In the long run such reduction may be unsustainable, as eventually the unit cost, and ultimately the Cumulative Average

Trendsetting with some Simple Moving Measures | 101

cost, will become negative! Our conclusion might be that the apparent straight line is merely a short-term approximation of a relationship which is potentially non-linear in the longer term – or at least, has a discontinuity when the data hits the floor!

However, that does not imply that negative values of Cumulative Averages or Unit data are never acceptable. Consider a cash flow profile over a period of time. The Cumulative Average may be decreasing month on month and the cash balance at any point in time may be negative and getting worse. It only becomes unsustainable in the longer term. (*Yep, the company is likely to go broke.*)

3.6.2 Dealing with missing data

When it comes to missing data with Cumulative Averages, then it can go one of two ways:

- Cumulative Averages can be more forgiving than Moving Averages or Exponential Smoothing.
- It can be terminal – stopping us in our tracks.

If we begin by considering the case where we only have the most recent data – we cannot find any reliable ancient history, for whatever reason. Whenever we take a Cumulative Average over time, we are starting from a fixed reference point i.e. a calendar date, month, year etc. We never go back to the beginning of time, whenever that was, do we?

Yes, I know that Microsoft have defined the beginning of time as 1st January 1900 in Excel (i.e. Day Number 1), but we have to accept that that was probably just a matter of convenience rather than any moment of scientific enlightenment.

On that basis then we can start a Cumulative Average sequence that is based on physical achievement, also from any fixed reference point, not necessarily 0 or 1. So, if we do not have the history of the first 70 units we can calculate the Cumulative Average from Unit 71. Naturally, we should state that assumption clearly for the benefit of others.

However, what if the missing data occurs in the middle of the series of data we are analysing?

If we happen to know the cumulative value of the units that we are missing, or the cumulative value over the periods of time we are missing, then the good news is that we can deal with it. Unfortunately, if we do not know the cumulative value missing, or the number of missing units, then we will be unable to continue with a Cumulative Average. There are only two solutions in this latter case:

- Replace the entire analysis with a Moving Average (leaving a gap plus an interval for all the missing data points, as we discussed in Section 3.2.8)
- Consider using a Curve Fitting routine (see Chapters 6 and 7)

102 | Trendsetting with some Simple Moving Measures

An example of the former might be where current records are being maintained monthly, but more historical records are only available to us at quarterly intervals, and 'ancient' history might only be available as annual records. However, we can still create a Cumulative Average as illustrated in Table 3.17 and Figure 3.27. (*We could, of course, still calculate a Moving Average over 12 months but it would be very 'staccato' like and may not be the most appropriate interval to consider.*)

In order to make a visual comparison of the raw data with the Cumulative Average, we will take the liberty of representing the raw data as an average value for the period it represents i.e. 12 months or 3 months, but only plot the Cumulative Average for the periods that we know to be correct, in other words the end of those points where only the cumulative data has been retained, (see Figure 3.27).

Table 3.17 Cumulative Average with Incomplete Historical Records

Date	End Month Number	Complaints Received	Cumulative Complaints	Cumulative Average Complaints	Moving Average (Interval 12)	End Month Number	Moving Average (Interval 12)
2006	12	108	108	9.00	9.00	12	9.00
2007	24	118	226	9.42	9.83		
2008	36	107	333	9.25	8.92	24	9.83
Q1 2009	39	39	372	9.54			
Q2 2009	42	15	387	9.21		36	8.92
Q3 2009	45	22	409	9.09			
Q4 2009	48	36	445	9.27	9.33	48	9.33
Q1 2010	51	25	470	9.22	8.17		
Q2 2010	54	29	499	9.24	9.33	51	8.17
Q3 2010	57	29	528	9.26	9.92		
Q4 2010	60	21	549	9.15	8.67	54	9.33
Jan-11	61	8	557	9.13			
Feb-11	62	17	574	9.26		57	9.92
Mar-11	63	16	590	9.37	10.00		
Apr-11	64	11	601	9.39		60	8.67
May-11	65	8	609	9.37			
Jun-11	66	3	612	9.27	9.42	63	10.00
Jul-11	67	12	624	9.31			
Aug-11	68	9	633	9.31		66	9.42
Sep-11	69	6	639	9.26	9.25		
Oct-11	70	13	652	9.31		69	9.25
Nov-11	71	5	657	9.25			
Dec-11	72	10	667	9.26	9.83	72	9.83
Jan-12	73	8	675	9.25	9.83	73	9.83
Feb-12	74	3	678	9.16	8.67	74	8.67
Mar-12	75	14	692	9.23	8.50	75	8.50
Apr-12	76	9	701	9.22	8.33	76	8.33
May-12	77	15	716	9.30	8.92	77	8.92
Jun-12	78	9	725	9.29	9.42	78	9.42
Jul-12	79	13	738	9.34	9.50	79	9.50
Aug-12	80	6	744	9.30	9.25	80	9.25
Sep-12	81	8	752	9.28	9.42	81	9.42

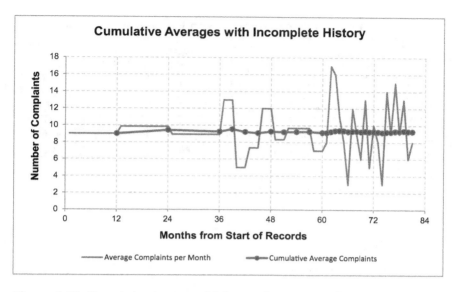

Figure 3.27 Cumulative Average with Incomplete Historical Records

3.6.3 Cumulative Averages with batch data

This leads on naturally to using Cumulative Averages where we only have batch averages, and that these might be incomplete records.

If we know the cumulative values and corresponding denominator values (i.e. the quantities that we are dividing by to get the Cumulative Averages, e.g. time or units produced) then we do not need every single value to calculate and plot the Cumulative Average. This is extremely useful if we only know batch costs and batch quantities.

Table 3.18 illustrates the purchase costs we may have discovered for a particular component. Unfortunately, we only seem to know the total for batches 4 and 5 combined. If we consider the batch number merely to be a label for correct sequencing, we can still calculate a Cumulative Average that is of use to us.

Bizarrely, we can now plot the data in Figure 3.28 that appears to show that we can conjure up one more average point than we have raw data to do! (*Now what were we saying about number jugglers, conjurers and a sleight of hand?*)

3.6.4 Being slightly more creative – Cumulative Average on a sliding scale

We said earlier that Cumulative Average denominators usually increment by one for every successive term, the implication being that Cumulative Averages are discrete functions. However, we do not need to be that prescriptive; we are estimators – we have a

Table 3.18 Cumulative Averages with Partial Missing Data

Batch No	Batch Quantity	Batch Cost	Batch Average Unit Cost	Cumulative Cost	Cumulative Quantity	Cumulative Average Cost
1	5	£62	£12.40	£62	5	£12.40
2	15	£122	£8.13	£184	20	£9.20
3	12	£104	£8.67	£288	32	£9.00
4 and 5				£592	72	£8.22
6	8	£80	£10.00	£672	80	£8.40
7	20	£152	£7.60	£824	100	£8.24

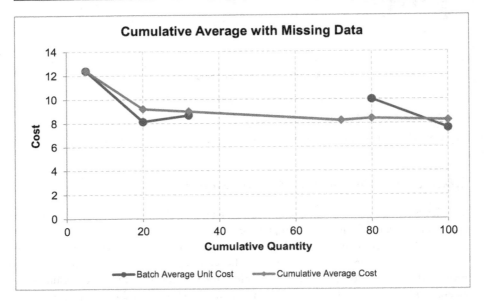

Figure 3.28 Cumulative Average with Partial Missing Data

responsibility to consider anything that we can do legitimately to identify a pattern of behaviour in data that we can then use to aid or influence the judgement process used to predict or estimate a future value.

A lesser used derivative of the traditional Cumulative Average is one that is based on Cumulative Achievement as a continuous function; in other words, we can consider the Cumulative Average based on non-integer denominators. For instance:

- We might collect the cost of successive units completed and calculate a Cumulative Average per unit to remove any unseemly variation in costs between units (see Volume IV Chapter 2 on Basic Learning Curves). This is a discrete function, dividing by a number of units completed (see previous Figure 3.25).

Trendsetting with some Simple Moving Measures | 105

- We might also the collect output per unit of input cost over time, and calculate a Cumulative Average productivity or efficiency measure over time. This again is a discrete function, dividing by a number of time periods that have elapsed (see previous Figure 3.24).
- Then again, we might want to be a little bit more creative with the data we have and calculate the Cumulative Cost per time period divided by the Cumulative Achievement realised. This is a continuous function, dividing by the cumulative equivalent number of units completed (see Table 3.19 and Figure 3.29)

The topic of Equivalent Units will be discussed further in Volume IV Chapter 5 on Equivalent Unit Learning Curves.

3.6.5 Cumulative Smoothing

A natural corollary to Cumulative Average smoothing is Cumulative Smoothing. In general, we may be better looking at Cumulative Smoothing under the section on Curve Fitting (Chapters 6 and 7) but as a simple taster of what is to come, we might like to

Table 3.19 Cumulative Average Equivalent Unit Costs

			Equivalent Unit Cost			
Month	Hours per Month	Units Achieved per Month	Cumulative Hours per Month	Cum Units Achieved per Month	Equivalent Unit Cost per Month	Cum Ave Equivalent Units Completed
0	-	-				
1	576	0.10	576	0.10	5,760	5,760
2	573	0.15	1,149	0.25	3,820	4,596
3	1,279	0.23	2,428	0.48	5,561	5,058
4	1,275	0.27	3,703	0.75	4,722	4,937
5	1,791	0.39	5,494	1.14	4,592	4,819
6	2,741	0.68	8,235	1.82	4,031	4,525
7	3,207	0.82	11,442	2.64	3,911	4,334
8	4,155	1.11	15,597	3.75	3,743	4,159
9	4,212	1.27	19,809	5.02	3,317	3,946
10	5,328	1.69	25,137	6.71	3,153	3,746
11	5,313	1.81	30,450	8.52	2,935	3,574
12	5,070	1.91	35,520	10.43	2,654	3,406
13	4,874	1.90	40,394	12.33	2,565	3,276
14	5,172	2.10	45,566	14.43	2,463	3,158
15	4,449	1.81	50,015	16.24	2,458	3,080
16	3,601	1.52	53,616	17.76	2,369	3,019
17	2,549	1.10	56,165	18.86	2,317	2,978
18	1,914	0.81	58,079	19.67	2,363	2,953
19	723	0.33	58,802	20.00	2,191	2,940
20	-	-				

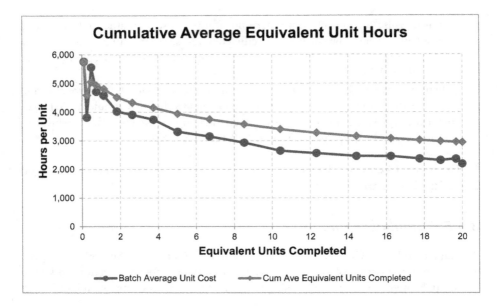

Figure 3.29 Cumulative Average Equivalent Unit Costs

consider an example where neither Moving Average nor Cumulative Average appear to be much help to us in answering a couple of simple questions such as '*what is the average component delivery rate, and at what point in time did it begin?*'

Consider actual component deliveries over a period of time (Table 3.20 and Figure 3.30). We can see that we have done some preliminary analysis of the data, and calculated the Cumulative Average Deliveries over time, and also a number of Moving Averages with increasing intervals.

In our example we can observe that:

- The actual deliveries per month are erratic with wild swings (*so, true to life, you will agree no doubt?*)
- There is no clear interval or repeating pattern on which to base the Moving Average calculation, but from the table higher order intervals yield smoother data (*we said they would!*)
- The Cumulative Average appears to be rising unabated
- Our best estimate of a nominal steady state for component deliveries per month would be in the region of 3 to 3.25 per month (*. . . but are we comfortable with that? Probably not*)
- The nominal steady state rate commences at Month 11 if we accept that a Moving Average of Interval 6 is appropriate.

Table 3.20 When Moving Averages and Cumulative Averages do not Appear to Help

Month	Units Delivered per Month	Cumulative Deliveries	Cumulative Average Deliveries	Moving Average (Interval 2)	Moving Average (Interval 3)	Moving Average (Interval 4)	Moving Average (Interval 5)	Moving Average (Interval 6)	Moving Average (Interval 7)	Moving Average (Interval 8)	Moving Average (Interval 9)
1	1	1	1.00								
2	0	1	0.50	0.50							
3	1	2	0.67	0.50	0.67						
4	2	4	1.00	1.50	1.00	1.00					
5	2	6	1.20	2.00	1.67	1.25	1.20				
6	1	7	1.17	1.50	1.67	1.50	1.20	1.17			
7	3	10	1.43	2.00	2.00	2.00	1.80	1.50	1.43		
8	1	11	1.38	2.00	1.67	1.75	1.80	1.67	1.43	1.38	
9	2	13	1.44	1.50	2.00	1.75	1.80	1.83	1.71	1.50	1.44
10	0	13	1.30	1.00	1.00	1.50	1.40	1.50	1.57	1.50	1.33
11	3	16	1.45	1.50	1.67	1.50	1.80	1.67	1.71	1.75	1.67
12	2	18	1.50	2.50	1.67	1.75	1.60	1.83	1.71	1.75	1.78
13	3	21	1.62	2.50	2.67	2.00	2.00	1.83	2.00	1.88	1.89
14	4	25	1.79	3.50	3.00	3.00	2.40	2.33	2.14	2.25	2.11
15	3	28	1.87	3.50	3.33	3.00	3.00	2.50	2.43	2.25	2.33
16	4	32	2.00	3.50	3.67	3.50	3.20	3.17	2.71	2.63	2.44
17	5	37	2.18	4.50	4.00	4.00	3.80	3.50	3.43	3.00	2.89
18	1	38	2.11	3.00	3.33	3.25	3.40	3.33	3.14	3.13	2.78
19	4	42	2.21	2.50	3.33	3.50	3.40	3.50	3.43	3.25	3.22
20	2	44	2.20	3.00	2.33	3.00	3.20	3.17	3.29	3.25	3.11
21	3	47	2.24	2.50	3.00	2.50	3.00	3.17	3.14	3.25	3.22
22	2	49	2.23	2.50	2.33	2.75	2.40	2.83	3.00	3.00	3.11
23	5	54	2.35	3.50	3.33	3.00	3.20	2.83	3.14	3.25	3.22
24	5	59	2.46	5.00	4.00	3.75	3.40	3.50	3.14	3.38	3.44
Min				1.00	1.00	1.50	1.40	1.50	1.57	1.50	1.33
Max				5.00	4.00	4.00	3.80	3.50	3.43	3.38	3.44
Max-Min				4.00	3.00	2.50	2.40	2.00	1.86	1.88	2.11

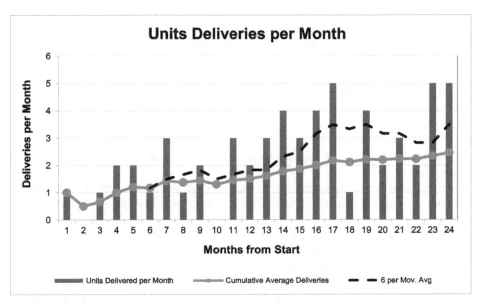

Figure 3.30 When Moving Averages and Cumulative Averages do not Appear to Help

OK, time for an honesty session now. How many of you thought that 'Month 11' was a misprint and should have read 'Month 16'? We must not forget that the Moving Average here has been depicted at the end of the cycle for which the average calculated is relevant; the start of the cycle for that average is given by the endpoint minus the interval plus one, hence 16−6 + 1 = 11.

It's not that intuitive immediately from the graph, I must admit? Even plotting the data at the interval mid-point doesn't really help. What we need is something simple that really makes it obvious . . . Perhaps we have just the thing.

For the Formula-phobes: Why do we not just deduct the interval to find the start point?

Consider a line of 5 black telegraph poles and spaces, followed by 5 white ones:

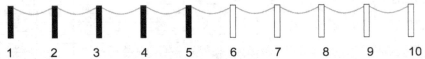

If we just deduct half the total number of poles from the last white one, we will end up with the last of the black ones. By adding 1 back on, we get the first of the white ones; we have to go in-between

---

If we take a step back from the Cumulative Average and plot the Cumulative data that we have available (Figure 3.31), we can immediately confirm the following:

- After Month 10 the cumulative number of units delivered per month follows a straight line − indicative of a nominally constant rate per month over Months 11 to 24
- The average rate per month can be determined from the difference between the cumulative deliveries at Month 24 and Month 10 divided by the number of months:

The average delivery rate is (59−13)/14 = 3.28 units per month after Month 10, i.e. beginning with the first delivery in Month 11. *Clearly, the end of Month 10 is the start of Month 11.*

The eagle-eyed will have spotted that the calculation gives 3.28 but the graph shows a straight line slope of 3.25. We are estimators, and precision of this nature is probably unnecessary and inappropriate, that's why the straight line on the graph has been rounded to increments of 3.25 or three and a quarter deliveries per month.

The curious amongst us who do not like loose ends, might ask how this all stacks up with Moving Averages and Cumulative Averages:

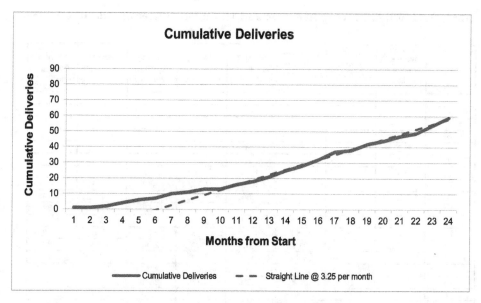

Figure 3.31 Cumulative Smoothing where Moving Averages and Cumulative Averages Fail

- In our example, we detected by observation that the Cumulative Deliveries followed a straight line from Month 10 to Month 24 – a difference of some 14 months.
- 14 is the product of 2 and 7, so it might be reasonable to try Moving Averages for intervals of 2 or 7
- We have also remarked that Cumulative Averages are slow to respond to changes in the underlying trend. However, we have also discussed that Cumulative Averages have to begin at a fixed reference point – usually observation point 1, but that does not have to be the case. In this instance we could begin the Cumulative Average at Month 10.

With this in mind, let's revisit the data from Table 3.20 in Table 3.21 but use higher order Intervals than previously used. We will consider the Cumulative Average only after Month 10 (i.e. Month 10 will become the new zero start point.)

From the table we can see that the difference between the Minima and Maxima is at it lowest level for a Moving Average of Interval 7, but only just ahead of a Moving Average of Interval 8.

Figure 3.32 shows us that the Cumulative Average between Months 10 and 24 is much more stable than the full holistic calculation. Also, the Cumulative Average provides an earlier indicator of there being a nominally steady state delivery rate per month over this period than the Moving Averages. The Moving Average can also be reconciled with the start point of the steady state average rate by deducting the interval minus 1

110 | Trendsetting with some Simple Moving Measures

Table 3.21 When Moving Averages and Cumulative Averages do not Appear to Help – Revisited

Month	Units Delivered per Month	Cum Deliveries after Month 10	Cum Ave Deliveries after Month 10	Moving Average (Interval 5)	Moving Average (Interval 6)	Moving Average (Interval 7)	Moving Average (Interval 8)	Moving Average (Interval 9)
1	1							
2	0							
3	1							
4	2							
5	2			1.20				
6	1			1.20	1.17			
7	3			1.80	1.50	1.43		
8	1			1.80	1.67	1.43	1.38	
9	2			1.80	1.83	1.71	1.50	1.44
10	0			1.40	1.50	1.57	1.50	1.33
11	3	3	3.00	1.80	1.67	1.71	1.75	1.67
12	2	5	2.50	1.60	1.83	1.71	1.75	1.78
13	3	8	2.67	2.00	1.83	2.00	1.88	1.89
14	4	12	3.00	2.40	2.33	2.14	2.25	2.11
15	3	15	3.00	3.00	2.50	2.43	2.25	2.33
16	4	19	3.17	3.20	3.17	2.71	2.63	2.44
17	5	24	3.43	3.80	3.50	3.43	3.00	2.89
18	1	25	3.13	3.40	3.33	3.14	3.13	2.78
19	4	29	3.22	3.40	3.50	3.43	3.25	3.22
20	2	31	3.10	3.20	3.17	3.29	3.25	3.11
21	3	34	3.09	3.00	3.17	3.14	3.25	3.22
22	2	36	3.00	2.40	2.83	3.00	3.00	3.11
23	5	41	3.15	3.20	2.83	3.14	3.25	3.22
24	5	46	3.29	3.40	3.50	3.14	3.38	3.44
Min				1.60	1.67	1.71	1.75	1.67
Max				3.80	3.50	3.43	3.38	3.44
Max-Min				2.20	1.83	1.71	1.63	1.78

from the start month indicated by the plot. This gives us the offset lag commencing at Month 11 as indicated by Table 3.22:

Ah, the estimator's equivalent of Utopia – compatible answers from more than one method!

3.7 Chapter review

We can apply basic trend smoothing techniques to any natural sequence whether time or achievement based. Moving Averages are simple to use and are ideal for estimating Steady State values, or changes in the Steady State value. However, they need to be adjusted where there is an upward or downward trend to take account of the inherent lag between the true underlying trend and the traditional practice of plotting the Moving Average at the end of the cycle. We can make this lag adjustment by plotting the Moving Average earlier in the middle of its Interval or cycle. Weighted Moving Averages

Trendsetting with some Simple Moving Measures

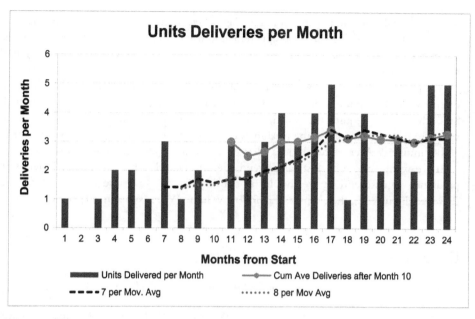

Figure 3.32 Cumulative Smoothing where Moving Averages and Cumulative Averages Fail – Revisited

Table 3.22 Moving Average Steady State – Applying the Offset Lag

Moving Average Base in Months Giving Steady State	Steady State Commences at Moving Average Endpoint Month	Moving Average Lag in Months (Base -1)	Steady State Commences at Month (Endpoint – Lag)
7	17	6	11
8	18	7	11

can be adjusted also to take account of the inherent lag; the adjustment will depend on the weightings used.

We can use other Moving Measures such as Maxima and Minima, Factored Standard Deviations or Percentiles to reflect the uncertainty around the steady state or upwards or downwards trend. However, Moving Averages can be unduly affected by extreme values or outliers but Moving Medians can provide a more stable alternative provided the interval is greater than two.

We have seen that Exponentially Weighted Moving Averages (Exponential Smoothing) is also useful for determining steady state values but need more complex analysis

112 | Trendsetting with some Simple Moving Measures

in order to take account of the inherent lag where there is an upward or downward trend. We have also discussed the apparent confusion and interpretation of the terms 'Smoothing Constant' and 'Damping Factor' and that it is important that we recognise how others use them.

We have also explored the use of Cumulative data and Cumulative Averages as an often overlooked smoothing mechanism. Whereas Moving Averages and Exponential Smoothing both require the full data history that is equally spaced with no missing values in order to maximise their value as trend smoothing techniques, the Cumulative and Cumulative Average techniques allow us to use a reduced number of observations.

If we want a more sophisticated trend smoothing technique that is also inherently predictive in nature, then we should be considering line or curve fitting techniques; these are discussed in the chapters that follow. Volume IV deals specifically with Learning Curves.

References

Brown RG (1963) *Smoothing Forecasting and Prediction of Discrete Time Series*, Englewood Cliffs, New Jersey, Prentice-Hall.

Casaregola, V (2009) *Theaters of War: America's Perceptions of World War II,* New York, Palgrave MacMillan, p. 87.

Holt, CC (1957) 'Forecasting Seasonals and Trends by Exponentially Weighted Averages', *Office of Naval Research Memorandum,* 52, reprinted (2004) in *International Journal of Forecasting*, 20 (1) pp. 5–10.

ONS, (2017) *MM23 Consumer Price Inflation time series dataset*, Newport, Office of National Statistics, 14 February 2017, [online] Available from: https://www.ons.gov.uk/economy/inflationandpriceindices/datasets/consumerpriceindices [Accessed 14–02–2017].

Winters, PR (1960) 'Forecasting sales by exponentially weighted moving averages', *Management Science*, 6 (3), pp. 324–342.

Younge, G (1999) 'Jesse Jackson: Power, politics and the preacher man', London, *The Guardian*, 17th April [online] Available from: https://www.theguardian.com/world/1999/apr/17/uselections2000.usa [Accessed 12/01/2017].

4 Simple and Multiple Linear Regression

Estimators and other number jugglers like us normally like to keep things simple – it's the rest of the world and life that makes things complicated for us. Wherever practical or justifiable, we are likely to try to put things into the perspective of linear relationships (i.e. to think in straight lines.) Sometimes this is a valid approximation to make; at other times, it would be naïve, so we need to keep our options open. We'll be looking at these other options in the next couple of chapters.

In general, the assumption of a Linear Relationship will raise a number of issues:

1. We're deluding ourselves if we think that everything can be predicted appropriately by a straight line.
2. What straight line? Especially when we only have a single reference point!
3. Which straight line do we take when our data is scattered around an invisible straight line? (*It's invisible because we haven't worked it out yet.*)
4. Which straight line do we take if we think we have more than one valid driver?

We'll deal with the first issue in Chapters 5–8 and we have already dealt with the second when we discussed the topic of Analogy in Volume I Chapters 2 and 5.

To save you looking back, every analogy is based on the implied assumption that there is a straight-line relationship through the Reference Point and the Origin (0, 0).

We will deal with the last two points here under Simple Linear Regression and Multiple Linear Regression (often referred to as Multi-Linear Regression or Multi-Variate Linear Regression.)

4.1 What is Regression Analysis?

on our perspective, we can define or explain Regression as:

1. An act of reflecting and recalling memories from an earlier stage of life, or an alleged previous life

114 | Simple and Multiple Linear Regression

2. A statistical technique that determines the 'Best Fit' relationship between two or more variables

Although we may be tempted to consider Regression as a means of entering a trance like state where we can reflect on life before we became estimators, it is, of course, the second definition that is of relevance to us here. However, this definition for me (although correct) does not fully convey the process and power of Regression Analysis as it misses the all-important element of what determines 'best fit'. Instead, we will revise our definition of Regression to be:

Definition 4.1 Regression Analysis

Regression Analysis is a systematic procedure for establishing the Best Fit relationship of a predefined form between two or more variables, according to a set of Best Fit criteria.

Note that Regression only assumes that there is a relationship between two or more variables; it does not imply causation. It also assumes that there is a continuous relationship between the dependent variable, i.e. the one we are trying to predict or model, and at least one of the independent variables used as a driver or predictor.

One of the primary outputs from a Regression Analysis is the Regression Equation, which we would typically use to interpolate or extrapolate in order to generate an estimate for defined input values (drivers). The technique has a very wide range of applications in business and can be used to identify a pattern of behaviour between one or more estimate drivers (the independent variables) and the thing or entity we want to estimate (the dependent variable). Examples might include the relationship between cost and a range of physical parameters, sales forecasts and levels of marketing budgets, Learning Curves, Time Series ... the list goes on.

We can define the estimating relationship to be any form we choose, whether it is linear or non-linear. We will see in Chapter 5 that there are occasions when we can *mathe-magically* transform a non-linear relationship into a linear one, in order to exploit the properties of a straight line. For now though we are going to stick with exploring the basic principles through Linear Regression.

In terms of defining what we mean by 'Best Fit', the most commonly used criteria is that based on 'Least Squares'. There are others, such as 'Least Absolute Deviations', which we could find using our favourite internet search engine, but we are only going to consider the Least Squares technique here (except for a brief mention in Section 4.1.1 to highlight the difference in relation to Least Squares).

4.1.1 Least Squares Best Fit

In order to define what we mean by 'Least Squares Best Fit' let's first consider the simple example of a straight line through a set of data points, as depicted in Figure 4.1. In the majority of cases, we would stipulate that the Best Fit Line must pass through the Arithmetic Mean of the data so that it is representative of the data set as a whole, i.e. the Mean of the Line is the same as the Mean of the Data, in terms of both the dependent and independent variables – it is unbiased towards any particular point. We might like to think of the Mean as being the centre of gravity of the data.

We can calculate the vertical difference between the each point and the corresponding position on the straight line through the data mean. These vertical differences are usually referred to as the 'Error Values' or 'Residuals'. The problem with the residuals is that some will be positive and some will be negative as they fall above or below the line, but their average will always be zero because the line passes through the Mean of the data. We could define the 'Best Fit' straight line as the line that minimises the total of the absolute errors or residuals between the points and the line. Unless you skipped over Volume II Chapter 3 on Measures of Dispersion and Shape, you may recall that statisticians are a bit like bank managers and don't like ignoring the negative signs. Instead, they prefer that we get rid of these inconvenient negative signs by squaring the differences

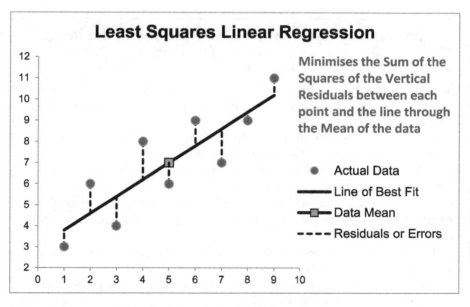

Figure 4.1 Determining the Line of Best Fit by Least Squares

116 | Simple and Multiple Linear Regression

instead (and later taking the positive square root to reverse it.) The straight line that returns the smallest value for the sum of the squares of the residuals is then considered to be the Least Squares Line of Best Fit through the Mean.

> ### Definition 4.2 Least Squares Regression
>
> Least Squares Regression is a Regression procedure which identifies the "Best Fit" of a pre-defined functional form by minimising the Sum of the Squares of the vertical difference between each data observation and the assumed functional form through the Arithmetic Mean of the data.

The most commonly sought functional function is the Linear Relationship, or straight line (*we do like to keep estimating as simple as possible wherever we can – honestly!*) but we can apply this principle of Least Squares Regression to any assumed form i.e. a 'curve', with any number of dimensions or variables, (*but let's not run before we know whether we even want to walk; we will return to Non-linear Regression and Curve Fitting in Chapters 6 and 7.*) For purposes of illustration we will stick with a simple straight line for the time being.

For the Formula-phobes: Why don't we take the Minimum Absolute Deviation?

Some of us may be wondering if the Least Squares Technique gives the same answer as that of the Minimum Absolute Deviation, and if not, why we don't take the latter instead.

The answer to the first question is that they don't give the same answer as a general rule. The answer to the second is more complicated.

Let's consider it from the perspective of 'maximum inclusion'. The Least Squares Technique will give greater emphasis to minimising the difference of points further from a line in comparison with those closer to it (squares of differences greater than one get bigger more quickly; squares of differences less than one get even smaller.)

Moreover, there may be more than one solution that gives us the Minimum Total Absolute Deviation. For instance, consider four points:

Any line between the two dotted lines passing through the Mean gives us the Minimum Total Absolute Deviation of 2.

	x-value	y-value
	1	1
	2	2
	3	2.5
	4	1.5
Mean	2.5	1.75

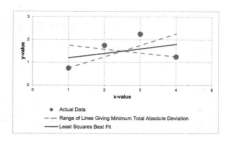

One extreme gives total emphasis on fitting the first observed data point, whereas the other extreme puts total emphasis on fitting the last observed data point. In this case, the Least Squares technique puts the emphasis on fitting the points further from any line at the sacrifice of some minor increase in the distance of points close to that line.

Let's consider our example in Figure 4.1 further. In Figure 4.2, we have drawn three lines at random through the Mean. For those of us who like to see numbers (*and which estimator doesn't?*), Table 4.1 calculates the sum of squares of the residuals for each of the three lines in the example. Visually, even though all three lines pass through the Mean of the data, when we compare each line with where it intersects the Lower and Upper Data Half Means, Line 1 appears to start too low and end too high, whereas Line 3 appears to start too high and end too low. On the other hand, Line 2 looks more reasonable, passing fairly evenly through the middle of the spread of data points.

Lower and Upper Data Half Means: The Mean of the first half of the data range and the Mean of the second half of the data range when the data is arranged in ascending order of the x-value.

OK, I confess, the lines weren't really drawn at random – I chose the middle one because it was the Least Square Line of Best Fit – having a near miss would not have been helpful in providing this explanation. What's more, the first line is one that returns the Minimum Total Absolute Deviation.

The good news is that we don't have to work out the sum of squares of the residuals or errors long-hand like this, and then iterate the procedure until we find a solution every time we want to perform a regression. The slope and intercept can be determined by a standard formula – even better news, Microsoft Excel will do all the calculations for us with functionality that comes as standard!

However, if we were to calculate the Least Squares Line of Best Fit iteratively for the data in our example, we could create a plot of the slope of the assumed line for each iteration against the Sum of Squares of the Residuals (or Sum of Squares

Simple and Multiple Linear Regression

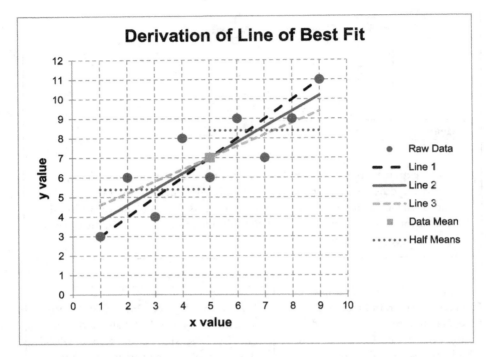

Figure 4.2 Derivation of Line of Best Fit

Table 4.1 Determining the Line of Best Fit by Least Squares

Raw Data		Line 1			Line 2			Line 3		
		Slope	1		Slope	0.8		Slope	0.6	
		Intercept	2		Intercept	3		Intercept	4	
x	y	y value	Residual	Residual Squared	y value	Residual	Residual Squared	y value	Residual	Residual Squared
1	3	3	0	0	3.8	0.8	0.64	4.6	1.6	2.56
2	6	4	-2	4	4.6	-1.4	1.96	5.2	-0.8	0.64
3	4	5	1	1	5.4	1.4	1.96	5.8	1.8	3.24
4	8	6	-2	4	6.2	-1.8	3.24	6.4	-1.6	2.56
5	6	7	1	1	7	1	1	7	1	1
6	9	8	-1	1	7.8	-1.2	1.44	7.6	-1.4	1.96
7	7	9	2	4	8.6	1.6	2.56	8.2	1.2	1.44
8	9	10	1	1	9.4	0.4	0.16	8.8	-0.2	0.04
9	11	11	0	0	10.2	-0.8	0.64	9.4	-1.6	2.56
Sum	45	63	63	0.00	16.00	63	0.00	13.60	63	0.00	16.00
Average	5	7	7	0.00	1.78	7	0.00	1.51	7	0.00	1.78

Error). We would get the result in Figure 4.3. We can compare this with the result we would get if we were to minimise the Sum of the Absolute Error. It is important that we appreciate that these will not necessarily be the same result i.e. 'Line 1' in Figure 4.2 and Table 4.1.

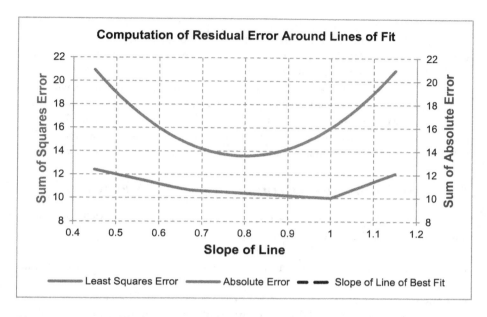

Figure 4.3 Line of Best Fit Defined by Least Squares Error cf. Minimum Absolute Error

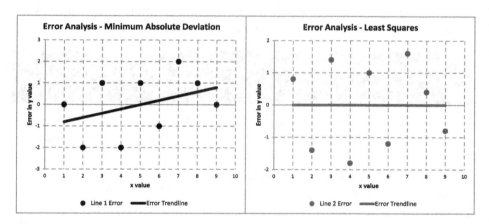

Figure 4.4 Error Distribution

So why do we use the Least Squares technique rather than the Minimum Absolute Deviation? Let's look at it from the perspective of the error. In Figure 4.4 we consider the error or residual between the first two of our lines and the observed data. Line 1 gives us the Minimum Absolute Deviation, and Line 2 provides the Least Squares Line of

120 | Simple and Multiple Linear Regression

Best Fit. There is an upward trend in the error values for the Minimum Absolute Deviation, indicating that the error is biased (it should be Normally Distributed around zero) whereas the Least Squares' trend is flat. Incidentally if we were to calculate the error for the third line, the trend would be decreasing.

Note that we should not misconstrue a flat line trend as being synonymous with only a Normally Distributed error term around a constant of zero – it could be arced – higher at the ends and lower in the middle or vice versa. This would indicate that the assumed functional relationship, i.e. linear, may not be correct; it may be non-linear. Later, in Section 4.5.7, we will classify these error distributions into two groups (*I know, I can hardly wait for it myself.*)

4.1.2 Two key sum-to-zero properties of Least Squares

There are two key properties implied by Least Squares to which we will refer back in understanding our Regression results later:

1. The sum of the errors between the Observed data and the Line of Best Fit (measured vertically) is zero.
2. The sum of the products of these errors and their associated independent variables is also zero! We might call this the Position-Weighted Sum of Errors.

In Table 4.2 we can demonstrate these two sum-to-zero properties where the error is the difference between the observed y value and the Regression Line, and the Position Weighted Error is the error multiplied by the Observed x value:

For the Formula-philes: Two Sum-to-Zero properties of Least Squares Error

For n paired observations x_1, x_2, x_3 ... x_n and y_1, y_2, y_3 ... y_n with corresponding Arithmetic Means $\bar{x}$, $\bar{y}$, we can fit a Line of Best Fit using Linear Regression:

Let β_0 and β_1 be the Intercept and Slope of the Regression Line of Best Fit, giving predicted values $\hat{y}_i$ corresponding to each x_i :

$$\hat{y}_i = \beta_0 + \beta_1 x_i \qquad (1)$$

The error term ε_i is the difference between the observed data and the Line of Best Fit:

$$\varepsilon_i = y_i - \hat{y}_i \qquad (2)$$

Substituting (1) in (2):

$$\epsilon_i = y_i - \beta_0 - \beta_1 x_i \qquad (3)$$

Sum of Squares Error SS_{err}

$$SS_{err} = \sum_{i=1}^{n} \left(y_i - \beta_0 - \beta_1 x_i \right)^2 \qquad (4)$$

Simple and Multiple Linear Regression | 121

Partial Derivative of SS_{err} with respect to β_0 :

$$\frac{\partial\ SS_{err}}{\partial\beta_0} = 2\sum_{i=1}^{n}\left(y_i - \beta_0 - \beta_1 x_i\right) \quad (5)$$

Partial Derivative of SS_{err} with respect to β_1 :

$$\frac{\partial\ SS_{err}}{\partial\beta_1} = 2\sum_{i=1}^{n} x_i\left(y_i - \beta_0 - \beta_1 x_i\right) \quad (6)$$

Using the standard property of calculus, in order to minimise SS_{err}, (i.e. Least Squares), both partial derivatives in (5) and (6) must be zero:

$$\frac{\partial SS_{err}}{\partial\beta_0} = \frac{\partial SS_{err}}{\partial\beta_1} = 0 \quad (7)$$

Applying (7) to (5), substituting (1) and (2), and then simplifying:

$$\sum_{i=1}^{n}\varepsilon_i = 0 \quad (8)$$

Applying (7) to (6), substituting (1) and (2), and then simplifying:

$$\sum_{i=1}^{n} x_i\varepsilon_i = 0 \quad (9)$$

. . . which are the two sum-to zero properties of Least Squares Error to which we referred

Table 4.2 Two Sum–to–Zero Properties of Least Squares Errors

Best Fit Line				
Slope	0.8			
Intercept	3			
Observed Data			Sum-to-Zero	
y	x	Regression $\hat{y}$ value	Error	Position Weighted Error
3	1	3.8	0.8	0.8
6	2	4.6	-1.4	-2.8
4	3	5.4	1.4	4.2
8	4	6.2	-1.8	-7.2
6	5	7	1	5
9	6	7.8	-1.2	-7.2
7	7	8.6	1.6	11.2
9	8	9.4	0.4	3.2
11	9	10.2	-0.8	-7.2
Sum	63	45	63	0.00	0.00
Average	7	5	7		

4.2 Simple Linear Regression

We mentioned in passing in the last section that we don't have to iterate manually in order to find the Least Squares Line of Best Fit as there is a standard analytical result that provides it. If we cast our minds back to Volume II Chapter 2 (unless we felt the need to resist that opportunity), we highlighted the relationship between the Slope of the Line of Best Fit, the covariance of the two variables, and the variance of the independent x-value. The relevant illustration is reproduced and annotated at Figure 4.5 to save you looking for it (*see how kind and caring I have become?*) The term Simple Linear Regression refers to the regression relationship assuming a simple straight line. We may hear of Simple Linear Regression being referred to as Ordinary Least Squares Regression in some texts.

For the Formula-philes: Slope and intercept for Simple Linear Regression

Consider a range of *n* paired observations $x_1, x_2, x_3 \ldots x_n$ and $y_1, y_2, y_3 \ldots y_n$ with Means $\bar{x}$ and $\bar{y}$ respectively

Let the Line of Best Fit for y have a slope of β_1 and an intercept of β_0 such that $\hat{y}_i$ for each corresponding value on the Line of Best Fit for each x_i :

$$\hat{y}_i = \beta_1 x_i + \beta_0$$

From Figure 4.5 the slope can be expressed as the ratio of the Covariance of x and y over the Variance of x:

$$\beta_1 = \frac{\sigma_{xy}}{\sigma_x^2}$$

Expanding σ_{xy} and σ_x^2 the Line of Best Fit Slope, β_1, can be calculated by the expression:

$$\beta_1 = \frac{\sum_{i=1}^{n}(x_i - \bar{x})(y_i - \bar{y})}{\sum_{i=1}^{n}(x_i - \bar{x})^2}$$

Substituting the Mean values of x_i and $\hat{y}_i$ the intercept, β_0, can be calculated as:

$$\beta_0 = \bar{y} - \beta_1 \bar{x}$$

Rule of Thumb

Recall (from Volume II Chapter 5) that we can use Linear Correlation as a measure of the degree of Linearity to see if it is worth running a full Linear Regression. As a rule of thumb, if the correlation is less than 0.71, then less than half of the variation in one variable can be explained by the linear relationship with the other variable. If that is the case perhaps we should be considering more variables or a non-linear relationship.

Simple and Multiple Linear Regression | 123

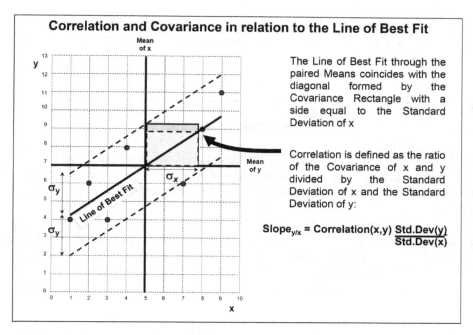

Figure 4.5 Correlation and Covariance in Relation to the Line of Best Fit

In Microsoft Excel, there are a number of functions and facilities that allow us to calculate the Line of Best Fit through Least Squares Linear Regression. There are advantages and disadvantages to each. We will review the merits of each in the following sub-sections.

4.2.1 Simple Linear Regression using basic Excel functions

There is a simple function in Excel which calculates and returns the line of Best Fit:

- **TREND(*known_ys, known_xs, new_xs, const*)** where ***known_ys*** and ***known_xs*** refer to the location of the y and x range of values; the location must be a set of contiguous cells (i.e. adjacent to each other) in either a row or a column format, but not mixed. Excel assumes (logically) that the values are paired together in the order that they are presented. The range ***new_xs*** refers to the location of a set of values for which we want to create the Line of Best Fit values; this could be one of the ***known_xs*** or a new value. In theory, this value accepts a range input which can be enabled by Ctrl+Shift+Enter but, as we said in Volume I, this is not recommended practice. It works well for a single cell that we can then copy across a range. The final parameter, ***const*** allows us to force the regression through the

124 | Simple and Multiple Linear Regression

origin rather than the Mean of the data. As a general rule, we should have a good reason for doing this; to force it through the origin set *const* to FALSE; otherwise set it to TRUE.

Caveat augur

As a general rule, array formulae or array functions such as **TREND** are discouraged (see Volume 1 Chapter 3) as they do not function correctly if the user forgets to enable them with Ctrl+Shift+Enter instead of the usual Enter key.

Failure to use Ctrl+Shift+Enter may not necessarily return an error. To avoid this it is recommended that the function's range parameter *new_xs* is restricted to a single cell, and that the formula is copied to adjacent cells to avoid this. That way it will work with the normal Enter key.

If we want to know the slope and intercept of the Least Squares Line of Best Fit, then there are two other functions available to us in Excel. The full syntax for each parameter is identical to the beginning of the **TREND** function i.e. contiguous cells in either a row or a column. Again Excel assumes that the values are paired together in the order that they are presented.

- **SLOPE**(*known_ys, known_xs*)
- **INTERCEPT**(*known_ys, known_xs*)

If we inadvertently include a non-numeric value, or a different number of cells, then Excel will return an error. In this case we can even mix columns and rows for the x and y ranges, but this is not recommended from a Best Practice Spreadsheet design perspective. (*Just because we can, doesn't mean we should!*) If we applied these two functions to the data in Table 4.1 we would get the results for Line 2 i.e. Slope = 0.8 and Intercept = 3.

In terms of using a regression output to interpolate or extrapolate the Best Fit Straight Line for any x-value position, Microsoft Excel provides a simple function which returns the value of y on the straight line corresponding to the requested value x, which may or may not have been included in the input range:

- **FORECAST**(x, *known_ys, known_xs*)

This function is largely redundant as the **TREND** function gives us the same functionality plus the ability to suppress the intercept. However, this is the safeguard option

Simple and Multiple Linear Regression | 125

Caveat augur

The major downside of these very simplistic functions is that there is no measure of how good, bad or indifferent the 'Best Fit' Straight Line actually is.
'Best Fit' does not necessarily imply a 'good fit'.

---

to use in preference to TREND if we want to stick with regular functions rather than risk using Array Functions.

The greatest strength of a Linear Regression is also its greatest weakness. Regardless of whether we use the Microsoft Excel in-built functions, or perform the calculations long hand, we will always be able to determine the Best Fit Slope and Intercept – even when there is no justification for doing so!

Ideally, we need a means of determining whether a regression line is statistically valid. The next two options in Excel allow us to do this, although it will not be until Section 4.5 that we will discuss what those measures might be.

4.2.2 Simple Linear Regression using the Data Analysis Add-in Tool Kit in Excel

'Data Analysis' is a very useful free add-in in Microsoft Excel that provides a range of statistical analysis routines, one of which is Linear Regression.

*   To load the add-in in Excel 2007 and above, go to **File/Options/Add-ins/Manage Excel Add-ins** and select the '**Analysis ToolPak**' option. Once enabled it will appear as 'Data Analysis' under Data.
*   In Excel 2003 and earlier, go to **Tools/Add-ins** and select '**Analysis ToolPak**'. Once enabled it will appear as 'Data Analysis' under Tools.
*   Once loaded on your computer, it should always be available (*unless you choose to unload it, of course.*)

The regression routine requires us to complete a dialogue box to define the location of the inputs and where to paste the standardised output, which includes multiple statistics relating to the analysis. (*We'll not trouble ourselves with these little delights for the moment – we'll leave that little pleasure until Section 4.5.*)

Let's consider some cost data for a range of parts of different weights, shown in Figure 4.6. We will assume that the costs are at the same economic conditions, and have been manufactured using comparable processes and technologies in the same factory. The Figure also shows the corresponding Excel Regression dialogue box.

Simple and Multiple Linear Regression

	A	B	C
1		y	x1
2	Obs	Cost £	Weight Kg
3	1	5123	10
4	2	6927	21
5	3	4988	34
6	4	9253	42
7	5	7022	59
8	6	10182	63
9	7	8348	71
10	8	10702	85
11	9	12092	97
13	Average	8293.000	53.556

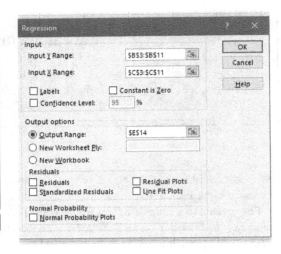

Figure 4.6 Data Input Using Microsoft Excel's Data Analysis Regression Add-In

Input options and limitations include:

- Location of the vertical range of cells containing the y-values or dependent data. The data must be in a column of contiguous cells (no gaps between cells and not in a row)
- Location of the vertical range of cells containing the x-values or independent data. The data must be in a column of contiguous cells (no gaps between cells and not in a row)
- We can elect to include a header cell for the x and y data columns (but it has to be both or neither.) Remember to tick the check box.
- We can elect to force the regression through the origin rather than the data mean, i.e. we can make the intercept equal to zero. We will return to this point later under 'Six key measures' in Section 4.3. Don't do it initially as a general rule!
- We can also tell Excel where we want it to paste the answer (*we only have to tell it the top left-hand corner position – Excel knows how many cells it needs* for the rest of the output.) The Regression routine will warn us if the output report is going to overwrite any existing data (*and obviously give us the chance to put it somewhere else!*)
    o Now, you may have noticed that I keep saying 'paste' in relation to the output; this is because Excel performs all its calculations in the background and pastes the numerical values in the output report (equivalent to Cut/Paste Special/Values); it doesn't put any formulae in the report.) This is a limitation of the Regression Routine; if we change the input data we have to re-run the routine manually.

Simple and Multiple Linear Regression | 127

Table 4.3 Determining the Line of Best Fit Using Microsoft Excel's Data Analysis Add-In

SUMMARY OUTPUT

Regression Statistics	
Multiple R	0.846881686
R Square	0.717208589
Adjusted R Square	0.676809816
Standard Error	1409.571009
Observations	9

ANOVA

	df	SS	MS	F	Significance F
Regression	1	35273716.98	35273716.98	17.75322707	0.003968678
Residual	7	13908233.02	1986890.431		
Total	8	49181950			

	Coefficients	Standard Error	t Stat	P-value	Lower 95%	Upper 95%	Lower 95.0%	Upper 95.0%
Intercept	4427.86612	1030.660497	4.296144204	0.003583531	1990.741314	6864.990925	1990.741314	6864.990925
Weight Kg	72.17054963	17.12857994	4.213457852	0.003968678	31.66789412	112.6732051	31.66789412	112.6732051

- There are other options available by way of tick boxes, which relate largely to whether we want tabular and graphical reports on residuals, and the ability to select a Confidence Interval. These are useful also in aiding our understanding of the output.

Table 4.3 illustrates the output of the Excel Regression Tool using the data from Figure 4.6.

In the bottom left hand corner of the Summary Output, we will see under the heading 'Coefficients' that the Intercept has been calculated as 4427.9 (rounded), and the slope (depicted here as the Coefficient of 'X Variable 1') is approximately 72.17. These coefficient values are the same as those we would get if we used the functions **SLOPE** and **INTERCEPT** (*as if you ever doubted it!*)

If we had included the header columns of Cost and Weight in the input range and ticked the input dialogue box to include 'Labels', then 'Weight' would have appeared in the Summary Output table instead of 'X Variable 1'. (You have probably guessed by implication that we can have 'X Variable 2' and 'X Variable 3', etc. We will get back to this in Section 4.3.) You may also have noticed that in the bottom right hand corner of the Summary Output there are two columns headed 'Lower 95%' and 'Upper 95%' that appear to have been repeated. This is not an error; we will get back to this one in Section 4.6 on Prediction and Confidence Limits.

4.2.3 Simple Linear Regression using advanced Excel functions

Now this option is the slightly scary one! People of a nervous or sensitive disposition may prefer to stick with the previous option. However, for the stout of heart, this option

128 | Simple and Multiple Linear Regression

in Microsoft Excel does offer two main advantages over the Data Analysis Add-in option:

- Data does not have to be in columns (greater flexibility)
- If we change our data then the regression calculations will automatically be updated (useful if we are performing routine analysis of large volumes of data every month or so; all we have to do is re-import the data)

This option utilises an advanced array function in Microsoft Excel called **LINEST**. (*Yes, it's one of those types of functions that we advised you against in Section 4.2.1.*) Before you berate me for dual standards, there is a way of avoiding some of the pitfalls of array functions; it requires us to combine the function with another function **INDEX**. Firstly, we need to consider the syntax for **LINEST**; it has four input fields or parameters:

- **LINEST(*known_ys, known_xs, const, stats*)** where:
  - o *known_ys* and *known_xs* refer to the location of the y and x range of values; the location must be a set of contiguous cells (i.e. adjacent to each other) in either a row or a column format, but not mixed. Excel assumes (logically) that the values are paired together in the order that they are presented. The parameter
  - o *const* allows us to force the regression through the origin rather than the Mean of the data. As a general rule, we should have a good reason for doing this; to force it through the origin set *const* to FALSE; otherwise set it to TRUE.
  - o The final parameter *stats* is asking us to specify whether we want Excel to calculate the supporting statistics (TRUE) or not (FALSE). To be honest, if we set the parameter to FALSE, we might as well use the simpler function **TREND** or **FORECAST**. We will be using the statistics later.

In the background, if we were to enable this array function with Ctrl+Shift+Enter then it will produce an array of 5 rows by 2 columns of values. However, instead, in order to use this simply, we can read this 'invisible' array with the function **INDEX(*array, row_num, column_num*)**. Here, the *array* is the **LINEST** function output and *row_num* and *column_num* are the row and column number that we are interested in extracting from the 'invisible' array. We can therefore generate any value we want from the invisible array using the formula:

INDEX(LINEST(known_ys, known_xs, Const, Stats), *row_num, column_num*)

I did warn you that this was going to get a bit scary!

For the time being we should assume that parameters **Const** and **Stats** are both set to TRUE, and that *row_num* can take any value from 1 to 5, and for Simple Linear Regression *column_num* takes the value 1 or 2. Table 4.4 illustrates this for our previous sample data from Figure 4.6.

The two values in the **LINEST** Output table of immediate use to us are the top row where the *row_num* equals 1. No doubt you will have spotted that when the *column_num*

Table 4.4 Determining the Line of Best Fit Using Microsoft Excel's Advanced Functions

	y	x
Obs	Cost £	Weight Kg
1	5123	10
2	6927	21
3	4988	34
4	9253	42
5	7022	59
6	10182	63
7	8348	71
8	10702	85
9	12092	97

LINEST Output Data			
		column_num	
		1	2
row_num	1	72.1705496	4427.86612
	2	17.1285799	1030.6605
	3	0.71720859	1409.57101
	4	17.7532271	7
	5	35273717	13908233

Table 4.5 LINEST Output Data for a Simple Linear Regression

		column_num	
		1	2
row_num	1	Slope	Intercept
	2	Standard Error for the Slope	Standard Error for the Intercept
	3	Coefficient of Determination (R-Square)	Standard Error for the y Estimate
	4	F-statistic for the Regression Model	Number of degrees of freedom in the Residuals
	5	Sum of Squares for the Regression Model	Sum of Squares for the Residual Error

takes the value of 1 or 2, we get the Best Fit Line's slope and intercept. At the moment, the rest of the data probably looks like random numbers. We will be discussing these and relating them back to the Summary Output Table from the Data Analysis Add-in tool later in Section 4.5. For completeness Table 4.5 summarises what they are.

4.3 Multiple Linear Regression

Sometimes we may have the situation where we believe that we have more than one driver that affects the outcome we want to estimate. For example, we might believe that

130 | Simple and Multiple Linear Regression

the time (effort or duration) required to assemble or construct a product will be dependent on either, or both, of its weight and the number of parts in the finished product:

- We could have two products of different weights with the same number of parts that take different times to assemble or construct due to the weight of the individual components and the need to provide additional physical support during assembly
- We could have two products that weigh the same, but one has more small components, and the other has a small number of larger, heavier components

So, instead of creating a two-dimensional model with just one driver or independent variable (*to use the technical term*), we can create a three-dimensional model that includes both drivers. The underlying assumption of a multi-linear model is that each driver or independent variable makes a linear contribution to the output. In other words, if we held the other driver as a constant (or inactive), then we could use a simple linear model to describe the relationship using the remaining active driver. The inactive driver in effect is contributing to the value of the constant (i.e. intercept) term. Similarly, the same is true if we reversed the roles of the two drivers in terms of being active or inactive. In this way we can add as many drivers as we want, each making a linear contribution to the final answer. However, just as with the Simple Linear Regression, where we had to have more data than variables, the same is true for a Multiple Linear Regression Model, so the more variables or drivers we add, the more data we will need to perform a regression analysis. Despite the contribution of multiple linear models, we only need one constant or intercept in our model (*the sum of several constituent constants is still a constant.*)

In a three-dimensional model, instead of a straight line we are defining a plane that slopes in two directions. Multiple dimensions become harder to visualise.

We should also think carefully about whether any additional driver is actually adding anything to the model. We will probably recall that we also refer to drivers as independent variables. If two potential drivers are highly correlated together then we can infer that they are not truly independent, so it is always worth checking that the so-called independent variables are 'sufficiently' independent of each other. We can do this in Microsoft Excel using the function **CORREL(*range1, range2*)**. (Alternatively, we can

---

For the Formula-philes: Multiple Linear Regression Equation

Consider a range of n drivers x_1, x_2, x_3 ... x_n all of which have a linear effect on the value of y, the entity which we wish to estimate. Let β_i represent the linear contribution (slope) of the corresponding x_i variable, and β_0 is the Intercept Constant of the model.

Multiple Linear Model:

$$y = \beta_0 + \beta_1 x_1 + \beta_2 x_2 + \beta_3 x_3 + \ldots + \beta_n x_n$$

Simple and Multiple Linear Regression | 131

For the Formula-phobes: Give me an example of what multi-linear means

Consider a football team (*that will be soccer if you are reading this in America*) playing in a progressive league. The league is progressive in that in order to promote attacking attractive football, teams are rewarded with a bonus point for scoring three or more goals in a match – even if they lose! That's on top of an award of 3 points for a win and one point for a draw. A team gets nothing for losing, unless they score three goals or more, in which case they get one bonus point.

We can calculate the number of points (the dependent variable) that they have amassed at any time based on the number of the number of wins, draws and games in which they score 3+ (independent variables):

Matches played	Wins (3 pts)	Draws (1 pt)	Losses (0 pt)	Score bonus (1 pt)	Total points
30	26	2	2	11	91
30	21	5	4	18	86
32	17	10	5	21	82

use **PEARSON(range1,range2)**, which has exactly the same functionality.) If the value is greater than 0.71 then this is indicative of at least half the variance of one being associated or linked with the variance in the other (see Volume II Chapter 5.)

Don't worry if we forget to do this in our excitement to complete the analysis; thankfully, the Regression technique will reject one of the variables as being statistically not significant if that were to be the case. It will also do this if the variable is random and not correlated with the dependent variable. However, we are getting ahead of ourselves again; we'll be covering that in Section 4.5.

In Microsoft Excel, we can perform a Multiple Linear Regression in only two ways:

1. Using the Data Analysis Add-in feature
2. Using the Advanced **LINEST** function

We will examine these in Sections 4.3.2 and 4.3.3. Unfortunately there are no basic functions in Microsoft Excel equivalent to **SLOPE** and **INTERCEPT** that will give us a short-cut answer. (*You look so disappointed!*)

4.3.1 Using categorical data in Multiple Linear Regression

Whereas with Simple Linear Regression both the x and y variables have to be 'continuous' functions, we can relax this requirement for Multiple Linear Regression. The

132 | Simple and Multiple Linear Regression

y-variable and at least one of the x variables must be continuous, or at least have a sufficiently distinct range (possibly unlimited) of discrete values such as a Build Sequence Number. In the case of Multiple Linear Regression, the other x-variables can be categorical data, i.e. data that defines or characterises data into groups.

> Examples of this include gender, test results (pass or fail), seasonal data (spring, summer, autumn, winter), types of product, classification of items into similar sizes or levels of complexity, . . . *the list goes on.*

By far the simplest way of dealing with these in Multiple Linear Regression is using a Dummy Variable or Binary On/Off switch i.e. 1 for 'On', and 0 for 'Off'. This works for any category where there are only two values – and it doesn't matter which way round we assign the On/Off switch.

Where we have three or more potential values in a single category e.g. small, medium, large then it is often easier to contemplate three switches, one for each value so long as we ensure that no more than one is switched on at any one time as illustrated in Table 4.6:

However, in truth, one of these Dummy Variables or switches is redundant as we can identify a base case when both Dummy Variables are switched off, as illustrated in Table 4.7, where medium size is taken as the base case.

If we 'forget' this option, don't worry, regression is such a 'forgiving' technique that it will point this out to us as part of the analysis. In Table 4.8 we have modelled our data with three Dummy Variables.

In this example, we have modelled three categories using three Dummy Variable On/Off Switches, whereas we only needed two Dummies (*Yes, I agree, I was the other dummy!*)

Table 4.6 Example 1 of Using Binary Switched for Multi–Value Categorical Data

Category Value	Small = x1	Medium = x2	Large = x3
Small	1	0	0
Medium	0	1	0
Large	0	0	1

Table 4.7 Example 2 of Using Binary Switched for Multi–Value Categorical Data

Category Value	Small = x1	Medium = x2	Large = x3
Small	1	0	0
Medium	0	0	0
Large	0	0	1

Simple and Multiple Linear Regression | 133

Table 4.8 Regression will Eliminate any Redundant Dummy Variable

Obs	y	x_1	x_2	x_3	x_4	Cat
1	4	1	0	0	1	Large
2	4	2	0	1	0	Medium
3	3	3	1	0	0	Small
4	8	4	0	0	1	Large
5	6	5	1	0	0	Small
6	8	6	0	1	0	Medium
7	9	7	0	1	0	Medium
8	11	8	0	0	1	Large
9	9	9	1	0	0	Small

Sum	62	45	3	3	3
Average	6.889	5	0.333	0.333	0.333

SUMMARY OUTPUT

Regression Statistics	
Multiple R	0.989135301
R Square	0.978388644
Adjusted R Square	0.76542183
Standard Error	0.513009059
Observations	9

ANOVA

	df
Regression	4
Residual	5
Total	9

	Coefficients
Intercept	3.408914729
x Variable 1	0.98255814
x Variable 2	-2.976744186
x Variable 3	-1.321705426
x Variable 4	0

Dummy Variable x_4 is redundant, making zero contribution >>>

Regression has calculated that one of them (x_4 in this case) is zero i.e. it is not adding anything to the model in comparison to the other independent variables. We should re-run the analysis without the redundant Dummy Variable – the resulting Coefficients will be the same (or equivalent) but it will correct an error generated in the statistics that we will be covering in Section 4.5 – *in case you're curious, it has identified that the total number of degrees of freedom equals the number of observations, whereas there are really only ever one less than this!*

The impact of using Dummy Variables like this is to determine a value that would be added or subtracted from the Model Base Case to take account of the different groups. If we restricted ourselves to one variable it would require us to either:

- Accept that the difference between category values was the same (i.e. switch values could be 0, 1, 2 ... all of which are multiplied by a constant)
- Require us to imply in advance (in the input) the relative scale of difference (e.g. Small = 0, Medium = 1, Large = 2.5), (*which somewhat defeats the objective of using Regression to some extent*) in which case we could normalise the data first and factor them to a common base.

4.3.2 Multiple Linear Regression using the Data Analysis Add-in Tool Kit in Excel

The way we use the Data Analysis Add-in Tool with Multiple Linear Regression is very similar to its use for Simple Linear Regression, but with one major difference. In the input dialogue box, for the '*Input x-range*' field we must specify the location of the independent variables (drivers) as a set of contiguous cells in contiguous columns. In effect we must define this as the top-left cell address to the bottom-right cell address e.g. B4:D13 for nine values (rows 4 to 12) covering three variables (in columns B, C and D.)

Everything else is the same:

- Data must be in columns, not rows
- We can include a header cell for the y column and each x column (but it has to be all or none). This feature is **highly recommended** so that we don't lose track of which variable is contributing what values when we interpret the results.
- We can elect to force the regression through the origin rather than the data mean, i.e. we can make the intercept equal to zero. We will return to this point later under 'Measures of Relevance' in Section 4.5.3.
- We can also tell Excel where to stick its output (*now, now, there's no need to be like that!*)
- We can also select the various graphical outputs and residual reports

In Figure 4.7 we have taken our original data from Figure 4.6 and added an extra variable for the number of Tests that have to be performed on each product. When we run the regression now with two independent variables (Weight and Tests), we get the results in Table 4.9. The layout of the Summary Output looks very similar to that for Simple Linear Regression, but with one difference ... the additional row at the bottom where the coefficients are given. In fact, if we were to run a regression with multiple drivers it will add an extra row for each additional driver.

Yes, you are correct; I have not practised what I preach in this example – I have not given the variables separate identifiable labels. This is to illustrate what happens when you don't define

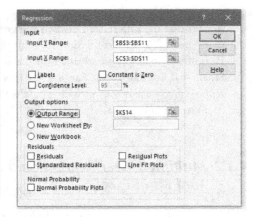

Figure 4.7 Multi-Linear Regression Input Example

Simple and Multiple Linear Regression | 135

Table 4.9 Determining the Plane of Best Fit Using Microsoft Excel's Data Analysis Add-In

SUMMARY OUTPUT

Regression Statistics	
Multiple R	0.969622971
R Square	0.940168705
Adjusted R Square	0.92022494
Standard Error	700.3118051
Observations	9

ANOVA

	df	SS	MS	F	Significance F
Regression	2	46239330.25	23119665.13	47.14098414	0.000214183
Residual	6	2942619.746	490436.6243		
Total	8	49181950			

	Coefficients	Standard Error	t Stat	P-value	Lower 95%	Upper 95%
Intercept	2600.419117	641.5343816	4.053436872	0.006700029	1030.641036	4170.197198
X Variable 1 - Weight	56.05580324	9.166965988	6.114978862	0.000872896	33.62504553	78.48656096
X Variable 2 - Test	1513.395674	320.0571051	4.72851766	0.003229465	730.2441503	2296.547198

Table 4.10 Regression Residuals

		Simple Linear Regression		Multi-Linear Regression	
	y	ŷ	e	ŷ	e
Obs	Cost £	Model £	Error £	Model £	Error £
1	5123	5,149.57	-26.57	4,674.37	448.63
2	6927	5,943.45	983.55	6,804.38	122.62
3	4988	6,881.66	-1893.66	6,019.71	-1031.71
4	9253	7,459.03	1793.97	9,494.95	-241.95
5	7022	8,685.93	-1663.93	7,421.11	-399.11
6	10182	8,974.61	1207.39	9,158.73	1023.27
7	8348	9,551.98	-1203.98	8,093.78	254.22
8	10702	10,562.36	139.64	10,391.95	310.05
9	12092	11,428.41	663.59	12,578.02	-486.02
Average	8293.000	8293.000	0.000	8293.000	0.000
Root Mean Square Error			1243.12		571.80
Max Absolute Residual			1893.66		1031.71

them and allow them to flow through to the Summary Output. We (and others who read our analysis) have to remember what each x variable is, and the order in which we specified them. Not too bad with just two, but if we include more then it becomes more of a memory challenge – a bit like Kim's Game *(Kipling, 1901).*

Here, the Coefficients of the two variables are seen to be approximately 56.1 and 1513.4 respectively with an intercept of 2600.4. Table 4.10 compares the result of

136 | Simple and Multiple Linear Regression

> ## For the Formula-phobes: Why does adding a new variable improve the 'Best Fit'?
>
> Suppose we have found the primary driver for something, say weight as an indicator of cost, but we are unhappy with the residual variation in that simple relationship, and we conclude, therefore, that there is at least one secondary driver at work. The Regression routine will always find the Least Squares Best Fit – even where the relationship is tenuous. It is just a dumb calculation – no artificial intelligence involved; the estimator or forecaster has to provide that.
>
> For almost anything we choose to add, the Regression technique will try to fit the data to the residuals. It will even rob some of the relationship already lined up for the primary driver in order to minimise the total error or residual. That's why we will be going to look at the 'S' word (statistics) in Section 4.5 to make sense of it all.

this Regression model with the observed data for the dependent y variable (cost) and that of the equivalent result we got for the Simple Regression Model based on just weight. The Root Mean Square Error (or Quadratic Mean – Volume I Chapter 2) and the Maximum Absolute Error indicate that the multi-linear model reduces the error. This is not untypical; including additional drivers or variables often improves the fit of the model to the data. As we will see in Section 4.5, this does not necessarily mean that it is the better model! Regression will always find the best fit, but it may not be credible or supportable.

4.3.3 Multiple Linear Regression using advanced Excel functions

The second option in Microsoft Excel is that slightly scary one using the advanced function **LINEST** in combination with **INDEX**. In principle, it works exactly like it does for Simple Linear Regression, but with an added complication. (*That groan was really quite audible from here!*)

The syntax of the main function is the same:

* **LINEST(*known_ys, known_xs, const, stats*)** where
  o *known_ys* and *known_xs* refer to the location of the y and x range of values. Here, the *known_xs* refers to a set of contiguous cells in contiguous columns i.e. a solid block of data, just as with the Data Analysis Add-in.
  o The *const* parameter should be set to TRUE unless we have a really good reason why we should force our relationship through the origin (FALSE).
  o Again, the parameter *stats* should be enabled to TRUE so that we can assess how good our 'Best Fit' really is.

So, where does that added level of complication come in? Answer: in the Output Array.

Simple and Multiple Linear Regression | 137

Table 4.11 Determining the Plane of Best Fit Using Microsoft Excel's Advanced Functions

Obs	y Cost £	x1 Weight Kg	x2 Tests
1	5123	10	1
2	6927	21	2
3	4988	34	1
4	9253	42	3
5	7022	59	1
6	10182	63	2
7	8348	71	1
8	10702	85	2
9	12092	97	3

LINEST Output Data			
	column_num		
	1	2	3
1	1513.395674	56.05580324	2600.419117
2	320.0571051	9.166965988	641.5343816
3	0.940168705	700.3118051	#N/A
4	47.14098414	6	#N/A
5	46239330.25	2942619.746	#N/A

Table 4.12 LINEST Output Data for a Multiple Linear Regression with Two Independent Variables

		column_num		
		1	2	3
row_num	1	Slope for Variable 1	Slope for Variable 2	Intercept
	2	Standard Error of the Slope for Variable 1	Standard Error of the Slope for Variable 2	Standard Error for the Intercept
	3	Coefficient of Determination (R-Square)	Standard Error for the y Estimate	Not Applicable
	4	F-statistic for the Regression Model	Number of degrees of freedom in the Residuals	
	5	Sum of Squares for the Regression Model	Sum of Squares for the Residual Error	

At first glance from Table 4.11, it may appear that the Output Array has just added an extra column to report the Coefficient of the additional independent variable. *If only it were that simple!*

For row numbers of 3 to 5 in combination with a column number greater than 2, the array is invalid and returns a 'not applicable' error. This in itself is not the issue. Table 4.12 illustrates the idiosyncrasy of **LINEST** in that the values in rows 1 and 2 take a side-step of one cell to the right for every additional independent variable we add to our model; bizarrely, it adds each new variable to the column on the left rather than the right as we might reasonably expect. They are in reverse order in relation to the Data Analysis Regression Tool and also in relation to the column_num!

138 | Simple and Multiple Linear Regression

4.4 Dealing with Outliers in Regression Analysis?

Having spent quite a bit of time in Volume II Chapter 7 discussing Outliers, we should really take time here to discuss briefly what constitutes an outlier in relation to Linear Regression. In Volume II we looked at outliers as those values that were relatively extreme in relation to all others against effectively a horizontal axis. Here though we are looking at relationships between a number of variables that we visualise as lines, planes, curved surfaces, and multi-dimensional entities that we can't begin to visualise.

We can reduce all of these simple and complex relationships 'mathe-magically' to a simple spread against a horizontal axis:

We already know that by definition the Least Squares Plane of Best Fit passes through the Mean of the data, and that the sum (and therefore the average) of the errors or residuals will be zero (Section 4.1.2). If the relationship that we are trying to calibrate through Regression is genuine, we can also probably assume that the spread of the Residuals around the Regression model will be Normally Distributed for data samples greater than 30 or distributed in accordance with a Student's t-Distribution for smaller samples (see Volume II Chapter 4) which is, shall we say, 'Normal-esque'.

The procedure is an iterative one:

1. Run a preliminary Regression Analysis and calculate the Residuals between the Regression model and the observed values.
2. Ideally, create a plot of the Residuals as a frequency diagram against a horizontal scale, and identify any that appear to be uncomfortably distant than the rest
3. Apply one (*or more if you have masochistic tendencies*) of the Outlier tests discussed in Volume II Chapter 7 to determine whether any extreme residual values might be considered to be Outliers.
4. If we detect potential Outliers, we should remove the 'worst' data point first from our dataset before repeating the procedure until we are happy that there are no Outliers.

Remember to record the removal of any Outliers in the Basis of Estimate in the spirit of making our estimates TRACEable (Transparent, Repeatable, Appropriate, Credible and Experientially-based).

Note that in many instances we will be looking at Simple Linear Regression, in which case a simple Linear Trendline through the data will often be sufficient to identify whether one or more points are sufficiently distant from the line to make us think '*Ahh, potential outlier!*' If it doesn't, then we might assume that there are none. However, we should not make a definitive decision based on this visual test alone!

The example in Figure 4.8 and Table 4.13 illustrates the procedure. Here we have chosen to use the simplistic slope and intercept functions for the preliminary regression run (*but this is simply a space saving measure for this book, as we are not ready yet to use the all-important statistical analysis.*)

Simple and Multiple Linear Regression

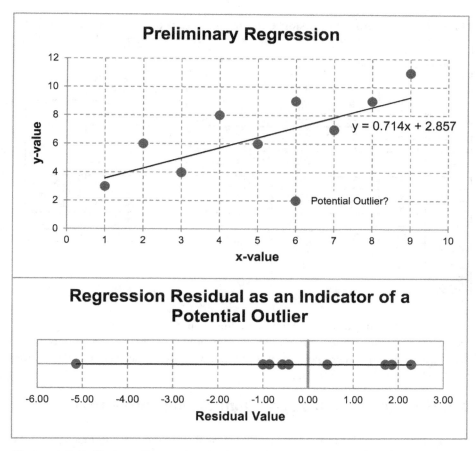

Figure 4.8 Preliminary Regression Analysis to Detect Presence of Outliers

The upper plot in Figure 4.8 highlights that we may suspect that the value y = 3.5 at x = 7.5 is a potential outlier, falling visually further away from the Line of Best Fit than all other points. The lower plot emphasises this; relative to the Mean of the Residuals (which is zero as a fundamental principle of our definition of 'Best Fit'), the point lies more than twice as far away as any other point on either side of the line.

Table 4.13 confirms our suspicions using either the Modified (SSS) Chauvenet's Criterion or Grubbs' Test for a Sample Size of 10, that we would be justified in excluding it from our final analysis. (Note that this point is not indicated as a potential outlier by some other more 'conservative' outlier tests such as the traditional Tukey's Fences, but would be flagged if we had used the more 'liberal' Slimline version of Tukey's Fences that we developed in Volume II Chapter 7.)

140 | Simple and Multiple Linear Regression

Table 4.13 Example Regression Outlier Test Based on Regression Residuals

Regression Input Data			Regression Output ($\hat{y}$)	Residual ($y - \hat{y}$)	Z-Score: Residual/ Std Dev	Modified Chauvenet Test: # Obs Expected in Sample of 10 with $\geq$ Z-Score		
Obs	x	y						
1	1	3	3.57	-0.57	0.260	8		
2	2	6	4.29	1.71	0.780	5		
3	3	4	5.00	-1.00	0.455	7		
4	4	8	5.71	2.29	1.041	3		
5	5	6	6.43	-0.43	0.195	8		
6	6	9	7.14	1.86	0.845	4		
7	7	7	7.86	-0.86	0.390	7	Grubbs Test
8	8	9	8.57	0.43	0.195	8	Critical Value
9	9	11	9.29	1.71	0.780	5	for Z
10	6	2	7.14	-5.14	2.341	0	2.176
Sum	51	65	65	0	↳	↳ Outlier ↵	
Average	5.1	6.5	6.50	0			
Slope			0.714				
Intercept			2.857				
Std Dev				2.197			

Note that if we decide to reject the data value as a potential outlier, the average of the remaining data points would change, and consequently, so would the slope, intercept and Residuals of the final regression. In other words, we must re-calculate the Regression results when we remove outliers.

4.5 How good is our Regression? Six key measures

Generating a Line of Best Fit using Simple Linear Regression is pretty much a straight forward task (*Did you spot the pun there? Ah, you were trying to ignore it; fair enough.*) It all sounds too good to be true, doesn't it? Maybe it is:

- The main advantage to us of Linear Regression Analysis is that it will always calculate the Line of Best Fit.
- The main disadvantage to us of Linear Regression Analysis is that it will always calculate the Line of Best Fit, even where there is no justification for it!

The same can be said for Multi-Linear Regression – it will always find the best fit relationships to our multiple dimensions so long as we have more data than variables.

Wouldn't it be great if we had a measure of just how good a fit the Regression actually was? We can only dream of having a single measure. However, we do have a set of six measures that we should use to inform our judgement.

- R-Square (and Adjusted R-Square)
- The F-Statistic
- The t-Statistic

Simple and Multiple Linear Regression | 141

- Homoscedasticity (*yes, really! Don't worry, we'll define what that is shortly)*)
- Coefficient of Variation (CV)
- Common Sense

Ok, perhaps I'm stretching it a bit in calling common sense a measure, as that is somewhat more subjective, but it is valid requirement in making that final judgement call on any analysis we do.

4.5.1 Coefficient of Determination (R-Square): A measure of linearity?!

Unless we avoided it like the proverbial plague, we may recall from Volume II Chapter 5 that the Coefficient of Determination is a measure of linearity, or more precisely, a measure of how much of the variance in one variable can be explained by a linear change in the other variable.

The way we measure this 'goodness of fit' is by looking at the proportion of the total variation or deviation that can be attributed to the variation which we can explain by the Straight Line Best Fit. It's no good doing it for single data points, we need to look at the overall data sample and take a holistic perspective of this proportion. To do this we measure:

$$\text{Goodness of Fit} = \frac{\text{Sum of Squares of Explained Variation around the Mean}}{\text{Sum of Squares of Total Variation around the Mean}}$$

This measure turns out to be the Coefficient of Determination.

For the Formula-philes: Coefficient of Determination as a measure of goodness of fit

For n paired observations x_1, x_2, x_3 ... x_n and y_1, y_2, y_3 ... y_n with corresponding Arithmetic Means $\overline{x}$, $\overline{y}$, we can fit a Line of Best Fit using Linear Regression:

Let β_0 and β_1 be the Intercept and Slope of the Regression Line of Best Fit, giving predicted values $\hat{y}_i$ corresponding to each x_i :

$$\hat{y}_i = \beta_0 + \beta_1 x_i \tag{1}$$

As the Line of Best Fit passes through the Mean of the data, we can conclude that:

$$\overline{y} = \beta_0 + \beta_1 \overline{x} \tag{2}$$

Let SS_{tot} be the Sum of Squares of the deviations of the observed data y_i from their Mean:

$$SS_{tot} = \sum_{i=1}^{n} \left(y_i - \overline{y} \right)^2 \tag{3}$$

(Continued)

142 | Simple and Multiple Linear Regression

Let SS_{reg} be the Sum of Squares of the deviations of the predicted values $\hat{y}_i$ on the regression line of Best Fit corresponding to each y_i :

$$SS_{reg} = \sum_{i=1}^{n} (\hat{y}_i - \overline{y})^2 \tag{4}$$

Substituting (1) and (2) in (4):

$$SS_{reg} = \sum_{i=1}^{n} (\beta_0 + \beta_1 x_i - \beta_0 - \beta_1 \overline{x})^2 \tag{5}$$

Simplifying (5):

$$SS_{reg} = \beta_1^2 \sum_{i=1}^{n} (x_i - \overline{x})^2 \tag{6}$$

From Section 4.1.1

$$\beta_1 = \frac{\sum_{i=1}^{n} (x_i - \overline{x})(y_i - \overline{y})}{\sum_{i=1}^{n} (x_i - \overline{x})^2} \tag{7}$$

Substituting (7) in (6) and simplifying:

$$SS_{reg} = \frac{\left(\sum_{i=1}^{n} (x_i - \overline{x})(y_i - \overline{y})\right)^2}{\sum_{i=1}^{n} (x_i - \overline{x})^2} \tag{8}$$

Dividing (8) by (3):

$$\frac{SS_{reg}}{SS_{tot}} = \frac{\left(\sum_{i=1}^{n} (x_i - \overline{x})(y_i - \overline{y})\right)^2}{\sum_{i=1}^{n} (x_i - \overline{x})^2 \sum_{i=1}^{n} (y_i - \overline{y})^2} \tag{9}$$

... which is the Coefficient of Determination, R^2 derived in Volume II Chapter 5

Pictorially, using the simple data example from Table 4.1 that we introduced in Section 4.1.1, we can illustrate what it is measuring in Figure 4.9. The Coefficient of Determination, R^2 is the Sum of Squares of the upper graph arrows expressed as a ratio of the Sum of Squares of the middle graph arrows. As usual, the squaring function

For the Formula-phobes: What does this all mean in (almost) plain English?

Suppose we think we have a linear relationship between two variables. Both variables will have a variation or scatter around their respective means. If we have a perfect straight line then the scatter around the two Means will be directly proportional to the slope of the line. However, where we don't have a perfect straight line, there will be a residual scatter around the line.

Consider any one point. The Best Fit straight line will account for the variation from the mean that can be associated purely with the general linear trend, i.e. the line 'explains' that as we increase or decrease one variable there will be a corresponding change in the other. It doesn't explain why some points are above or below the trendline; this is the unexplained variation.

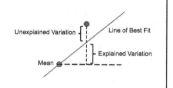

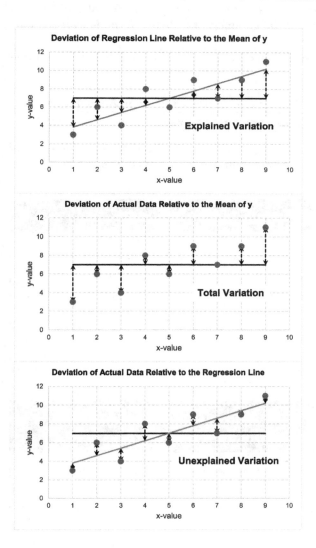

Figure 4.9 Explained and Unexplained Variation in Linear Regression

144 | Simple and Multiple Linear Regression

is to avoid the problem of positive and negative arrows cancelling each other out (as they always will!).

If the data falls exactly on the Best Fit straight line, i.e. forming a perfect straight line, then the ratio will be one. If the data is scattered closely around the Best Fit straight line, then the ratio will be less than but close to one. If the data is scattered at random, and the Best Fit straight line has a slope that is close to zero then the ratio will be close to zero. As the term is a function of squared values it can never be less than zero, so the Coefficient of Determination gives us an index of between zero and one as an expression of the goodness of fit between the Regression Line and the data.

We may find an alternative definition of R^2 as a measure of goodness of fit in relation to the unexplained variation in Regression:

$$\text{Goodness of Fit} = 1 - \frac{\text{Sum of Squares of Unexplained Variation Around the Mean}}{\text{Sum of Squares of Total Variation aound the Mean}}$$

The Unexplained Variation relates to the difference between the Regression Line of Best Fit and the observed data, as illustrated in the lower graph of Figure 4.9.

For the Formula-philes: R-Square

Definition 1:

$$R^2 = \frac{SS_{reg}}{SS_{tot}} \qquad (1)$$

Definition 2:

$$R^2 = 1 - \frac{SS_{err}}{SS_{tot}} \qquad (2)$$

Eliminating R^2 betweeen (1) and (2) and simplifying:

$$SS_{tot} = SS_{reg} + SS_{err} \qquad (3)$$

For the Formula-philes: Analysis of Variance (ANOVA) Partitioning the Sum of Squares

Consider a Linear Regression through a series of n paired data points (x_i, y_i) generating corresponding values (x_i, y_i).

Let β_0 and β_1 be the parameters of the Regression Line giving pred icted values $\hat{y}_i$ corresponding to each x_i :

$$\hat{y}_i = \beta_0 + \beta_1 x_i \qquad (1)$$

Simple and Multiple Linear Regression | 145

The difference δ_i between the true value of y_i and their Mean $\overline{y}$ is:

$$\delta_i = y_i - \overline{y} \qquad (2)$$

The difference r_i between the regression value of $\hat{y}_i$ and their Mean $\overline{y}$ is:

$$r_i = \hat{y}_i - \overline{y} \qquad (3)$$

The difference ε_i between the true value of y_i and the Regression Line $\hat{y}_i$ is:

$$\varepsilon_i = y_i - \hat{y}_i \qquad (4)$$

Expressing (2) as a function of (3) and (4):

$$\delta_i = \varepsilon_i + r_i \qquad (5)$$

Squaring and expanding (5):

$$\delta_i^2 = \varepsilon_i^2 + 2\varepsilon_i r_i + r_i^2 \qquad (6)$$

Taking the Sum of (6) across all the data points:

$$\sum_{i=1}^{n}\delta_i^2 = \sum_{i=1}^{n}\varepsilon_i^2 + 2\sum_{i=1}^{n}\varepsilon_i r_i + \sum_{i=1}^{n}r_i^2 \qquad (7)$$

Consider the middle term of the right hand side of (7) and substitute r_i with (3):

$$\sum_{i=1}^{n}\varepsilon_i r_i = \sum_{i=1}^{n}\varepsilon_i \left(\hat{y}_i - \overline{y}\right) \qquad (8)$$

Expanding (8) and substituting (1):

$$\sum_{i=1}^{n}\varepsilon_i r_i = \sum_{i=1}^{n}\varepsilon_i \left(\beta_0 + \beta_1 x_i\right) - \overline{y}\sum_{i=1}^{n}\varepsilon_i \qquad (9)$$

Re-arranging (9):

$$\sum_{i=1}^{n}\varepsilon_i r_i = \left(\beta_0 - \overline{y}\right)\sum_{i=1}^{n}\varepsilon_i + \beta_1 \sum_{i=1}^{n}\varepsilon_i x_i \qquad (10)$$

Using the two Sum-to-Zero Properties derived in Section 4.1.2:

$$\sum_{i=1}^{n}\varepsilon_i r_i = 0 \qquad (11)$$

Substituting (11) in (6):

$$\delta_i^2 = \varepsilon_i^2 + r_i^2 \qquad (12)$$

... which is the Partitioned Sum of Squares, often expressed as thus justifying the alternative definitions for R-Square

This result may seem either amazing or intuitive depending on how our individual brains are wired. For the doubters and algebraically resistant amongst us, Table 4.14 demonstrates that it works (again using the data from Table 4.1.)

Enough rambling, where were we? Oh, yes!

R-Square, therefore, is used as a measure of goodness of fit of our Regression equation. Now, that is all well and good we could argue (*we are estimators, it's in our DNA to argue*) if we are thinking purely of a Simple Linear Regression, but what about a Multi-linear Regression where we're talking about relationships in multiple dimensions. However, the key to understanding this is in the name – we are talking about Multi-*linear* Regression. If we were to hold all our independent variables except one as being temporarily fixed or constant, then the relationship defaults to a Simple Linear Regression. This is true for every one of our multiple dimensions.

146 | Simple and Multiple Linear Regression

Table 4.14 Example of Analysis of Variance Sum of Squares

Regression Input			Best Fit	Regression Variations			Sums of Squares of Variations		
Obs	Y	X	$\hat{Y}$	$Y - \bar{Y}$	$Y - \hat{Y}$	$\hat{Y} - \bar{Y}$	$(Y - \bar{Y})^2$	$(Y - \hat{Y})^2$	$(\hat{Y} - \bar{Y})^2$
1	3	1	3.8	-4	-0.8	-3.2	16	0.64	10.24
2	6	2	4.6	-1	1.4	-2.4	1	1.96	5.76
3	4	3	5.4	-3	-1.4	-1.6	9	1.96	2.56
4	8	4	6.2	1	1.8	-0.8	1	3.24	0.64
5	6	5	7	-1	-1	0	1	1	0
6	9	6	7.8	2	1.2	0.8	4	1.44	0.64
7	7	7	8.6	0	-1.6	1.6	0	2.56	2.56
8	9	8	9.4	2	-0.4	2.4	4	0.16	5.76
9	11	9	10.2	4	0.8	3.2	16	0.64	10.24

Sum	63	45	63	0.00	0.00	0.00			
Average	7	5	7						

Sum of Squares	52	13.6	38.4
			52

However, as we have already indicated in Section 4.3, if we add another data variable then Regression would try its level best to account for the residual error with the new variable – even if the link is decidedly tenuous. The trade-off for this so-called improvement is that we lose a degree of freedom from our model (see Section 4.5.2 for a discussion on degrees of freedom); . . . but, is that trade-off worth it? So, the question then becomes 'Is the Coefficient of Determination reliable for anything other than

---

Definition 4.3 Adjusted R-Square

Adjusted R-Square is a measure of the "goodness of fit" of a Multi-Linear Regression model to a set of data points, which reduces the Coefficient of Determination by a proportion of the unexplained variance relative to the Degrees of Freedom in the model, divided by the Degrees of Freedom in the Sum of Squares Error.

For the Formula-philes: Calculation of Adjusted R-Square

For a Multi-linear Regression based on a sample size of n observations and v independent variables, the Adjusted R-Square, R_{adj}^2 is calculated as:

$$R_{adj}^2 = R^2 - \left(1 - R^2\right)\frac{v}{n - v - 1} \qquad (1)$$

Where, by definition:

$$R^2 = \frac{SS_{reg}}{SS_{tot}} = 1 - \frac{SS_{err}}{SS_{tot}} \qquad (2)$$

Substituting (2) in (1):

$$R_{adj}^2 = \frac{SS_{reg}}{SS_{tot}} - \frac{SS_{err}}{SS_{tot}}\frac{v}{n - v - 1}$$

straight lines?' For this reason, a statistician named Wherry (Field, 2007, p.172) invented the Adjusted R-Square to take account of its predictive power based on the population rather than a sample. (The adjustment concept is similar to using Bessel's Correction Factor that we discussed in Volume II Chapter 3 to remove bias in sample variance relative to the population variance. The formula is quite complex.

Figure 4.10 illustrates the impact on the Adjusted R-Square of increasing the number of independent variables for a given number of observations. A marginal increase in R-Square may see a reduction in the Adjusted R-Square. Interestingly (*to some*) we can replicate the exact values in this graph (*if we wanted*) for 19 observations with 2, 4, 6 and 8 variables respectively. (*Yes, you're right; for me to know that this combination gives exactly the same values then I was one of the few of us who just 'wanted' to do it. Don't worry though, I've been booked in for some therapy.*)

Unlike R-Square, the adjusted R-Square increases only if the new term improves the model fit to the data more than we would expect to get by chance. We will note from Figure 4.10 that the adjusted R-Square can take a negative value, but will always be less than or equal to R-Square. In the case illustrated, we have a benefit from adding the extra variable in that an increase in R-Square also yields an increase in Adjusted R-Square.

Adjusted R-Square does not have the same interpretation as R^2 in terms of being a Measure of Linearity. As such, care must be taken in interpreting and reporting this statistic. Adjusted R-Square will be more helpful to us when we are choosing which variables to include and exclude from our model. We will cover this in more

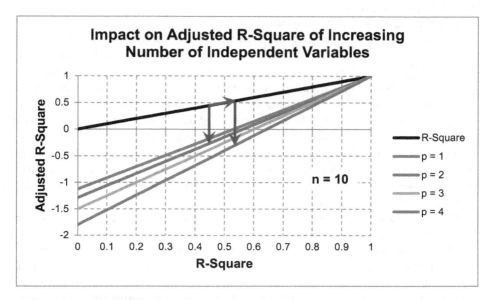

Figure 4.10 Adjusted R-Square Decreases as the Number of Independent Variables Increases

148 | Simple and Multiple Linear Regression

Table 4.15 Example 1 – Impact of an Additional Variable on R-Square and Adjusted R-Square

Simple Linear Regression		Multiple Linear Regression		Change
Independent Variable	Weight	*Independent Variable*	Weight	
			N° Tests	< variable

SUMMARY OUTPUT   SUMMARY OUTPUT

Regression Statistics		*Regression Statistics*	
Multiple R	0.846881686	Multiple R	0.969622971	
R Square	0.717208589	R Square	0.940168705	0.223
Adjusted R Square	0.676809816	Adjusted R Square	0.92022494	0.243
Standard Error	1409.571009	Standard Error	700.3118051	-709.259
Observations	9	Observations	9	

ANOVA			ANOVA		
		df			*df*
Regression		1	Regression		2 < variables
Residual		7	Residual		6
Total		8	Total		8

depth in Section 4.7. Andy Field (2005, p.171) neatly describes Adjusted R-Square as an indicator of '*the loss of predictive power or shrinkage*' inherent in the model. Let's look at that by way of an example.

In Table 4.15 we compare the results of our two Regression examples from Table 4.3 and Table 4.9, the difference being the addition of a second driver (Number of Tests). Here we can see the impact on R-Square and Adjusted R-Square of adding the second variable. As predicted, R-Square increases as the Regression procedure tries its level best to fit the extra driver to the data. More importantly, the Adjusted R-Square increases by even more, telling us that the extra variable is having a positive impact on the overall fit.

Compare this with the results in Table 4.16. Here we have replaced the second driver with a different variable (Number of Circuits). Again, we will observe that R-Square has increased but the Adjusted R-Square has now reduced with the additional 'driver' relative to the Simple Linear Regression Model; in other words, the predictive power of the model has reduced with the additional variable. Also, the Standard Error has increased.

Our conclusion should be that this alternative model is not as good as either our original Simple Linear Regression Model or our first Multi-Linear Model.

The next two paragraphs should be read in conjunction with Section 4.5.3 on t-Statistics.

We will often find that if the Adjusted R-Square value reduces, then the p-value of the added variable *or* for one of the other variables is not significant, and we would reject the model as it stands. If the latter happens then we should consider the replacement driver as the primary driver in place of the original one.

Simple and Multiple Linear Regression | 149

Table 4.16 Example 2 – Impact of an Additional Variable on R-Square and Adjusted R-Square

Simple Linear Regression		Multiple Linear Regression		Change
Independent Variable	Weight	*Independent Variable*	Weight	
			Circuits	< variable
SUMMARY OUTPUT		SUMMARY OUTPUT		

Regression Statistics		*Regression Statistics*		
Multiple R	0.846881686	Multiple R	0.862994614	
R Square	0.717208589	R Square	0.744759704	0.028
Adjusted R Square	0.676809816	Adjusted R Square	0.659679605	-0.017
Standard Error	1409.571009	Standard Error	1446.444808	36.874
Observations	9	Observations	9	

ANOVA		ANOVA		
	df		*df*	
Regression	1	Regression	2	< variables
Residual	7	Residual	6	
Total	8	Total	8	

There may be situations where the addition of another variable causes the Adjusted R-Square to increase but at the same time the p-value or significance of that variable (or another variable) is or becomes marginal, i.e. it is slightly to the wrong side of the notional cut-off value (the critical value). This is where we can exercise our judgement in trading off a slightly higher risk of making an error in accepting a parameter we should really reject, against improved predictability power in the model.

4.5.2 F-Statistic: A measure of chance occurrence

In Volume II Chapter 6 we referred to the F-Statistic as a means of comparing the unexplained variance with the explained variance around a regression relationship, and as such it provides us with a measure of chance occurrence – in other words, what's the chance that the regression relationship has occurred by fluke. (*This is one way of remembering the purpose of the F-statistic here – F for fluke – it actually stands for Fisher, but that probably doesn't help us remember what it's used for.*)

As always, we will try to visualise what it is doing for us, and for simplicity we will look at a Simple Linear Regression – the same sample we've been looking at throughout this chapter.

Going back to Figure 4.9 and using the simple data example from Table 4.1 as an aid to readability, the F-Statistic is looking at the ratio of the Explained Variation to the Unexplained Variation. To save us looking back we have extracted the two relevant graphs and reproduced them in Figure 4.11; the F-Statistic is essentially the ratio of the Sum of Squares of the upper graph over the lower graph.

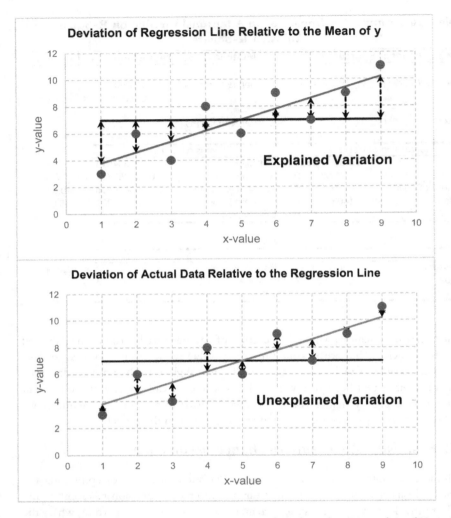

Figure 4.11 F-Distribution as a Comparison of Explained to Unexplained Variation

Yes, well spotted – 'essentially' implies that it is not simply the raw ratio of the Sum of Squares. The F-statistic is actually measuring the Ratio of the Mean Sum of Squares of the explained to the unexplained where the Mean is based on the relevant number of *degrees of freedom*:

> The number of degrees of freedom in the overall data sample (assuming that the sample mean is representative of the population mean) is one less than the sample size
> In the Explained Variation, the number of degrees of freedom is the number of independent or predictor variables in the regression model. The Total Variation

(Explained plus Unexplained) is one less than the number of observations in order to be able to ensure that the sum of the error terms is zero, i.e. there is one constraint on the number of degrees of freedom.

Therefore, for the Unexplained Variation, the number of degrees of freedom is one less than the number of observations or data points less the number of independent variables.

Let's consider a Simple Linear Regression. The closer that the data is scattered around the Regression Line, then the smaller the Unexplained Variation (the denominator or bottom line) will be, and consequently the greater the F-Statistic becomes. If we refer to Volume II Chapter 6 we will see that the probability of getting a very large F-Statistic by chance is very remote.

Intuitively, this feels right as well. The chance of a tight fit around a Regression Line occurring by fluke chance is a bit like our winning the lottery jackpot — not impossible but highly unlikely. (*Disappointingly, understanding all this doesn't improve our chances of winning the lottery, just lowers our expectations of it so that we are not quite as disappointed when we don't win.*)

The precise shape of the F-Distribution changes depending on the number of degrees of freedom in the Explained and Unexplained Variation. An example for ten observations, and two independent variables is shown in Figure 4.12.

The actual value of the F-Statistic will depend on the number of data points and the number of independent variables in our model, and to some extent the actual value

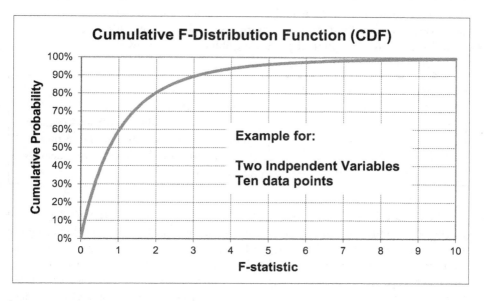

Figure 4.12 Example F-Distribution CDF

For the Formula-phobes: Run that 'degrees of freedom' thing by me again

Consider a simple Straight Line Regression through our data. It has to pass through the Mean in order to meet our Best Fit Criterion. Suppose our line is a little out of kilter because our data is just a sample – but we are comfortable that the Mean is representative of the data overall. We can vary the tilt of the line so that it still passes through the Mean. This will change both the slope and the intercept so we only have one degree of movement or freedom to move, but in doing so it affects two parameters.

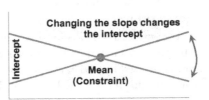

If we change the data scatter around the Line of Best Fit, then we can change every point until the last one – this one will be constrained so that the sum of all the residual values scattered around the line equals zero. In this case, we have freedom to move all but one of the data points if we are to maintain the constraint stipulated.

is not really important; it is the level of significance that is important. In other words what is the probability that we could get a value of F greater than that generated by our model? The good news is that Excel does this for us if we are using the Data Analysis Add-in for Regression. In Table 4.17 we have extracted the relevant ANOVA (Analysis of Variance) sections from our earlier examples in Tables 4.3 and 4.9.

In Excel's ANOVA section, the following abbreviations are used:

df	degrees of freedom	as defined above
SS	Sum of Squares	of the Explained (Regression), Unexplained (Residual) and Total Variation in the dependent variable around its mean
MS	Mean Square	Sum of Squares divided by the degrees of freedom

The F-Statistic is calculated as the Mean Square for the Regression Model divided by the Mean Square for the Residual. The Significance of F can be verified using the Excel Function **F.DIST.RT**(x, *deg_freedom1*, *deg_freedom2*), where x is the Regression's F-Statistic, ***def_freedom1*** and ***def_freedom2*** relate to the degrees of freedom for the Regression and Residuals respectively. (*Go on, try it, you know you want to!*) Note that in earlier versions of Microsoft Excel this can be verified using **FDIST**(x, *deg_freedom1*, *deg_freedom2*)

Simple and Multiple Linear Regression | 153

Table 4.17 Two Examples of the Significance of the F-Statistic

Simple Linear Regression				
ANOVA				
	df	SS	MS	F	Significance F
Regression	1	35273717	35273717	17.7532271	0.003968678
Residual	7	13908233	1986890.43		
Total	8	49181950			

Multiple Linear Regression				
ANOVA				
	df	SS	MS	F	Significance F
Regression	2	46239330.3	23119665.1	47.1409841	0.000214183
Residual	6	2942619.75	490436.624		
Total	8	49181950			

The higher proportion of the Sum of Squares in Table 4.17 tell us that our enhanced Multi-Linear Regression Model accounts for more of the total variation in the data than the Simple Linear Regression. Note that the total variation is the same for both models as the dependent variable data (cost) is the same.

For the Formula-philes: Calculation of the F-Statistic

Regression Partitioned Sum of Squares is:

$$SS_{tot} = SS_{reg} + SS_{err} \tag{1}$$

Dividing (1) by SS_{tot}

$$1 = \frac{SS_{reg}}{SS_{tot}} + \frac{SS_{err}}{SS_{tot}} \tag{2}$$

By definition:

$$R^2 = \frac{SS_{reg}}{SS_{tot}} \tag{3}$$

Substituting (2) in (3):

$$R^2 = 1 - \frac{SS_{err}}{SS_{tot}} \tag{4}$$

By definition F is the Ratio of the two Mean Squares:

$$F = \frac{MS_{reg}}{MS_{err}} \tag{5}$$

Where:

$$MS_{reg} = \frac{SS_{reg}}{p} \tag{6}$$

and:

$$MS_{err} = \frac{SS_{err}}{n - p - 1} \tag{7}$$

(Continued)

154 | Simple and Multiple Linear Regression

From (5), (6) and (7):	$$F = \frac{SS_{reg}}{SS_{err}} \frac{(n-p-1)}{p}$$	(8)
Inverting (8)	$$\frac{1}{F} = \frac{SS_{err}}{SS_{reg}} \frac{p}{(n-p-1)}$$	(9)
Substituting (4) and (3) in (9)	$$\frac{1}{F} = \left(\frac{1}{R^2} - 1\right)\frac{p}{(n-p-1)}$$	(10)
Multiplying (10) through by R^2	$$\frac{R^2}{F} = \left(1 - R^2\right)\frac{p}{(n-p-1)}$$	(11)
By definition	$$R_{adj}^2 = R^2 - \left(1 - R^2\right)\frac{p}{(n-p-1)}$$	(12)
Substituting (11) into (12)	$$R_{adj}^2 = R^2\left(1 - \frac{1}{F}\right)$$	(13)
Rearranging (13)	$$F = \frac{R^2}{R^2 - R_{adj}^2}$$	

Those of us who are still with us and have not yet lost the will to live, or have nodded off, may have realised that there is some relationship between the F-Statistics, R-Square and the Adjusted R-Square:

- R-Square is based on two elements of the Partitioned Sum of Squares.
- The Adjusted R-Square is derived from R-Square and takes account of those degrees of freedom that are floating around.
- The F-Statistic is based on the Ratio of two Mean Squares which in turn are based on elements of the Partitioned Sum of Squares and the degrees of freedom.

It can be shown that provided we have not forced a regression through the origin (see Section 4.5.4, or done anything silly like including a redundant independent variable (see Section 4.3.3) that is perfectly correlated to another independent variable creating terminal collinearity within the model, then the F-Statistic is equal to the Ratio of R-Square divided by the difference between R-Square and the Adjusted R-Square.

Consequently, as a rule of thumb for Multi-Linear Models, we will find that if the F-Statistic reduces in comparison to an alternative model then so would the Adjusted R-Square, and *vice versa*. (*I can just hear the cynics amongst us commenting on how rare it is for two statistics not to contradict each other!*)

If we have been brave enough to use the **LINEST** function in Excel then we can generate the F-statistic using the expression:

F = **INDEX(LINEST(known_ys, known_xs, TRUE, TRUE), 4, 1)**

Getting the significance of the F-Statistic is even more involved.

We know how many degrees of freedom there are in the Regression model from the number of columns in the **known_xs** (*and being able to count is not an unreasonable expectation for an estimator or other number juggler.*)

LINEST can give us the degrees of freedom for the residuals using:

df_{res} = **INDEX(LINEST(known_ys, known_xs, TRUE, TRUE), 4, 2)**

However, in truth we can just as easily get this from the number of observations *minus* the number of independent variables *minus* one.

Has anyone ever told you that you have a very expressive face? If you are trying this at home, please ensure that a responsible adult is available to administer the safe amount of pain relief medication for the headache.

In determining what might be a reasonable level of significance for us to accept the regression model, we should be looking for values as small as possible. In respect of both our examples in Table 4.17 then the chance of getting such a good fit to the data by fluke is very remote. In many applications, the default position is to reject the model if the significance is greater than 5%, but this should really be decided on the context in which the model is to be used. Where accepting a model that we should have rejected is an untenable situation (i.e. a Type I Error), then we might want to tighten the threshold to the 1% level. On the other hand, there are situations where we might relax the threshold to 10%. We can make this judgement call, but we should really have the acceptable threshold determined in advance and not wait until we get a regression result – as it might suggest a degree of conscious bias (to ensure that we get the result that we want.)

Incidentally, if we were to calculate the F-statistic for the revised Multi-Linear Regression we used in Table 4.16 where we saw the Adjusted R-Square fall, then we would get a value of F=8.75 with a Significance Level of 1.66%. This would compare with the equivalent Simple Linear Model value in Table 4.17 of F=17.75 with a Significance

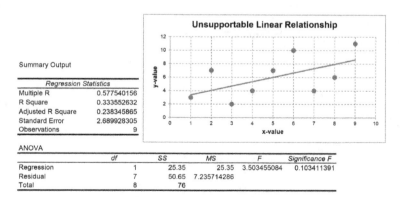

Figure 4.13 Example of a Questionable Linear Regression

of 0.4%. Despite the reduction in the F-Statistic it is reasonable to conclude that the assumed relationship in either model has not occurred by fluke.

Consider instead the regression formed by data in Figure 4.13. Here the F-Statistic is very low, and its significance is calculated to be 10.34%. Visually, we wouldn't have much faith in the relationship being valid; the F-Statistic gives us some moral support to reject the model. In this particular case, we will note also that we would have rejected the model on the basis of R-Square being only 0.33, indicating a poor fit to the data.

4.5.3 t-Statistics: Measures of Relevance or Significant Contribution

If we could re-run all our previous projects for which we have historical evidence that we are trying to find the Best Fit linear or multi-linear relationship, then we would very probably get slightly different results due to the underlying uncertainty, risks and opportunities that underpin everything. For instance, a regular commute to work does not always take exactly the same time, does it? We would hope that the data we get would be consistent with that which we already have but there would be some variation. For instance, our Best Fit Line through some data might me slightly steeper or shallower in slope with a slightly different intercept. In a perfect world, all our data points would fall precisely on a pre-defined relationship such as a straight line. (*We don't live in a perfect world, that's why it needs estimators like us.*)

The best we can hope for in our imperfect world is that the data we have is representative and that the Best Fit Slope and Intercept etc are therefore representative also. We can use the Student's t-Test to test the Null Hypothesis that the true value of each Best Fit parameter could be zero.

Let's consider the last example from Figure 4.13, which we would have rejected anyway based on the R-Square and F-Statistic, but the t-statistics might help us appreciate why the overall regression should be rejected. In Figure 4.14 we have plotted the range

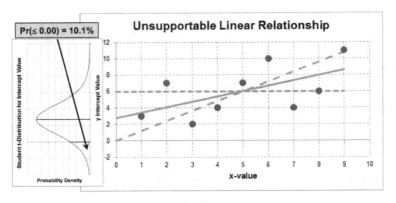

Figure 4.14 Example of a Questionable Linear Regression

Simple and Multiple Linear Regression | 157

of potential values for the Intercept based on the degree of scatter in the actual data as a Student's t-Distribution based on a Mean value of 2.75, which is the calculated Intercept for the Best Fit Line through the data. The long-dashed line represents a line through both the origin (Intercept = 0) and the mean of the data. In this case, this indicates that there is a 10.1% chance that the intercept could be zero or less. Conversely, we could plot same t-Distribution centred on an intercept of zero, and this would tell us that the chance of getting a value of 2.75 or greater is 10.1%.

In Table 4.18 we see the output from Excel's Data Analysis Add-in which calculates the Intercept's *t-Stat* and advises us that the Probability (*P-value*) of getting that *t-Stat* is 20.2%. This is the two-tailed version of the test as any error or variation could be in either direction, giving 10.1% on either side.

For the Formula-philes: Calculation of the t-Statistic

The t-statistic for the Coefficient of any Regression Variable or Intercept can be calculated as:

$$t = \frac{Coefficient}{Standard\ Error} \tag{1}$$

For a Regression of n observations using v independent variables, the residual degrees of freedom df can be expressed as:

$$df = n - v - 1$$

The Two-Tailed Cumulative Probability of getting that t-Statistic can be calculated in Excel by:

$$\Pr(t) = \mathbf{T.DIST.2T}(t, df) \tag{2}$$

Alongside this we have the **Lower 95%** and **Upper 95%** values. These represent the values of the intercept and slope that would give us a 95% Confidence Interval [2.5%, 97.5%] in which zero falls in both cases. The conclusion must be that it is distinctly possible that the true values of the intercept and slope are zero, and the usual interpretation would be that the calculated Best Fit Line is not significant and should be considered to be questionable. (*It doesn't mean that we have to reject the model, but we should acknowledge the health risk associated with using it!*)

We can repeat the exercise for the data from Figure 4.15. This is a tighter spread of data, and the probability of the true value of the intercept being zero, based on the evidence available, is only 1.05%., or 2.1% based on a two-tailed test.

Table 4.19 calculates the range of values of the t-Statistic based on a 95% Confidence Interval, which does not include an intercept or slope value of zero. The conclusion

Table 4.18 Example Excel Output for a Questionable Linear Regression

	Coefficients	Standard Error	t Stat	P-value	Lower 95%	Upper 95%
Intercept	2.75	1.954187608	1.407234387	0.202175279	-1.870919409	7.370919409
X Variable 1	0.65	0.347268251	1.871751876	0.103411391	-0.171158928	1.471158928

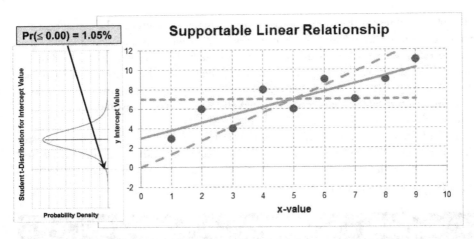

Figure 4.15 Example of a Supportable Linear Regression

Table 4.19 Example Excel Output for a Supportable Linear Regression

	Coefficients	Standard Error	t Stat	P-value	Lower 95%	Upper 95%
Intercept	3	1.012618796	2.962615362	0.021026869	0.605537038	5.394462962
X Variable 1	0.8	0.179947082	4.445751442	0.002986904	0.374492766	1.225507234

would be that it is extremely unlikely that the true value of the intercept should be zero, and the usual interpretation would be that the calculated Best Fit Line is statistically sound. (*It doesn't mean that the model is right, but it is statistically supportable.*)

What we have done here for the intercept is essentially what Excel is doing in the bottom right hand corner of its Data Analysis Output Table for both the intercept and any slope coefficients.

- We will see two sets of columns there labelled 'Lower 95%' and 'Upper 95%' (*I'll get come back to the question of 'why twice?' shortly.*)
- Excel calculates the Upper and Lower Limits of a 95% Confidence Interval around each Coefficient parameter of the Best Fit. The 'Lower 95%' equals the 2.5% Confidence Level, and the 'Upper 95%' equals the 97.5% Confidence level for the parameter.
- We are not bound to a 95% Confidence Interval (*yes, the pun was intended*) which is why we have two sets of columns. In the Data Analysis Regression Input dialogue box there is an option to specify a 'Confidence Level'. This is a misnomer as it actually means 'Confidence Interval'. It defaults to 95% unless we specify otherwise (hence why there are two repeated columns.)
- We can specify any other Confidence Interval that suits our needs e.g. 90%, or 80% (or 86.4% *if we so wished*). The second set of columns will be headed 'Lower 90%'

Simple and Multiple Linear Regression | 159

and 'Upper 90%' (or whatever Interval we requested). By default we will always get the 95% Interval as well in the first pair of columns

- If the Lower limit is less than zero for our chosen Confidence Interval then the Null Hypothesis should be accepted that the coefficient's true value could be zero.

> **Note:** For a Simple Linear Regression, the significance of the F-Statistic is always identical to the t-value significance for the slope parameter, i.e. the two probabilities will always be equal. This is not the case for Multi-linear Regression where the F-Statistic really comes into its own and 'comments' on the relationship overall.

For the Formula-phobes: Why do these F and t measures give the same probability?

Let's just take a few moments out to think about why the significance of the Regression F-Statistic is the same as the probability of the 'true' value of the slope coefficient being zero.

Well, in essence, they are saying the same thing in two different ways.

Remember from Section 4.5.2 that the F-Statistic is measuring the ratio of the Explained Variance in our y-value to the Unexplained Variance, and that the former is directly proportional to the Variance in our x-value on the assumption of an underlying straight line. If the true value of the slope coefficient is zero, then we do not have a linear relationship between our two variables, and as a consequence, there is no element of Explained Variance in the y-value attributable to the changes in the x-value.

So the probability of an Explained Variance being low is the same as the probability that the slope of the linear relationship is zero.

The implication of accepting a Null Hypothesis that one of the regression coefficients is potentially zero, is that we then exclude it from the model, and re-run the Regression. In essence, what we are doing here is a Stepwise Regression procedure that we will be covering in a little more detail in Section 4.7. In the meantime, let's look at what that implies, summarised in Table 4.20:

The obvious question this probably prompts us to ask is:

'*What if more than one parameter fails its Null Hypothesis Test, do we reject them all out of hand?*'

The answer is: '*No, we should take one step at a time! Read Section 4.7 first.*'

160 | Simple and Multiple Linear Regression

Table 4.20 Implications of Null Hypothesis that the Parameter Value may be Zero

Parameter Coefficient	Simple Linear Regression	Multiple Linear Regression
Intercept	The model passes through the origin – there is no apparent fixed term for the data range that is being modelled (but read Section 4.5.4)	
Slope	i. Using a constant value may be better than using the current predicator variable or driver	i. It is questionable whether the parameter is adding anything material to the relationship
	ii. Somewhere we need to find a better predicator variable than the one we have used. (*Yes, shock, horror! It's true . . . sometimes we don't get it right first time!*)	ii. If by rejecting the parameter, we are left with a model that has a large degree of unexplained variation as indicated by a small F-Statistic then somewhere we need to find a better predicator variable than the one we have rejected

Let's amuse ourselves with the example in Figure 4.16, which is the model we rejected in favour of a Simple Linear Regression back in Table 4.16 (Section 4.5.1) based on the reduction in Adjusted R-Square. Here we have a Multi-Linear Regression of the Cost to manufacture, assemble and test a number of products in relation to two Cost Drivers: Finished Product Weight and the Number of Electrical Circuits.

The rationale to support these as Cost Drivers is that the Cost and the Finished Product Weight are 85% Correlated, and that Cost is also 75% correlated with the Number of Circuits in each product, and that potentially these are a differentiator of complexity. Let's validate that assumption.

If we follow through our tests, we will observe that the R-Square value is quite acceptable, and that the significance of the F-Test suggests that this apparent multi-linear relationship has not occurred by chance. The Null Hypotheses that the true value of the Intercept or the Slope Coefficient for Weight are zero are unlikely. However, there is a c.45% probability that the Number of Circuits does not add any significant support to understanding the product cost, and the true value of its coefficient is highly likely to be zero.

So where did it all go wrong? The answer is Collinearity. In other words, as well as checking the correlation between each potential independent variable (weight or

Simple and Multiple Linear Regression 161

	INPUT DATA >		y	x1	x2
		Obs	Cost £	Weight Kg	Circuits
		1	5123	10	0
		2	6927	21	2
		3	4988	34	1
		4	9253	42	1
		5	7022	59	2
		6	10182	63	2
		7	8348	71	1
		8	10702	85	3
		9	12092	97	3
		Average	8293.000	53.556	1.667

SUMMARY OUTPUT

Regression Statistics	
Multiple R	0.862994614
R Square	0.744759704
Adjusted R Square	0.659679605
Standard Error	1446.444808
Observations	9

ANOVA

	df	SS	MS	F	Significance F
Regression	2	36628734.51	18314367.25	8.753629983	0.016628295
Residual	6	12553215.49	2092202.582		
Total	8	49181950			

	Coefficients	Standard Error	t Stat	P-value	Lower 95%	Upper 95%
Intercept	4253.803965	1079.511651	3.940489167	0.00762035	1612.334114	6895.273817
Weight Kg	55.89853767	26.79118351	2.086452719	0.081999597	-9.657126765	121.4542021
Circuits	627.3112769	779.4935141	0.804767795	0.45165584	-1280.040641	2534.663194

Figure 4.16 Example of an Unsupportable Multi-Linear Regression – Insignificant Parameter

circuits) and the dependent variable (cost), we should also have tested the correlation between the Product Weight and the Number of Circuits. Had we tested this we would have discovered a 75% Correlation. In other words, there is a large degree of commonality between the two variables in terms of their driver or predictive power – we don't need both! The variable that the model rejects is the one with the lower correlation with the dependent variable, in this case, Number of Circuits.

The really good news for us is that, we don't really have to worry about it anyway. Simple Linear Regression and Multi-Linear Regression will do all the donkey work for us and highlight the net result . . . we just have to apply the intelligence in interpreting and understanding the result.

Where we have a high degree of Linear Correlation between two candidate independent variables, we should not be surprised that the Regression Statistics say 'yes' to a linear model with either variable, but 'no' to a Multi-Linear Model using both; there will be a high degree of duplication and therefore predictive redundancy between the two variables.

If we were to re-run the regression without the second driver, we would get the valid result we had back in Table 4.3. The F-Statistic has also increased, suggesting that there is less Unexplained Variance in the Model than there was when we tried to use two drivers.

In Section 4.3, we did introduce a second driver or independent variable (Number of Tests). Had we checked at the time we would have found that it too was some 75% correlated with Cost . . . just like Number of Circuits! However, the Correlation between Finished Product Weight and the Number of Tests was only 50%, leaving a little more

162 | Simple and Multiple Linear Regression

Table 4.21 Regression of a Secondary Driver Only

Summary Output

Regression Statistics	
Multiple R	0.75318686
R Square	0.56729044
Adjusted R Square	0.50547479
Standard Error	1743.62103
Observations	9

ANOVA

	df	SS	MS	F	Significance F
Regression	1	27900450	27900450	9.17713272	0.019131371
Residual	7	21281500	3040214.29		
Total	8	49181950			

	Coefficients	Standard Error	t Stat	P-value	Lower 95%	Upper 95%
Intercept	4309	1437.826733	2.99688405	0.02003023	909.0800374	7708.91996
Tests	2241	739.7557512	3.02937827	0.01913137	491.7556105	3990.24439

in the way of 'independent elbow room'. Suppose out of curiosity we were to run the regression of Cost with the Number of Tests but without the Weight (Table 4.21).

Interestingly (*well, it is to me, but then I don't get out much*) the secondary driver provides a model with mixed messages. The F-Statistic and t-Statistics are sending positive messages, but the R-Square is saying that the driver or predictor variable only accounts for around half of the variation in Cost. There is a moral here. If we find that a potential driver is only telling us half the story we need to hang on to it but keeping looking for one or more other drivers to fill in the rest of the story for us.

Caveat augur

If the Null Hypothesis is accepted that the true value of the Intercept is zero, then we can elect to re-run the regression, forcing the Regression Line to pass through the origin.

Many commercial off-the-shelf software applications such as Microsoft Excel allow us to do this. However, *all is not what it seems* . . . Doing this creates a number of changes, some of which are not always obvious or intended.

4.5.4 Regression through the origin

It must be said that in some circles (e.g. ICEEA Training Materials), rejecting the Intercept is strongly discouraged. However, rejecting the intercept should not be misconstrued

Simple and Multiple Linear Regression | 163

as implying that there is no fixed element to the relationship, but more a matter of convenience, given the scope and range of the input variables, that this is a better approximation to our perception of reality than one with a fixed element. Potentially the true underlying relationship may only exhibit an approximation to a linear relationship in the region of our data (i.e. it is locally linear).

In the case of a Multiple Linear Regression in particular, the Intercept implies the value that would remain unexplained should all the independent variable values be zero at the same time. This can sometimes be a purely theoretical concept. However, if we can conceive of a situation where all the independent variables can be zero simultaneously, then we have to decide in the logical scheme of things whether or not it is appropriate to force a regression through the Origin. If we take this argument to the next level and take the view that if all the drivers (independent variables) are zero then perhaps the intercept is just a reminder that in the black art and white science world of estimating, there is always something that hasn't been taken account of fully, and that the intercept is an approximation of all those variables that we weren't smart enough to include or able to gather data on, and on that basis we should ignore the t–Statistic for the Intercept and not force the regression through the origin. It could also mean, of course, that there is a fixed value, but we should not assume this to be the default interpretation.

Putting all that to one side, what if we do anyway?

Forcing a regression through the Origin does have one major impact on our key regression assumptions – it drives the proverbial bus through them! If we feel we have to force the Regression Line through the Origin, and yet keep the premise that it must also go through the average of the data, then we have a straight line fixed by two points. In this case, in terms of a Simple Linear Regression, there is no latitude to calculate the Least Squares.

For a multi-linear regression, we can perform a restricted Least Squares routine by allowing the model to 'rotate around a diagonal axis' (to use a 3D analogy.) In fact, by implying that we must pass the relationship through both the Origin and the average, would we not in effect be describing the conditions implied by an analogical method (see Volume 1 Chapters 2 and 5)?

To avoid this in the case of a Simple Linear Regression, in which we compare everything to the average or mean value, we have to amend our thinking – we must assume that the sample average is not representative of the population average, and therefore, remove that criterion from our definition of 'Best Fit'. In short, we calculate the least squares based on a line through the Origin instead.

For example, in Figure 4.17, we have a Linear Regression in which the probability of the Null Hypothesis (p-value) that the intercept is zero is some 35.6% – quite high.

If we accept the Null Hypothesis and re-run the Regression procedure but this time tick the box that says 'constant is zero' in the Microsoft Excel Data Analysis Add-in tool for regression. This time we will get the result in Figure 4.18. (*Note that the Data Analysis Add-in will always report the Intercept – even when we have forced it through the Origin. The #N/A is not an error flag in this case.*)

INPUT DATA >

Obs	y	x
1	1	1
2	4	2
3	2	3
4	6	4
5	4	5
6	7	6
7	5	7
8	7	8
9	9	9
Average	5	5

SUMMARY OUTPUT

Regression Statistics	
Multiple R	0.8593378
R Square	0.7384615
Adjusted R Square	0.7010989
Standard Error	1.3938641
Observations	9

ANOVA

	df	SS	MS	F	Significance F
Regression	1	38.4	38.4	19.764706	0.002986904
Residual	7	13.6	1.9428571		
Total	8	52			

	Coefficients	Standard Error	t Stat	P-value	Lower 95%	Upper 95%
Intercept	1	1.012618796	0.9875385	0.3562763	-1.39446296	3.394463
X Variable 1	0.8	0.179947082	4.4457514	0.0029869	0.374492766	1.2255072

Figure 4.17 Example of a Linear Regression with an Intercept Close to Zero

INPUT DATA >

Obs	y	x
1	1	1
2	4	2
3	2	3
4	6	4
5	4	5
6	7	6
7	5	7
8	7	8
9	9	9
Average	5	5

SUMMARY OUTPUT

Regression Statistics	
Multiple R	0.9716287
R Square	0.9440623
Adjusted R Square	0.8190623
Standard Error	1.3917047
Observations	9

ANOVA

	df	SS	MS	F	Significance F
Regression	1	261.5052632	261.50526	135.0163	7.8903E-06
Residual	8	15.49473684	1.9368421		
Total	9	277			

	Coefficients	Standard Error	t Stat	P-value	Lower 95%	Upper 95%
Intercept	0	#N/A	#N/A	#N/A	#N/A	#N/A
X Variable 1	0.9578947	0.082437475	11.619652	2.739E-06	0.76779358	1.1479959

Figure 4.18 Example of a Linear Regression with an Intercept Constrained to Zero

When we compare the two outputs, we will see a number of differences, some obvious, and others less so, until we look more closely:

- The Slope Coefficient changes (*not surprisingly*)
- The P-value of the Null Hypothesis reduces dramatically (*as perhaps we would hope*)
- The Significance of the F-Statistic reduces also ... but this time it no longer matches the P-value for the Slope Coefficient (*Oops, that doesn't sound right, does it? After all the rationale for why they should be the same hasn't changed, or has it?*)
- Multiple R, R-Square and Adjusted R-Square have both improved significantly, which on the face of it sounds good, until we think about it and then the seeds of doubt begin to fall around us – why does forcing a line through the Origin improve the fit wen we have moved away from the best fit?
- The whole basis of the ANOVA Sum of Squares (SS) and Mean Square (MS) appears to have changed as the values returned are an order of magnitude different.

The fundamental reason why we are observing these large changes for what we might see as a relatively small shift in the position of the Regression Line is that we have moved the proverbial goalposts. To keep the sport analogy going, we are measuring things from the corner post and not the centre spot in the middle of the pitch. Figure 4.19 illustrates the fundamental change in relation to a single point (*not a goalpost or corner post in sight.*)

So, that proverbial bus ride through our fundamental assumptions means that we cannot make a direct comparison with the statistics reported by a 'normal' regression (i.e. one that does not force the Best Fit Line through any particular point other than the Mean of the data) and one that is forced through the Origin. To use a more scientific analogy rather than sport, it's like switching from Fahrenheit to Centigrade but only reading the numbers and not the scale!

Table 4.22 summarises what's happening with all the statistics being reported in Microsoft Excel's Data Analysis Regression Output when we force a Regression line through the Origin, i.e. set the intercept to zero.

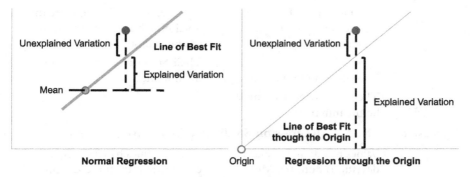

Figure 4.19 Regression Through the Origin Changes the Basis of Measuring the Goodness of Fit

Table 4.22 Regression Through the Origin Changes the Basis of Measuring the Goodness of Fit

Entity	"Unconstrained" Least Squares Linear Regression		Least Squares Linear Regression Through the Origin
Total SS	... these values could be an order of magnitude different		
	Sum of the Squares of the Deviations of the Observed y values from their Mean	<	Sum of the Squares of the Deviations of the Observed y values from the Origin
Regression SS	... these values could be an order of magnitude different		
	Sum of the Squares of the Deviations of the Predicted y values from their Mean	<	Sum of the Squares of the Deviations of the Predicted y values from the Origin
Residual SS	... these values will be a similar order of magnitude		
	Sum of the Squares of the Deviations between the Observed y-values and those predicted by the unconstrained regression line	≈	Sum of the Squares of the Deviations between the Observed y-values and those predicted by the constrained regression line
Regression df	... these values will be the same		
	One Degree of Freedom to allow the Line of Best Fit to swivel around the Mean	=	One Degree of Freedom to allow the Line of Best Fit to swivel around the Origin
Residual df	... these values will differ by one.		
	The Degrees of Freedom are the Total Number of Observations less two constraints: Their Sum must be zero Their Sum of Squares must be the minimum	>	The Degrees of Freedom are the Total Number of Observations less one constraint: Their Sum of Squares must be the minimum
Regression MS	... Regression SS divided by the Regression df		
	Values based on Squares of deviations from the Mean	<	Inflated values based on Squares of deviations from the Origin

Entity	"Unconstrained" Least Squares Linear Regression	Least Squares Linear Regression Through the Origin
Residual MS	. . . **Residual SS divided by the Residual _df_**	
	Residual SS divided by 1 »	**Residual _SS_** divided by 2
F-Statistic	. . . **Regression MS divided by the Residual _MS_**	
	Value is driven by the Squares of the Deviations from the Mean	< Value is inflated by the Squares of the Deviations from the Origin
R-Square (R^2)	. . . **Regression SS divided by the Total _SS_**	
	$\dfrac{\text{Regression } SS}{\text{Total } SS} = 1 - \dfrac{\text{Residual } SS}{\text{Total } SS}$ <	**Total _SS_** inflated but **Residual _SS_** is largely unaltered
Multiple R	Square Root of **R-Square** – no difference other than the values involved	

Note: If the data is scattered around the Origin (with positive and negative values) and the Mean of the observed y-values is close to the Origin, then these comments are somewhat exaggerated . . . _but hey, how often do we get data like that?_

There is also an anomaly (_some might say error, but I don't want to upset Microsoft_) in the Significance of F that is reported in Excel. If we calculate the value 'long-hand' using Excel's **F.DIST.RT(_x, deg_freedom1, deg_freedom2_)** function, where x is the Regression's F-Statistic as calculated in Table 4.22, and **_def_freedom1_** and **_def_freedom2_** relate to the degrees of freedom for the Regression and Residuals respectively in the same table, then the value for the Unconstrained Regression can be verified accordingly, but for the Regression through the Origin it gives a slightly inflated answer. It appears to use the original degrees of freedom rather than revised ones have been used (_values have been rounded here_):

Expected:	**F.DIST.RT(_135.0163043, 1, 8_)** = 2.7393E-06
Reported:	**F.DIST.RT(_135.0163043, 1, 7_)** = 7.8903E-06

. . . at least the anomaly is on the right side, increasing the chance of a Type I Error rather than a Type II Error (see Volume II Chapter 6.)

Let's look at another example. In Figure 4.20 we have performed a Simple Linear Regression of some data, for which the Line of Best Fit passes through the Origin, which coincidentally is also the Mean point of the data. Figure 4.21 provides a plot of this data

SUMMARY OUTPUT

INPUT DATA >

Obs	y	x
1	-3	-4
2	-1	-3
3	-3	-2
4	1	-1
5	-1	0
6	2	1
7	-1	2
8	2	3
9	4	4
Sum	0	0
Average	0	0

Regression Statistics	
Multiple R	0.7994563
R Square	0.6391304
Adjusted R Square	0.5875776
Standard Error	1.5399443
Observations	9

ANOVA

	df	SS	MS	F	Significance F
Regression	1	29.4	29.4	12.39759	0.009714308
Residual	7	16.6	2.3714286		
Total	8	46			

	Coefficients	Standard Error	t Stat	P-value	Lower 95%	Upper 95%
Intercept	0	0.51331478	0	1	-1.21379658	1.2137966
X Variable 1	0.7	0.198805959	3.5210212	0.0097143	0.229898607	1.1701014

Figure 4.20 Example of a Linear Regression with a Natural Intercept of Zero

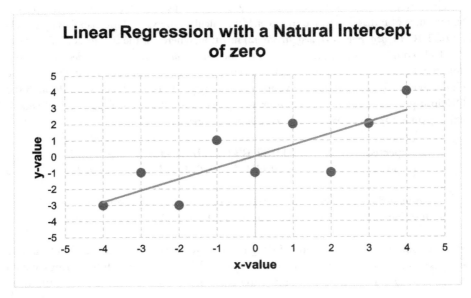

Figure 4.21 Linear Regression with a Natural Intercept of Zero

Simple and Multiple Linear Regression | 169

The Regression Line shows that there is still some uncertainty around its true slope, but suggests also that there is no uncertainty whatsoever in the value of the intercept, i.e. the Null Hypothesis that the true value of the Intercept is zero appears to have a probability of one! On one level this is counter-intuitive until we realise that a 'normal' Regression forces the Line of Best Fit through the Mean of the data, which in this case happens to be the Origin, and hence the probability will be one.

We might wonder what would happen if we were to run another regression through the data, but this time constrain it to pass through the Origin. Let's compare the results of the two regressions, one that passes through the Origin naturally (Figure 4.21) and the other with the same data that is forced through the Origin (Table 4.23). (*We would expect them to be the same, wouldn't we . . . or would we? After all we're talking statistics here!*) Hmm, they are not identical; so, what is happening here?

For the calculated values of the Coefficients we get the same; so too for the Multiple R, R-Square, and for the Sum of Squares for the Regression, Residual and total.

Not so though, for the residual degrees of freedom, which have increased by one, which then has a domino effect on the Mean Square of the Residual, the F-Statistic and ultimately the Significance of the F-Statistic.

Even when we correct for the anomaly in the calculation of the latter, we still have a difference. One crumb of comfort to us may be that once we have corrected the anomaly in Excel regarding the calculation of the Significance of F, we get the same probability as that for the t-Statistic associated with the Slope Coefficient . . . but it is still less in Table 4.23 than the corresponding p-value in Figure 4.20.

So, whilst the two lines of Best Fit appear to be identical, the removal of the intercept as a parameter makes a difference in that we are allowing the degrees of freedom

Table 4.23 Example of a Linear Regression with a Natural Intercept of Zero but Constrained to Zero

SUMMARY OUTPUT

Regression Statistics	
Multiple R	0.79945634
R Square	0.63913043
Adjusted R Square	0.51413043
Standard Error	1.44048603
Observations	9

ANOVA

	df	SS	MS	F	Significance F
Regression	1	29.4	29.4	14.1686747	0.007036749
Residual	8	16.6	2.075		
Total	9	46			

	Coefficients	Standard Error	t Stat	P-value	Lower 95%	Upper 95%
Intercept	0	#N/A	#N/A	#N/A	#N/A	#N/A
X Variable 1	0.7	0.185965947	3.76413001	0.00551169	0.271161758	1.12883824

170 | Simple and Multiple Linear Regression

in the Residual to increase by one, which then cascade through to the calculation of the F-Statistic and its significance. Furthermore, the probability of the t-Statistics is also based on the number of degrees of freedom in the Residual and the Mean Square for the Residuals, which itself is based on those same degrees of freedom. By removing the Intercept as a Parameter we are in effect reducing any 'slack' in the system giving us a narrower range of movement in the slope.

As a consequence of moving the basis of measuring the goodness of fit when we force a Regression through the Origin we cannot make an easy direct comparison between our three basic measures of R-Square, F-Statistic and t-Statistics.

However, not wanting you to feel let down by an inability to make a direct comparison between models, with and without an Intercept, perhaps we should consider the following as the next best thing. By forcing the Regression through the Origin we are to all intents and purposes saying that we do not think that the Sample Mean is a fair reflection of the Population Mean. By implication then we might surmise that the value of the Regression function at the point where we achieve the Mean of our Predicator Variables is a better estimate of the Population Mean.

Table 4.24 illustrates a procedure to make a fairer comparison between the Constrained and Unconstrained Regression through the origin using our example from Figure 4.18: The procedure we might like to consider is:

1. Take the Regression Equation through the Origin, and calculate the Predicted Value, $\hat{y}$ for each x-value in Column (F)
2. Determine the Mean value of the Regression function in Cell F15 that corresponds to the x-value Mean (or Means in the case of a Multi-linear Regression). We'll call this the 'Implied Mean'
3. In Column G, determine the deviation between Observed y in Column B and the Implied Mean in Cell F15
4. In Column H, determine the deviation between the Regression value $\hat{y}$ in Column F and the Implied Mean in Cell F15
5. In Column I, determine the deviation between the Observed y in Column B and the Predicted $\hat{y}$ in Column F
6. In Rows 14–16, determine the Sum, Mean and Sum of Squares for each column using Excel functions **SUM**, **AVERAGE** and **SUMSQ**
7. In Cells G17:H17, calculate the R-Square value by dividing Cell G16 by Cell H16
8. In Cells H18 and I18, enter the degrees of freedom for the Regression Model and the Residuals from previous Figure 4.18
9. In Cells H19 and I19, calculate the Mean Square for the Regression and Residuals by Dividing Cells H16 and I16 by Cells H18 and I18 respectively
10. In Cells H20:I20, calculate the F-Statistic by dividing Cell H20 by I20
11. In Cell I21, we can calculate the Standard Error as the Square Root of Cell I19
12. Finally, in Columns K and L we can compare key statistics with those from our previous results from Figures 4.15 and 4.16

Simple and Multiple Linear Regression — 171

Table 4.24 Comparison of a Regression through the Origin with a Natural Regression Through the Origin

	Input Data		Regression through the Origin ($\hat{y}$)	Deviation in y from Mean of $\hat{y}$	Deviation in $\hat{y}$ from Mean of $\hat{y}$	Deviation in y from $\hat{y}$ (Residual)		Unconstrained Regression	Regression Through the Origin
Obs	y	x							
1	1	1	0.958	-3.789	-3.832	0.042		Using Excel Data Analysis standard Regression Output	
2	4	2	1.916	-0.789	-2.874	2.084			
3	2	3	2.874	-2.789	-1.916	-0.874			
4	6	4	3.832	1.211	-0.958	2.168			
5	4	5	4.789	-0.789	0.000	-0.789			
6	7	6	5.747	2.211	0.958	1.253			
7	5	7	6.705	0.211	1.916	-1.705			
8	7	8	7.663	2.211	2.874	-0.663			
9	9	9	8.621	4.211	3.832	0.379			
Sum	45	45	43.105	1.895	0.000	1.895			
Mean (Average)	5	5	4.789	0.211	0.000	0.211			
Sum of Squares	277	285	261.505	52.399	55.054	15.495			
R-Square	Ratio of Sum of Squares >			0.952				0.738	0.944
Deg Freedom				9	1	8			
Mean Square	Sum of Squares divided by Deg Freedom >				55.054	1.937			
F-Statistic	Ratio of Mean Squares >				28.424			19.765	135.016
Standard Error	Square Root of Mean Square Error >					1.392		1.394	1.392

Whilst many of us may not choose to make this comparison as a matter of routine practice (*we've all got busy lives after all*) its inclusion here is justified on the basis that it illustrates that whilst some of the Regression Statistics reported as standard by Excel's Data Analysis facility are comparable between an Unconstrained Natural Regression and Regression through the Origin, the F-Statistic is not . . . it is an order of magnitude different!

Later, in Section 4.7 we will be reviewing an incremental approach to Multi-Linear Regression and we will revisit the issue of forcing a Regression through the Origin and the risk and implications of doing so 'too early' in our decision-making process. (*I can't wait either! Oh! You were just being sarcastic again; I should have known.*)

4.5.5 Role of common sense as a measure of goodness of fit

Earlier we alluded to the importance of applying 'common sense' to interpreting the output of a regression analysis (*something that cynics would say is much less common than its name suggests.*).

If there is no obvious rationale linking the predicator variable or driver and the entity that we are trying to estimate, then we should really question why we are considering it at all (*just because we can doesn't mean we should, does it?*) This doesn't mean to say that we should not entertain an indirect relationship between variables.

For instance, let's say that we believe that the number of parts is a good indicator of the time to assembly a product but until the design is complete we cannot easily predict the number of parts. However, we may have an early indication of the target weight and from previous products we have found this to be a reasonable indicator of the number

172 | Simple and Multiple Linear Regression

> ## A word (or two) from the wise?
>
> *"What nature hath joined together, multiple regression analysis cannot put asunder."*
>
> **Richard E. Nisbett**
> Professor of social psychology
> and author
> b.1941

of parts. Consequently, whilst there is no direct association between weight and assembly time, we can still use weight as an indirect driver because of its known association with a more direct driver, the number of parts assembled.

A date (year, month, etc.) may be a reasonable indicator of the underlying technology that underpins some other performance characteristic. The date variable is not the Primary Driver (see Volume I Chapter 4) but may turn out to be valid as a Secondary Driver.

After we have performed a Regression Analysis and ascertained that the parameters are all statistically significant, we should still take a step back and ask ourselves "Does this make sense?" As Richard Nisbett (2009, p.18) astutely points out, multiple regression is not the 'be all and end all'; just because multiple regression has found what appears to be a statistically good relationship, does not necessarily make it valid in real life. If it appears to be nonsensical, then more than likely it is nonsensical.

We should be very careful especially in respect of negative values:

It is said by some that 'size matters'. In relation to estimating, this is often the case, although not always in the same direction! Suppose we have a regression in which the cost to design, develop and construct the First of Class of a new ship is shown to be dependent on the ship's 'gross tonnage' (*a measure of capacity or volume, not weight*), but in our particular model, the Best Fit has a negative contribution ... the implication being that as we increase the size of the ship it gets cheaper (. . . *I don't think so*). The model therefore would fail the 'common sense' test.

However, in the context of miniaturisation, increasing costs may well be associated with diminishing size! A smaller watch may be costlier to make than a larger watch.

In that case, it may be possible that one variable makes an over-contribution to the final estimate whereas the other variable makes some adjustment or negative contribution. For this type of rationale to be supportable, we cannot have the scenario where the first variable makes no contribution or a less contribution that the compensatory variable i.e. a 'major' cannot be less than 'minor'. We cannot have a zero value for either driver.

4.5.6 Coefficient of Variation as a measure of tightness of fit

The Coefficient of Variation is a dimensionless measure that can be used to compare alternative Regression models. In terms of a regression it can be calculated as the

Simple and Multiple Linear Regression | 173

Regression Standard Error divided by the Mean of the Dependent Variable values (i.e. the observed y-values.)

The Coefficient of Variation (CV) gives us a comparative measure of the tightness of fit of the data, i.e. the scatter around the regression relationship, and is often expressed as a percentage. However, its usefulness as an absolute measure is questionable as its value increases as the Mean tends towards zero, which could be the case if the entity being estimated can take negative values. CV as a comparative measure between alternative regressions is more useful when we have only positive or negative values but not both, as we might get with a measure of process tolerance.

Let's look at our previous Multi-linear Regression example from Section 4.5.3. There we consider the relationship between Assembly Time, Product Weight and Number of Component Parts. We may recall that we ran three Regressions: one with both predictor variables, and two others using each predictor variable on its own in turn. To save you looking back (*I'm too good to you really*), Table 4.25 summarises some of the key statistics from Tables 4.9, 4.15 and 4.21.

Here we see that the Multi-Linear Regression model using both Weight and Parts gives the lowest level of relative scatter, but for the preferred (*in this case*) Simple Linear Regression model, using Weight only, the CV is not too far behind in relative terms. The lesser preferred but acceptable Simple Linear Regression model using Number of Parts gives us a higher degree of scatter or unexplained variation in the data; this is something that the F-Statistic can also give us

If we are to use the Coefficient of Variation as a measure of tightness of fit then the question this begs is '*What is an acceptable value?*' to which there is no easy answer, because it is an imperfect measure, depending as it does on the value of the

Table 4.25 Coefficient of Variation as a Measure of Tightness of Fit

Regression Statistic	Predicator Variables		
	Table 4.9	*Table 4.15*	*Table 4.21*
	Weight and Tests	*Weight only*	*Tests only*
R-Square	0.940	0.717	0.567
Adjusted R-Square	0.920	0.677	0.505
F	47.141	17.753	9.177
Standard Error	700.3	1409.6	1743.6
Mean of Cost	8293	8293	8293
Coefficient of Variation	8.4%	17.0%	21.0%

174 | Simple and Multiple Linear Regression

Mean in relation to the Origin. For example, if we were to offset the entire dataset vertically up or down by a constant amount we would not change the standard deviation of the dependent y-variable, or the standard error of the Regression but the Mean of the y-variable would change by the constant amount, thus affecting the CV. From a practical perspective, the CV is best kept as a relative measure to compare alternative models for the same observed y-values as we have here, rather than as an absolute benchmark.

> Note: If we have constrained a regression to pass through origin then we have effectively said that we do not believe the sample Mean to be a good estimate of the population Mean. Therefore, we should determine the Estimated Mean to be that of the Predicted Values as illustrated in Section 4.5.4, Table 4.24, and use this as the basis of the CV calculation.

4.5.7 White's Test for heteroscedasticity . . . and, by default, homoscedasticity

In Section 4.1.1 we specified that the error or scatter around the regression line should sum to zero, and that it was expected to be Normally Distributed. What we neglected to mention was that the error term should have constant variance around and along the regression line. If it does this, then we can say that the relationship displays 'homoscedasticity'. (*Although personally, I have difficulty in saying it. It doesn't easily trip off the tongue.*)

If the scatter is not homoscedastic, then it is heteroscedastic. Figure 4.22 illustrates the principle. (*Sometimes these are spelt with a kicking 'kuh' rather than a curly 'cuh' as 'homoskedasticity' and 'heteroskedasticity'.*)

Definition 4.4 Homoscedasticity, or homoskedasticity

Data is said to exhibit homoscedasticity if data variances are equal for all data values.

Definition 4.5 Heteroscedasticity, or heteroskedasticity

Data is said to exhibit heteroscedasticity if data variances are not equal for all data values.

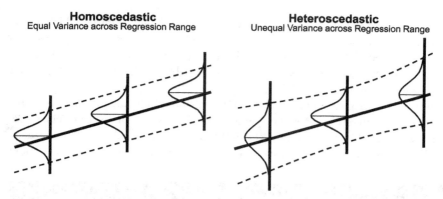

Figure 4.22 Homoscedastic and Heteroscedastic Error Distribution

In reality, most people just run the regression and don't even think about the equal variance requirement. Others may take a cursory look at the error plot to see is there is any pattern such as arcing or divergence/convergence that would suggest that the model is not truly linear. So, in terms of a formal test, we can introduce a change to our normal mantra:

Just because we don't, it doesn't mean we shouldn't!

As highlighted, many of us may often do a simple visual test of whether the scatter appears to be evenly distributed, i.e. no obvious narrowing, widening or arcing of the scatter. This may be sufficient to highlight heteroscedasticity when we have a large number of data points, but where we have only a few data points, we need a 'proper test' to support our paradigm of TRACEability (Volume I Chapter 3). There are several such tests from which we can choose, but the White Test (White, 1980), which emanates from the heady world of econometrics, is probably the most popular. (*Popular is a relative term in this context implying that it is probably used less infrequently than others! Sorry, is my cynicism beginning to show again?*)

White's Test is an unusual one; we might call it a double negative test. It requires us to set up a Multi-Linear Model of our data that implies that our error data is heteroscedastic, and then tests whether the data supports that Null Hypothesis, by performing another regression, referred to as the Auxiliary Regression. For this we need to take the square of the residuals (or error terms0 from our Primary Regression as our dependent variable. This is taken to be our estimate of the error variance with a Mean Variance of zero. We then perform a Multi-Linear Regression against the original independent variables (x values), the squares of these x-values, and the pairwise cross-products if we have more than one independent variable. White's Test is unusual because we are looking for the model to fail, i.e. to demonstrate that the assumption of heteroscedasticity cannot be supported.

176 | Simple and Multiple Linear Regression

From this Auxiliary Regression, we can create the White Statistic, which is the R-Square for the Auxiliary Regression multiplied by the number of data points. This statistic is assumed to have a chi-squared distribution with v degrees of freedom (see Volume II Chapter 6), where v is the number of variables in the Auxiliary Regression (including squares and cross-products.)

However, there is a downside to White's Test in that it is only intended for large sample sizes ... and with relatively few independent variables. As we can see from Table 4.26, the number of variables required to perform White's test grows quite quickly (quadratically). This means that when we have multiple independent variables, we need more

---

For the Formula-philes: Limitations of White's Test for heteroscedasticity

For a Primary Regression with independent variables, the number of pseudo variables,, in the Auxiliary Regression required to perform White's Test is a quadratic function of the form

$$a = \frac{p(p+3)}{2}$$

Caveat augur

Just like any other statistical test, White's Test is not infallible.

With a small number of data points (e.g. 10), a Simple Linear Regression through data that is known not to be linear can pass the White's Test for homoscedasticity.

Sometimes all we have to do is look at the error pattern created. As a backup, we should also perform a F-Test on the Auxiliary Regression.

Table 4.26 Variables Required for White's Test

Primary Regression	Ind Var	White's Test Auxiliary Regression (Error Squared as a Function of Variables)	Ind Var & deg. freed.
$y = f(x)$	1	$\varepsilon^2 = f(x, x^2)$	2
$y = f(x_1, x_2)$	2	$\varepsilon^2 = f(x_1, x_2, x_1^2, x_2^2, x_1 x_2)$	5
$y = f(x_1, x_2, x_3)$	3	$\varepsilon^2 = f(x_1, x_2, x_3, x_1^2, x_2^2, x_3^2, x_1 x_2, x_1 x_3, x_2 x_3)$	9
$y = f(x_1, x_2, x_3, x_4)$	4	$\varepsilon^2 = f(x_1, x_2, x_3, x_4, x_1^2, x_2^2, x_3^2, x_4^2, x_1 x_2, x_1 x_3, x_1 x_4, x_2 x_3, x_2 x_4, x_3 x_4)$	14

Simple and Multiple Linear Regression | 177

data points in order to test for homoscedasticity than we do to perform the primary regression! For example, in theory, we could run a Multi-Linear Regression with three independent variables using just four data points (*but probably not a good idea!*). However, we cannot run White's Test without at least ten data points, and therefore we would be challenged to demonstrate homoscedasticity in our data.

The procedure for White's Test is as follows:

1. Perform the primary Regression Analysis as normal and plot the Residuals between the Regression Model and the observed values. We will use our Multi-Linear Regression of Cost as a function of Finished Weight and Number of Tests from Section 4.3.2 as our example.
2. Perform all the usual 'safety checks': R2, Adjusted R2, F Statistic significance and the significance of the parameter t-statistics (p-values). A quick look back at Table 4.9 suggests that we are good to go.
3. Assuming that we give ourselves the green light at Step 2, in relation to the Measures of Linearity and Significance, calculate the Regression Line (which we did for our example in Table 4.10)
4. Calculate the Residual as the difference between the observed or actual y-values and the regression line of predicted y-values (also done in Table 4.10)
5. We now square the residuals. This will be the dependent variable of our Auxiliary Regression. We have included this in a new Table 4.27
6. Set up an array in consecutive (contiguous) columns of the independent variables for White's Test, as we have in Table 4.27:

    *   Each independent variable (x-value columns from the primary regression)
    *   The square of each independent x-variable from the primary regression
    *   The cross-products of each pair of independent x-variables from the primary regression

Table 4.27 Example of White's Test – Auxiliary Regression Input Data

	Auxiliary Regression Inputs					Primary Regression Error	
	Independent Variables Derived form Primary Regression						
	x1	x2	Squares		Cross-Product	ε	ε^2
Obs	Weight Kg	Tests	Weight^2	Test^2	Weight x Test	Error £	Error^2
1	10	1	100	1	100	448.63	201266.343
2	21	2	441	4	1764	122.62	15035.092
3	34	1	1156	1	1156	-1031.71	1064429.860
4	42	3	1764	9	15876	-241.95	58539.742
5	59	1	3481	1	3481	-399.11	159286.543
6	63	2	3969	4	15876	1023.27	1047089.537
7	71	1	5041	1	5041	254.22	64629.425
8	85	2	7225	4	28900	310.05	96128.683
9	97	3	9409	9	84681	-486.02	236214.520
	Aux-x1	Aux-x2	Aux-x3				Aux-y

178 | Simple and Multiple Linear Regression

7. Run the auxiliary regression. Table 4.28 shows the Regression Output
8. Extract the R-Square statistic (not the Adjusted version) from this Auxiliary Regression (i.e. 0.6807 in Table 4.28) and multiply it by the number of data points (9) to create the White Statistic of 6.126 in our example
9. We can now generate the Critical Value for the White Statistic using Microsoft Excel's Chi-Squared Distribution, **CHISQ.INV.RT(*significance, v*)** with our chosen *significance* level (say 5%), and *v* is the number of degrees of freedom equal to the number of 'independent' variables in the Auxiliary Regression (including squares and cross-products.) If our White's Test Statistic is greater than our Critical Value then we reject the hypothesis that the data is homoscedastic, and therefore it can be assumed to be heteroscedastic, and as a consequence, our primary Regression would not be valid. Fortunately, in our example, we have shown that the White Statistic is comfortably below our Critical Value of 11.070, so our data is Homoscedastic. Alternatively, we can determine the significance of the White Statistic calculates in step 8 using **CHISQ.DIST.RT(*whitestatistic, v*)**; if this is greater than our chosen significance level then the data is heteroscedastic.
10. To be on the safe side, if the first test shows that the data may be homoscedastic, we should also look at the F-Statistic of the Auxiliary Regression . . . *and perhaps have a mild panic.* In our example in Table 4.28, we will notice that we have a very high F-Statistic Significance, and also high P-values for most if not all of t-Statistics for the input variables. **Don't worry, this is exactly what we would be hoping for!** It is the exact opposite of what we would normally be looking for in a Regression, in which we are trying to establish a relationship. Here, we are looking to disprove a relationship. The model assumes heteroscedasticity in the square of the error terms.

Table 4.28 Example of White's Test Output Data and Test Result

SUMMARY OUTPUT

Regression Statistics		White's Test - Chi.Square Dist with Reg *df*		
		Significance Level	Critical Value	
Multiple R	0.825056723			
R Square	**0.680718596**	5%	11.070	
Adjusted R Square	0.148582922		Observed White's Statistic	Significance Level
Standard Error	386834.8077			
Observations	9	R Square x Observations >	6.126	29.4%

ANOVA

	df	SS	MS	F	Significance F
Regression	5	9.5712E+11	1.91424E+11	1.279219998	0.447027664
Residual	3	4.48924E+11	1.49641E+11		
Total	8	1.40604E+12			

	Coefficients	Standard Error	t Stat	P-value	Lower 95%	Upper 95%
Intercept	-1512925.455	1163999.33	-1.299764885	0.284538235	-5217290.823	2191439.912
Weight Kg	73037.72195	32277.70662	2.262791555	0.108645448	-29684.34624	175759.7901
Tests	1495984.234	1207308.856	1.239106486	0.303412414	-2346211.373	5338179.841
Weight^2	-956.2144542	400.9254257	-2.385018242	0.097176806	-2232.138094	319.7091856
Test^2	-579228.4586	347656.7707	-1.666092846	0.194285106	-1685627.464	527170.547
Weight x Test	52.72982258	23.07022848	2.285622036	0.10638361	-20.68994079	126.149586

Simple and Multiple Linear Regression | 179

By rejecting this error model as being heteroscedastic, we are accepting that it must be homoscedastic! In other words, the distribution of the variance is independent of the position in the data range. (*Was that a 'whoopee!' I heard, or perhaps it was just my echo?*)

There is a risk that these two test statistics contradict each other. If they do contradict, we should err towards caution and assume the data is heteroscedastic. It could be an indication that a linear model is not appropriate and that a Non-Linear Model should be considered (see Chapters 5 to 7).

Caveat augur

We may discover an alternative procedure on the internet to perform White's Test using the Regression's Predicted Values and their Squares as the only two sets of variables in the Auxiliary Regression, regardless of the number of primary variables. On the face of it this seems much simpler and more flexible than the technique we have described here that appears to circumvent the quadratic growth in auxiliary variables.

However, this only works for a Simple Linear Regression with a single independent variable. It does not work for Multi-Linear Regression.

Whilst by inference all the squares and pairwise cross-products can be accounted for in the square of the Predicted Values using this alternative procedure, it also places constraints on their inter-relationships, creating different and potentially inconsistent results with the non-abbreviated test described here.

4.6 Prediction and Confidence Intervals – Measures of uncertainty

Wouldn't it be really useful if we could run a Regression and calculate the degree of accuracy or uncertainty in the values it predicts . . . and be able to express that degree of uncertainty in terms of levels of confidence or probability?

Well, the good news is that we can (*OK, calm down, put a hold on those celebrations, there's no gain without pain . . .*) The bad news is that Microsoft Excel doesn't do it for us automatically – it could, but unfortunately, it doesn't. However, the slightly better news is that Excel does give us all the building blocks we need to calculate them for ourselves. (*OK, perhaps now is the time for some restrained celebration, but let's not get too excited just yet.*)

4.6.1 *Prediction Intervals and Confidence Intervals: What's the difference?*

Prediction Intervals and Confidence Intervals are very closely related, both conceptually and mathematically, but they are different.

Prediction Intervals, defined by a Lower and Upper Limit, relate to the range of uncertainty in the individual point predictions indicated by the Regression Equation. In other words, there is a degree of scatter in the observed data around the Best Fit, and so it is only reasonable that a degree of scatter around the regression line should be expected when the estimated values are eventually actualised. This is an expression of the uncertainty in the eventual outcome.

Confidence Intervals, also defined by a Lower and Upper Limit, relate to the range of uncertainty in the Regression Equation due to the inherent uncertainty in the Regression Coefficients calculated. If we had had one or more additional or fewer data observations to drive the Regression, we would have had a slightly different answer – unless the additional or fewer points all fell on the Regression Best Fit Relationship (*which is unlikely*).

Definition 4.6   Regression Prediction Interval

A Regression Prediction Interval of a given probability is an expression of the Uncertainty Range around future values of the dependent variable based on the regression data available. For a known value of a single independent variable, or a known combination of values from multiple independent variables, the future value of the dependent variable will occur within the Prediction Interval with the probability specified.

Definition 4.7   Regression Confidence Interval

The Regression Confidence Interval of a given probability is an expression of the Uncertainty Range around the Regression Line. For a known value of a single independent variable, or a known combination of values from multiple independent variables, the Mean of all future values of the dependent variable will occur within the Confidence Interval with the probability specified.

We might like to think of a Confidence Interval as an expression of the reliability of the regression equation as an estimate of a future value, and a Prediction Interval as a measure of the uncertainty around that future value.

Simple and Multiple Linear Regression | 181

Both Intervals assume a Student's t-Distribution, and by implication the Regression Equation is the 50% Confidence Level of that distribution, which of course is symmetrical, giving us Median (50%) = Mean = Mode or Most Likely value (see Volume II Chapter 2.)

The boundaries of the Confidence Interval are referred to as the Upper and Lower Confidence Limits (UCL and LCL); similarly the Prediction Interval is bounded by the Upper and Lower Prediction Limits (UPL and LPL). It is general custom and practice that both Prediction and Confidence Intervals assume symmetrical Limits. For instance:

Interval	Lower Limit	Upper Limit
90%	5%	95%
80%	10%	90%

For a matching pair of intervals (e.g. 90%) the Confidence Interval will always sit within the confines of the equivalent Prediction Interval but will always appear to be more curved, tending asymptotically to the same Upper and Lower Limit values of the Prediction Interval.

Figure 4.23 shows how the Confidence Interval is contained within

> **Asymptote:** A straight line that tends continually closer in value to that of a given curve as they tend towards infinity (positively or negatively). The difference between the asymptote and the associated curve reduces towards zero, but never reaches it at any finite value.

the bounds of the Prediction Interval, and that our 'confidence' in both the individual point predictions and the Best Fit line overall wanes as we extrapolate further from the observed data we used to perform the analysis, causing it to fan or funnel out.

Prediction and Confidence Intervals can be generated for both Simple and Multiple Linear Regressions, and importantly can be extrapolated or interpolated for any valid point indicated by that relationship. An easy way to remember the difference is that Confidence Interval relates to our Confidence in the values calculated for the Regression Coefficients e.g. Slope and Intercept, whereas the Prediction Interval relates to the inevitable scatter in the Data Points:

C is for our **Confidence** in the **Coefficients Calculated**

P is for the **Potential** range of **Points Predicted**

So, for a Prediction Interval of 90% and a data sample of ten we can expect one point to fall outside the Interval. If we widen the Interval to 95% then we will get one outside the Interval if we are unlucky or none if we are lucky. (*By Chauvenet's Criterion we would expect there to be around half an observation outside the interval, which is nonsensical, of course, it will either round up or round down in reality; we can't have half a point!*)

182 | Simple and Multiple Linear Regression

Rule of Thumb

As a rule of thumb, in order to verify that the Prediction Interval has been calculated appropriately, we can apply the fundamental principle of Chauvenet's Criterion (see Volume II Chapter 7) to count the number of observed points that fall inside or outside the Prediction Interval. For instance if we have a 90% Prediction Interval then by implication we should find 90% of points lying within the interval. (*That's lying in the sense of 'resting', and not in the sense of 'being untruthful' – just in case your cynical side was wondering!*)

Some food for thought . . .

We can generate Prediction and Confidence Intervals for any range of probability e.g. 10%–90%, 5%–95%, 1%–99% or even 1.2345%–98.7655% if we wanted. However, if we are prepared to accept or reject regression parameter values based on a particular level of significance, say 5%, which implies an interval of 2.5%–97.5%, why then would we ever be interested in Prediction or Confidence Intervals that fall outside those same values of probability? Should we not be consistent? We can go narrower, but logically, we should not go wider.

4.6.2 Calculating Prediction Limits and Confidence Limits for Simple Linear Regression

Without a standard function or facility in Microsoft Excel then some of us may be dismayed to learn that calculating the Prediction and Confidence Intervals for a Simple Linear Regression (SLR) is one for the formula-philes.

For the Formula-philes: Prediction and Confidence Intervals for Simple Linear Regression

Consider a range of n paired observations $x_1, x_2, x_3 \ldots x_n$ and $y_1, y_2, y_3 \ldots y_n$ with Means $\bar{x}$ and $\bar{y}$ respectively, for which we generate a Simple Linear Regression Let the Line of Best Fit for y have a slope of β_1 and an intercept of β_0 then the Line of Best Fit is:
$$\hat{y}_i = \beta_1 x_i + \beta_0$$

The Standard Error, ε, of the Regression
$$\varepsilon = \sqrt{\sum_{i=1}^{n} \frac{(y_i - \hat{y}_i)^2}{n-2}}$$

Simple and Multiple Linear Regression

The p% Confidence Interval, $(LCL_i, UCL_i) = \gamma \hat{y}_i \mp \varepsilon t \sqrt{\dfrac{1}{n} + \dfrac{(x_i - \bar{x})^2}{\sum_{i=1}^{n}(x_i - \bar{x})^2}}$

(LCL_i, UCL_i), around $\hat{y}_i$ is:
The p% Prediction Interval, $(LCL_i, UCL_i) = \hat{y}_i \mp \varepsilon t \sqrt{1 + \dfrac{1}{n} + \dfrac{(x_i - \bar{x})^2}{\sum_{i=1}^{n}(x_i - \bar{x})^2}}$
(LPL_i, UPL_i), around $\hat{y}_i$ is:
Where t is the two-tailed Student's t-Distribution value for a p% Confidence Interval that can be calculated in Microsoft Excel as: $t = \mathbf{T.INV.2T}(1 - p\%, n - 2)$

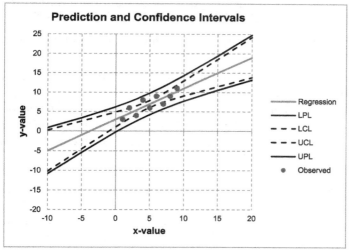

Figure 4.23 Prediction and Confidence Intervals

Table 4.29 Calculation of Prediction and Confidence Intervals in Microsoft Excel

	A	B	C	D	E	F	G	H	I	J	K	L	M	N	O	P	Q
1					Confidence & Prediction Interval >			90%									
2	Input Data			Δ(x)²	Extended	Extended	Δ(x)²		Std Error	Obs	Half	Half	Regresssion	Confidence Interval		Prediction Interval	
3	Obs	y	x	$(x_i - \bar{x})^2$	x	Δ(x)²	ΔΣ(x)²	t	ε	n	CI	PI	Output (ŷ)	LCL	UCL	LPL	UPL
4					-10	225	3.7500	1.8946	1.3939	9	5.1891	5.8224	-5	-10.1891	0.1891	-10.8224	0.8224
5					-5	100	1.6667	1.8946	1.3939	9	3.5210	4.4013	-1	-4.5210	2.5210	-5.4013	3.4013
6					0	25	0.4167	1.8946	1.3939	9	1.9185	3.2641	3	1.0815	4.9185	-0.2641	6.2641
7	1	3	1	16	1	16	0.2667	1.8946	1.3939	9	1.6231	3.0997	3.8	2.1769	5.4231	0.7003	6.8997
8	2	6	2	9	2	9	0.1500	1.8946	1.3939	9	1.3494	2.9656	4.6	3.2506	5.9494	1.6344	7.5656
9	3	4	3	4	3	4	0.0667	1.8946	1.3939	9	1.1135	2.8659	5.4	4.2865	6.5135	2.5341	8.2659
10	4	8	4	1	4	1	0.0167	1.8946	1.3939	9	0.9440	2.8044	6.2	5.2560	7.1440	3.3956	9.0044
11	5	6	5	0	5	0	0.0000	1.8946	1.3939	9	0.8803	2.7836	7	6.1197	7.8803	4.2164	9.7836
12	6	9	6	1	6	1	0.0167	1.8946	1.3939	9	0.9440	2.8044	7.8	6.8560	8.7440	4.9956	10.6044
13	7	7	7	4	7	4	0.0667	1.8946	1.3939	9	1.1135	2.8659	8.6	7.4865	9.7135	5.7341	11.4659
14	8	9	8	9	8	9	0.1500	1.8946	1.3939	9	1.3494	2.9656	9.4	8.0506	10.7494	6.4344	12.3656
15	9	11	9	16	9	16	0.2667	1.8946	1.3939	9	1.6231	3.0997	10.2	8.5769	11.8231	7.1003	13.2997
16					10	25	0.4167	1.8946	1.3939	9	1.9185	3.2641	11	9.0815	12.9185	7.7359	14.2641
17					15	100	1.6667	1.8946	1.3939	9	3.5210	4.4013	15	11.4790	18.5210	10.5987	19.4013
18					20	225	3.7500	1.8946	1.3939	9	5.1891	5.8224	19	13.8109	24.1891	13.1776	24.8224
19	Sum	63	45	60													
20	Average	7	5														

184 | Simple and Multiple Linear Regression

SUMMARY OUTPUT

Regression Statistics	
Multiple R	0.859337849
R Square	0.738461538
Adjusted R Square	0.701098901
Standard Error	1.393864105
Observations	9

INPUT DATA >

Obs	Y	X
1	3	1
2	6	2
3	4	3
4	8	4
5	6	5
6	9	6
7	7	7
8	9	8
9	11	9

ANOVA

	df	SS	MS	F	Significance F
Regression	1	38.4	38.4	19.76470588	0.002986904
Residual	7	13.6	1.942857143		
Total	8	52			

	Coefficients	Standard Error	t Stat	P-value	Lower 95%	Upper 95%
Intercept	3	1.012618796	2.962615362	0.021026869	0.605537038	5.394462962
X Variable 1	0.8	0.179947082	4.445751442	0.002986904	0.374492766	1.225507234

Figure 4.24 Calculation of Prediction and Confidence Intervals in Microsoft Excel

These formulae can be verified from a number of reliable and authoritative sources, (e.g. Walpole et al., 2012, pp. 409–411).

In Table 4.29 we work though the example from Table 4.1 and Table 4.2 in calculating the two intervals shown in Figure 4.23. Figure 4.24 summarises the inputs and basic regression Output. In this example, we have data for a range of x-values and wish to extrapolate forwards and backwards.

1. In Column D of Table 4.29, we calculate the square of the deviation of each Observed x from their Mean in Cell C20
2. In Column E we specify the range of x-values for which we wish to show the two Intervals around the Regression Lines
3. In Column F, we repeat the calculation from Column D but this time extending it to the wider x-values. It is important that the difference still relates to the average of the Observed x-values only in Cell C20. (*This could have been down in a single column but it was important to emphasise this latter point; visually it is clearer in two steps.*)
4. Column G divides Column F by the Sum of Squares of Observed x-value deviations from their Mean i.e. Cell D19
5. Column H calculates the t-Distribution Statistic for the two-tailed Confidence Interval specified in Cell H1 using Excel Function **T.INV.2T(*1-$H$1, Residual df*)** where the *Residual df* can be referenced from the Data Analysis Output Report and equals the number of observations minus two (*i.e. we have two parameters*)
6. The Standard Error from the Data Analysis Output Report is copied into Column I
7. The number of Observations is input into Column J
8. Column K depicts the calculation for the Half Confidence Interval (CI). Note: the Full Interval is this Half Interval value either side of the Regression Line. For example, the Column K = Column H multiplied by Column I multiplied by the

Square Root of (Column G plus the reciprocal of Column J). We can use the Excel function **SQRT** here.

9. Column L depicts the calculation for the Half Prediction Interval (PI). For instance the Column L = Column H multiplied by Column I multiplied by the Square Root of (Column G plus the reciprocal of Column J plus 1); the only difference to the formula for a Confidence Interval above is the addition of 1 inside the Square Root function.
10. The Regression Line based on the Coefficients of the Slope and Intercept is shown in Column M
11. Column N = Column M − Column K and is the Lower Confidence Limit (LCL)
12. Column O = Column M + Column K and is the Upper Confidence Limit (UCL)
13. Column P = Column M − Column L and is the Lower Prediction Limit (LPL)
14. Column Q = Column M + Column L and is the Upper Prediction Limit (UPL)

It's not pretty, but it works!

There is a more elegant albeit more challenging solution to this using some advanced functionality in Microsoft Excel; we will deal with this in the next section on Multi-Linear Regression ... *We have no real choice there but to use it.*

4.6.3 *Calculating Prediction Limits and Confidence Limits for Multi-Linear Regression*

For those of a sensitive disposition, be warned, this section (~~might~~) will get a little scary, especially for the formula-phobes amongst us!

There's no real way to break this gently, so let's just wade in.

The bad news is: If we want to generate Prediction and/or Confidence Intervals as Measures of Uncertainty in the case of Multiple Linear Regression, we need to use Matrix Multiplication.

The good news is: We can do this in Microsoft Excel using special Array Formula Functions

The bad news is: The formulae look horrendous!

The good news is: They're not quite as bad as they look!

For the Formula-philes: Prediction and Confidence Intervals for multi-linear regression

Consider a range of n observations $y_1, y_2, y_3 \ldots y_n$ and v associated driver values $x_{11}, x_{12}, x_{13} \ldots x_{1n}$ to $x_{v1}, x_{v2}, x_{v3} \ldots x_{vn}$ for which we generate a Multi-linear Regression

Let be the vertical array of the regression parameters :
$$[\beta] = \begin{bmatrix} \beta_0 \\ \beta_1 \\ \ldots \\ \beta_n \end{bmatrix}$$

(Continued)

Simple and Multiple Linear Regression

Let $[X]$ represent the array of driver values for the associated dependent variable value y_i with a leading column of unity values

$$[X] = \begin{bmatrix} 1 & x_{11} & x_{21} & \cdots & x_{v1} \\ 1 & x_{12} & x_{22} & \cdots & x_{v2} \\ \cdots & \cdots & \cdots & \cdots & \cdots \\ 1 & x_{1n} & x_{2n} & \cdots & x_{vn} \end{bmatrix}$$

The Regression Output function can be depicted by either an array or algebraic equation:

$$\hat{y}_i = [X][\beta]$$
$$\hat{y}_i = \beta_0 + \beta_1 x_{1i} + \beta_2 x_{2i} + \ldots + \beta_v x_{vi}$$

Let $[x]_i$ represent the array of driver values associated with variable value with a leading cell of value 1:

$$[x]_i = \begin{bmatrix} 1 & x_{1i} & x_{2i} & \cdots & x_{vi} \end{bmatrix}$$

Let Z_i be the value calculated by the Matrix Product

$$Z_i = [x]_i \left[[X]^T [X] \right]^{-1} [x]_i^T$$

From Walpole (2012, pp. 457–458), the Confidence and Prediction Intervals are given by:

The p% Confidence Interval, (LCL_i, UCL_i), around $\hat{y}_i$ is:

$$(LCL_i, UCL_i) = y_i \mp t\varepsilon \sqrt{Z_i}$$

The p% Prediction Interval, (LPL_i, UPL_i), around $\hat{y}_i$ is:

$$(LPL_i, UPL_i) = y_i \mp t\varepsilon \sqrt{1 + Z_i}$$

... where ε is the standard error of the regression given by:

$$\varepsilon = \sqrt{\sum_{i=1}^{n} \frac{\left(y_i - \hat{y}_i \right)^2}{n - v - 1}}$$

... and t is the two-tailed Student's t-Distribution value for a p% Confidence Interval that can be calculated in Microsoft Excel as:

$$t = T.INV.2T \left(1 - p\%, n - v - 1 \right)$$

If only books could take a picture; the look of horror on your faces now would make a credible rival for Edvard Munch's iconic painting The Scream ...

In my defence, I did try to warn you in advance.

For the Formula-phobes: So, what's all this Matrix Multiplication malarkey?

Basically, mathematicians like to try to be efficient. (*And we may have heard it said that one definition of 'efficiency' is 'intelligent laziness'!*) Mathematicians have developed a shorthand for a series of repetitive calculations to avoid writing things out

Simple and Multiple Linear Regression | 187

long-hand, and (perhaps surprisingly) it makes it less unintelligible to the formula-phobes. (*Well, perhaps the jury is still out on that one.*)

You can always tell a Matrix; they arrange the values in a block and put square brackets around them to show that they are 'special'. In short Matrix Multiplication is equivalent to a block of Microsoft Excel **SUMPRODUCT** calculations . . . two or more series of paired numbers multiplied together and added up! For instance, the hours per year could be one matrix and the charging rates per year could be in another matrix. Multiplying them together in pairs with matching corresponding years, and aggregating them together gives us the total cost.

The weird thing is that the accepted convention is to multiply the rows of the first matrix by the columns of the second matrix, which is probably not the most intuitive thing to do . . . *but no-one said that Mathematicians were always logical, all of the time, did they?*

$$\begin{bmatrix} 1 & 2 & 3 \\ 4 & 5 & 6 \end{bmatrix} \begin{bmatrix} 1 & 4 \\ 2 & 5 \\ 3 & 6 \end{bmatrix} = \begin{bmatrix} 1x1 + 2x2 + 3x3 & 1x4 + 2x5 + 3x6 \\ 4x1 + 5x2 + 6x3 & 4x4 + 5x5 + 6x6 \end{bmatrix} = \begin{bmatrix} 14 & 32 \\ 32 & 77 \end{bmatrix}$$

The resulting matrix is always a square matrix with the number of rows equal to the number of rows in the left-hand matrix and the columns equal to the number of columns in the right-hand matrix.

Despite its appearance, it is something that we can do relatively easily in Microsoft Excel so long as we are happy to use Array Functions (and all that Ctrl+Shift+Enter malarkey) . . . what's more it can be used for Simple Linear Regression also (*so we can forget all about the previous section and remember just the one technique!*)

Table 4.30 Input Array in Microsoft Excel for Prediction and Confidence Intervals

	A	B	C	D	E	F	G
1							
2		**Input Data and Prediction Range**					Z
3	Obs	y	Unity	x_1	x_2		
4			1	0	1		1.050
5			1	0	3		0.553
6			1	1	1		0.849
7	1	7	1	1	3		0.416
8	2	6	1	2	2		0.305
9	3	9	1	3	2		0.207
10	4	11	1	4	3		0.222
11	5	9	1	5	1		0.402
12	6	12	1	6	3		0.272
13	7	11	1	7	2		0.178
14	8	13	1	8	3		0.467
15	9	12	1	9	1		0.532
16			1	9	3		0.618
17			1	12	2		0.953
18			1	15	1		1.807

188 | Simple and Multiple Linear Regression

Why do I say that it is so easy? With the exception of a column of 1s, we will already have the grid of independent x-variables set out in Excel alongside the corresponding values for the dependent y variable that we are trying to predict. Table 4.30 highlights the key elements.

For ease of description we will refer to the array of data covering the Unity Column and the x-values in Cells C3:E11 as the '*FullArray*'; any row of data across Columns C:E will be referred to as a '*RowArray*' e.g. C3:E3 through to C11:E11.

In order to calculate the somewhat anti-social formulae for Prediction and Confidence Intervals in Microsoft Excel we need to use some of its advanced Array Functions, even though we discouraged their use without good reason under our Good Practice Spreadsheet Principles in Volume I. This is one of those good reasons; it is not practical another way:

- Matrix Multiplication **MMULT(*array1,array2*)**

   Ordinary multiplication is 'commutative' i.e. you can switch the order and it doesn't change the result e.g., but this is not the case with Matrix Multiplication.

   A [2x3] Matrix multiplied by a [3x2] Matrix gives us a [2x2] Matrix whereas if we switched the order a [3x2] Matrix multiplied by a [2x3] Matrix gives us a [3x3] Matrix. **Order Matters!**

- Matrix Inverse   **MINVERSE(*array*)**

   The Matrix Inverse Function gives that array which when multiplied by the original array, gives us the 'Identity Array' i.e. a Square Array with 1s on the top-left to bottom-right diagonal and 0s elsewhere.

   We can only use the **MINVERSE** function with a **square array.**

- Matrix Transposition **TRANSPOSE(*array*)**

   The transposition of an array or matrix swaps the rows and columns around, leaving the top-left to bottom-right corners unchanged.

Caveat augur

Array functions will not work if the squiggly brackets are typed in manually! It will interpret it as a text string.

If we forget to use a combination of the keys '**Ctrl+Shift+Enter**' and just press enter then we may get a value, if we are 'unlucky' (*because we may not realise that it is in error*), whereas if we are 'lucky', we will get a #VALUE error (*at least we will know that we have done something wrong!*)

Simple and Multiple Linear Regression | 189

For any Excel Array function we have to activate it in a special way using a combination of the keys '**Ctrl+Shift+Enter**'. This action inserts 'squiggly' brackets or braces { } around the formula (including the equal sign) to tell Excel (*and us*) that it as an array function.

We can use these functions to generate the Z_i value in the Confidence and Prediction Interval calculation defined in the formula-phile box. As you can imagine with a nasty looking formula such as this, the finished cell formula is quite horrendous, so we'll take it one step at a time:

1. The transposition of our '*FullArray*' is **TRANSPOSE(*FullArray*)**
2. The product of our transposed '*FullArray*' and the '*FullArray*' itself is the Matrix Multiplication **MMULT(TRANSPOSE(*FullArray*),*FullArray*)** which is a square array
3. The inverse of this product of the transposed '*FullArray*' and the '*FullArray*' is **MINVERSE(MMULT(TRANSPOSE(*FullArray*),*FullArray*))**
4. The product of one of the *RowArray*s and this Inverted Matrix is another Matrix Multiplication **MMULTI(*RowArray*,MINVERSE(MMULT(TRANSPOSE(-*FullArray*),*FullArray*)))**
5. Using Matrix Multiplication again we can find product of the above with the transposition of the *RowArray* that we have just used to get the wonderful expression **MMULT(MMULT(*RowArray*,MINVERSE(MMULT(TRANSPOSE(*FullArray*),*FullArray*))),TRANSPOSE(*RowArray*))**
6. All we need to do now is to replace the '*FullArray*' text with the Cell Range references or a Range Name (unless we've already called the range of cells by that name.) Note that if we are using absolute cell references then we need to lock them with $ signs. We need to replace the 'RowArray' text with the Cell Range references, but

---

Z_i is like a 'Multi-dimensional Quadratic Function'

If you've never heard of one, it's probably because I just made the term up. However, by way of explanation, if we were to hold all but one of our independent x-variables constant and plot the value of Z_i as the remaining x-variable changes we will get a perfect quadratic function.

Why is this? By holding all but one independent x-variable constant we are effectively reducing our multi-linear regression to a Simple Linear Regression. From Section 4.6.2 we then have the situation where Z_i can be expressed in the more tangible form of:

$$Z_i = \frac{(x_i - \bar{x})^2}{\sum_{i=1}^{n} (x_i - \bar{x})^2}$$

The denominator (bottom line) is fixed and the numerator (top line) is a quadratic equation in x_i

190 | Simple and Multiple Linear Regression

this time we are not going to lock them as we want this to be a dynamic range that changes with the row number we are on.

7. Finally, we can activate the calculation by **Ctrl+Shift+Enter**

After the first cell has been created without an error, the calculation can then be copied to all other relevant cells like a normal calculation.

Note: If we have used the dummy strings FullArray and RowArray in setting this up, then the length of the formula at step 7 as a text string is 90 characters – this may help to validate our typing ability! We can test this in Excel using **LEN**(*text*) function using the calculation without the equal sign.

We can extend the Z_i value to cover any combination of the independent x-variables for which we want a prediction by allowing the *RowArray* to vary. However, the *FullArray* must be locked to the Observed x-values that we have used in the Regression Analysis Input data.

In Figure 4.25 and Table 4.31 we complete the example we started in Table 4.30.

1. Columns A to G reproduce Table 4.30 with the addition of the Regression Output Coefficients for Observations 1 to 9.
2. In Column G, it is important that the Z-value's 'FullArray' still relates to the Observed x-values only in Cells F7:H15.
3. Column H calculates the t-Distribution Statistic for the two-tailed Confidence Interval specified in Cell H1 using Excel Function **T.INV.2T**(*1–H1, Residual df*) where the *Residual df* can be referenced from the Data Analysis Output Report in Figure 4.25 and equals the number of observations minus the number of regression parameters (in this case 3 i.e. the coefficients of the 2 independent variables plus the intercept)
4. The Standard Error from the Data Analysis Output Report (Figure 4.25) is copied into Column I
5. Column K depicts the calculation for the Half Confidence Interval (CI). Note: the Full Interval is this Half Interval value either side of the Regression Line. For example, the Column K = Column H multiplied by Column I multiplied by the Square Root of Column G. We can use the Excel function **SQRT** here.
6. Column L depicts the calculation for the Half Prediction Interval (PI). For instance, the Column L = Column H multiplied by Column I multiplied by the Square Root of (Column G plus 1); the only difference to the formula above is the addition of 1 inside the Square Root function.
7. The Regression Line based on the Coefficients of the Slope and Intercept are shown in Column M. The easiest way to do this is to use Microsoft Excel's **SUMPRODUCT** function to multiply the range $C$1:$E$1 (Coefficients) with each row in turn below, e.g. **M4=SUMPRODUCT$C$1:$E$1,C4:E4)**
8. Column N = Column M – Column K and is the Lower Confidence Limit (LCL)
9. Column O = Column M + Column K and is the Upper Confidence Limit (UCL)

Simple and Multiple Linear Regression 191

SUMMARY OUTPUT

Regression Statistics	
Multiple R	0.954261475
R Square	0.910614963
Adjusted R Square	0.88081995
Standard Error	0.827819597
Observations	9

INPUT DATA >

Obs	y	Unity	x_1	x_2
1	7	1	1	3
2	6	1	2	2
3	9	1	3	2
4	11	1	4	3
5	9	1	5	1
6	12	1	6	3
7	11	1	7	2
8	13	1	8	3
9	12	1	9	1

ANOVA

	df	SS	MS	F	Significance F
Regression	2	41.88828829	20.94414414	30.56266433	0.000714158
Residual	6	4.111711712	0.685285285		
Total	8	46			

	Coefficients	Standard Error	t Stat	P-value	Lower 95%	Upper 95%
Intercept	3.171171171	1.135918982	2.79172302	0.031505848	0.391677552	5.95066479
X Variable 1	0.861261261	0.111119227	7.750785208	0.000242422	0.589362308	1.133160215
X Variable 2	1.135135135	0.365175043	3.108468544	0.020889591	0.241583994	2.028686276

Figure 4.25 Calculation of Prediction and Confidence Intervals in Microsoft Excel using Array Formulae

Table 4.31 Simple Linear Regression Prediction and Confidence Intervals Using Array Formulae

	A	B	C	D	E	F	G	H	I	J	K	L	M	N	O	P	Q
1	Regression Coeffs >		3.1712	0.8613	1.1351				95%		< Confidence & Prediction Interval						
2	Input Data and Prediction Range						Z	t	Std Error		Half	Half	Regresssion	Confidence Interval		Prediction Interval	
3	Obs	y	Unity	x_1	x_2				ε		CI	PI	Output ($\hat{y}$)	LCL	UCL	LPL	UPL
4			1	0	1		1.050	2.447	0.828		2.076	2.901	4.306	2.230	6.382	1.406	7.207
5			1	0	3		0.553	2.447	0.828		1.507	2.524	6.577	5.070	8.083	4.052	9.101
6			1	1	1		0.849	2.447	0.828		1.866	2.754	5.168	3.302	7.034	2.413	7.922
7	1	7	1	1	3		0.416	2.447	0.828		1.307	2.411	7.438	6.131	8.745	5.027	9.848
8	2	6	1	2	2		0.305	2.447	0.828		1.118	2.314	7.164	6.046	8.282	4.850	9.478
9	3	9	1	3	2		0.207	2.447	0.828		0.922	2.226	8.025	7.103	8.947	5.800	10.251
10	4	11	1	4	3		0.222	2.447	0.828		0.954	2.239	10.022	9.068	10.975	7.783	12.260
11	5	9	1	5	1		0.402	2.447	0.828		1.284	2.398	8.613	7.329	9.897	6.214	11.011
12	6	12	1	6	3		0.272	2.447	0.828		1.057	2.285	11.744	10.688	12.801	9.460	14.029
13	7	11	1	7	2		0.178	2.447	0.828		0.856	2.199	11.470	10.615	12.326	9.271	13.669
14	8	13	1	8	3		0.467	2.447	0.828		1.384	2.453	13.467	12.083	14.850	11.014	15.920
15	9	12	1	9	1		0.532	2.447	0.828		1.477	2.507	12.058	10.581	13.534	9.551	14.564
16			1	9	3		0.618	2.447	0.828		1.592	2.577	14.328	12.736	15.920	11.751	16.905
17			1	12	2		0.953	2.447	0.828		1.978	2.831	15.777	13.799	17.754	12.946	18.607
18			1	15	1		1.807	2.447	0.828		2.723	3.394	17.225	14.502	19.948	13.831	20.619

10. Column P = Column M − Column L and is the Lower Prediction Limit (LPL)
11. Column Q = Column M + Column L and is the Upper Prediction Limit (UPL)

That wasn't so bad after all, was it? Steady! I know what you're thinking; I can read minds, even from this distance!

As this technique works just as well for Simple Linear Regression as Multiple Linear Regression we can replicate the previous results of the example in Table 4.29 using the Z_i technique as illustrated in Table 4.32.

$$\frac{\left(x_i - \bar{x}\right)^2}{\sum_{i=1}^{n}\left(x_i - \bar{x}\right)^2} + \frac{1}{n} = Z_i$$

Column G + Reciprocal of Column J of Table 4.29 equals Column G of Table 4.32.

192 | Simple and Multiple Linear Regression

Table 4.32 Simple Linear Regression Prediction and Confidence Intervals Using Array Formulae

	A	B	C	D	E	F	G	H	I	J	K	L	M	N	O	P	Q
1	Regression Coeffs >		3.0	0.8					90%		< Confidence & Prediction Interval						
2	Input Data and Prediction Range							z	t		Std Error ε	Half CI	Half PI	Regression Output (ŷ)	Confidence Interval LCL UCL		Prediction Interval LPL UPL
3	Obs	y	Unity	x													
4			1	-10				3.861	1.895		1.394	5.189	5.822	-5.000	-10.189	0.189	-10.822	0.822
5			1	-5				1.778	1.895		1.394	3.521	4.401	-1.000	-4.521	2.521	-5.401	3.401
6			1	0				0.528	1.895		1.394	1.918	3.264	3.000	1.082	4.918	-0.264	6.264
7	1	3	1	1				0.378	1.895		1.394	1.623	3.100	3.800	2.177	5.423	0.700	6.900
8	2	6	1	2				0.261	1.895		1.394	1.349	2.966	4.600	3.251	5.949	1.634	7.566
9	3	4	1	3				0.178	1.895		1.394	1.113	2.866	5.400	4.287	6.513	2.534	8.266
10	4	8	1	4				0.128	1.895		1.394	0.944	2.804	6.200	5.256	7.144	3.396	9.004
11	5	6	1	5				0.111	1.895		1.394	0.880	2.784	7.000	6.120	7.880	4.216	9.784
12	6	9	1	6				0.128	1.895		1.394	0.944	2.804	7.800	6.856	8.744	4.996	10.604
13	7	7	1	7				0.178	1.895		1.394	1.113	2.866	8.600	7.487	9.713	5.734	11.466
14	8	9	1	8				0.261	1.895		1.394	1.349	2.966	9.400	8.051	10.749	6.434	12.366
15	9	11	1	9				0.378	1.895		1.394	1.623	3.100	10.200	8.577	11.823	7.100	13.300
16			1	10				0.528	1.895		1.394	1.918	3.264	11.000	9.082	12.918	7.736	14.264
17			1	15				1.778	1.895		1.394	3.521	4.401	15.000	11.479	18.521	10.599	19.401
18			1	20				3.861	1.895		1.394	5.189	5.822	19.000	13.811	24.189	13.178	24.822

A Side Benefit of Adding a 'Onesie' Column

By adding a column of 1s to the left of our Regression Input Array in Microsoft Excel (i.e. the Unity column) we get an added benefit. If we were to copy and transpose the Regression Coefficients from the Data Analysis Regression Output Table to a row above the Input Array such that the intercept was above the Unity column and the Regression 'Slope' Coefficients were above their corresponding Input columns then we can create the Regression Predicted Value for y using a simple **SUMPRODUCT** function:

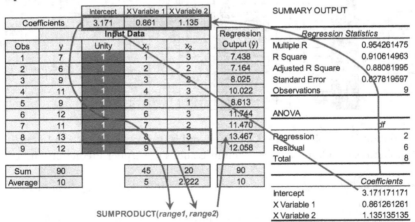

For the more adventurous we can assign the Regression Coefficients as a Named Range such as 'RegCoeffs' and insert **TRANSPOSE(RegCoeffs)** into one of the **SUMPRODUCT** arrays.

However, we mustn't forget to enable it with Ctrl+Shift+Enter as it is an Array Function. Otherwise we will get one of those rather unpleasant #VALUE error messages.

Simple and Multiple Linear Regression | 193

4.7 Stepwise Regression

Got too many potential variables and don't know where to start? Stepwise Regression may help!

The really good news is that this is not a different kind of Regression technique to tax our brain cells. Stepwise Regression is a recognised procedure used with Multi-Linear Regression that takes incremental steps in selecting appropriate predictor x-variables or drivers from a range of potential ones. Essentially, there are two procedures, but the principles are the same; it's more a question of the direction of approach like Top-down or Bottom-up Estimating. In this case we have:

- **Forward Selection**, which progressively adds variables to the mix until no further improvement can be made to the Regression Best Fit that is statistically supportable.
- **Backward Elimination**, which throws everything into the pot and removes one at a time until no further improvement can be made to the Regression Best Fit that is statistically supportable.

Definition 4.8 Stepwise Regression by Forward Selection

Stepwise Regression by Forward Selection is a procedure by which a Multi-Linear Regression is compiled from a list of independent candidate variables, commencing with the most statistically significant individual variable (from a Simple Linear Regression perspective) and progressively adding the next most significant independent variable, until such time that the addition of further candidate variables does not improve the fit of the model to the data in accordance with the accepted measures of goodness of fit for the Regression.

Definition 4.9 Stepwise Regression by Backward Elimination

Stepwise Regression by Backward Elimination is a procedure by which a Multi-Linear Regression is compiled commencing with all potential independent candidate variables and eliminating the least statistically significant variable progressively (one at a time) until such time that all remaining candidate variables are deemed to be statistically significant in accordance with the accepted measures of goodness of fit.

194 | Simple and Multiple Linear Regression

There is a halfway house, of course. If we have an existing acceptable model but we wish to improve it, we can add additional variables and test for significance. This opens up the possibility that existing variables may also become unsupportable where there is an overlap or correlation with a new variable. (*Thoughts of two steps forwards and one step back comes to mind!*)

Whilst the two approaches often give the same result, there are occasions where the final Regression models are different. This is usually because one (or possibly both) of the two approaches has reached an acceptable conclusion (statistically speaking), but not necessarily the optimum solution.

The criteria we use for making our step changes to the model will be based on the variables' t-Statistics, the overall F-Statistic, R-Square and the Adjusted R-Square ... and of course, common sense. It is important that the criteria for acceptance or rejection of candidate variables to or from the model is decided in advance.

In order to compare the two approaches, we will be considering the relatively simple example in Table 4.33.

The example relates to the time taken for a number of journeys by road. The mileage is split down into four categories:

- Motorways, for which the speed limit is 70 mph (unless there are road works, or Variable Speed Controls in place).
- Open A-roads, for which the speed limit can be anything from 30 mph to 70 mph in 10 mph increments.
- Open B-roads, for which the speed limit can be anything from 30 mph to 60 mph in 10 mph increments,

Table 4.33 Example Data for Stepwise Regression

	Input Data						
Obs	Trip Hours	Number of Driving Breaks	Motorway Mileage	Open A-road Mileage	Open B-road Mileage	Urban Road Mileage	Total Mileage
1	1.45	0	10	19	4	5	38
2	1	1	29	4	3	3	39
3	2.9	0	29	20	8	14	71
4	2.75	1	68	33	3	3	107
5	3.1	0	112	18	4	5	139
6	4.5	1	105	43	11	8	167
7	4.7	0	72	99	8	5	184
8	5	1	184	11	2	9	206
9	6.75	2	91	105	13	11	220
10	5.1	2	224	10	0	5	239
Sum	37.25	8	924	362	56	68	1410
Average	3.725	0.8	92.4	36.2	5.6	6.8	141

Simple and Multiple Linear Regression | 195

- Urban roads, which relate to roads in built-up areas such as towns and cities where the speed limit is generally 30 mph but could be 20 mph in places. However, due to the proliferation of traffic lights, pedestrian crossings, parked vehicles, congestion etc, the average speed is likely to be somewhat less,

In reality we would not achieve the speed limit as an average value as that would imply that there would be periods when we exceeded the speed limit. The data also takes account of the number of breaks that were taken, whether for fuel, leg-stretching or fluid management (*i.e. loos and brews*).

At this stage of developing our model we may even take the view that the overall mileage may be a good enough indicator or driver of the total trip time. In Tables 4.34 and 4.35 we have run a Simple Linear Regression using the Total Mileage as the independent Predicator Variable as a benchmark to which we can compare a more refined model with multiple variables.

The Unconstrained Regression (Table 4.34) highlights that there is a 28% probability that the Null Hypothesis of the Intercept being zero is true. This could lead us to re-run the regression forcing it through the origin (Table 4.35) on the logical basis that if we travel zero miles we would expect the trip to last zero hours. (*We wouldn't claim to have started if we began with a break, would we?*)

However, because we have a more detailed breakdown (synonymous with a Bottom-up Estimate), we may be able to get a better model. If we follow our first principles, we should test that there is a 'reasonable' linear relationship between the Trip Hours (our y-value) and at least one of our multiple x-values. We should also check that there is no substantive linear correlation between the various x-values within the model (a property known as 'multicollinearity').

The Correlation Matrix for our example is shown in Table 4.36:

Table 4.34 Unconstrained Simple Linear Regression Based on Total Mileage

SUMMARY OUTPUT

Regression Statistics	
Multiple R	0.938367703
R Square	0.880533945
Adjusted R Square	0.865600688
Standard Error	0.656272652
Observations	10

ANOVA

	df	SS	MS	F	Significance F
Regression	1	25.39569965	25.39569965	58.96462866	5.85766E-05
Residual	8	3.445550354	0.430693794		
Total	9	28.84125			

	Coefficients	Standard Error	t Stat	P-value	Lower 95%	Upper 95%
Intercept	0.536687606	0.46418376	1.156196429	0.280959249	-0.533722065	1.607097276
Total Mileage	0.022612145	0.002944733	7.678842924	5.85766E-05	0.015821577	0.029402712

Simple and Multiple Linear Regression

Table 4.35 Simple Linear Regression Through the Origin Based on Total Mileage

SUMMARY OUTPUT

Regression Statistics	
Multiple R	0.987930271
R Square	0.97600622
Adjusted R Square	0.864895109
Standard Error	0.66843911
Observations	10

ANOVA

	df	SS	MS	F	Significance F
Regression	1	163.5762024	163.5762024	366.0972084	5.76805E-08
Residual	9	4.02129759	0.446810843		
Total	10	167.5975			

	Coefficients	Standard Error	t Stat	P-value	Lower 95%	Upper 95%
Intercept	0	#N/A	#N/A	#N/A	#N/A	#N/A
Total Mileage	0.025657603	0.001340966	19.13366688	1.34132E-08	0.022624127	0.02869108

Definition 4.10 Collinearity and multicollinearity

Collinearity is an expression of the degree to which two supposedly independent predicator variables are correlated in the context of the observed values being used to model their relationship with the dependent variable that we wish to estimate. Multicollinearity is an expression to which collinearity can be observed across several

---

Caveat augur

Multicollinearity is not a reason to reject a model out of hand. The apparent correlation between two independent variables may be entirely coincidental. However, it may result in their being some cross-over between the variables in terms of their contribution to the value of the dependent variables. So long as the apparent relationship continues, then the model results may still be valid.

However, we should be aware that reliance on the impact of individual changes (e.g. driver sensitivity analysis) in any one variable may not be valid.

Where there is a high degree of multicollinearity, one or more of the independent variables may be rejected statistically because their apparent contribution to the model is insignificant based on the data we have; if we had different data, we may have had a different conclusion. (*No-one said life as an estimator was straightforward.*)

Simple and Multiple Linear Regression | 197

Table 4.36 Correlation Matrix for Example Stepwise Regression

Correlation Matrix	Number Driving Breaks	Motorway Mileage	Open A-road Mileage	Open B-road Mileage	Urban Road Mileage	Total Mileage
Number Driving Breaks	1	0.570	0.121	0.007	0.023	0.586
Motorway Mileage		1	-0.126	-0.341	0.000	0.839
Open A-Road Mileage			1	0.758	0.217	0.428
Open B-Road Mileage				1	0.573	0.142
Urban Road Mileage					1	0.188
Total Mileage						1
Trip Hours	0.563	0.642	0.614	0.428	0.439	0.938

From this we can conclude that there appears to be some relationship between each x-variable and the Trip Hours (y-variable), but also that there is a high degree of correlation between the Total Mileage and the Motorway Mileage, but also between the mileage on Open A-roads and Open B-roads. To a lesser extent there is some correlation between Open B-roads and Urban Roads, and the Number of Driving Breaks and the Total Mileage (and therefore Motorway Mileage also. On this basis we should question whether we should be including the Total Mileage and Open B-road Mileage.

4.7.1 Backward Elimination

Just for now, in order to demonstrate what happens if we forget or don't spot this multi-collinearity within the model, we will run the regression first with all candidate variables included. Table 4.37 illustrates what we get.

Table 4.37 Example Backward Elimination – Step 1

SUMMARY OUTPUT

Regression Statistics	
Multiple R	0.998397056
R Square	0.996796682
Adjusted R Square	0.742792534
Standard Error	0.151976722
Observations	10

ANOVA

	df	SS	MS	F	Significance F
Regression	6	28.7488623	4.791477051	248.9410463	0%
Residual	4	0.092387697	0.023096924		
Total	10	28.84125			

	Coefficients	Standard Error	t Stat	P-value	Lower 95%	Upper 95%
Intercept	-0.203694056	0.135885341	-1.499014196	21%	-0.580972246	0.173584135
Total Mileage	0.137000813	0.020530249	6.673119784	0%	0.079999703	0.194001924
Number Driving Breaks	0.220146309	0.082207476	2.677935392	6%	-0.008098235	0.448390852
Motorway Mileage	-0.119197872	0.020991357	-5.678426141	0%	-0.177479223	-0.060916521
Open A-road Mileage	-0.108265879	0.019440068	-5.569212971	1%	-0.162240161	-0.054291597
Open B-road Mileage	-0.11275526	0.045784969	-2.462713488	7%	-0.239874713	0.014364192
Urban Road Mileage	0	0	65535	#NUM!	0	0

198 | Simple and Multiple Linear Regression

What do our 'goodness of fit' measures tell us?

✓	**R-Square:**	Very high value (*giving us that feel good factor*)
✓	**F-Statistic:**	Very high value (*suggesting that there is virtually no chance of a random fluke relationship*)
?	**t-Statistics:**	21% chance that the Intercept is really zero, but more alarming is the error message against Urban Road Mileage!
✗	**Common Sense:**	The whole model is questionable on the basis of the multiple negative contributions to a Trip time which cannot be less than zero

There are two clear signs here that we have a multicollinearity issue to deal with:

1. The error message against Urban Road Mileage, which is telling us that something is wholly duplicated within our model
2. The nonsensical negative contributions indicating that they are reversing the effect of something else. (Note: negatives in themselves are not bad if the logic supports that.)

Let's re-run the Regression without the Total Mileage on the basis that it is not independent, being the exact sum of a combination of other drivers as we showed in Table 4.33. We get the result in Table 4.38 instead.

Some very interesting points emerge here (*well, some of us may find them interesting*) when we look at our key statistical measures in relation to the previous invalid model:

Table 4.38 Example Backward Elimination – Step 1 Revisited

SUMMARY OUTPUT

Regression Statistics	
Multiple R	0.998397056
R Square	0.996796682
Adjusted R Square	0.992792534
Standard Error	0.151976722
Observations	10

ANOVA

	df	SS	MS	F	Significance F
Regression	5	28.7488623	5.749772461	248.9410463	0%
Residual	4	0.092387697	0.023096924		
Total	9	28.84125			

	Coefficients	Standard Error	t Stat	P-value	Lower 95%	Upper 95%
Intercept	-0.203694056	0.135885341	-1.499014196	21%	-0.580972246	0.173584135
Number Driving Breaks	0.220146309	0.082207476	2.677935392	6%	-0.008098235	0.448390852
Motorway Mileage	0.017802941	0.001094149	16.27104653	0%	0.014765098	0.020840785
Open A-road Mileage	0.028734934	0.002496996	11.50780329	0%	0.021802163	0.035667705
Open B-road Mileage	0.024245553	0.028806384	0.841672905	45%	-0.055733791	0.104224897
Urban Road Mileage	0.137000813	0.020530249	6.673119784	0%	0.079999703	0.194001924

Simple and Multiple Linear Regression | 199

- R-Square is unchanged, thus underlining that the Total Mileage was not in the least bit independent of the other drivers. The Adjusted R-Square increases because it is based on one less variable's worth of adjustment!
- The F-Statistic is identical for the same reason as above.
- The Coefficient, t-Statistic and associated p-value for the Intercept and the Number of Driving Breaks have not changed because their implied association with the dependent variable though their correlation with Total Mileage is picked up by the sum of the others.
- The Coefficients, t-Statistics and associated p-values for the other drivers have all changed because they were not independent of the Total Mileage. However, if we were to add the coefficient for each driver in Table 4.37 to that for the Total Mileage then we would get the value calculated in Table 4.39. For instance:

Motorway mileage: $0.017802941 = 0.137000813 - 0.119197872$

This is confirming that the negative values in Table 4.35 were compensating for an overstatement by Total Mileage.

Perhaps not surprisingly, based on our observation that there was a high correlation between the Open A-road and Open B-road mileages and a moderate correlation between Open B-road and Urban Road mileages, that there is a good chance (45%) that the contribution to the model made by the Open B-road Mileage variable is zero (i.e. the Null Hypothesis is probably true.)

As a consequence, we would now step back one variable and remove the Open B-road Mileage variable from our model and re-run the Regression, the results of which are shown in Table 4.39.

What do our 'goodness of fit' measures tell us this time?

Table 4.39 Example Backward Elimination – Step 2

SUMMARY OUTPUT

Regression Statistics	
Multiple R	0.998112901
R Square	0.996229364
Adjusted R Square	0.993212855
Standard Error	0.14747872
Observations	10

ANOVA

	df	SS	MS	F	Significance F
Regression	4	28.73250014	7.183125034	330.2590342	0%
Residual	5	0.108749864	0.021749973		
Total	9	28.84125			

	Coefficients	Standard Error	t Stat	P-value	Lower 95%	Upper 95%
Intercept	-0.179560436	0.128894488	-1.393080788	22%	-0.510894266	0.151773395
Number Driving Breaks	0.23420709	0.078109855	2.998431992	3%	0.033419314	0.434994865
Motorway Mileage	0.017316963	0.000901888	19.20078696	0%	0.014998586	0.019635341
Open A-road Mileage	0.030429869	0.001432691	21.23965637	0%	0.026747019	0.034112719
Urban Road Mileage	0.149345016	0.013941142	10.71253856	0%	0.113508171	0.185181861

200 | Simple and Multiple Linear Regression

✓ **R-Square:** Very high value – marginally less than before. Still strong support for a linear relationship

✓ **Adjusted R-Square:** Very high value – significantly better than before giving strong support for a multi-linear relationship

✓ **F-Statistic:** Very high value, even better than before – again, highly unlikely to have occurred by chance or fluke

? **t-Statistics:** 22% chance that the Intercept is really zero – very similar to the previous step

✗ **Common Sense:** Logic would suggest that the negative Intercept is not logical. We would have to drive 10 miles on a motorway or nearly 6 miles on an A-road or more than 1 mile on an Urban Road just to start the clock; either that or begin our trip with a break ... which is somewhat of an oxymoron!

Logically, we would now take another step back and force the Intercept to be zero and re-run the Regression again, giving us the results in Table 4.40.

This gives us the following:

✓ **R-Square:** Very high value – better than before. Strong support for a linear relationship

✓ **Adjusted R-Square:** High value – not as good as before but still giving strong support for a multi-linear relationship

✓ **F-Statistic:** Very high value, but if we recall Section 4.5.4 this cannot be compared with Unconstrained Regressions. However,

Table 4.40 Example Backward Elimination – Step 3 (Intercept = 0)

SUMMARY OUTPUT

Regression Statistics	
Multiple R	0.999549535
R Square	0.999099274
Adjusted R Square	0.831982244
Standard Error	0.158618764
Observations	10

ANOVA

	df	SS	MS	F	Significance F
Regression	4	167.4465405	41.86163513	1663.822785	0%
Residual	6	0.150959473	0.025159912		
Total	10	167.5975			

	Coefficients	Standard Error	t Stat	P-value	Lower 95%	Upper 95%
Intercept	0	#N/A	#N/A	#N/A	#N/A	#N/A
Number Driving Breaks	0.231681057	0.08398737	2.758522591	3%	0.026171366	0.437190748
Motorway Mileage	0.016779615	0.000876821	19.13687325	0%	0.014634111	0.018925119
Open A-road Mileage	0.029863904	0.00147766	20.21027245	0%	0.026248201	0.033479607
Urban Road Mileage	0.137007739	0.011580125	11.83128305	0%	0.108672193	0.165343284

Simple and Multiple Linear Regression | 201

		a big increase when compared with the Simple Linear Regression through the Origin in Table 4.35. Conclusion: virtually no chance of a fluke occurrence
✓	**t-Statistics:**	No issues. Null Hypotheses for all other drivers being zero are rejected
✓	**Common Sense:**	No issues with the logic of the model

We can now use this model with some degree of confidence (95% to be precise because that was our selection criterion.)

4.7.2 Forward Selection

The alternative Stepwise approach of Forward Selection is more of a minimalist approach, starting with the best two independent variables (when considered in isolation of all others) and only adding others if they improve the model by reducing the unexplained variance. Using the same data as in the Backward Elimination example, we begin by examining and ranking all the candidate variables or drivers so that we can trial them in the model in a logical sequence. Let us assume that we have already realised that the total mileage is not independent of the other mileage variables.

We can rank the likely contribution to the model in a couple of ways, and it doesn't really matter which one we do:

- Based on the Coefficient of Determination or R-Square of the independent variable with the dependent variable. We can do this in Excel using the function **RSQ** or using either function **CORREL** or **PEARSON** with the two ranges and squaring it
- Based on the F-Statistic that a Simple Linear Regression would return using either the Data Analysis Wizard, or the advanced composite function:

INDEX(LINEST(*y-range,x-range*, TRUE, TRUE),4,1) *plus Ctrl+Shift+Enter*

The first is the easier, so why over-complicate matters. Table 4.41 summarises the Rankings in question for the example.

The second step is to run a Simple Linear Regression for the highest ranked variable, in this case Motorway Mileage, and note the key statistics (Table 4.42.) The R-Square does not particularly fill us with any great confidence being less than our 0.5 rule of thumb value (from Section 4.5.1.) The significance of the F-statistic and the Slope's t-Statistic's p-value do just get under the 5% wire, so we wouldn't reject the variable out of hand

We can then add the second highest ranked variable, Open A-road Mileage, and re-run the Regression (Table 4.43) to compare our key statistics indicating 'goodness of fit'. In this case we will note that there is evidence to support the acceptance of the

Table 4.41 Ranking of Independent Data's Likely Contribution to the Model for Trip Hours – Step 1

	Number Driving Breaks	Motorway Mileage	Open A-road Mileage	Open B-road Mileage	Urban Road Mileage	Total Mileage
R-square	0.32	0.41	0.38	0.18	0.19	0.88
F	3.70	5.61	4.85	1.79	1.91	58.96
Rank	3	1	2	5	4	Excluded

Table 4.42 Example of Forward Selection – Step 2 (Highest Ranked Variable)

SUMMARY OUTPUT

Regression Statistics	
Multiple R	0.642061383
R Square	0.412242819
Adjusted R Square	0.338773171
Standard Error	1.455663586
Observations	10

ANOVA

	df	SS	MS	F	Significance F
Regression	1	11.8895982	11.8895982	5.611062965	4.53%
Residual	8	16.9516518	2.118956475		
Total	9	28.84125			

	Coefficients	Standard Error	t Stat	P-value	Lower 95%	Upper 95%
Intercept	2.172100327	0.801043787	2.71158751	2.66%	0.324890042	4.019310611
Motorway Mileage	0.016806274	0.007094942	2.368768238	4.53%	0.000445308	0.033167239

Table 4.43 Example of Forward Selection – Step 3 (Two Highest Ranked Variables)

SUMMARY OUTPUT

Regression Statistics	
Multiple R	0.950819974
R Square	0.904058623
Adjusted R Square	0.876646801
Standard Error	0.628725383
Observations	10

ANOVA

	df	SS	MS	F	Significance F
Regression	2	26.07418075	13.03709037	32.9806103	0.03%
Residual	7	2.76706925	0.395295607		
Total	9	28.84125			

	Coefficients	Standard Error	t Stat	P-value	Lower 95%	Upper 95%
Intercept	0.700176527	0.424361498	1.649953001	14.29%	-0.303278962	1.703632017
Motorway Mileage	0.019147103	0.003089239	6.198000648	0.04%	0.011842215	0.026451992
Open A-road Mileage	0.034685943	0.005790369	5.990282243	0.05%	0.020993896	0.048377989

Null Hypothesis that the Intercept is zero. However, although this may be logical in this context for reasons already discussed, as a general rule we should resist the temptation to force the model through the Origin too early as it may be compensating for another, as yet missing, variable. If we have the time, then we can run parallel models (i.e. with and without an intercept) for a period of time to see if they stay consistent.

What do our 'goodness of fit' measures tell us about adding the extra variable?

✓	**R-Square:**	Very high value (giving us that feel good factor)
✓	**Adjusted R-Square:**	Significant increase in value – giving strong support for a multi-linear relationship
✓	**F-Statistic:**	A high value – suggesting that the added variable does make a difference making it worth our while to lose a degree of freedom
?	**t-Statistics:**	14% chance that the Intercept is really zero
?	**Common Sense:**	There is a case that the model should return a zero time for zero miles travelled. For short journeys the model is not valid unless we reject the Intercept

In order not to make a hasty decision with regards to the intercept, we can run two Stepwise Regressions in parallel to see if the pattern continues – it's only a question of ticking a box in Excel and pointing the output to a different cell. In this case we will set the intercept to zero (Table 4.44.)

Step 5 sees the introduction of the next highest ranked variable – in this case the Number of Driving Breaks (Table 4.45.)

Comparing Tables 4.44 and 4.45, we would reject the addition of this new variable on several counts:

Table 4.44 Example of Forward Selection – Step 4 (Constrained Through the Origin)

SUMMARY OUTPUT

Regression Statistics	
Multiple R	0.98846794
R Square	0.977068869
Adjusted R Square	0.849202478
Standard Error	0.693108959
Observations	10

ANOVA

	df	SS	MS	F	Significance F
Regression	2	163.7542998	81.87714988	170.4353558	0.00%
Residual	8	3.843200234	0.480400029		
Total	10	167.5975			

	Coefficients	Standard Error	t Stat	P-value	Lower 95%	Upper 95%
Intercept	0	#N/A	#N/A	#N/A	#N/A	#N/A
Motorway Mileage	0.022894127	0.002308741	9.916279925	0.00%	0.017570159	0.028218094
Open A-road Mileage	0.040217913	0.005204356	7.727740664	0.01%	0.028216647	0.05221918

204 | Simple and Multiple Linear Regression

Table 4.45 Example of Forward Selection – Step 4 (3 Highest Ranked Variables with Intercept=0)

SUMMARY OUTPUT

Regression Statistics	
Multiple R	0.988982023
R Square	0.978085441
Adjusted R Square	0.828966995
Standard Error	0.724354424
Observations	10

ANOVA

	df	SS	MS	F	Significance F
Regression	3	163.9246747	54.64155823	104.1407837	0.00%
Residual	7	3.672825322	0.524689332		
Total	10	167.5975			

	Coefficients	Standard Error	t Stat	P-value	Lower 95%	Upper 95%
Intercept	0	#N/A	#N/A	#N/A	#N/A	#N/A
Motorway Mileage	0.02138024	0.003588834	5.957433862	0.06%	0.012893996	0.029866483
Open A-road Mileage	0.03926865	0.005688359	6.903335355	0.02%	0.025817818	0.052719482
Number Driving Breaks	0.21853663	0.383506331	0.56983839	58.66%	-0.688311741	1.125385001

✓ **R-Square:** Very high value – marginally higher than before. Strong support for a linear relationship

✗ **Adjusted R-Square:** Very high value – but reduced compared to the previous step suggesting that the loss of a degree of freedom may not be worth our while

✗ **F-Statistic:** Very high value, but significantly reduced from the previous step

✗ **t-Statistics:** 59% chance that the contribution of the Number of Driving Breaks is zero

Now we could give up there but that would be premature of us. We should continue and see if the next variable adds any further value to our model. Step 5 sees us trial the variable Urban Road Mileage. Table 4.46 shows that the ends justified the means, with all our key statistics showing improvement.

That leaves us with only one variable left to try – the least ranked in terms of its potential contribution on its own – open B-road mileage. Step 6 is shown in Table 4.47. Based on the key statistics we would reject this final step:

✓ **R-Square:** Very high value – marginally higher than before. Strong support for a linear relationship

✗ **Adjusted R-Square:** High value – but reduced compared to the previous step suggesting that the loss of a degree of freedom may not be worth our while

✗ **F-Statistic:** Very high value, but significantly reduced from the previous step

Simple and Multiple Linear Regression | 205

Table 4.46 Example of Forward Selection – Step 5 (One Step Back One Step Forward)

SUMMARY OUTPUT

Regression Statistics	
Multiple R	0.998977945
R Square	0.997956936
Adjusted R Square	0.85451606
Standard Error	0.221169776
Observations	10

ANOVA

	df	SS	MS	F	Significance F
Regression	3	167.2550875	55.75169584	1139.741932	0.00%
Residual	7	0.342412488	0.04891607		
Total	10	167.5975			

	Coefficients	Standard Error	t Stat	P-value	Lower 95%	Upper 95%
Intercept	0	#N/A	#N/A	#N/A	#N/A	#N/A
Motorway Mileage	0.018398466	0.00090838	20.25414982	0.00%	0.016250489	0.020546444
Open A-road Mileage	0.030899093	0.001992821	15.50520178	0.00%	0.02618682	0.035611366
Urban Road Mileage	0.136585181	0.016145313	8.459741881	0.01%	0.098407582	0.17476278

Table 4.47 Example of Forward Selection – Step 6 (Last Potential Variable Added)

SUMMARY OUTPUT

Regression Statistics	
Multiple R	0.999063242
R Square	0.998127362
Adjusted R Square	0.830524377
Standard Error	0.22870994
Observations	10

ANOVA

	df	SS	MS	F	Significance F
Regression	4	167.2836506	41.82091265	799.5091293	0.00%
Residual	6	0.313849419	0.052308237		
Total	10	167.5975			

	Coefficients	Standard Error	t Stat	P-value	Lower 95%	Upper 95%
Intercept	0	#N/A	#N/A	#N/A	#N/A	#N/A
Motorway Mileage	0.018794418	0.001081428	17.37925954	0.00%	0.016148259	0.021440577
Open A-road Mileage	0.028578348	0.003756325	7.608059951	0.03%	0.019386951	0.037769745
Urban Road Mileage	0.118912767	0.0291667	4.077004495	0.65%	0.047544423	0.190281111
Open B-road Mileage	0.030657834	0.041488151	0.738953984	48.78%	-0.070860013	0.132175681

✗ **t-Statistics:** 49% chance that the contribution of the Open B-road Mileage is zero

In conclusion we would reject step 6 and return to step 5 which would give us our best model and ticks all the right boxes ... *Hang on, isn't that different from the model we created using Backward Elimination approach?*

Well spotted, prompting the question ...

206 | Simple and Multiple Linear Regression

4.7.3 Backward or Forward Selection – Which should we use?

Backward Elimination and Forward Selection approaches to Stepwise Regression will often give the same result, which is quite re-assuring, but as this example shows we can get differences. Where we have alternative end-points we cannot say that either model is wrong, only that one model perhaps better accounts for the variance in the observed values. In effect, at least one of these approaches has reached a 'statistical cul-de-sac' with no means of progressing without backing up!

In Table 4.48 we compare the final position of the two approaches in relation to the example we used.

Table 4.48 Comparison of Stepwise Regression Results by Forward Selection and Backward Elimination

Measure		Backward Elimination	Forward Selection	Comments
R–Square		**0.999**	0.998	Marginally in favour of Backward Model, but both are excellent
Adjusted R–Square		0.832	**0.855**	Forward Model makes better us of the Degrees of Freedom available
F-Statistic (Both Models assume an Intercept of 0)		**1663.8**	1139.7	Backward Model better explains its Variance
Standard Error		**0.159**	0.221	Supports the F-Statistic that the Backward Model better explains the Variance, due largely but not entirely down to the Degrees of Freedom
Residual Sum of Squares		**0.151**	0.342	Demonstrates that the Backward Model better accounts for the Variance in the Observed Data
Degrees of Freedom	Regression	4	3	For Info
	Residuals	`6	7	
t-Statistic p-Values		Worst is c.3%	**All less than 1%**	

Table 4.49 compares the two models in terms of the Predicted Values and the Observed Values and, in reality, there is very little to choose between them in terms of the predicted values for the observed values. However, in using either model as a predictive tool, the Backward Model is more sensitive to the Number of Driving Breaks taken, each break contributing on average some 14 minutes (0.23 hours) to the journey time; in the Forward Model, this is merely implied by the distance travelled, that 'on average' a break would be taken after a set time (which is implied itself by the distance travelled overall). Note that the averages of the Regression Line in this instance do not match the average of the input data because we have rejected he intercept.

There are some advantages and disadvantages of each approach, and these are summarised in Table 4.50.

To get the best of both worlds we can take the truly Stepwise approach and commence with the Backward Elimination, and re-introduce variables eliminated at an earlier stage to test their significance in the context of a smaller number of variables. We can do this also by commencing with a Forward Selection Process.

The approach we take is largely down to a matter of personal preference, and in terms of my personal preference, I would usually go for the Backward Elimination approach. (*Go on, say it – you always knew I was a little bit backward in my thinking. Thanks for the vote of confidence!*)

Table 4.49 Comparison of Regression Residuals Using Forward Selection and Backward Elimination

Obs	Observed Trip Hours	Backward Stepwise Regression	Forward Stepwise Regression	Backward Stepwise Residual	Forward Stepwise Residual
1	1.45	1.42	1.45	0.03	0.00
2	1	1.25	1.07	-0.25	-0.07
3	2.9	3.00	3.06	-0.10	-0.16
4	2.75	2.77	2.68	-0.02	0.07
5	3.1	3.10	3.30	0.00	-0.20
6	4.5	4.37	4.35	0.13	0.15
7	4.7	4.85	5.07	-0.15	-0.37
8	5	4.88	4.95	0.12	0.05
9	6.75	6.63	6.42	0.12	0.33
10	5.1	5.21	5.11	-0.11	-0.01
Sum	37.25	37.485	37.473	-0.235	-0.223
Average	3.725	3.749	3.747	-0.024	-0.022
Sum of Squares	167.598	167.447	167.255	0.151	0.342

208 | Simple and Multiple Linear Regression

Table 4.50 Advantages and Disadvantages of Forward Selection and Backward Elimination Approaches

	Advantages	Disadvantages
Backward Elimination Approach	• Generally converges more quickly to a statistically valid and acceptable model where there are several candidate drivers in the final model • Easier to "spot" the impact of multicollinearity (correlation between candidate variables)	• Increased risk of developing a more complex model where a simpler model is adequate
Forward Selection Approach	• Generally converges more quickly to a statistically valid and acceptable model where there are few relevant drivers amongst a number of potential candidate drivers available	• Generally a more time-consuming procedure to converge to a statistically valid and acceptable model where there are several candidate drivers in the final model • Risk of premature acceptance of a sub-optimal model • Can lead to premature rejection of a variable

4.7.4 Choosing the best model when we are spoilt for choice

Just suppose we have the luxury (*although some might say 'curse'*) of more than one Linear or Multi-Linear Regression Model that has passed all the form, fit and function tests we can throw at it in terms of R-Square, F-Statistic and various t-Statistics, Sensibility and Homoscedasticity (*I still can't say that*), and we want to choose the best of the Best Fits; which do we choose?

There is an argument that if we think that they all have merit then we should use them all to test the sensitivity or robustness of the estimate for which we use it. However, most people would want to know which model is the best, statistically speaking.

For Linear Models, we should use either the R-Square or Adjusted R-Square, depending on the number of independent Variables we have:

• Use R-Square as the differentiator when:

Simple and Multiple Linear Regression | 209

- o We want to compare alternative Simple Linear Models with different independent variables or drivers
- o We want to compare alternative Multi-Linear Models which have the same number of independent variables or drivers as each other. Some of these variables may be common but at least one will be different
- Use Adjusted R-Square as the differentiator when:
  - o We want to choose between alternative Multi-Linear Regression Models, which have a different number of independent variables or drivers. Some of the variables may be common across the models.

In both cases the higher the statistic's value, the better the model fit.

4.8 Chapter review

Let's go back over what we've just explored – *a kind of 'Regression' regression, we might call it!*

We began by defining Simple Linear Regression as a means of determining the line of best fit through paired data points, derived by minimising the square of the differences between the line and the observed data points. We extended this concept to multiple dimensions where each dimension or driver has a linear relationship with the dependent variable that we are trying to estimate. (*Well, they would be linear if all those other pesky variables stayed still!*). This allowed us to derive the concept of Multiple Linear Regression or Multi-linear Regression.

In order to determine whether 'Best Fit' was actually a 'good fit' (*as opposed to being just the best of a bad bunch*) we explored a number of statistical measures to guide us:

- R-Square as a Measure of Linearity
- F-Statistic as a measure of chance occurrence (has the linear relationship occurred by fluke)
- t-Statistics to test the Null Hypotheses that any (or all) of the calculated Coefficients are really no better than zero.
- The Coefficient of Variation can be used as a Measure of Tightness of Fit, or how closely (or otherwise) the data is scattered around the 'Line of Best Fit'.

We turned to White's Test to determine whether the residual values of our linear or multi-linear model was homoscedastic (equal variance) or heteroscedastic (unequal variance) as this was important to eliminate inadvertent bias in our models.

We also discussed that all-important use of 'common sense' in reviewing and interpreting the model's statistics. Common sense is an important element in deciding whether we should or should not force a regression through the origin, and that if we

210 | Simple and Multiple Linear Regression

did, there would be a fundamental change in some of the key statistics to measure the 'goodness of fit'.

Any decision on whether to accept the Null Hypothesis that any Coefficient Value is really zero is done on the basis of the assumed probability distribution of the parameter values around the mean calculated values. This thought led naturally to the generation of uncertainty ranges around the regression line called the Confidence Interval. The uncertainty range for individual points scattered around the regression line is called the Prediction Interval.

Whilst Microsoft Excel provides two principle ways of performing Simple and Multiple Linear Regression through its Data Analysis Add-in, or by using advanced Array Functions, it does not directly support the generation of Confidence or Prediction Intervals. However, so long as we are not too put off by horrendous looking Array Formulae, we can easily derive these for ourselves long-hand in Excel, even though this means sacrificing a Good Practice Spreadsheet Principle we introduced in Volume I.

Finally, we considered a procedure called Stepwise Regression which allows us to select or de-select variables from a model based on their statistical impact. The two basic approaches are Forward Selection where variables are added one at a time, and Backward Elimination where we commence with all candidate variables and progressively delete variables where the Null Hypothesis of the variable's parameter being zero is significant.

I have fond memories of learning regression . . . It takes me back to a former life in my youth. Happy days!

References

Field, A (2005) *Discovering Statistics Using SPSS, 2nd Edition*, London, Sage, p.730 and p. 739.

Kipling, R (1901) *Kim*, London, Macmillan & Co, Ltd.

Nisbett, RE (2009) *Intelligence and How to Get It: Why Schools and Cultures Count*, New York, WW Norton & Company.

Walpole, RE, Myers, RH, Myers, SL & Ye, K (2012) *Probability & Statistics for Engineers and Scientists*, 9th Edition, Boston, Pearson.

White, H (1980) 'A Heteroskedasticity-Consistent Covariance Matrix Estimator and a Direct Test for Heteroskedasticity', *Econometrica*. 48 (4): 817–838.

5 Linear transformation: Making bent lines straight

A straight line is just a curve that never bends!

Estimating would be so much more straightforward if our estimating or forecasting technique was, well, straight forward ... or backward, i.e. just a question of projecting values along a straight line. That way we only need to concern ourselves with one input variable and two constants (an intercept and a slope or gradient) in order to get an output variable:

Output = slope × Input + Intercept

Unfortunately, in reality many estimating relationships are not linear. However, there are some instances (especially where we need to interpolate rather than extrapolate) a linear approximation may be adequate for the purposes of estimating. As George Box pointed out (Box, 1979, p. 202; Box & Draper, 1987, p. 24), just because the model is not right in an absolute sense, it does not mean that it is not useful in helping to achieve our aims.

Even if a linear approximation is not appropriate, it still does not mean that we have to throw in the proverbial towel. There are many estimating relationships that follow generic patterns and can be considered to be part of one of a number of families of curves or functions. Whilst some of these may be distinctly nonlinear in reality, they can often be transformed into linear relationships with just a little touch of a '*mathe-magical sleight of hand*' called a linear transformation (*a case perhaps of unbending the truth!*)

In this chapter, we will be dealing in the main with the transformation of the perfect relationship. In the next chapter, we will be looking at finding the 'best fit' linear transformation where we have data scattered around one of these nonlinear relationships or models.

Whilst there are some very sophisticated linear transformations in the dark and mysterious world of mathematics, the three groups that are probably the most useful to estimators are the relatively simple groups of curves which can be transformed

> **A word (or two) from the wise?**
>
> *"Essentially, all models are wrong, but some are useful."*
>
> **George Box**
> 1919–2013
> Statistician

using logarithms. As we will see later this also includes the special case of reciprocal values. Before we delve into the first group, we are probably better reminding ourselves what logarithms are, and some of their basic properties. (*What was that? Did someone say that they have never taken the logarithm of a value before? Really? Well you're in for a treat – welcome to the wacky weird world of logarithms!*)

5.1 Logarithms

Logarithms – another one of those things we may (*or may not*) have done in school just to pass an exam (*or at least we may have thought that at the time.*) Some of us may have tried to bury the event as a mere bad memory. We probably just know them by their shortened names of logs, and anti-logs when we want to reverse them.

Those of us who were lucky enough to own a slide rule were using logarithms to the base 10. (*Yes, you've spotted it, there's more than one base that can be used for logarithms.*) As some of us may never have seen a slide rule, the principle is illustrated in Figure 5.1 – it's an ingenious device, but now largely obsolete, that converts multiplication into addition. (*From an evolutionary computational aid perspective, a slide rule falls between an abacus and a calculator.*)

Definition 5.1   Logarithm

The Logarithm of any positive value for a given positive Base Number not equal to one is that power to which the Base Number must be raised to get the value in question.

For the Formula-philes: Definition of a logarithm to any given base

Consider any two numbers N and B:
The logarithm of N to the base of B is defined notationally as: $N = \log_B(B^N)$

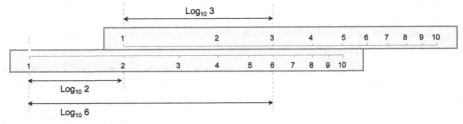

Figure 5.1 Slide Rules Exploit the Power of Logarithms

Linear transformation: Making bent lines straight | 213

Many people may look at the definition of a logarithm, shake their heads and mutter something like, '*What's the point of raising a number to a power of another number, just to get the number you first thought of?*' Well, sometimes it does feel like that, but in truth logarithms are just exploiting the power of powers, so let's take a short detour into the properties of powers ... it should help later!

5.1.1 Basic properties of powers

Let's consider powers from a first principles perspective:

When we take a number raised to a power, we are describing the procedure of taking a number of instances (equal to the power) of a single number, and multiplying them together. For instance:

Ten to the power of 3:	10^3	=	10 x 10 x 10	=	1000
Two to the power of 5:	2^5	=	2 x 2 x 2 x 2 x 2	=	32
Four to the power of 2:	4^2	=	4 x 4	=	16

This gives us an insight into one of the basic properties of powers:

In the special case of powers of ten, the value of the power tells us how many zeros there are in the product following the leading digit of one. By implication, ten raised to the power of zero has no zeros following the digit one. We can conclude that ten raised to the power of zero is one.

In the case of two values that are the powers of the same number, when we multiply them together, this is equivalent to raising that number to the power of the sum of the individual powers. This is known as the '**Additive Property of Powers**'.

By implication, any value raised to the power of zero takes the value of one. If we multiply a number by one, we leave it unaltered. This is the same as adding zero to a power; it leaves it unaltered. (*So, it is not just for power of ten that this works.*)

If we extend this thinking further to square roots, the definition of a square root of a number is that which when multiplied by itself gives the original number. By implication, this means that the square root of a number can be expressed by raising a number to a power of a half, i.e. when multiplied by itself, this is equivalent to doubling the power of a half to get one, and any number raised to the power of 1 is simply the number itself.

The other major property that we can derive from our example is the '**Multiplicative Property of Powers**'. Consider a value that is the power of a number. If we then raise that value to a power then the result would be another value which can be expressed as the original number raised to the power of the product of the two constituent powers.

214 | Linear transformation: Making bent lines straight

For the Formula-phobes: Example of the Additive Property of Powers

This is one area where we cannot get away totally from using formulae, but we can avoid symbolic algebra.

Let's consider powers of 2:

2^1	2^2	2^3	2^4	2^5	2^6
2	4	8	16	32	64

The Additive Property of Powers:

The value 32 can be expressed as the product of other numbers:

$$32 = 16 \times 2 = 8 \times 4$$
$$2^5 = 2^4 \times 2^1 = 2^3 \times 2^2$$
$$2^5 = 2^{4+1} = 2^{3+2}$$

The value 2 can be expressed as the product of its square root multiplied by itself:

$$2^1 = 2^{0.5+0.5} = 2^{0.5} \times 2^{0.5}$$
$$2 = \sqrt{2} \times \sqrt{2}$$

For the Formula-phobes: Example of the Multiplicative Property of Powers

Again, let's consider powers of 2:

2^1	2^2	2^3	2^4	2^5	2^6
2	4	8	16	32	64

The Multiplicative Property of Powers:

The value 64 can be expressed in a number of ways:

$$64 = 4 \times 4 \times 4 = 8 \times 8$$
$$2^6 = 2^2 \times 2^2 \times 2^2 = 2^3 \times 2^3$$

This is the same as saying two squared cubed, or two cubed squared

$$2^6 = (2^2)^3 = (2^3)^2$$

Linear transformation: Making bent lines straight | 215

For the Formula-philes: The Additive and Multiplicative Properties of Powers

Consider n, p and q which are all real numbers:

The Additive Property of Powers $\qquad n^p n^q = n^{p+q}$

The Multiplicative Property of Powers $\qquad \left(n^p\right)^q = n^{pq}$

The flipside to these properties is that their inverses relate to the properties of subtraction and division. If we take a number raised to a power and divide it by that same number raised to another power, this can be simplified by subtracting the latter power from the former. This implies (correctly) that a negative power is equivalent to a 'real division'.

Also, raising a value to a fractional power is equivalent to taking the appropriate root of the value.

For the Formula-phobes: Examples of subtracting and dividing powers

Again, using powers of 2:

2^1	2^2	2^3	2^4	2^5	2^6
2	4	8	16	32	64

Power subtraction

We can express 32 divided by 8 in powers, and subtract the powers to get the result:

$$32 \div 8 = 2^5 \div 2^3 = 2^{5-3} = 2^2 = 4$$

Fractional or decimal powers

Also, raising a value to a fractional power is equivalent to taking the appropriate root of the value. For instance, using the multiplicative property of powers:

$$8 = 8^{\frac{1}{3} \times 3} = \left(8^{\frac{1}{3}}\right)^3 = \left(\sqrt[3]{8}\right)^3 = 2^3$$

Or to put it another way:

$$8 = 2^3 \implies 2 = \left(2^3\right)^{\frac{1}{3}}$$

We can extend the principle to integer and decimal combinations. We can use a calculator to check the following example:

$$4^{2.5} = 4^{2+0.5} = 4^2 \times 4^{0.5} = 4^2 \times 4^{\frac{1}{2}} = 4^2 \times \sqrt{4} = 16 \times 2 = 32$$

216 | Linear transformation: Making bent lines straight

So where does all this leave us? Well, if only we could express all multiplication and division calculations in terms of the powers of a common number, we could just use addition and subtraction operators. Oh, guess what! That's where logarithms come to the fore, courtesy of mathematician John Napier in 1614.

5.1.2 Basic properties of logarithms

The word 'logarithm' is a bit of a mouthful to say every time (even to ourselves), so we normally just say 'log' or to be correct we should say 'log to the base n' where n is whatever number is being used as the basis. Notationally, we should use Log_n. However, in cases where we omit mentioning the base, we are usually implying that we are using Common Logs, although some people (such as Engineers) prefer to use Natural Logs:

- Common Logs are logs to the base 10, usually abbreviated to **Log** or Log_{10}
- Natural Logs, sometimes known as Naperian Logs, and usually abbreviated to **Ln**, are logs to the base e, where e is the 'Exponential Number' or 'Euler's Number' and is approximately 2.71828182845905.

Caveat augur

Even though 10 is considered to be a Natural Number, and is the base for counting in many cultures, Natural Logs are *not* logs to the base 10. Natural Logs are those logs based on Euler's Number, e.

Euler's Number, e, does occur in nature and not just in the minds or virtual realities of mathematicians and other scientists, but it's choice as the base for Natural Logs may be construed as an oxymoron, as there appears to be nothing 'natural' about it to the lay person. Even in mathematics a 'natural number' is defined as one of the set of positive integers used in counting, not some irrational number such as e. On that basis, we could argue that the more obvious choice for the base of Natural Logs would have been the number ten – we have ten digits across our two hands, and ten more across our two feet. *However, we are where we are; we won't change centuries of convention by having a whinge here!*

As the constant e is transcendental, with decimal places that go on infinitely and cannot be substituted by a fraction, mathematicians would call it an 'irrational' number. Many non-mathematicians would probably agree.

In truth, it the majority of cases it does not make any difference what number base we choose to use as our preference for logarithms, so long as we are consistent in what

Linear transformation: Making bent lines straight | 217

we use within a single relationship or equation. Until more recent versions were released, the logarithmic graph scale in Microsoft Excel assumed logs to the base 10; now we can use whatever base we want. The beauty of logarithms is that their properties transcend the choice of base and can be related back to the properties of powers:

The Additive Property of Logs: The log of the product of two numbers is equal to the sum of the logs of the two individual numbers

The Multiplicative Property of Logs: The log of a number raised to a power is equal to that power being multiplied by the log of the number

However, the first property that we should recognise is based on the definition of a log, i.e. the log of a value is that power to which the Base Number must be raised to get the value in question. If we were to raise the Base Number to the power of one, we would simply get the Base Number – nothing would change. Therefore, for any Base Number, the log of the Base Number to that base is simply, and invariably, 1.

For the Formula-philes: The Additive Property of Logs

Consider n, b, p and q which are all real numbers such that:

$$m = b^p$$
$$n = b^q \tag{1}$$

Taking the product of m and n from (1)

$$mn = b^p b^q \tag{2}$$

Applying the Additive Property of Powers to (2):

$$mn = b^{p+q} \tag{3}$$

By the definition of a log , p and q are:

$$p = \log_b\left(b^p\right)$$
$$q = \log_b\left(b^q\right) \tag{4}$$

Similarly, $p+q$ can be expressed as:

$$p + q = \log_b\left(b^{p+q}\right) \tag{5}$$

Substituting (1) and (2) in (4) and (5):

$$p = \log_b m \tag{6}$$
$$q = \log_b n \tag{7}$$
$$p + q = \log_b\left(mn\right) \tag{8}$$

From (6), (7) and (8) we get:

$$\log_b m + \log_b n = \log_b\left(mn\right)$$

. . . which is known as the **Additive Property of Logs** and is true for any positive logarithmic base not equal to one

218 | Linear transformation: Making bent lines straight

Table 5.1 Example of the Additive Property of Logarithms

Row	x	y	xy	$Log_{10} x$	$Log_{10} y$	$Log_{10} (xy)$
1	2	3	6	0.301	0.477	0.778
2	2	5	10	0.301	0.699	1.000
3	4	15	60	0.602	1.176	1.778
4	10	10	100	1.000	1.000	2.000
5	20	30	600	1.301	1.477	2.778
6	6	100	600	0.778	2.000	2.778

Table 5.1 illustrates an example of the use of the Additive Property:

- The left-hand set of columns calculate the products of x and y; the right-hand set of columns show the Common Log values for x, y and xy; the last of these columns is the sum of the previous two columns.
- Similarly, on the left-hand side the third row is the product of the first and second rows. In terms of corresponding log values, the third row is the sum of the previous two rows.
- Rows 1, 3 and 5 show that as the product of x and y increases by a factor of ten, the log values of 6, 60 and 600 all increase by 1, which is the Common Log of 10. Row 4 shows that the Common Log of 100 is 2.

We can see from this that multiplication operations in terms of 'real space numbers' are replaced by addition operations in 'log space'. The inverse of a multiplication operation in 'real space' is a division operation; the equivalent inverse operation in 'log space' is subtraction.

However, we do have multiplication and division operations with logarithms; these are equivalent to powers and roots in 'real space'.

For the Formula-philes: The Multiplicative Property of Logs

Consider n, b, p and q which are all real numbers such that:

$$m = b^p$$
$$n = m^q \tag{1}$$

Applying the Multiplicative Property of Powers to (1)

$$n = \left(b^p\right)^q$$
$$n = b^{pq} \tag{2}$$

But, by the definition of a log:

$$p = \log_b\left(b^p\right)$$

$$pq = \log_b\left(b^{pq}\right) \tag{3}$$

Substituting (1) and (2) in (3)

$$p = \log_b m$$

$$pq = \log_b n \tag{4}$$

From (4) and (1), eliminating p:

$$\log_b\left(m^q\right) = q\log_b m \tag{5}$$

... which is known as the **Multiplicative Property of Logs** and is true for any positive logarithmic base not equal to one

Table 5.2 Example of the Multiplicative Property of Logarithms

Row	x	n	x^n	$Log_{10}\ x$	$Log_{10}\ (x^n)$
1	2	3	8	0.301	0.903
2	4	3	64	0.602	1.806
3	3	2	9	0.477	0.954
4	10	2	100	1.000	2.000
5	16	0.25	2	1.204	0.301
6	4	0.5	2	0.602	0.301

Table 5.2 illustrates the Multiplicative Property of Logs:

- The left-hand set of columns are examples of the calculation that raise **x** to the power of **n**
- The right hand two columns are the logs of **x** and **x^n**. In all cases we can arrive at the log of the latter by multiplying the former by the power, **n**

There is one other noteworthy property of logs and that is the **Reciprocal Property**, which allows us to switch easily and effortlessly between different bases.

The log of one number to a given base of another number is equal to the reciprocal of the log of the second number to the base of the first number.

Some examples of the Reciprocal Property are provided in Table 5.3 in which for any pair of values for b and c, we calculate the log of one value to the base of the other value. If we multiply these two logs together, we always get the value 1.

220 | Linear transformation: Making bent lines straight

Table 5.3 Example of the Reciprocal Property of Logarithms

Row	b	c	$\log_b c$	$\log_c b$	Product
1	0.25	0.8	0.161	6.213	1.000
2	0.5	1.25	-0.322	-3.106	1.000
3	2	3	1.585	0.631	1.000
4	2	4	2.000	0.500	1.000
5	2.5	8	2.269	0.441	1.000
6	3	5	1.465	0.683	1.000

Now that's what I call a bit of mathe-magic! I can feel a cold shower coming on to quell my excitement.

For the Formula-philes: The Reciprocal Property of Logs

Consider b, c, p and q which are all real number s such that:

$$b^p = c^q \tag{1}$$

Taking logs to the base b and base c:

$$\log_b \left(b^p \right) = \log_b \left(c^q \right)$$

$$\log_c \left(b^p \right) = \log_c \left(c^q \right) \tag{2}$$

Applying the Multiplicative Property of Logs to (2):

$$p \log_b b = q \log_b c$$

$$p \log_c b = q \log_c c \tag{3}$$

Substituting and, in (3):

$$p = q \log_b c$$

$$p \log_c b = q \tag{4}$$

Eliminating p and q from (4)

$$\log_c b = \frac{1}{\log_b c}$$

... which is known as the **Reciprocal Property of Logs** and is true for any pair of positive logarithmic bases not equal to one

The good news is that we don't have to start using Log Tables out of a reference book, or learn how to use a slide rule, Microsoft Excel is log friendly, providing three different complementary (and complimentary) functions as standards:

LOG(*number, base*) We can specify any base so long as the value is positive and not equal to 1. The base is an optional parameter; if it is omitted it defaults to 10

LOG10(*number*) We might be forgiven for thinking that this is a superfluous function given the above. All we can say is that it provides an option for improved transparency in a calculation

Linear transformation: Making bent lines straight | 221

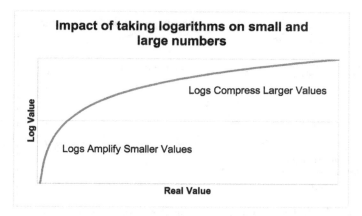

Figure 5.2 Impact of Taking Logarithms on Small and Large Numbers

For the Formula-phobes: Logarithms are a bit like vision perspective

Think of looking at a row of equally spaced lamp-posts on a straight road. The nearest to you will appear quite far apart, but the ones furthest away, nearer the horizon will seem very close together. Similarly, with speed: distant objectives appear to travel more slowly than those closer to us travel more slowly than those closer to us

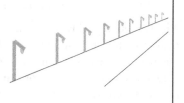

For the Formula-phobes: How can logs turn a curve to a line?

In effect that is exactly how they work . . . by creating a 'Turning Effect'. This is achieved as the natural outcome of the compression of large value and amplification of smaller values.

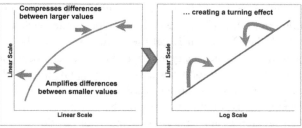

(Continued)

If there is already a curve in the direction of the turning effect, taking logarithms will emphasise the curvature.

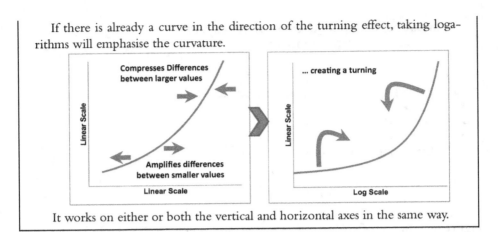

It works on either or both the vertical and horizontal axes in the same way.

LN(*number*)   This function returns the Natural or Naperian Log of a number. Again, given the first function, this option is not strictly speaking necessary but without it, we would have to use another function embedded within the first to generate Euler's Number **LOG(*number*, EXP(1))** for the ***base***, which is not a particularly neat way of doing things.

The last and arguably the most important property of logarithms (to any base) for estimators is that they amplify smaller numbers in a range and compress larger numbers relative to their position in the range, as illustrated in Figure 5.2 – as the real value gets bigger, the rate of change of the log value diminishes. It is this property that we will exploit when we consider linear transformations in the next section.

5.2 Basic linear transformation: Four Standard Function types

OK, now we can get to the nub of why we have been looking at the property of logarithms. As we discussed in Volume I Chapter 2, there are a number of estimating relationships that are power based rather than linear. Typical examples include Learning Curves, Cost-weight relationships, Chiltern Law, etc.

There are three basic groups of functions that we can transform into straight lines using logs depending on whether we are taking the log of the vertical y-axis, the horizontal x-axis, or both. The Microsoft Excel users amongst us may be familiar with the graph facility where we can define the type of Trendline (Linear, Logarithmic, Exponential and Power) that we may want to fit through some data. The names relate to the basic groups or Function Types we will consider initially.

Note that in all instances that follow, the specific shape and direction of curvature of the transformations depend on the parameter values of the input functions (e.g. slope and intercept), and that logarithmic values can only be calculated for positive values of x or y. (*Strictly*

Linear transformation: Making bent lines straight 223

speaking, we can take logs of negative numbers, but this needs us to delve into the weird and wonderful world of Complex and Imaginary Numbers – and you probably wouldn't thank me for going there!)

> ### Definition 5.2   Linear Function
>
> A Linear Function of two variables is one which can be represented as a monotonic increasing or decreasing straight line without any need for mathematical transformation.

5.2.1 *Linear functions*

Just to emphasise a point and to set a benchmark for the other Function Types, we'll first look at what happens if we try to transform something using logs when it is already a simple linear function. (*We just wouldn't bother in reality, but it may help to understand the dynamics involved.*)

The 2x2 array of graphs in Figure 5.3 show what happens if we were to calculate the log values for either or both the x and y axes. And plot the four possible combinations:

Transformation graph	*Characteristics*
Linear x by Linear y	The input function is already a straight line, so no transformation is required.
Log x by Linear y	If we take the log of the x-value we will create a curve because we have 'stretched' the distance between low values and 'compressed' the distance between high values relative to each other. The degree of curvature will depend on the value of the intercept used in the input function (bottom left quadrant) relative to its slope, and the direction of the input slope.
Linear x by Log y	If we take the log of the y-value we will create a curve because we have 'stretched' the distance between low values and 'compressed' the distance between high values relative to each other. The degree of curvature again will depend on the value of the intercept used in the input function (bottom left quadrant) relative to its slope.
Log x by Log y	If we take the log of both the x-value and the y-value we will usually create a curve because of the 'stretching' and 'compressing' property inherent in taking log values. The degree and direction of the curvature will depend largely on the value of the intercept used in the input function (bottom left quadrant). In the special case of a zero intercept, we will get a Log-Log transformation that is also linear. So the smaller the intercept relative to the slope, the straighter the Log-Log Curve will appear

Linear transformation: Making bent lines straight

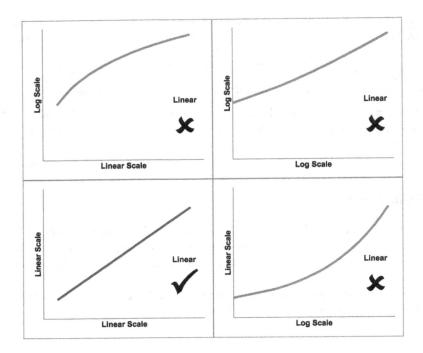

Figure 5.3 Taking Logarithms of a Linear Function

Unsurprisingly, the family name under which we classify straight lines is **Linear Function**. As a general rule, taking the log of a Linear Function distorts its shape into a curve, so we would not apply a transformation in this case. The only exception would be in the case where the straight line passes through the origin, i.e. the intercept equals zero. In this case alone, the straight line remains a straight line if we transform both the x and y-values (*but only for positive values, as we cannot take the log of zero or a negative value.*)

In terms of the special case of a zero intercept, in the majority of circumstances, we just wouldn't bother making the transformation, even though we could (*why create work for ourselves?*) However, there is the exception where the relationship is just part of a more

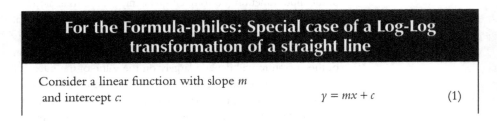

Taking the log of both sides (any base):	$\log y = \log(mx + c)$ (2)
If $c = 0$, (2) becomes:	$\log y = \log(mx)$ (3)
Expanding (4) using the Additive Property of Logs:	$\log y = \log m + \log x$ (4)
Let X and Y be the log values of x and y:	$Y = \log y$
	$X = \log x$ (5)
Substituting (5) in (4):	$Y = \log m + X$
... which is a linear equation with intercept of Log m and slope of 1	

complex model involving multiple drivers. We will discuss this at the appropriate time in Chapter 7.)

We will find that Linear Functions can increase or decrease as illustrated in Figure 5.4 in which we might consider the time from departure or to arrival if we travel at a constant or average speed:

5.2.2 Logarithmic Functions

Despite its generic sounding name (implying that it may cover all the transformations using logarithms that we are considering), a Logarithmic Function is actually a specific

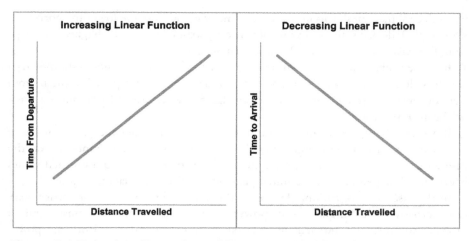

Figure 5.4 Examples of Increasing and Decreasing Linear Functions

Linear transformation: Making bent lines straight

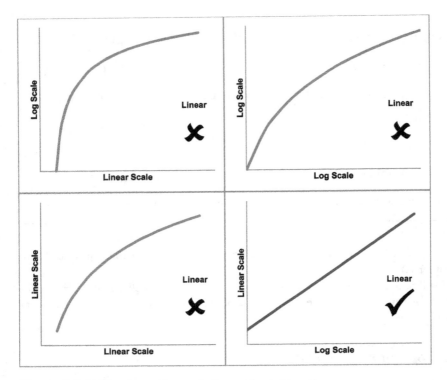

Figure 5.5 Taking Logarithms of a Logarithmic Function

type of curve that transforms into a straight line when we take the log of the horizontal x-value plotted against the real value of vertical y-axis. Figure 5.5 highlights what happens if we take the log of either the x or y variable, or both.

In our example, the bottom left quadrant again depicts the raw untransformed data which is clearly nonlinear. If our 'Log x by Linear y' graph, as depicted in the bottom right quadrant, is a straight line then we can classify the raw data relationship as a **Logarithmic Function**.

If we take the upper quadrants with the log of y with either the Linear or log of x, it will still leave us with a curve as shown in the top two quadrants. As was the case with our Linear Function, the degree of curvature in these transformed values is very much dependent on the initial parameters of the raw data function.

In the case of a Logarithmic Function, the dependent variable, y or vertical axis, can alternatively be described as the power to which a given constant must be raised in order to equal the independent variable, x or horizontal axis. Both the power and constant parameters of the function must be positive values if we are to take logarithms.

Linear transformation: Making bent lines straight | 227

Definition 5.3 Logarithmic Function

A Logarithmic Function of two variables is one in which the dependent variable on the vertical axis produces a monotonic increasing or decreasing straight line, when plotted against the Logarithm of the independent variable on the horizontal axis.

For the Formula-philes: Linear transformation of a Logarithmic Function

Consider a function with positive constants m and c:

$$c^y = mx \qquad (1)$$

Taking the log to the base b of both sides:

$$\log_b\left(c^y\right) = \log_b\left(mx\right) \qquad (2)$$

Expanding (2) using the Additive and Multiplicative Property of Logs:

$$y \log_b c = \log_b x + \log_b m \qquad (3)$$

Rearranging (3):

$$y = \frac{\log_b x + \log_b m}{\log_b c} \qquad (4)$$

Let X be the log value of x to the base, b:

$$X = \log_b x \qquad (5)$$

Substituting (5) in (4):

$$y = \frac{X}{\log_b c} + \frac{\log_b m}{\log_b c}$$

... which is a linear equation for y with an intercept of $\dfrac{\log_b m}{\log_b c}$ and a slope of $\dfrac{1}{\log_b c}$

Note: In most cases it is probably convenient to use logs to the base of 10 or e

Logarithmic Graphs can express positive or negative trends yielding convex or concave curves, as illustrated in Figure 5.6. For a decreasing trend, the power parameter is greater than zero but less than one; for an increasing trend the power parameter is greater than one. The function cannot be transformed if the power parameter is one, zero or negative. The intercept must also always be a positive value.

Concave Curve: one in which the direction of curvature appears to bend towards a viewpoint on the x-axis, similar to being on the inside of a circle or sphere

Convex Curve: one in which the direction of curvature appears to bend away from a viewpoint on the x-axis, similar to being on the outside of a circle or sphere

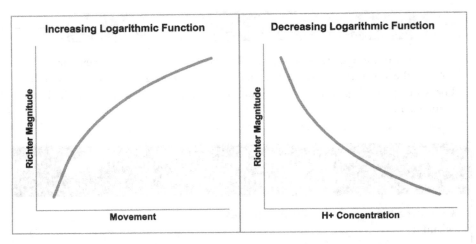

Figure 5.6 Examples of Increasing and Decreasing Logarithmic Function

When we come to the generalised version of these standard functional forms, we will see that a simple vertical shift can change a concave function into a convex one and vice versa.

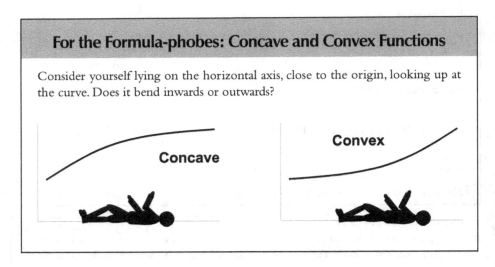

The **Richter Magnitude Scale** (usually abbreviated to 'Richter Scale') is used to measure the intensity of earthquakes and is an example of a Logarithmic Function. Each whole unit of measure on the Richter Scale is equivalent to a tenfold increase in intensity, i.e. an earthquake of magnitude 4.3 is ten times more powerful than one of magnitude 3.3. (The vertical y-axis is the Richter Magnitude and the horizontal x-axis is the 'Amplified Maximum Ground Motion', as measured on a seismograph a specified distance from the earthquake's epicentre.)

On the other hand (*the right one in this case*), **pH** is a Logarithmic Function with a decreasing rate. pH is a measure of acidity or alkalinity of an aqueous solution that some of us may have tested using litmus paper in school; it is actually measuring the concentration of hydrogen ions which are present in a liquid. Another well-known Logarithmic Function is the decibel scale used as a measure of sound intensity or power level of an electrical signal.

With decreasing Logarithmic Functions, there is a likelihood that the output y-value may be negative for extreme higher values of x. We need to ask ourselves whether this passes the 'sense check' that all estimators should apply – the C in TRACE (Transparent, Repeatable, Appropriate, Credible and Experientially based, see Volume I Chapter 3).

Caveat augur

It is quite commonplace to depict a date along the horizontal x-axis. However, we should never take the log of a date! It is meaningless. Date is a relative term, a label, but in reality, it is transcendental as we do not know when Day 1 occurred, and any attempts to define a specific start date is inherently arbitrary.

Microsoft Excel makes the assumption that Day 1 was 1st January 1900, but this is more a matter of convenience. We can take the log of an elapsed time, duration or time from a start point, but not a date. Just because we can, does not mean we should! (Dates can be used as an indicator of technology used.)

Having determined that a curve falls into the genre of a Logarithmic Function, we will probably conclude that the transformed linear form is the easier to use in interpolating or extrapolating some estimated value because its output will give us an answer for the 'thing' we are trying to estimate (usually the vertical y-scale). However, there may be occasions when we want to 'reverse engineer' an estimate insomuch that we want to know what value of a driver (or x-value) would give us a particular value for the thing we want to estimate (the y-value). For that we need to know how to reverse the transformation back to a real number state. In simplistic terms, we need to raise the answer to the power of the log base we used. (We will look more closely at examples of these in Chapter 7.)

We could simply use the Logarithmic Trendline option in Microsoft Excel and select the option to display the equation on the graph. For those of us who are uncomfortable with using Natural or Naperian Logs, you'll be disappointed to hear that is how Excel displays the equation (i.e. there is no option for Common Logs.)

For the Formula-philes: Reversing a linear transformation of a Logarithmic Function

Suppose our transformed curve has a linear equation with constants M and C:	$y = MX + C$	(1)
Where X is the log to the base b of x:	$X = \log_b x$	(2)
Substituting (2) in (1):	$y = M \log_b x + C$	(3)
Dividing through by M:	$\dfrac{y}{M} = \log_b x + \dfrac{C}{M}$	(4)
Applying both sides of the equation as equal powers of b:	$b^{y/M} = b^{\log_b x} b^{C/M}$	(5)
Substituting for x using the definition of a log:	$\left(b^{1/M}\right)^y = \left(b^{C/M}\right)x$	(6)
Simplifying (6) with alternative constants, m and c:	$m^y = cx$	
... which is a Logarithmic Function with constants	$m = \left(b^{1/M}\right)$ and $c = \left(b^{C/M}\right)$	

Note: In most cases, it is probably convenient to use logs to the base of 10 or e

5.2.3 Exponential Functions

In Figure 5.7, we can do exactly the same process of taking the log of either or both of the axes for another group of curves called **Exponential Functions**. These are those

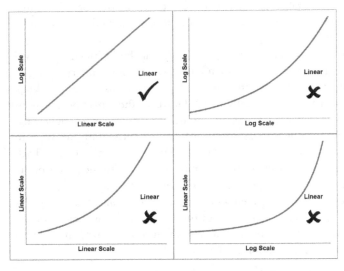

Figure 5.7 Taking Logarithms of an Exponential Function

Linear transformation: Making bent lines straight 231

relationships which gives rise to expressions such as 'exponential growth' or 'exponential decay'. Commonly used Exponential Functions in cost estimating, economics, banking and finance are 'steady state' rates of escalation or inflation, e.g. 2.5% inflation per year, discounted cash flow and compound interest rates.

Exponential Functions are those that produce a straight line when we plot the log of the vertical y-scale against the linear of the horizontal x-scale.

> ## Definition 5.4 Exponential Function
>
> An Exponential Function of two variables is one in which the Logarithm of the dependent variable on the vertical axis produces a monotonic increasing or decreasing straight line when plotted against the independent variable on the horizontal axis.

For the Formula-philes: Linear transformation of an Exponential Function

Consider a function with positive constants m and c:

$$y = cm^x \qquad (1)$$

Taking the log to the base, b of both sides:

$$\log_b y = \log_b \left(cm^x \right) \qquad (2)$$

Expanding (2) using the Additive Property of Logs:

$$\log_b y = \log_b c + \log_b \left(m^x \right) \qquad (3)$$

Expanding (3) using the Multiplicative Property of Logs

$$\log_b y = \log_b c + x \log_b m \qquad (4)$$

Let Y be the log to the base b of y:

$$Y = \log_b y \qquad (5)$$

Substituting (5) in (4):

$$Y = x \log_b m + \log_b c \qquad (6)$$

... which is a linear equation with an intercept of $\log_b c$ and a slope of $\log_b m$

Note: In the majority of cases it is probably convenient to use logs to the base of 10 or e.

In common with Logarithmic Functions, the two parameters of the Exponential Function must both be positive otherwise it becomes impossible to transform the data, as we cannot take the log of zero or a negative number. (*I'm beginning to sound like a parrot!*) This may lead us to think '*What about negative inflation?*' This is not a problem when inflation is expressed as a factor relative to the steady state, so 3% escalation will be expressed as a recurring factor of 1.03; consequently 2% deflation would be expressed as a recurring factor of 0.98.

232 | Linear transformation: Making bent lines straight

For any Exponential Function, a recurring factor:

- greater than zero but less than one, will give a monotonic decreasing function
- greater than one will result in a monotonic increasing function
- equal to one gives a flat line constant

Examples of increasing (positive slope) and decreasing (negative slope) Exponential Functions are shown in Figure 5.8. Steady-state Inflation or Economic Escalation is probably the most frequently cited example of an Increasing Exponential Function. Radioactive Decay is an example of a decreasing Exponential Function, as are discounted Present Values (see Volume I Chapter 6). Interestingly all Exponential Functions are Convex in shape.

There may be occasions when we want to estimate a date based on some other parameters. If we follow the normal convention then the date (as the entity we want to estimate) will be plotted as the vertical y-axis (rather than the horizontal x-axis which we would use if date was being used as a driver. In this case, we can read across the same caveat for dates that we stated for Logarithmic Functions, for the same reason:

Never take the log of a date!

As we will see from the example in Figure 5.8, we can use date in an Exponential Function so long as it is the x-scale or independent input value, but not if it is the y-scale output value.

Where we have used the linear form of an Exponential Function to create an estimate we will need to reverse the transformation to get back to a real space value for the number. To do this we simply take the base of logarithm used and raise it to the power of the value we created for the transformed y-scale.

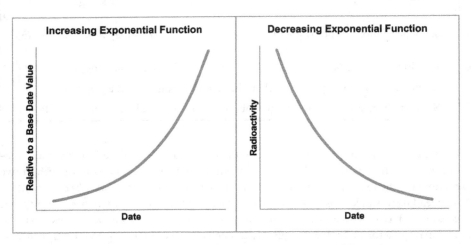

Figure 5.8 Examples of Increasing and Decreasing Exponential Functions

Linear transformation: Making bent lines straight | 233

For example, if we have found that the best fit straight line using Common Logs is:

$$\log_{10} y = 2x + 3$$
$$\text{Then, } y = 10^{2x+3}$$
$$\text{or, } y = 1000 \text{ x } 10^{2x}$$
$$\text{or, } y = 1000 \text{ x } 100^{x}$$

Sorry, we just cannot avoid the formula here.

For the Formula-philes: Reversing linear transformation of an Exponential Function

Consider a linear function with constants M and C: $\qquad Y = Mx + C \qquad$ (1)

Where Y is the log to the base, b of y: $\qquad Y = \log_b y \qquad$ (2)

Substituting (2) in (1): $\qquad \log_b y = Mx + C \qquad$ (3)

Applying both sides of the equation as equal powers of b: $\quad b^{\log_b y} = b^{Mx+C} \quad$ (4)

Substituting for y using the definition of a log: $\qquad y = \left(b^M\right)^x b^C \qquad$ (5)

Simplifying (5) with alternative constants, m and c: $\qquad y = cm^x$

...which is an Exponential Function with constants $m = b^M$ and $c = b^C$ (*Here, the power C is upper case. It is not easy to differentiate between c and C.*)

Note: In most cases, it is probably convenient to use logs to the base of 10 or Euler's Number, e

5.2.4 Power Functions

Definition 5.5 Power Function

A Power Function of two variables is one in which the Logarithm of the dependent variable on the vertical axis produces a monotonic increasing or decreasing straight line when plotted against the Logarithm of the independent variable on the horizontal axis.

In Figure 5.9, we can repeat process of taking the log of either or both of the axes for final group of curves in this section called **Power Functions.** These are those curves which mathe-magically transform into straight lines when we take the log of both the horizontal x-axis and the vertical y-axis, as illustrated in the top-right quadrant. Taking

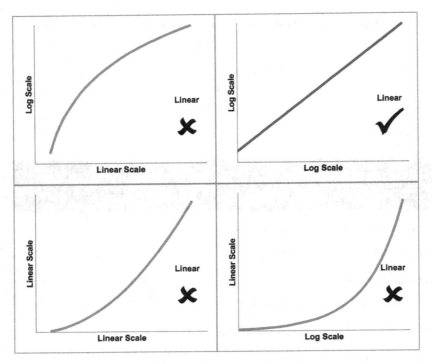

Figure 5.9 Taking Logarithms of a Power Function

the log of only one of these exacerbates the problem of the curvature, although the precise degree is down to the parameters involved.

For the Formula-philes: Linear transformation of a power function

Consider a function with constants m and c:	$y = cx^m$	(1)
Taking the log to the base, b of both sides:	$\log_b y = \log_b\left(cx^m\right)$	(2)
Expanding (2) using the Additive Property of Logs:	$\log_b y = \log_b c + \log_b\left(x^m\right)$	(3)
Expanding (3) using the Multiplicative Property of Logs:	$\log_b y = \log_b c + m \log_b x$	(4)
Let X be the log to the base b of x and Y be the log to the base b of y:	$X = \log_b x$ $Y = \log_b y$	(5)

Substituting (5) in (4): $Y = mX + \log_b c$

...which is a linear equation with an intercept of $\log_b c$ and a slope of
Note: In the majority of cases it is probably convenient to use logs to the base of 10 or e.

For exactly the same reasons as we discussed under Logarithmic and Exponential Functions:

Never take the log of a date!

(*Yes, I know what you're thinking – that I'm beginning to turn into a bit of an old nag – well, just be a little more charitable and think of it as 'positive reinforcement'.*)

There are many examples of Power Functions being used in estimating relationships such as:

- Chilton's Law (Turré, 2006) is used in the petrochemical industry to relate cost with size (). This is represented in the left-hand graph of Figure 5.10 by the increasing slope.
- Learning Curves or Cost Improvement Curves are used in the manufacturing industry to relate the reduction in effort or cost or duration observed as the cumulative number of units produced increases, as illustrated in the right-hand graph of Figure 5.10 by the decreasing slope.

The Power Function is the most adaptable of the set of logarithmic-based transformations depending on the value of the Power/Slope parameter; the Power parameter in real space is equivalent to the slope parameter in log space. Figure 5.11 illustrates the range of different shapes created by the different parameter values:

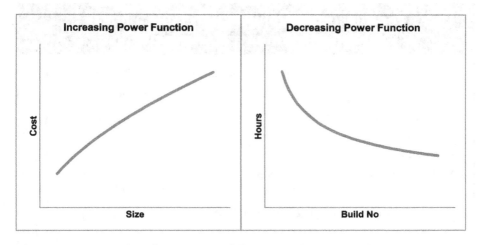

Figure 5.10 Examples of Increasing and Decreasing Power Functions

Linear transformation: Making bent lines straight

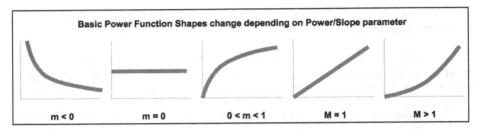

Figure 5.11 Basic Power Function Shapes Change Depending on Power/Slope Parameter

- Less than zero: Convex and decreasing like a playground slide
- Equal to zero: Returns a constant value
- Greater than zero but less than one: Concave and increasing (up then right)
- Equal to one: Increasing straight line (i.e. a Linear Function)
- Greater than one: Convex and increasing (right then up)

Where the Power/Slope parameter equals -1, we have the special case Power Function that we normally call the Reciprocal.

In order to use the output of a Power Function that has been transformed into Linear form for the purposes of interpolating or extrapolating a value we wish to estimate, we need to be able to convert it back into 'real space' values. To do this we must raise the logarithm base to the power of the predicted y-value of the transformed equation:

- The real intercept is the base raised to the power of Transformed Linear Intercept.
- The x-value must be raised to the power of the Transformed Linear Slope.
- The y-value is then the product of the above elements.

For the Formula-philes: Reversing a linear transformation of a Power Function

Consider a Linear Function with constants M and C:	$Y = MX + C$	(1)
Where X is the log to the base b of x and Y is the log to the Base b of y	$X = \log_b x$ $Y = \log_b y$	(2)
Substituting (2) in (1):	$\log_b y = M \log_b x + C$	(3)
Applying both sides of the equation as equal powers of b:	$b^{\log_b y} = b^{M \log_b x + C}$	(4)
Re-arranging (4) and applying the Multiplicative and Additive Properties of Logs:	$b^{\log_b y} = \left(b^{\log_b x}\right)^M b^C$	(5)

Linear transformation: Making bent lines straight | 237

Substituting for x and y in (5) using the definition of a log:

$$y = (x)^M b^C \qquad (6)$$

Simplifying (6) with alternative constants, m and c:

$$y = cx^m$$

... which is a Power Function with constants m = M and $c = b^c$ (*Here, the power C is upper case. It is not easy to differentiate between c and C.*)

Note: *In the majority of cases it is probably convenient to use logs to the Base of 10 or e.*

5.2.5 Transforming with Microsoft Excel

When it comes to real life data and looking for the best fit curve to transform, we are likely to find that the decisions are less obvious than the pure theoretical curves that we have considered so far. This is because in many situations the data will be scattered in a pattern around a line or curve. Consequently, we need an objective measure of 'best fit'. Microsoft Excel's basic Chart utility will provide us with some 'first aid' here.

In essence we are looking for the straight line best fit after we have transformed the data. If we recall (*unless you opted to skip that chapter, of course*) in Volume II Chapter 5 on Measures of Linearity, Dependence and Correlation, we discussed how Pearson's Linear Correlation Coefficient would be a helpful measure. Unfortunately, Excel's Chart Utility will not generate this directly; however, it will generate the Coefficient of Determination or R-squared (R^2), which is the square of Pearson's Linear Correlation. Consequently, the highest value of R^2 is also the highest absolute value of the Correlation Coefficient. (*It has the added benefit being a square that we don't have to worry about the sign!*)

Furthermore, unless we really want to go to the trouble of changing the axes of the charts, we can trust Excel to do it all in the background by selecting a Trendline of a particular type:

*   A Linear Trendline will do exactly what we would expect and fit a line through the data
*   A Logarithmic Trendline will fit a straight line through Log x and Linear y data, but then display the resultant trend as the equivalent 'untransformed' curve of a Logarithmic Function
*   An Exponential Trendline will fit a straight line through Linear x and Log y data, but then display the resultant trend again as the equivalent 'untransformed' curve of an Exponential Function
*   *No prizes for guessing what a Power Trendline does.* It calculates the best fit through the Log x and Log y data and then transforms the Trendline back to the equivalent Power Function in real space.

In each case, we will notice that there is a tick box to select whether we want to display the R^2 value on the chart. We do.

Figures 5.12 to 5.15 show examples of the above. The four quadrants in each of these are equivalent to the four graphs discussed in the previous sections, but in these latter

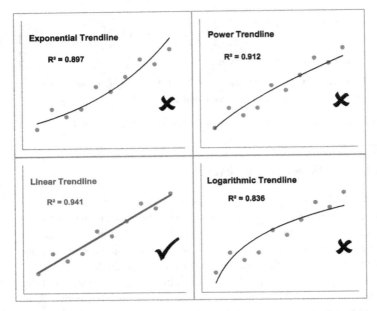

Figure 5.12 Examples of Alternative Trendlines Through Linear Data

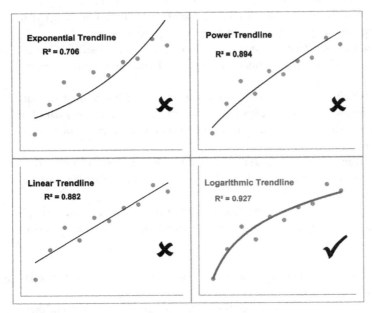

Figure 5.13 Examples of Alternative Trendlines Through Logarithmic Data

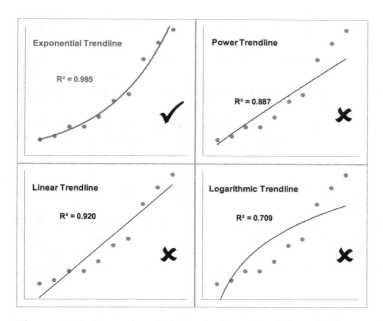

Figure 5.14 Examples of Alternative Trendlines Through Exponential Data

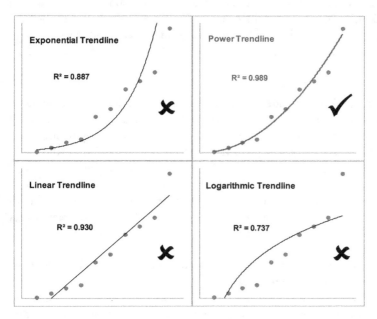

Figure 5.15 Examples of Alternative Trendlines Through Power Data

240 | Linear transformation: Making bent lines straight

cases we have chosen not to change any of the scales to logarithmic values as the different trendlines do this 'in the background'. In all cases the best fit Trendline returns the highest R^2 (i.e. closer to the maximum value of 1 – see Volume II Chapter 5), but also the distribution of our data points around the Trendline is more evenly scattered than in the other three cases and not arced. Values of R^2 less than 0.5 are indicative that the function type is a poor fit.

We can also observe that it is not unusual in the case of increasing data for a Power Function to look very similar to a Linear Function through the origin, especially where the power value is close to one.

- We will observe that the Exponential and Logarithmic Trendlines curve in opposite directions to each other.
- We might like to consider a Power Trendline to be a weighted average of these two. For this reason, it can often require a judgement call by the estimator whether the relationship is really a Linear Function or a Power Function. We will address this issue in Chapter 7 on Nonlinear Regression and also later in this Chapter in Section 5.3

Caveat augur

We should note that just because we can find the best fit trendline using Excel does not mean that it hasn't occurred by chance. This is especially the case where there are equally acceptable levels of R^2 returned for the other Trendlines.

We should always look at the pattern of data around the Trendline. Any 'bowing' of the data around the selected Trendline, characterised by data being on one side of the line or curve at both 'ends' but on the other side in the middle area, may be indicative that the selected Trendline is not a particularly good fit. We will look into this more closely in Chapter 7 on Nonlinear Regression. (*I know . . . Have a cold shower if the thought is getting you all excited with anticipation – I find that it usually helps.*)

If we prefer, we can quite easily check in Microsoft Excel whether it is possible to transform a curve into a straight line by changing either or both of the variables or axes to a Logarithmic Scale. We can do this either:

- By making a conversion using one of the in-built Functions (**LOG, LOG10 or LN**) and plotting the various alternatives
- By plotting the untransformed data and fitting the four Trendlines in turn to determine whether a Linear, Exponential, Logarithmic or Power Trendline fits the data best, using R^2 as an initial differentiator of a 'Best Fit'

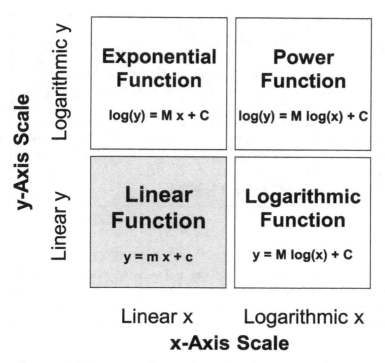

Figure 5.16 Summary of Logarithmic-Based Linear Transformations

- By plotting the untransformed data and changing the scale of each or both axes in turn to a Logarithmic scale

Figure 5.16 summarises the three standard logarithmic-based linear transformations. (*Yes, you're right, the use of the word 'standard' does imply that there are some non-standard logarithmic-based linear transformations – just when you thought it couldn't get any better! . . . or should that be worse?*) We will consider the case of 'near misses' to straight line transformations in Section 5–3 that follows where we will consider situations where there appears to be a reasonable transformation apart from at the lower end of the vertical or horizontal scale, or both.

In the majority of cases we will probably only be considering values where both horizontal and vertical axis values are positive. However, it is possible for negative values of the vertical y-axis to be considered in the case of Linear and Logarithmic Functions, and for negative values of the horizontal x-axis to be considered in the case of Linear or Exponential Functions. We must ask ourselves whether such negative values have any meaning in the context of the data being analysed. Where Excel cannot calculate the Trendline or the log value because of zeros or negatives, it will give us a nice helpful reminder that we are asking for the impossible!

242 | Linear transformation: Making bent lines straight

> ## Caveat augur
>
> Microsoft Excel provides us with an option to fit a polynomial up to an order 6. The higher the order we choose, the better the fit will be if measured using the Coefficient of Determination R^2.
>
> However, fitting a Polynomial Curve without a sound rationale for the underlying estimating relationship it represents, arguably breaks the Transparency and the Credibility objectives in our TRACEability rule, *and if we are honest with ourselves, it is a bit of an easy cop-out!*
>
> **Just because we can doesn't mean we should!**

There are some exceptions to our *Caveat augur* on not using Polynomial Curve Fitting. One of them we have already discussed in Chapter 2 on Linear Properties which we will revisit briefly in Section 5.6, in that the cumulative value of a straight line is known to be a quadratic function (polynomial of order 2.) We will discuss others in Chapter 7 on Nonlinear Regression.

5.2.6 Is the transformation really better, or just a mathematical sleight of hand?

Having used this simple test to determine which function type appears to be the best fit for our data, we should really now stand back and take a reality check. We wouldn't dream of comparing the length of something measured in centimetres with something measured in inches without normalising the measurements, would we? Nor would we compare the price of something purchased in US dollars with the equivalent thing measured in Australian dollars, or pounds' sterling, or euros without first applying an exchange rate.

The same should be true here. Ostensibly, the R-Square or Coefficient of Determination, is a measure of linearity, but in the case of Exponential or Power Functions, we are measuring that linearity based on logarithmic units and comparing it with those based on linear units. *Oops!*

To ensure that we are comparing like with like, we must first normalise the data by reversing the log transformation of the model back into linear units. We can then compare the errors between the alternative functional best fit curve (exponential or power) with the best fit line (linear). We will be revisiting this in the next chapter, when we look at nonlinear regression, but just for completion, we will finish the last example we started.

Linear transformation: Making bent lines straight | 243

Table 5.4 Comparing Linear and Power Trendline Errors in Linear Space (1)

Linear Function				Power Function				
Linear Values				Logarithmic Values			Linear Values	
x	y	Line of Best Fit	Error	Log(x)	Log(y)	Line of Best Fit	Curve of Best Fit	Error
1	1.25	-15.25	16.50	0.0000	0.0969	0.1316	1.35	-0.10
2	6.5	0.55	5.95	0.3010	0.8129	0.7441	5.55	0.95
3	13	16.35	-3.35	0.4771	1.1139	1.1024	12.66	0.34
4	16.5	32.15	-15.65	0.6021	1.2175	1.3566	22.73	-6.23
5	44.75	47.95	-3.20	0.6990	1.6508	1.5538	35.79	8.96
6	54.5	63.75	-9.25	0.7782	1.7364	1.7149	51.86	2.64
7	78.75	79.55	-0.80	0.8451	1.8963	1.8511	70.97	7.78
8	89.25	95.35	-6.10	0.9031	1.9506	1.9691	93.13	-3.88
9	100	111.15	-11.15	0.9542	2.0000	2.0732	118.35	-18.35
10	154	126.95	27.05	1.0000	2.1875	2.1663	146.64	7.36

R2	0.930			0.989	
Slope	15.80			2.03	
Intercept	-31.05			0.13	
Sum of Errors		0.00			-0.54
Sum of Squares Error		1553,48			593.38

In Table 5.4 we have taken the data from which we generated Figure 5.15 and:

- Determined the Best Fit Lines in Linear and Log Space
- Reversed the transformation on the Best Fit Line for the Power Function by taking the anti-log of the Best Fit Line to get the Best Fit Curve. To do this we just raised 10 (the log base) to the power of the Best Fit Line, e.g. $1.35 = 10^{0.1316}$.
- Calculated the error between the raw data and the two alternative Best Fits. We make the comparison on the Sum of Square Errors

In this case, we see that the Power Function is the Best Fit Function for our data. So, what was all the fuss about? Well, let's just change the position of the last data point (Table 5.5) and re-do the calculations. The simple R-Square test still indicates that the Power Function is the best fit, but when we normalise the scales again, we find that the Linear Function is the one that minimises the Sum of Square Errors in linear space.

To be completely thorough, we should also check for homoscedasticity in the error terms for both models. We will not prove it here but both models pass do White's Test for homoscedasticity (see Chapter 4, Section 4.5.7 for further details.)

Where the difference in R-Square Values is large, it is reasonable expect that the R-Square test will suffice, but where it is relatively small, then we will need to verify that the normalised Sum of Square Errors has been minimised.

244 | Linear transformation: Making bent lines straight

Table 5.5 Comparing Linear and Power Trendline Errors in Linear Space (2)

Linear Function				Power Function				
Linear Values				Logarithmic Values			Linear Values	
x	y	Line of Best Fit	Error	Log(x)	Log(y)	Line of Best Fit	Curve of Best Fit	Error
1	1.25	-11.47	12.72	0.0000	0.0969	0.1434	1.39	-0.14
2	6.5	2.91	3.59	0.3010	0.8129	0.7468	5.58	0.92
3	13	17.30	-4.30	0.4771	1.1139	1.0998	12.58	0.42
4	16.5	31.68	-15.18	0.6021	1.2175	1.3502	22.40	-5.90
5	44.75	46.06	-1.31	0.6990	1.6508	1.5444	35.03	9.72
6	54.5	60.44	-5.94	0.7782	1.7364	1.7031	50.48	4.02
7	78.75	74.82	3.93	0.8451	1.8963	1.8373	68.76	9.99
8	89.25	89.20	0.05	0.9031	1.9506	1.9536	89.86	-0.61
9	100	103.59	-3.59	0.9542	2.0000	2.0561	113.79	-13.79
10	128	117.97	10.03	1.0000	2.1072	2.1478	140.54	-12.54

R2	0.967		0.988
Slope	14.38		2.00
Intercept	-25.85		0.14

Sum of Errors	0.00
Sum of Squares Error	589.35

-7.91
594.11

Note that this normalisation process only needs to be done where the Best Fit Curves appear to be Exponential or Power Functions. We don't need to worry about Logarithmic Functions because the error terms are still measured in linear space not log space.

5.3 Advanced linear transformation: Generalised Function types

Or what if our transformations are just a near miss to being linear?

When we try transforming our data, we may find that one or more of our four standard function types appear to give a reasonable fit apart from a tendency to veer away from our theoretical best fit line at the lower end (i.e. closest to the origin). This may be because the assumption of a perfect Function Type is often (but not always) an over-simplification of the reality behind the data. It may be that our data may not fit any one of the four standard Function Types, but before we give up on them, it may be that there is simply a constant missing from the assumed relationship.

If we are extrapolating to a point around which or beyond where the fit appears to be a good one, then it may not matter to us. Where we want to interpolate within or extrapolate beyond a value where the fit is not as good as we would like, we may be tempted to make a manual adjustment. This may be quite appropriate but it may be better to look for a more general linear transformation that better fits the data.

Linear transformation: Making bent lines straight | 245

Let's consider three more generalised versions of our three nonlinear function types:

- Generalised Logarithmic Functions
- Generalised Exponential Functions
- Generalised Power Functions

Linear Functions are already in a Generalised form compared with a straight line through the origin, which is equivalent to being the Standard form in this context.

The difference between these generalised forms and the Standard Transformable versions is that we have to make an adjustment to the values we plot for either or both the x and y values. The good news is that the adjustments are constant for any single case; the bad news is that we will have to make an estimate of those constants.

For the Formula-philes: Generalised Transformable Function types

Function Type	Standard Form	Generalised Form
Logarithmic	$m^y = cx$	$m^y = c(x + a)$
Exponential	$y = cm^x$	$y = cm^x - b$
Power	$y = cx^m$	$y = c(x + a)^m - b$

The Standard forms are those where the Constants a and b in the Generalised forms are equal to zero

---

Let's look at each in turn and examine whether there are any clues as to which generalised function we should consider. We will then look at a technique to estimate the missing constants for us in a structured manner. (*We could just guess and iterate until we're happy. We might also win the lottery jackpot the same way.*)

5.3.1 Transforming Generalised Logarithmic Functions

In Figures 5.17 and 5.18 we can compare the shape of the Generalised form of a Logarithmic Function with the Standard form both in Real Space and the appropriate

Linear transformation: Making bent lines straight

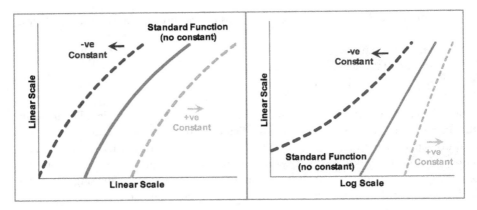

Figure 5.17 Generalised Increasing Logarithmic Functions

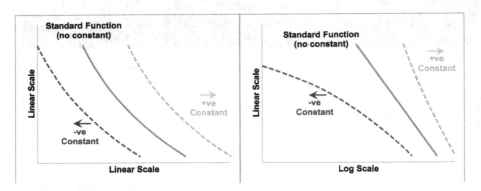

Figure 5.18 Generalised Decreasing Logarithmic Functions

Transformed Log Space. Here, the solid lines are the Standard form with the additional 'Offset Constant' term set to zero. The dotted lines show examples of the effect of introducing negative and positive offset constants. The constant simply adjusts the horizontal x-scale to the left or right for the same y-value. If we were to substitute x for a new left or right adjusted X value as appropriate (X = x + a), we would recover our linear transformation; the logarithmic scale is stretched for lower values of x and compressed for the higher values. Both the outer curves are asymptotic to the Standard form in the Linear-Log form if we were to extend the graph out to the right.

In the case of positive adjustments, despite being asymptotic to the Standard form, we may find that an approximation to a Standard form is a perfectly acceptable fit for the range of data in question (as it is in the cases illustrated.) In reality, with the natural scatter around an assumed relationship, where we have a fairly small range of x-values, we will probably never really know whether we should be using the Standard or Generalised form of the relationship. Only access to more extreme data would help us. However, putting things into perspective, we are not looking for precision; we are estimators and are looking instead for a relative level of accuracy. There's a lot of merit in keeping things simple where we have the choice. Both curves would generate a similar value within the expected range of acceptable estimate sensitivity.

Let's bear that in mind when we look at what happens when we consider our Generalised values in the four Linear/Log graph Matrix with a slightly different offset constant than in Figure 5.17. Figure 5.19 gives us a real dilemma. Here we are assuming that the true underlying relationship is a generalised Logarithmic Function, but for the data range in question for the negative offset constant, we would be forgiven for assuming that the relationship was a Standard Power Function as its transformation appears to be fairly linear in the top right-hand Log-Log Quadrant. In the case of a positive offset constant, we may well interpret it to be a Standard and not a Generalised Logarithmic Function as it could easily be taken as a straight line.

In the case of Figure 5.20, we have assumed that the underlying relationship is a decreasing Generalised Logarithmic Function. In the case of a negative offset constant,

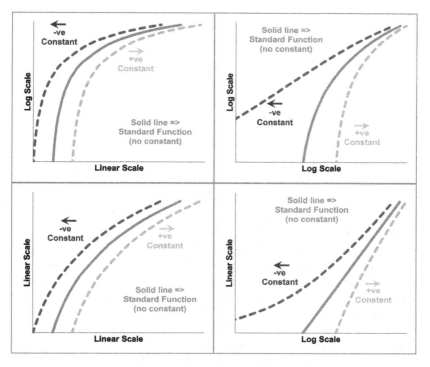

Figure 5.19 Generalised Increasing Logarithmic Functions

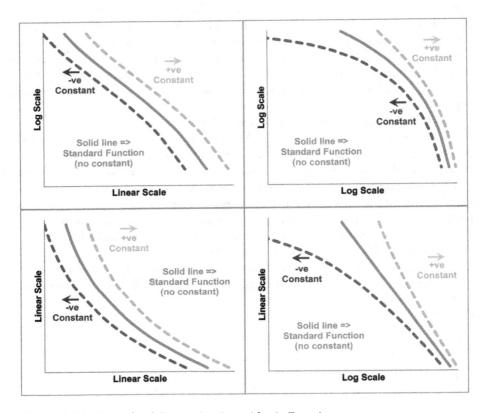

Figure 5.20 Generalised Decreasing Logarithmic Functions

we would probably not look much further than a Standard Exponential Function. With a positive offset constant in a decreasing Generalised Logarithmic Function we may be spoilt for choice between a Standard Exponential or Standard Logarithmic Function. (The slight double waviness of the transformed curves in the top left Log-Linear Quadrant may easily be lost in the 'noise' of the natural data scatter. Consequently we should always consider it to be good practice to plot the errors or residuals between the actual data and the assumed or fitted underlying relationship to highlight any non-uniform distribution of the residuals.

Note: It is not always the case that the Offset Constant term will create this dilemma, but we should be aware that certain combinations of the offset constant and the normal slope and intercept constants may generate these conditions – forewarned is forearmed, as they say.

5.3.2 Transforming Generalised Exponential Functions

If we repeat the exercise on the assumption of a true underlying Generalised Exponential Function with either positive or negative Offset Constants, we will discover (*perhaps to our continuing dismay*) that we may get other dilemmas of a similar nature. Unsurprisingly perhaps, the difficulties with the Generalised Exponential Function Types largely mirror those of the Generalised Logarithmic Function Types.

Figure 5.21 illustrates that we could be misled into thinking that an increasing Generalised Exponential Function with a positive offset constant was just a Standard form of Exponential Function (top left quadrant). Similarly Figure 5.22 suggests in the case of a decreasing Generalised Exponential Function, we could easily be convinced that (with the inherent data scatter characteristic of real life) that a positive offset function was a Standard Exponential Function. Furthermore, in the case of certain positive or negative offsets, we would be tempted to believe that the relationship was a Standard Logarithmic Function (bottom right quadrant). Again, the waviness would be lost in the natural data scatter.

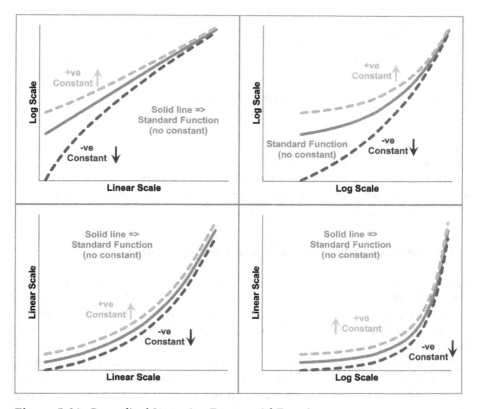

Figure 5.21 Generalised Increasing Exponential Functions

250 | Linear transformation: Making bent lines straight

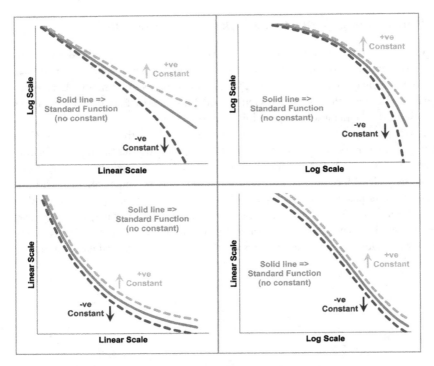

Figure 5.22 Generalised Decreasing Exponential Functions

Oh, the trials and tribulations of being an estimator!

Before we consider what we might do in these circumstances, let's complete the set and examine the possible conclusions we could make with Generalised Power Functions, depending on the value of the constants involved.

5.3.3 Transforming Generalised Power Functions

Figure 5.23 illustrates that in the case of a small positive offset adjustment in both the horizontal and vertical axes, we might accept that the result could be interpreted as just another Standard Power Function. If we did not have the lower values of x, we may conclude also that a Generalised Power Function could be mistaken for a Standard Exponential Function also.

However, where we have a small positive offset constant in either the horizontal or vertical axis, with a small negative offset constant in the other, then to all intents and purposes the effects of the two opposite-signed offsets exacerbate the effects of each other, as illustrated in Figure 5.24. However, we would be prudent to check what happens with a

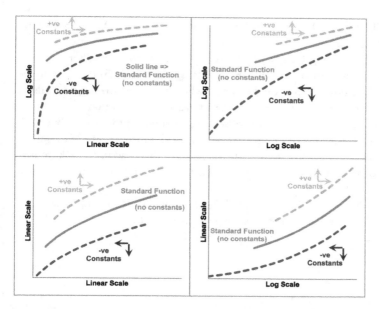

Figure 5.23 Generalised Increasing Power Functions with Like-Signed Offsets Constants

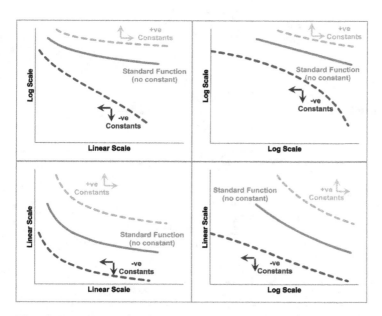

Figure 5.24 Generalised Increasing Power Functions with Unlike-Signed Offset Constants

decreasing Generalised Power Function first as illustrated in Figure 5.25 before making any premature conclusions.

Where we have a decreasing Generalised Power Function with both offset constants being positive and relatively small in value, the results (see Figure 5.25) could easily be mistaken for a Standard Power Function. However, where we have two negative offsets, we could also look at the result as being a Standard Logarithmic Function.

In the case where the signs of the two offsets are in similar proportions but with opposite signs, i.e. small negative of one paired with a small positive value of the other, we may find the result is similar to Figure 5.26 which again may lead us to look no further than a Standard Power Function, a Standard Logarithmic Function or possibly even a Standard Exponential Function!

The moral of these dilemmas is that sometimes 'being more sophisticated' can equal 'being too smart for our own good'! There is a lot of merit in keeping things simple.

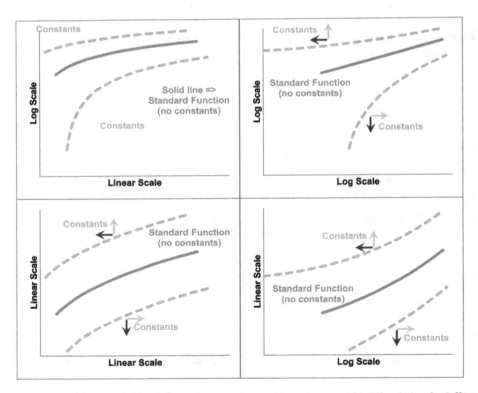

Figure 5.25 Generalised Decreasing Power Functions with Like-Signed Offset Constants

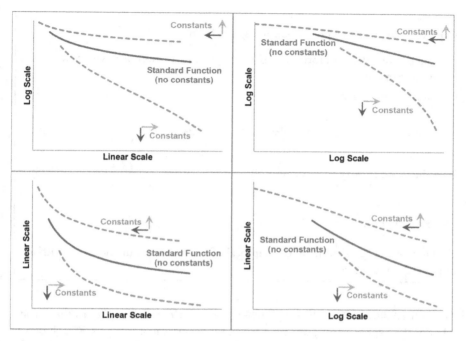

Figure 5.26 Generalised Decreasing Power Functions with Unlike-Signed Offset Constants

5.3.4 Reciprocal Functions – Special cases of Generalised Power Functions

There are two special cases of Generalised Power Functions – those that give us reciprocal relationships for either x or y. Where the power parameter is -1 (or very close to it), a Reciprocal Function of either x or y will transform our data observations into a linear function. A simple graphical test of plotting y against the reciprocal of x or vice versa is a simple option to try without digging into more sophisticated curve fitting routines required for these Generalised Functions.

> **For the Formula-philes: Reciprocals are special cases of Generalised Power Functions**
>
> Consider a Generalised Power Function with constants m, a, b and c:
>
> $$y = c(x + a)^m - b \qquad (1)$$
>
> Let $a = 0$ and $m = -1$. Substituting these in (1):
>
> $$y = cx^{-1} - b \qquad (2)$$
>
> *(Continued)*

Expressing the negative power of x in (2) as a fraction: $y = \dfrac{c}{x} - b$

... which is the Standard form of y as a Reciprocal Function of x with a slope of c and intercept of $-b$

Consider (1) in which $b = 0$ and $m = -1$: $\qquad y = c(x+a)^{-1}$ \hfill (3)

Expressing the negative power of x in (3) as a fraction: $y = \dfrac{c}{(x+a)}$ \hfill (4)

Inverting both sides of the equation (4): $\qquad \dfrac{1}{y} = \dfrac{(x+a)}{c}$ \hfill (5)

Or, expanding the terms: $\qquad \dfrac{1}{y} = \dfrac{x}{c} + \dfrac{a}{c}$

... which is the Standard form of reciprocal of y as a Linear Function of x with a slope of $1/c$ and intercept of a/c

Figure 5.27 illustrates the first type where the offset constant applies to the vertical axis. This is the expression of y as a function of the reciprocal of x. Figure 5.28 illustrates the second special case of the Generalised Power Function where the offset constant applies to the horizontal axis. This is the case where the Reciprocal of y is a function of x.

5.3.5 Transformation options

So if we suspect that we have a relationship that could be classified as one of these Generalised Function types, how can we find the 'Best Fit'? We could simply use the old 'Guess and Iterate' technique, but in reality that is only likely to give us a 'good fit' at best, and

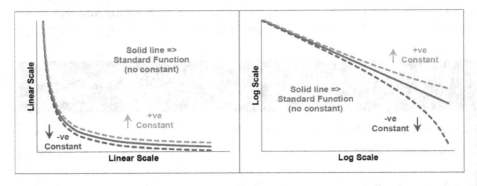

Figure 5.27 Reciprocal-x as a Special Case of the Generalised Power Function

Linear transformation: Making bent lines straight

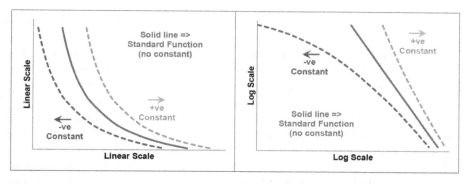

Figure 5.28 Reciprocal-y as a Special Case of the Generalised Power Function

not the 'Best Fit', prompting the question 'Is a "good fit" not a "good enough fit"?' If we are going down that route (*and it's not wrong to ask that question; in fact there is a strong case of materiality to support it in many instances*), we may as well look at whether we can approximate the Generalised form to one of the Standard forms for the range of data we are considering?

We can summarise the results of our observations from the discussion of each Generalised Function in Table 5.6. We should note that these are examples and not a definitive

Table 5.6 Examples of Alternative Standard Functions that might Approximate Generalised Functions

	Conditions when an alternative Standard Functional Form may be appropriate			
Function Type		Standard Logarithmic	Standard Power	Standard Exponential
Generalised Logarithmic	Increasing Trend	With +ve Offset	With -ve Offset	
	Decreasing Trend	With +ve Offset	With -ve Offset	With any offset (including zero)
Generalised Power	Increasing Trend		With any small Offsets	With any small Offsets
	Decreasing Trend	With any small Offsets	With any small +ve Offsets	With any small Offsets
Generalised Exponential	Increasing Trend			With +ve Offset
	Decreasing Trend	With any offset (including zero)		With +ve Offset

256 | Linear transformation: Making bent lines straight

list of all potential outcomes of using an alternative Standard Function type based on our discussions above.

The practical difficulties of transforming the Generalised Function Types into corresponding Standard Function Types so that we can subsequently transform them into Linear Functions can be put into context if we reflect on the words of George Box (1979, p.202) with which we opened this chapter: 'Essentially all models are wrong, but some are useful'.

If we find that a Standard form of a function appears to be a good fit, we would probably have no compunction in using the transformation – even though it may not be right in the absolute sense. So what has changed? Nothing – if we didn't know about the Generalised Function Types, we wouldn't have the dilemma over which to use. So, unless we know for certain, that the Standardised Function is not appropriate on grounds of Credibility, we should use that simpler form as it improves Transparency and Repeatability by others in terms of TRACEability. In other words, is it reasonable to assume that another estimator with the same data, applying the same technique, would draw the same conclusion?

Caveat Augur

Beware of unwarranted over-precision!
If we cannot realistically justify why we are making a constant adjustment to either our dependent or independent variables, then we should be challenging why we are making the adjustment in the first place. 'Because it's a better fit' is not valid justification.

However, in the majority of cases where we believe that the Generalised form of one of these Function Types is a better representation of reality (note that we didn't say 'correct'), we can still use the Standard form if we first estimate (and justify) the offset constant(s) involved. That way we can then complete the transformation to a linear format. The process of doing this is a form of Data Normalisation (see Volume I Chapter 6).

In many circumstances we may also feel that there is little justification in looking for negative offset constants, especially in the output variable (y-value) that we are trying to estimate. Positive offset constants can be more readily argued to be an estimate of a fixed element such as a fixed cost or process time. In truth, the model is more likely to be an approximation to reality, and the constant term may be simply that adjustment that allows us to use the approximation over a given range of data. Consequently, in principle a negative offset constant is often just as valid as a positive one in terms of allowing that approximation to reality to be made. However, in practice it presents the added difficulty that it may limit the choice of Generalised Function type that can be normalised into a Standard form if that then requires us to take the Log of a negative number.

Linear transformation: Making bent lines straight | 257

However, regardless of whether we have a positive or negative offset constant we should always apply the 'sense test' to our model, i.e. '*Is the model produced logical? Does the model make sense?*'

Hopefully, we will recall the argument we made in the previous section on the Standard Function types:

Never take the log of a date!

In theory, we could with the Generalised form but the positive Offset Constant would be so enormous as to render any adjustment pointless.

5.4 Finding the Best Fit Offset Constant

Before we rush headlong into the details of the techniques open to us, let's discuss a few issues that this gives us so that we can make an informed judgement on what we might do.

At one level we want to transform the data into a format from which we can perform a linear regression that we can feel comfortable in using as an expression of future values. The principles of linear regression lead us to define the Best Fit straight line such that:

- It passes through the arithmetic mean of the data.
- Minimises the Sum of Squares Errors (SSE) between the dependent y-variable and the modelled y-variable.

We already know that by taking the Logarithm of data then the Regression Line will pass through the arithmetic mean of the log data, which is equivalent to the Geometric Mean of the raw data.

If we add or subtract an offset value to transform a Generalised Function into a Standard Functional form then the 'adjusted' Sum of Squares Error and the R-Square value will change as the offset value changes. Ideally, we want to minimise the SSE or maximise the values of R-Square. Where we have an offset to the independent x-variable, these two conditions coincide. However, where we have an offset to the dependent y-variable, these two conditions will not coincide! (*You were right; you just sort of knew that it was all going too well, didn't you?*) In fact there are many situations where the Adjusted SSE reduces asymptotically to zero. Table 5.7 and Figure 5.29 illustrate such a situation.

Sometimes we may have three options with three different results. We can look to:

1. Maximise R-Square of the transformed offset data as a measure of the 'straightest' transformed line
2. Minimise the value of SSE of the transformed offset data to give us the closest fit to the line. As the example shows this is not always an option.
3. Minimise the value of the SSE of the untransformed data

Linear transformation: Making bent lines straight

Table 5.7 Non-Harmonisation of Minimum SSE with Maximum R2 Over a Range of Offset Values

Offset >			-45	-43	-42	-41	-40	-39	-35	-30	-25
x	y		Log(y-45)	Log(y-43)	Log(y-42)	Log(y-41)	Log(y-40)	Log(y-39)	Log(y-35)	Log(y-30)	Log(y-25)
1	131		1.9345	1.9445	1.9494	1.9542	1.9590	1.9638	1.9823	2.0043	2.0253
2	120		1.8751	1.8865	1.8921	1.8976	1.9031	1.9085	1.9294	1.9542	1.9777
3	117		1.8573	1.8692	1.8751	1.8808	1.8865	1.8921	1.9138	1.9395	1.9638
4	111		1.8195	1.8325	1.8388	1.8451	1.8513	1.8573	1.8808	1.9085	1.9345
5	98		1.7243	1.7404	1.7482	1.7559	1.7634	1.7709	1.7993	1.8325	1.8633
7	85		1.6021	1.6232	1.6335	1.6435	1.6532	1.6628	1.6990	1.7404	1.7782
12	67		1.3424	1.3802	1.3979	1.4150	1.4314	1.4472	1.5051	1.5682	1.6232
17	56		1.0414	1.1139	1.1461	1.1761	1.2041	1.2304	1.3222	1.4150	1.4914
22	55		1.0000	1.0792	1.1139	1.1461	1.1761	1.2041	1.3010	1.3979	1.4771
27	47		0.3010	0.6021	0.6990	0.7782	0.8451	0.9031	1.0792	1.2304	1.3424
LINEST Parameters											
Row	Col	R2	0.9641	0.9798	0.9818	0.9823	0.9820	0.9812	0.9759	0.9686	0.9620
5	2	SSE	0.0903	0.0367	0.0293	0.0256	0.0238	0.0228	0.0220	0.0216	0.0207

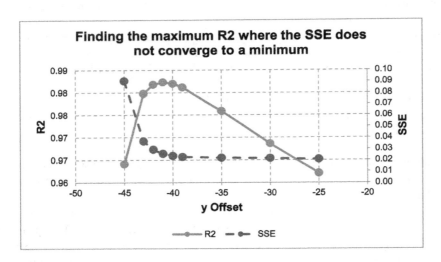

Figure 5.29 Finding the Maximum R2 Where the SSE does not Converge to a Minimum

It is worth noting that we can probably always 'improve' the linearity or fit of any curve by tweaking the offset value(s) — even those where we might have been happy to use a standard curve as if we'd never heard of the Generalised version! So, if it would have been good enough before what has changed now?

This last point is hinting that to make such a refinement in the precision of fit would be bordering on the '*change for change sake*' category, and in a real sense would often be pointless and relatively meaningless. So perhaps the sensible thing to do would be to ignore option 3 (we do at present with standard linear transformations) and consider majoring on the simpler option of maximising R-Square rather than go to the trouble of calculating the SSE, either by an incremental build-up of the square of each error, or taking a short-cut approach using the Excel calculation **INDEX(LINEST(*known_y's, known_x's*, TRUE, TRUE), 5, 2)** as we have done in Table 5.7.

If we are unhappy with this, the only other sensible option we could try would be to run it both ways and assess whether the difference was significant. If it was different, then we could use both results to derive a range estimate. In the example in question, this is not an option. Figure 5.30 illustrates the results of our endeavours choosing to maximise the R-Square value. What appears to be a relatively small change in R-Square from 0.94 to 0.982 we get a much better looking fit to the data by a relatively large offset value.

5.4.1 Transforming Generalised Function Types into Standard Functions

In terms of determining the offsets we should apply to give us the maximum R-Square there are two basic approaches:

- A logical rationalisation or estimate of what the adjustment may be, and then testing whether it is credible. This approach is often used in the analysis of Learning Curves (see Volume IV)
- 'Guess and Iterate' to find the 'Best Fit'

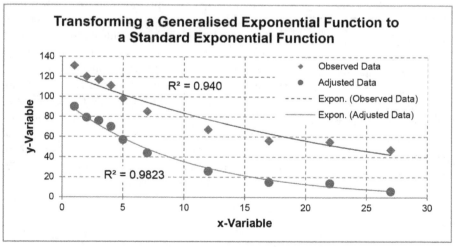

Figure 5.30 Transforming a Generalised Exponential Function to a Standard Exponential Function

260 | Linear transformation: Making bent lines straight

All models will carry some inaccuracy in relation to actual data. If there is a valid logical reason for a particular offset constant to be used this should be considered preferable to using a better fit that we may find through an iteration procedure or technique.

If we are going to use the 'Guess and Iterate' technique, we may as well do it in a structured manner using the 'Random-Start Bisection Method' or 'Halving Rule', or even better we can let Excel's Solver or Goal Seek facilities do all the work for us. However, as Excel is an algorithm, it can sometimes fail to converge. This might be interpreted as Solver or Goal Seek being 'a bit flaky' but, in reality, it is probably more likely the case that we have given it one of the following problems:

1. **Mission Impossible:**   There is no acceptable solution with the data and constraints we have given it, i.e. the data is not one of the Generalised Function Types

2. **Hog Tied:**   We have been too tight or restrictive in the constraints we have fed it – we need to widen the boundaries

3. **Needle in a Haystack:**   We have been too loose in the constraints we have fed it – it has effectively 'timed out'

4. **Wild Goose Chase:**   We gave it a false start point and set it off on in the wrong direction. We can try giving it a different starting position.

Before we look in more detail at how Excel can help us, let's look at the 'Random-Start Bisection Method'.

5.4.2 Using the Random-Start Bisection Method (Technique)

By our definition of the terms 'method' and 'technique' we would refer to it as a technique not a method (Volume I Chapter 2). However, it is referred to as the 'Bisection Method' in wider circles.

The principle of the Random-Start Bisection Method as an iteration technique is very simple. It is aimed at efficiency in reducing the number of iterations and dead-end guesses to a minimum. We take an initial guess at the parameters and record the result we want to test. In this case we would probably be looking to maximise the Coefficient of Determination or R^2 (Volume II Chapter 5). We then take another guess at the parameters and record the result. The procedure is illustrated in Figure 5.31 and an example given in Figure 5.32 for a Logarithmic Function.

Model Set-up

1. Choose a Generalised Function type for the data that we want to try to transform into a Standard form. (*We can always try all three if we have the time!*)

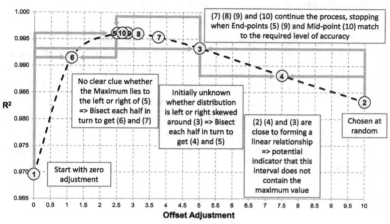

Figure 5.31 Finding an Appropriate Offset Adjustment Using the Random-Start Bisection Method

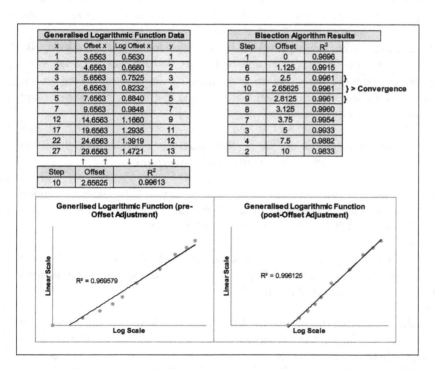

Figure 5.32 Example of Offset Adjustment Approximation Using the Random-Start Bisection Method

262 | Linear transformation: Making bent lines straight

2. Choose whether we want to apply the offset adjustment to either the x or y value, or both in the case of a Power Function (*we can always try another function later*)
3. Set up a spreadsheet model with an offset adjustment as a variable with columns for x, y, the adjusted x or y, and the corresponding log values of each depending on the Function Type chosen. The y-value should assume that it is the Standard Function chosen at Step 1.
4. Calculate the R^2 value for combinations of the adjusted x and y-values) for both their linear and log values, depending on the Function Type chosen. We can use **RSQ(*y-range, x-range*)** within Microsoft Excel to calculate R^2.
    o   In the example shown in Figure 5.32 we have calculated 'Offset x' as the value of x plus a 'fixed' Offset Constant, and the log of that value as we are considering a Generalised Logarithmic Function. The constant is 'fixed' for each iteration.
    o R^2 is calculated using the data for y and the log of the Offset x

Model Iteration

The circled numbers in Figure 5.31 relate to the following steps

1. We start with an assumption of zero adjustment (i.e. the Standard form of the Function Type). We calculate and plot the R^2 on the vertical y-axis against zero on the horizontal axis depicting the offset adjustment
2. Choosing a large positive (or negative) offset value at random, we can re-calculate R^2 and plot it against the chosen offset value. In this case we have chosen the value 10.
3. Now we can bisect the offset adjustment between the one chosen at random (i.e. 10) and the initial one (i.e. zero offset). Calculate and plot R^2 for the bisected offset adjustment value (in this case at 5)

As we do not know whether the optimum offset value lies to the left or right of the value from step 3, we may need to try iterating in both directions:

4. In this case we have guessed to the right first. Now bisect the offset adjustment between the one chosen at random in Step 2 and the previous bisection from step 3. Calculate and plot R^2 for the bisected offset adjustment value (at 7.5).

If the R^2 from step 4 is higher than both those for the offset values on either side of it (i.e. that led to the bisected value from Steps 2 and 3) we can skip step 5 and move to step 7. In the example illustrated this is not the case:

5. Repeat the last step but this time to the left of the value derived in step 3 and bisect the value between step 1 (zero offset) and that calculated for step 3. Calculate the R^2 and plot it. In this case we plot it at 2.5
6. Select the highest of the three R^2 values from steps 3–5 (or steps 1–4 if we have skipped step 5) and repeat the process of bisecting the offset value between the steps to the left or right. In this case step 6 has been chosen between steps 1 and 5.

Linear transformation: Making bent lines straight   263

7.   If the R^2 in step 6 is less than either value for the steps to the left or right of it, then we should try bisecting the offset on the opposite side of the highest R^2, in this case from step 5.

For steps 8 and up, we repeat the procedure until the R^2 does not change to a level of precision appropriate. The graph should steer us into which direction we should be trying next.

For the example summarised in Figure 5.32, we can derive an offset value of some 2.65625 after 10 steps. We should note that all the values of R^2 are high, and that this alone is an imperfect measure of the linearity of the Generalised Functions being discussed here. It is important that we observe the scatter of data (residual or error) around the best fit line to identify whether the points are evenly scattered. For the raw data shown in the left-hand graph of Figure 5.32, the data is more arced than it is for the adjusted data, but this only represents a relatively small numerical increase in the value of R^2 from 0.9696 to 0.9961.

5.4.3 Using Microsoft Excel's Goal Seek or Solver

Microsoft Excel has two useful features (OK, *let me re-phrase that . . . Microsoft Excel has many useful features, two of which that can help us here.*) that allow us to specify an answer we want and then get Excel to find the answer. These are **Goal Seek** and **Solver,** both of which are iterative algorithms. **Solver** is much more flexible than **Goal Seek**, which to all intents and purposes can be considered a slimmed down version of **Solver**:

• *Goal Seek*:

  o   Included as standard under the **Data/Forecast/What-If Analysis**
  o   Allows us to look for the value of a single parameter in one cell that generates a given value in a target cell (or as close as it can get to the specified value)
  o   It is a one-off activity that you must initialise every time you want to find a value

• *Solver*:

  o   Packaged as standard but must be loaded from the Excel Add-Ins. Once loaded it will remain available to us under **Data/Analysis** until we choose to unload it
  o   Allows us to look for the values of one or more parameters in different cells that generate a given value in a target cell, or minimises or maximises the value of that target cell
  o   It allows us to specify a range of constraints or limits on the parameters or other values within the spreadsheet
  o   It has 'memory' and will store the last used/saved parameters and constraints on each spreadsheet within a workbook, thus it is fundamentally consistent with the principles of TRACEability (Transparent, Repeatable, Appropriate, Credible and Experientially-based; see Volume I Chapter 3).

264 | Linear transformation: Making bent lines straight

Let's use Solver on the data and model we set up in the previous section for the Random-Start Bisection Model. The Model Set-up is the same as before and is not repeated here. The principle is the same in that we wish to maximise the Correlation between the data and the theoretical offset adjusted Logarithmic Function.

Running Solver

1. Start with an assumption of zero adjustment (i.e. the Standard Form of the Function Type).
2. Select Solver from the Excel Data Ribbon. A dialogue box will appear
3. Under '**Set Objective**' specify the Cell that contains the calculation with R^2
4. Below this, select the Radar Button to 'Max'
5. In '**By Changing Variable Cells**' identify the cell in which the Offset Constant is defined. (For Power Functions we can include additional parameters separated by commas.)
6. As an option, under '**Subject to the Constraints**' we can add the constraint that the Offset must never allow the situation where our model tries to calculate the log of a negative number or zero (e.g. in the case of a Generalised Logarithmic Function we can specify that the offset cell must be greater than the negative of the minimum of the x values). We can add as many constraints as we feel necessary to limit the range of iterations that Solver will consider, but if we add too many it may become restrictive leading to a 'Failure to converge' message. If we add too few, it may find an answer that it not expected or appropriate.

    Underneath the Constraints block ensure that the option to 'Make Unconstrained Variables Non-negative' is not checked. (*Yes, I spent twenty minutes trying to figure out why one of my models wasn't converging.*)

An example of the Model and Solver set up is illustrated in Figure 5.33.

7. When we click '**Solve**', another dialogue box will advise whether Excel has been able to determine an appropriate answer (Figure 5.34).

If we get a message that Solver cannot find an appropriate solution, then we may wish to try a different start point other than zero offset. Alternatively, we may have to consider an alternative transformation model.

Caveat Augur

There should be some bound on the degree of offset that is reasonable otherwise Solver will try to maximise it to an extent that any curvature in the source data becomes insignificant and indistinguishable from a very shallow almost flat line.

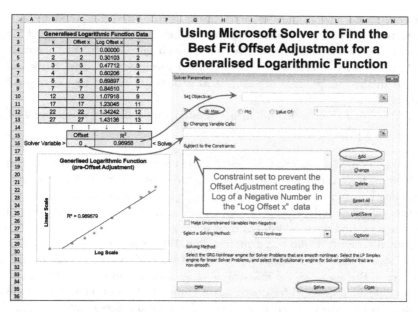

Figure 5.33 Using Microsoft Excel Solver to Find the Best Fit Offset Adjustment for a Generalised Logarithmic Function

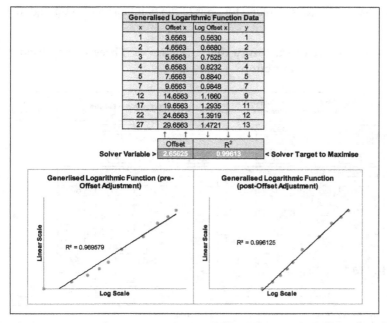

Figure 5.34 Finding an Appropriate Offset Adjustment Using Solver (Result)

266 | Linear transformation: Making bent lines straight

In this case we could have found the same answer as this using the Random-Start Bisection Method if we had continued to iterate further... assuming that we had the time, patience (*and nothing better to do*). **Solver** is much quicker than the manually intensive Random-Start Bisection Method. We could also have used **Goal Seek**, but that loses out in the transparency department as it does not record any logic used; **Solver** does.

Wouldn't it be good if we could find the best solution out of all three in one go? Well, we can, at least to a point ...

- If we set up our model with all three Generalised Functions side by side we can then sum the three R-Square values together. In its own right this statistic has absolutely no meaning but it can be used as the objective to maximise in **Solver** – the three R-Square values are independent of each other so the maximum of each one will occur when we find the maximum of their sum.
- A suggested template is illustrated in Table 5.8. Whatever template we use, it is strongly advised that each function type is plotted separately so that we can observe the nature of the scatter of the offset or adjusted data around each of the best fit trendlines provided in Excel for the Standard Function Types of interest to us; this will allow us to perform a quick sensibility check on the Solver solution rather than just being seduced into accepting the largest possible R-Square. In Volume II Chapter 5 a high value R-Square can indicate a good linear fit but does not measure whether a nonlinear relationship would be an even better fit; a graph will help us with this. In the graph, we are looking for signs of the trendlines arcing through the

Table 5.8 Using Solver to Maximise the R-Square of Three Generalised Functions – Set-Up

		Generalised Logarithmic Function		Generalised Exponential Function		Generalised Power Function				3 in 1 Solvers
		Offset a		Offset b		Offset a		Offset b		
Solver Variable >		0	< Truncated	0		0	< Truncated	0		
Minimum Offset >		0		-10.9999		0		-10.9999		
Maximum Offset >		54		46		54		46		
Raw Input Data		Transformed Data		Transformed Data		Transformed Data				
x	y	X = x+a	Log(x+a)	Y=y+b	Log(y+b)	X = x+a	Log(x+a)	Y=y+b	Log(y+b)	
1	11	1.0000	0.0000	11	1.0414	1.0000	0.0000	11	1.0414	
2	12	2.0000	0.3010	12	1.0792	2.0000	0.3010	12	1.0792	
3	13	3.0000	0.4771	13	1.1139	3.0000	0.4771	13	1.1139	
4	15	4.0000	0.6021	15	1.1761	4.0000	0.6021	15	1.1761	
5	16	5.0000	0.6990	16	1.2041	5.0000	0.6990	16	1.2041	
7	18	7.0000	0.8451	18	1.2553	7.0000	0.8451	18	1.2553	
12	19	12.0000	1.0792	19	1.2788	12.0000	1.0792	19	1.2788	
17	21	17.0000	1.2304	21	1.3222	17.0000	1.2304	21	1.3222	
22	22	22.0000	1.3424	22	1.3424	22.0000	1.3424	22	1.3424	
27	23	27.0000	1.4314	23	1.3617	27.0000	1.4314	23	1.3617	
		R^2 with y		R^2 with x		R^2 based on Log Transforms				Sum
R^2 to Maximise >		0.9775		0.8207		↳	0.9804	↵		19.7785

Notes: (1) Minimum Offset is fractionally more than -Min(Raw Data Range)
(2) Maximum Offset is set nominally at twice Max (Raw Data Range)

Linear transformation: Making bent lines straight | 267

data as a sign of a poor fit rather than one that passes through the points 'evenly' as an indicator of a good fit to the data. (Recall that big word, that's not too easy to say that we discussed in Chapter 3 . . . Homoscedasticity? We want our data errors to have equal variance i.e. homoscedastic not heteroscedastic.)

The set-up procedure for the template illustrated is:

- In terms of the R-Square we are looking to maximise (i.e. get as close to 1 as possible) by allowing the offset value to vary, we need to calculate R-Square for each function type using the **RSQ** function in Microsoft Excel based on any starting parameter for the offset; a value of zero is a logical place to start as that will return the standard logarithmic function. Note that R-Square is calculated on different data ranges for each Function Type:
  - o For a Logarithmic Function, R-Square should be calculated based on our raw y-variable with the log of the x-variable plus its offset parameter
  - o For an Exponential Function, R-Square must be calculated based on the log of the y-variable plus its offset parameter with our raw x-variable
  - o For a Power Function, R-Square is calculated based on the log of the y-variable plus its offset parameter with the log of the x-variable plus its offset parameter
- We can now open **Solver** and set the cell containing the sum of the three R-Square calculations as the **Solver** Objective to **Maximise**.
  - o By maximising R-Square we are inherently minimising the Sum of Square Error (SSE) as $R^2 = 1 - SSE/SST$ from Chapter 4 Section 4.5.1 (the Partitioned Sum of Squares)
- In terms of setting the constraints, the offset value chosen can be positive or negative but not too large in the context of the raw data itself.
  - o As a lower limit the offset cannot be less than or equal to the negative of the smallest x-value in our data range, otherwise we will cause Microsoft Excel to try to do the impossible and take the log of a negative value during one of its iterations; this will cause **Solver** to return a failure error.
  - o In our example in Table 5.8 we should consider setting a Solver Constraint that the Offset parameter is greater than -1. In reality, we can't do this in **Solver** as the option provided is for something that is 'greater than or equal to' some value. Recognising this we may want to pick a value arbitrarily close to the negative of the minimum, such as -10.9999. As we have done here for the x-value offset we have overwritten this minimum to say that we do not want it to be less than zero rather than a value of -0.9999. The argument here is that an offset less than one either way is probably not significant in the scheme of things.
  - o We should also set a maximum positive value to prevent **Solver** from trying to iterate towards infinity. (The closer the values are around a constant value, the closer to 1 that R-Square will get.) The choice of value is down to Estimator Judgement. As a starting point, we may want to consider a limit of approximately twice the maximum absolute value. It is often more useful to specify an

Linear transformation: Making bent lines straight

- integer value as the constraint, or one that creates an integer somewhere else in the template so that it is easier to spot if the constraint is invoked.
  - o For this example, we have also placed a constraint that the x-variable offset must be an integer value only
- Underneath the Constraints block ensure that the option to 'Make Unconstrained Variables Non-negative' is **not** checked. It may prevent our model from converging if we don't.
- Click 'Solve'. In the example in question, we will get the result in Table 5.9 and the three mini-plots in Figure 5.35

In their own right, all three values of R-Square are good, but a quick look at the mini-plots show that the data is not an Exponential Function as the data is Concave and exponential data is always Convex. The raw statistic can deceive — always plot the data!

Table 5.9 Using Solver to Maximise the R-Square of Three Generalised Functions

	Generalised Logarithmic Function		Generalised Exponential Function		Generalised Power Function			3 in 1 Solvers
	Offset a		Offset b		Offset a		Offset b	
Solver Variable >	1		46	< Maximum	1		46	< Maximum
Minimum Offset >	0		-10.9999		0		-10.9999	
Maximum Offset >	54		46		54		46	
Raw Input Data		Transformed Data	Transformed Data		Transformed Data			
x	y	X = x+a	Log(x+a)	Y=y+b	Log(y+b)	X = x+a	Log(x+a)	Y=y+b	Log(y+b)
1	11	2.0000	0.3010	57	1.7559	2.0000	0.3010	57	1.7559
2	12	3.0000	0.4771	58	1.7634	3.0000	0.4771	58	1.7634
3	13	4.0000	0.6021	59	1.7709	4.0000	0.6021	59	1.7709
4	15	5.0000	0.6990	61	1.7853	5.0000	0.6990	61	1.7853
5	16	6.0000	0.7782	62	1.7924	6.0000	0.7782	62	1.7924
7	18	8.0000	0.9031	64	1.8062	8.0000	0.9031	64	1.8062
12	19	13.0000	1.1139	65	1.8129	13.0000	1.1139	65	1.8129
17	21	18.0000	1.2553	67	1.8261	18.0000	1.2553	67	1.8261
22	22	23.0000	1.3617	68	1.8325	23.0000	1.3617	68	1.8325
27	23	28.0000	1.4472	69	1.8388	28.0000	1.4472	69	1.8388
R² to Maximise >		R² with y 0.9868	R² with x 0.8694		R² based on Log Transforms 0.9852			Sum R² 2.8415

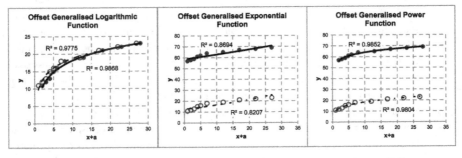

Figure 5.35 3-in-1 Solver Graphical Sensibility Check (Before and After)

Linear transformation: Making bent lines straight | 269

There is not much to choose between Logarithmic and Power Functions, in terms of the value of R-Square. This problem is not uncommon. However, a quick look at the **Solver** output summary shows us that the iterations truncated when the offset parameter, b, reached the maximum allowed value. (*Do you see what I mean about being easy to spot?*). If we want to be sure that this is a limiting factor and not just (coincidentally) the best fit result, then we can extend our maximum range, and re-run **Solver**.

In this particular case, our 'three-in-a-bed' Solver Model has found that a best fits our raw data if we transform our data by offsetting our x data by a value of 1 to the right. We still get an excellent fit for a Generalised Power Function if we are prepared to accept such a large adjustment to the y-variable. Each case must be looked at in its individual context.

Note that this example highlights that in practice this technique often offers us little clear difference to inform our choice of Generalised Logarithmic and Generalised Power Functions, as they are always concave or convex together. Consequently, the estimator must make that choice based on which model is more intuitively correct, or at least more supportable (i.e. in the spirit of the credibility principle of TRACE-ability.) This is particularly the case with small data samples as we have here. The technique is more powerful (and helpful) where we have larger data samples covering a wider range of data values.

However, there is another constraint we could apply . . .
If we were to take a step back and reflect on what we are doing here we could make the argument that we should only be prepared to offset the x-values for Logarithmic and Power Functions, and the y-values for Exponential and Power Functions by the same amount each (that is, one x offset value and one y offset value.)

The rationale for this approach would be that we are offsetting the observed data to take account of something tangible within the model that has a significant impact on the observed values – not just because we can get a better fit! It is far better to have a sound rationale for offsetting by a particular value than allowing one to be picked that we observe but cannot justify. It's the estimating equivalent of the age-old parental retort when a child questions a decision of:

> '*Why do I have to do that?*'
> '*. . . Because I say so!*'

An example of where we might justify such an adjustment could be observed reductions in production costs due to operator and organisational learning, but there may be an element of fixed material cost that cannot be reduced. Simply taking the Best Fit simple curve unchallenged may simply be smoothing over the cracks of a more complex relationship with greater unexplained variances.

Having said all that, if we applied this logic to our model, tying the two x-offset values together and likewise the two y-offset values, we would get the same results as we got in Table 5.9. *You're probably thinking, 'So, why mention it then?' . . . because we can! . . . No, seriously, we mention it because such duplication is not always the case!*)

> ## Caveat Augur
>
> As Microsoft Excel's Solver is an iterative algorithm, it may not always converge first time to the optimum solution. We should always try either:
>
> - Running Solver again from where it left off to see if it will converge further (which we did here because it 'timed out')
> - Or, we can give it a different set of initial parameter values to prevent it from getting stuck in a 'local turning point'
> - Or, we can set more stringent convergence criteria in the Solver Options (i.e. make it work harder)

> ## For the Formula-phobes: What are local turning points?
>
> Nothing to do with that moment when realisation and understanding sets in, necessarily, but things do turn around ...
>
> If we have identified a variable whose value changes when a parameter changes then we can choose to minimise (or maximise) that variable's value by choosing the appropriate value of the parameter. The relationship may not be a simple one. It becomes even more difficult to minimise the variable if there is more than one parameter that affects the variable. We may end up in what might be described as the mathematical equivalent of a cul-de-sac. The only way out is to go back out at the top and carry on down the main road.
>
>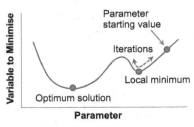

As a consequence, we may find that what we intuitively expect to be the rational model, does not necessarily return the best fit to the scatter of the data. In fact, the shape of some Generalised Functions may well masquerade as Standard form of themselves, or other functional forms or as Generalised forms of other Functional Types, as we demonstrated in Sections 5.3.1 to 5.3.3. Sometimes estimators have to trust their instincts more than the maths! (*Did I hear someone mutter something about heresy?*)

Linear transformation: Making bent lines straight | 271

5.5 Straightening out Earned Value Analysis . . . or EVM Disintegration

This section is not intended to be as defamatory or demeaning of EVM as some might think that the title suggests . . .

Earned Value Management (EVM) is all about an integrated approach to managing cost and schedule on a project. It is an invaluable tool for the cost estimator or schedule planner who can use it to track performance and achievement against a plan.

EVM graphical reports typically display three curves representing the Planned Cost, the Actual Cost and the Achievement on a single graph to give it that integrated perspective against the schedule; the horizontal scale is almost invariable a date scale. More often than not these three elements form a set of 'lazy S-curves' as depicted in Figure 5.36. (*Is it me, or do the majority of examples of EVM chart examples always show us to be late and overspent?*)

As the name suggests, EVM is a management technique encompassing the Plan-Do-Review Cycle. As part of the review cycle, there is an expectation that someone with the appropriate skills analyses the data. This element might be referred to as Earned Value Analysis (EVA).

One of the EVA challenges with which estimators are often tasked is in predicting an outturn value and date of completion. It is the practice of displaying the three graphs as a group of S-Curves against time that makes this so challenging because the integrated approach of EVM is both its strength and its weakness – it's hard to see beyond the three S-Curves. Projecting the direction of curves is difficult. We should fervently resist any temptation we might feel to use Microsoft Excel's Polynomial Trendlines; it is highly unlikely that there is a strong case to support that the underlying relationship for each line is a polynomial function of time (be it cubic, quartic, or any other higher order). So, **just because we can, does not mean we should!**

All is not lost! We can dissect the problem into its fundamental components (*we could call it our 'dis-integration' technique!*) It is the time scale that really causes us the difficulty. Take a moment to review Figure 5.36 and form an opinion on what the cost outturn might be and the date when it will be achieved by projecting the actual and achievement curves across time, and compare it with the answer we get at the end of Section 5.5.2. (*No cheating or pretending you forgot – write it down on a piece of paper.*) There, we are going to explore how we can look at the EVM problem as one that we can express in simpler terms, hopefully linear ones. Before that we will just remind ourselves of the basic EVM/EVA terminology.

5.5.1 EVM terminology

EVM is a TLA (Three-Letter Abbreviation) for Earned Value Management. EVM itself has a proliferation of two-, three- and four-letter abbreviations that 'describe' the

Linear transformation: Making bent lines straight

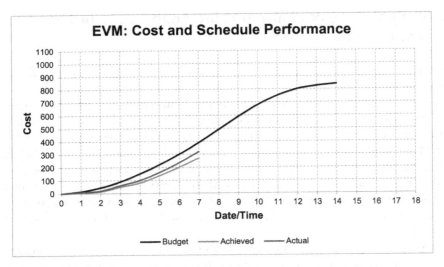

Figure 5.36 Lazy S-Curves Typical of EVM Cost and Schedule Performance

different elements of a Spend and Achievement, or Cost and Scheduling Monitoring system. EVM or its component EVA has its own terminology or language, which now comes in two different dialects – the old and the new, as various people and organisations have tried to simplify the terminology used:

Each point represents the cumulative budgeted value of the work authorised to be completed at that point in time. The curve represents the budget profile over time for a defined scope of work. Generally, it excludes any unallocated contingency. (*Note: As estimators are often keen to point out, this may or may not be the value they estimated.*)

- BCWS – Budgeted Cost of Work Scheduled, or more simply the Planned Value (PV)
- BCWP – Budgeted Cost of Work Performed, or simply the Earned Value (EV).
  Each point represents the cumulative budgeted value of the work completed at that point in time, including Work In Progress. The curve represents the profile by which the budgeted value has been consumed or achieved over time.
- ACWP – Actual Cost of Work Performed, or just the Actual Cost (AC)
  Each point represents the cumulative actual cost of the work completed or in progress at that point in time. The curve represents the profile by which the actual cost has been expended for the value achieved over time.
- BAC – Budget At Completion
  The last value on the BCWS or PV (and BCWP or EV) lines, equivalent to 100% of the defined scope of work
- EAC – Estimate At Completion

Linear transformation: Making bent lines straight | 273

This is the sum of the actual cost to date for the work achieved, plus an estimate of the cost to complete any outstanding or incomplete activity or task in the defined scope of work.

- CV – Cost Variance
  The difference between the value achieved (BCWP or EV) and the actual cost (ACWP or AC) at any point in time (CV = BCWP – ACWP or CV = EV - AC)
- CPI – Cost Performance Index
  The ratio of the value achieved (BCWP or EV) divided by the actual cost (ACWP or AC) at a point in time
- SV – Schedule Variance (cost impact)
  The difference between the value achieved (BCWP or EV) and the planned value (BCWS or PV) at any point in time (SV = BCWP - BCWS or SV = EV - PV)
- SPI – Schedule Performance Index
  The ratio of the value achieved (BCWP or EV) divided by the planned value (BCWS or PV) at a point in time
- ES – Earned Schedule
  This is a relatively new term in the vocabulary of EVM and EVA, coined by Lipke (2003). It represents the time at which the work achieved was scheduled to have been achieved. In old EVM dialect, it may have been referred to as the Budgeted Time of Work Performed (BTWP) (*but it wasn't around when the four-letter abbreviations were being defined*)
- AT – Actual Time
  This is another relatively new term to EVM, also created by Lipke (2003). It is the corresponding term to ES which denotes the time that work was actually completed. In old-style terminology, it may have been referred to as ATWP (*but again it hadn't been thought of at the time*)
- SV(t) – Schedule Variance (time impact)
  The difference between the Actual Time (AT) that an Earned Value (EV or BCWP) was achieved and the time when it should have been achieved – the Earned Schedule (ES) such that SV(t) = ES - AT
- SPI(t) – Schedule Performance Index-Time
  This is the ratio that Lipke (2003) created that compares the time at which work was actually performed with the time when it was planned to be performed. It is the measure ES divided by AT. (*I wonder why they never capitalised it all instead of putting the last letter in brackets? Some people just don't have a sense of humour.*)

These terms are illustrated in Figure 5.37. Despite its unfortunate name SPI(t) is not something we should look down on and avoid. The main reason that the SV(t) and SPI(t) variations were promoted (successfully) by Lipke was that there is a fundamental flaw in the original SPI measures in that it will always converge to 1 regardless of how late a project delivery is, because the Earned Value at Completion will always be the Planned Value at Completion.

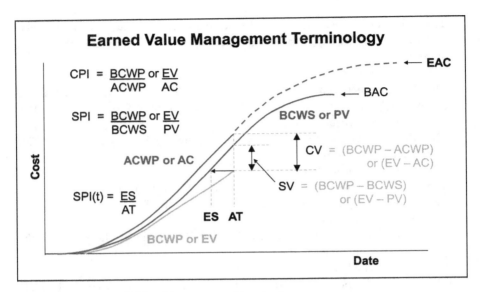

Figure 5.37 Earned Value Management Terminology

Furthermore, in EVM circles it is standard to depict cost on the vertical axis, and imply schedule by the horizontal date axis. If we think about it, we may conclude that it seems somewhat bizarre that traditional Schedule Variance (SV) and the corresponding SPI is measured by a vertical difference rather than a horizontal one i.e. measuring a schedule performance in terms of a cost measure. SV(t) and SPI(t) resolves this inconsistency.

5.5.2 Taking a simpler perspective

Let's take a step back and look at what we can conclude from this. In an ideal EVM world we would expect that the Budget against Time, Actual Spend against Time, and Achievement against Time would always be perfectly correlated. It is the activity ramp up and wind down that creates the characteristic lazy s-curves, the specific profile of which is dependent on the activity schedule, and progress (*or otherwise*) against it. It is this integrated view that makes forecasting more complicated than it needs to be, so let's consider Cost and Schedule separately for the moment (*let's 'dis-integrate' them!*)

As we expect Actual Cost and Achieved Value to be reasonably linearly correlated, why don't we simply plot one against the other to see whether it is true? The only role that the Date plays is in matching the data pairs.

Also, if our progress or achievement slips in time relative to the planned schedule that it may be considered reasonable to expect that the actual cost would also move to the right, assuming that we were managing these things appropriately. *(Good point, we don't always manage these things properly . . . that's why we have EVM.)*

Let's consider plotting the following matching pairs of data in Table 5.10 against our objective of determining cost and schedule outturn:

The slope of our Cost Performance graph in Figure 5.38 depicts the trend of Actual Cost in relation to Earned Value we have achieved. The average EVM Cost Performance Index (CPI) is the reciprocal of slope of the best fit line through the data.

As the horizontal scale is bounded by the minimum and maximum achievement (equivalent to 0% and 100%), any projection of the Cost Performance Trend cannot

Table 5.10 Key Component Elements of Earned Value Analysis

Requirement	Horizontal x-axis	Vertical y-axis	Comments
Cost Outturn or EAC	Achievement (BCWP or EV)	Actual Cost (ACWP or AC)	Graph will directly compare the actual data (ACWP or AC) against a measure of achievement (BCWP or EV)
Schedule Completion Date	Actual Time (AT)	Earned Schedule (ES)	Graph will directly compare the actual date achieved (AT) with the schedule's planned date (ES)

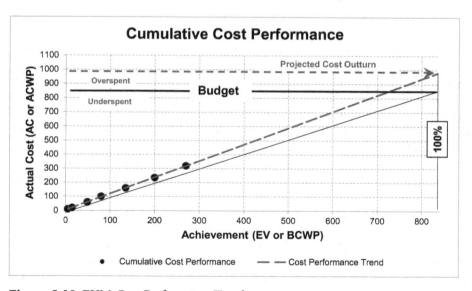

Figure 5.38 EVM Cost Performance Trend

276 | Linear transformation: Making bent lines straight

extend beyond that maximum Earned Value. Where any projection we make intersects that right hand maximum Earned Value boundary, this is a view of the Estimate At Completion. In answer to our unasked question 'What type of projection should we be making?' is that we should be informed by what we observe in the graph.

Theoretically speaking from an EVM First Principles perspective, the relationship of Actual Cost to Achievement should be fundamentally linear in nature. Unfortunately that inconvenient thing called reality keeps getting in the way. Instead, when we have plotted our data and analysed it, we may be faced with one of the four possible conclusions (*five if we count the silly or undesirable one at the end*). Our potential response may be as summarised in Table 5.11.

In terms of an example of where we may legitimately have a nonlinear Cost Performance trend would be in of a recurring production activity in which the budget has been based on an assumption of a given Learning Curve rate (see Volume IV Chapter 2), but that in reality the true Learning Curve rate turns out to be steeper or shallower than this. We would then observe that the Cost Performance trend is likely to be a Power Function. We will discuss this more fully in Volume IV Chapter 3.

We can repeat the process with the Schedule Performance Data as shown in Figure 5.39.

In this case the intercept of the Line of Best Fit through the Schedule Performance graph measures how early or late work commenced relative to the scheduled start. The slope of the Line of Best Fit depicts the trend of Earned Schedule in relation to the Actual Time we have taken. The average EVM Schedule Performance Index (time) or SPI(t) is the slope of the best fit line through the data. (*Note: Unlike Cost Performance it is not the reciprocal of the graph's slope this time around.*)

Depending on the pattern we observe in our Schedule Performance Trend we have all the options discussed in Table 5.11, with the exception that the trend should be projected to the maximum vertical scale value, equivalent to 100% of the Earned Schedule i.e. 100% achievement. By now we will be familiar with our well-voiced mantra '**Never take the log of a date!**' but in this case we can take the log of the horizontal axis because we are not considering Actual Time as a date as such, but as a duration of time that has elapsed from the start. So, in this case we do know when time began. Furthermore, by the very nature of a slip in the start time, it opens up the possibility of using the Generalised form of a Logarithmic or Power Function.

Our two plots can now be recombined to provide a forecast through to completion for both Cost and Schedule in the standard EVM Lazy S-curve format as shown in Figure 5.40.

Now doesn't that just give you a nice warm feeling? So how did we compare with the 'guessedimate' value we wrote down on that piece of paper just before Section 5.5.1?

We can, of course apply any of the transformations discussed in the earlier sections of this chapter in place of a simple linear extrapolation if the underlying trend is progressively getting better or worse. Note that in the case of the Schedule progress we are considering elapsed time from the start or planned start so the '*never take the log of a date*' rule does not apply here.

Table 5.11 Potential Response to EVM Cost Performance Trends

Observed Cost Performance Pattern	Potential Response
1 The graph is perfectly monotonic, increasing, and appears to be a Linear Function	We can project the best fir straight line through to the right hand axis to give us a Cost Estimate At Completion if we believe that the steady state cost performance to date will continue through to the end of the contact or project
2 The graph is perfectly monotonic, increasing, but appears to be a composite of two or more straight lines e.g. one straight line up to and including one level of achievement or Earned Value, breaking to another straight line afterwards	If we believe that the current (latter) steady state cost performance will continue, then we can project it through to the right hand axis. If we understand the reason why the cost performance trend changed, we can make a judgement whether that revised trend will continue, or change again, possibly back to the original steady state trend. Based on our decision we can project the trend through to the right hand axis to get the Cost Estimate At Completion
3 The graph is perfectly monotonic, increasing, but appears to be a nonlinear Function (i.e. it is curved)	Here, we might want to consider whether the observed pattern fits any of the three Standard Function Types i.e. Logarithmic, Exponential or Power (*as if you had forgotten them already*). It may be difficult to justify why we would want to consider the Generalised Forms of these functions in this case as inherently they should pass through the origin (*no cost implies no achievement and vice versa*). Regardless of the transformation type used, the EAC is the value equivalent to the right-hand side.
4 The data is generally, but not perfectly, monotonic and increasing	If we accept the fundamental principle on which EVM is based that Actual Cost is incurred in proportion to Earned Value achieved then the Cost Performance graph should always be monotonic and increasing. However, we will probably accept that mistakes do happen (*e.g. fat fingers inputting data*) and that it is right to make appropriate corrections. If on investigation we find, understand and accept the reason why Actual Cost or Earned Value has reduced, then in an ideal world we should either back flush the change to the point that the error occurred, or remove all the prior data where the error is *known* to exist. As the data is cumulative in nature, this should not adversely affect the underlying true trend. If we find we have this situation, don't forget to document the change in the Basis of Estimate and then check whether the data pattern now fits any of the three prior situations
5 The data is largely not monotonic and/or decreases	We have major data integrity problems, or a creative accountant, or a project manager in denial! If we genuinely do have this, then we need to understand why as it breaks the fundamental principles on which EVM is based

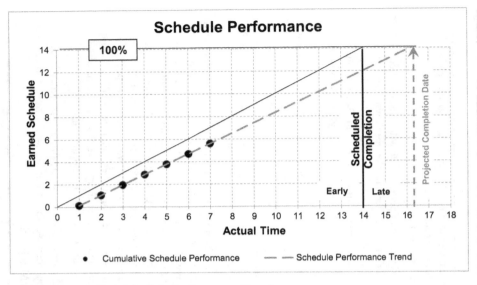

Figure 5.39 EVM Schedule Performance Trend

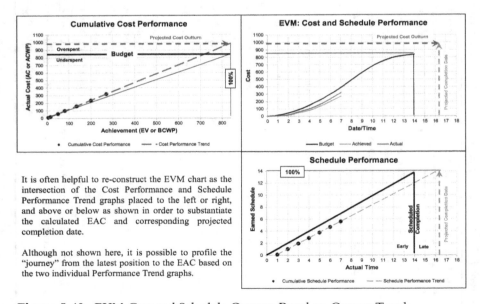

It is often helpful to re-construct the EVM chart as the intersection of the Cost Performance and Schedule Performance Trend graphs placed to the left or right, and above or below as shown in order to substantiate the calculated EAC and corresponding projected completion date.

Although not shown here, it is possible to profile the "journey" from the latest position to the EAC based on the two individual Performance Trend graphs.

Figure 5.40 EVM Cost and Schedule Outturn Based on Current Trends

5.6 Linear transformation based on Cumulative Value Disaggregation

As we saw in Chapter 2 on exploiting linear properties, the Cumulative Value of a straight line is a Quadratic through the origin. If we are confident that the cumulative values we have are correct, but that we have an incomplete history for the constituent units, then we can exploit this property. We can use a polynomial trendline of order 2 to identify the underlying quadratic equation through the cumulative values and then disaggregate it to find the Best Fit straight line through the unit data. (*Yes, I know we said earlier in Section 5.2.5 that we should resist the temptation to use Polynomial Trendlines in Microsoft Excel, but we did add the caveat of 'unless we have a sound rationale' for doing so; this is one of those sound rationales and not a hypocritical convenience!*)

To demonstrate the point we will again consider perfect world data before looking at a real world scenario. Consider a history of design effort to resolve a number of design queries we have received, as depicted in Table 5.12 and Figure 5.41. Plotting a Quadratic Trendline through the origin provides a perfect fit to the data, i.e. one in which the Coefficient of Determination, R^2 (see Volume II Chapter 5) equals 1. (*We did say we would be looking at perfect world scenario initially – don't expect it with the real world stuff!*)

The implication of this is that the unit data is linear and that we can transform the cumulative trend into a linear unit trend. The slope of the line is double the coefficient of the x^2 term, and the intercept is the coefficient of the x term minus that for the x^2 term (see Chapter 2 on Exploiting Linear Properties.) For the case in question:

*   Slope, m equals 0.3 (i.e. twice 0.15)
*   Intercept, c equals 4.85 (i.e. 5 minus 0.15)
*   Straight line is:

The implication is that the effort required in design, y, to resolve each new query gets progressively more as the cumulative number of queries, x increases, presumably because the easier or most obvious queries were raised first.

Although we have not shown it graphically here, we could use the Cumulative Average data in Table 5.12 to indicate that there is an underlying linear relationship; we may recall from Chapter 2 that the Cumulative Average of a discrete straight line is also a straight line of half the slope.

If we now look at an equivalent problem using data more representative of real life (in other words, there is some variation around a perfect 'Best Fit' curve, we can perform the same procedure. Table 5.13 and Figure 5.42 illustrates this. In this case the Trendline is not perfect, but due to the smoothing property of cumulative data, we can expect there to be a very high Coefficient of Determination, R^2 – assuming that the underlying unit relationship is linear, of course.

Figure 5.42 shows that there is a good fit to a Quadratic through the origin. Applying the standard conversion back to a linear function we get a straight line with a slope twice

Linear transformation: Making bent lines straight

Table 5.12 Design Effort in Response to Design Queries (Perfect World)

Month	Design Queries	Cum Queries	Design Hours	Cum Hours	Cum Ave Hours	Hours per Query
1	37	37	390	390	10.55	10.55
2	30	67	618	1008	15.05	20.60
3	31	98	922	1931	19.70	29.75
4	33	131	1299	3229	24.65	39.35
5	27	158	1305	4535	28.70	48.35
6	27	185	1524	6059	32.75	56.45
7	25	210	1606	7665	36.50	64.25
8	22	232	1569	9234	39.80	71.30
9	18	250	1391	10625	42.50	77.30
10	20	270	1660	12285	45.50	83.00
11	18	288	1597	13882	48.20	88.70
12	12	300	1118	15000	50.00	93.20

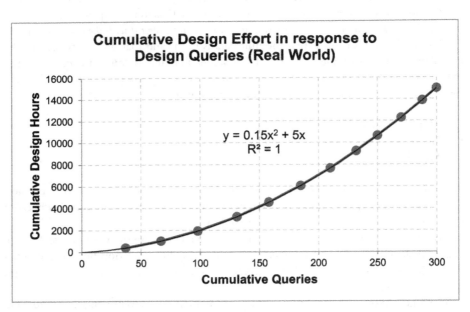

Figure 5.41 Cumulative Design Effort in Response to Design Queries (Perfect World)

that of the quadratic's x^2 term, or 0.29 and an intercept of the difference between the x and x^2 coefficients i.e. 4.965–0.145 = 4.82

In truth, the mathe-magical principle we are using here is differential calculus. (*Now that wasn't too scary, was it?*)

Table 5.13 Design Effort in Response to Design Queries (Real World)

Month	Design Queries	Cum Queries	Design Hours	Cum Hours	Cum Ave Hours	Hours per Query
1	37	37	394	394	10.65	10.65
2	30	67	635	1029	15.36	21.17
3	31	98	882	1911	19.50	28.45
4	33	131	1480	3391	25.89	44.85
5	27	158	735	4126	26.11	27.22
6	27	185	2127	6253	33.80	78.78
7	25	210	746	6999	33.33	29.84
8	22	232	1681	8680	37.41	76.41
9	18	250	1955	10635	42.54	108.61
10	20	270	1341	11976	44.36	67.05
11	18	288	1073	13049	45.31	59.61
12	12	300	1801	14850	49.50	150.08

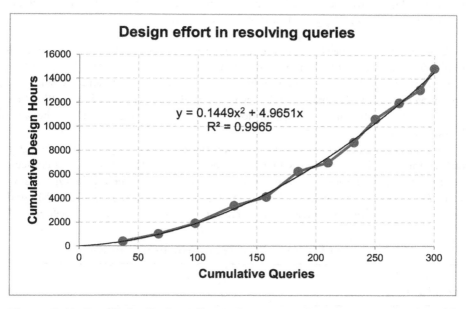

Figure 5.42 Cumulative Design Effort in Response to Design Queries (Real World)

5.7 Chapter review

In this chapter we considered the techniques we can exploit in some cases to transform curved estimating relationships into straight lines. Extrapolating and interpolating straight lines is easier than it is with curves, and arguable is less prone to error.

282 | Linear transformation: Making bent lines straight

Many of the transformations we can use, exploit the properties of logarithms, which themselves exploit the properties of powers. As we saw, logarithms (or logs) effectively compress the difference between large values and stretch the difference between small values. This stretching and compressing creates a bending effect on a line or curve. Obviously we are trying to find one that straightens out a curve. Incidentally, we also discussed that whilst Common Logs (those based on the number 10) and 'Natural' or Napierian Logs (those based on Euler's Number, commonly referred to as 'e') are the most frequently used logs, we can take the log to any base so long as we are consistent within a single transformation.

The properties of logs lead us to consider four Standard Function types:

- Linear Function: Those which are already straight lines and need no transformation
- Logarithmic Functions: Those which transform into straight lines when we take the log of the horizontal x-axis
- Exponential Functions: Those which transform into straight lines when we take the log of the vertical y-axis
- Power Functions: Those which transform into straight lines when we take the log of both the horizontal x-axis and vertical y-axis

The easiest technique for determining which transformation type is appropriate is to plot four versions of the data on graphs with different combinations of linear and log scales. If we select the appropriate trendlines in Excel's Chart utility and also select the option to display the Coefficient of Determination, R^2, we can identify visually which options appear to be linear and which don't. R^2 can be used to discriminate between 'close calls':

In addition to the latter three Standard Nonlinear Functions, there are also three generalised forms of these functions which effectively offset the standard ones by a constant. In order to transform these functions, we must first estimate a value for the Offset Constant. We can do this manually by iteration using the Random-Start Bisection Method, or we can exploit the Solver feature in Microsoft Excel to maximise the Coefficient of

Table 5.14 Function Type Transformation Summary

Function and Trendline Type:	Horizontal x-axis Scale	Vertical y-axis Scale
Linear	Linear	Linear
Logarithmic	Logarithmic	Linear
Exponential	Linear	Logarithmic
Power	Logarithmic	Logarithmic

Linear transformation: Making bent lines straight | 283

Determination (the closer it is to one, the stronger the degree of linearity). Once we have an estimate of the constant we can transform the curves using the technique discussed for Standard forms.

Reciprocal functions of x or of y are special cases of the Power Function.

We then looked at how we can apply these concepts to Earned Value Analysis in a two-step approach to straightening out the traditional Lazy S-curves typical of EVM Reports in order to determine Estimates At Completion and Completion Dates. In these circumstances, we can dis-integrate the EVM data into plots of Actual Cost versus Earned Value or Achievement and Actual Time versus Earned Schedule. We can then examine whether the data trends in either or both of these sub-ordinate plots fit any of the function types (standard or generalised) discussed previously.

Finally, we took a quick look at how we can transform the Cumulative Value of a Linear Function back into a straight line by differentiating the Best Fit Quadratic curve through the data to give us the straight line's slope and intercept from the Quadratic's coefficients.

Perhaps now after all that we can claim that we are truly all transformed characters at last!

References

Box, GEP (1979) 'Robustness in the strategy of scientific model building', in Launer, R. L.; Wilkinson, G. N., Robustness in Statistics, New York, Academic Press, pp. 201–236.

Box, GEP & Draper, NR (1987) *Empirical Model-Building and Response Surfaces*, Hoboken, NJ, John Wiley & Sons, p. 424.

Lipke, W (2003) 'Schedule is Different', *The Measurable News*, Project Management Institute, March.

Turré, G (2006) 'Plant Capacity and Load' in Foussier, P, *Product Description to Cost: A Practical Approach, Volume 1: The Parametric Approach*, London, Springer-Verlag, pp. 141–143.

6 Transforming Nonlinear Regression

A word (or two) from the wise?

"I have got this obsessive compulsive disorder where I have to have everything in a straight line, or everything has to be in pairs."

David Beckham
b.1975
Footballer

Many estimating relationships are not straight lines (even in the short term), but straight lines are easier to forecast (*well, a bit of a doddle really, if the truth be told.*) For this reason estimators like straight lines, and it appears that David Beckham does too (Dolan, 2006). Perhaps he should have become an estimator instead of a footballer! (*David Beckham's compulsion is a little bizarre as his trademark free kicks were usually bent or curled around a line of defenders.*)

In Chapter 4 we discovered the power of Linear Regression in determining a line or a multi-dimensional linear system of 'Best Fit' lines, and in Chapter 5 we looked at some common families of curves that can be transformed into a linear format. Guess what we are going to do here? That's right we are going to transform some of these curves into straight lines and then perform some Linear or Multi-linear Regression on them. The early part of this chapter in effect weaves these two themes together.

6.1 Simple Linear Regression of a linear transformation

In Chapter 5 we identified three basic function types that can be transformed simply into a linear form by taking the Logarithm of either the dependent or independent variable, or both. (*The fourth function type we looked at was the linear function which requires no transformation.*)

We are going to start with Simple Linear Regression of these functions. Before we do that, we must transform them into a linear form, as summarised in Table 6.1.

Transforming Nonlinear Regression | 285

Table 6.1 Basic Function Types

Function Type	Transform by Taking Logarithm of	Example
Logarithmic	Independent Variable, x	Richter Scale for Earthquakes
Exponential	Dependent Variable, y	Economic Inflation
Power	Both Variables, x and y	Learning Curve

The good news is that apart from the transformation bit at the beginning, and the reversal of the transformation at the end, everything in between is the same as we did for Linear Regression in the last but one chapter, so we won't need to repeat all the justification of why we test the output as we do. We will use the same statistics (R2, F and t-values) to decide whether our transformed regression is any good or not. On top of all that, (*you'll be pleased to hear*) we can still derive our Prediction and Confidence Intervals. However . . .

Caveat Augur

The fundamental assumptions of linear regression, i.e. that the Regression Linewill pass through the mean of the data, and that the data will be scattered Normall-yaround the Regression Line only holds true in the linear space, and not inthe logarithmic space.

The Regression will pass through the Arithmetic Mean of any Variable that is not transformed, and through the Arithmetic Mean of any logarithmic data. However, the Arithmetic Mean of logarithmic data is actually the Geometric Mean of the untransformed data in logarithmic dimensions.

For the Formula-philes: The Mean of the logs is the log of the Geometric Mean

Consider a series of values $v_1 \ldots v_n$ with $L_1 \ldots L_n$ being the associated Logarithmic Values to the Base b

By definition, L_i equals:

$$L_i = \log_b v_i \qquad (1)$$

(Continued)

286 | Transforming Nonlinear Regression

The Arithmetic Mean of the Log Values can be expressed as:

$$\frac{1}{n}\sum_{i=1}^{n}L_i = \frac{1}{n}\sum_{i=1}^{n}\log_b v_i \qquad (2)$$

Applying the Additive Property of Logs to (2) (see Chapter 5 Section 5.1.2):

$$\frac{1}{n}\sum_{i=1}^{n}L_i = \frac{1}{n}\log_b\left(\prod_{i=1}^{n}v_i\right) \qquad (3)$$

Applying the Multiplicative Property of Logs to (3) (see Chapter 5 Section 5.1.2):

$$\frac{1}{n}\sum_{i=1}^{n}L_i = \log_b\left(\sqrt[n]{\prod_{i=1}^{n}v_i}\right) \qquad (4)$$

Substituting (1) back into (4):

$$\frac{1}{n}\sum_{i=1}^{n}\log_b v_i = \log_b\left(\sqrt[n]{\prod_{i=1}^{n}v_i}\right)$$

... which demonstrates that the Arithmetic Mean of the Logs of set of values is equal to the Log of the Geometric Mean of the values themselves

Table 6.2 summarises how the regression is 'pegged' for the different function types, i.e. the point through which the Regression Line will pass.

Table 6.3 provides an example of this with a set of linear values and their logarithmic equivalents in Base 10 (*Common Logs*), Base e (*Natural Logs*) and Base 2, clearly showing that the choice of logarithmic base is irrelevant.

Does this idiosyncrasy matter, i.e. that the sum of the error data in logarithmic space will be zero, but in linear space it won't?

- In practice, most people ignore this fact; although in truth many are blissfully unaware of it!
- As estimators we may be reasonably comfortable with this 'anomaly' because in truth any difference in the error terms that this makes is probably no more or less significant than any inherent variance in the 'actual' data we have used.
- Who's to say that logarithmic space is not the true reality rather than linear space? OK, perhaps that was a bit philosophical, but there was a time that people thought

Table 6.2 Means Through which the Nonlinear Regression Passes

Function Type	Regression Line Passes Through	
	x-Variable	*y-Variable*
Linear	Arithmetic Mean	Arithmetic Mean
Logarithmic	Geometric Mean	Arithmetic Mean
Exponential	Arithmetic Mean	Geometric Mean
Power	Geometric Mean	Geometric Mean

Transforming Nonlinear Regression | 287

Table 6.3 Arithmetic Mean of Logarithmic Values Equals the Geometric Mean of the Linear Values

Obs	Value	LOG(Value)	LN(Value)	LOG$_2$(Value)
1	5	0.6990	1.6094	2.3219
2	3	0.4771	1.0986	1.5850
3	7	0.8451	1.9459	2.8074
4	3	0.4771	1.0986	1.5850
5	8	0.9031	2.0794	3.0000
6	9	0.9542	2.1972	3.1699
7	4	0.6021	1.3863	2.0000
8	8	0.9031	2.0794	3.0000
9	6	0.7782	1.7918	2.5850
10	2	0.3010	0.6931	1.0000

Arithmetic Mean	5.5	0.6940	1.5980	2.3054
Geometric Mean	4.9431			
10$^{\text{Arithmetic Mean}}$		4.9431		
e$^{\text{Arithmetic Mean}}$			4.9431	
2$^{\text{Arithmetic Mean}}$				4.9431

that the earth was flat. (*If you are a modern day flat-earther, no offence is intended and I hope that I haven't undermined your belief system!*)

If you are really uncomfortable with the potential bias in the linear error term not summating to zero, we consider an alternative in Chapter 7 (*but we should be asking ourselves whether it is really worth the trouble?*)

The procedure for running a Simple Linear Regression of a linear transformation is a four or five step procedure:

1. Identify the function type that best reflects our data, and convert the appropriate variable values into logarithmic values as appropriate (as per Table 6.1).
2. Perform the Regression on the transformed data
3. Interpret the results in terms of their goodness of fit in the 'transformed' Linear Space
4. If the Regression is valid, generate the Regression Line and create the Prediction and Confidence Intervals around it.
5. For Exponential and Power Functions, convert the results back into Linear Space; this is not required for Logarithmic Functions as they are already in linear output format.

We will look at an example of each in Sections 6.1.1 to 6.1.3

6.1.1 Simple Linear Regression with a Logarithmic Function

Let's consider a production system in which we have recorded the number of design queries raised as each successive production unit is completed. Intuitively we would expect the rate of query arising to fall off as we get further into production. In Figure 6.1 we have determined visually that the relationship appears to be a Logarithmic Function. For a Logarithmic Function we can perform a Linear Regression based on the original dependent variable (y) against the linear transformation of the Production Build Number (the logarithmic values) which constitutes the independent variable (x).

Note: Even though the example depicts unique sequential values of the independent variable this is not a requirement. There can be missing values or duplicated values in the data.

We will note that the standard Arithmetic Mean of the raw data does not lie on the Logarithmic Trendline but sits within the concave shape of the curve – in short, the trendline appears to be too low.

The Regression results, generated by the usual Data Analysis Add-in facility within Microsoft Excel, are shown in Table 6.4.

We can now generate the Regression results, based on the transformed values of the x-variable. As the dependent variable is still in 'linear space' we don't have to reverse the transformation. (*The sighs – and size – of relief were noted! Enjoy it while you can because this won't be the case when we consider Exponential and Power Functions! Forewarned is forearmed.*)

In Table 6.5 we have extended the analysis to generate the Lower and Upper Prediction Limits (LPL and UPL) for a 90% Prediction Interval. (The calculation details are a direct follow-on from those for Prediction Intervals in Chapter 5 for 'normal' Linear Regression, but applied here to the Logarithmic values of the independent variable. The Confidence Interval of the Regression Line can also

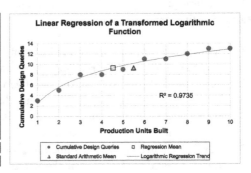

Figure 6.1 Linear Regression of a Transformed Logarithmic Function

Transforming Nonlinear Regression | 289

Table 6.4   Linear Regression of a Transformed Logarithmic Function

SUMMARY OUTPUT

Regression Statistics		
Multiple R	0.986663824	
R Square	0.973505502	
Adjusted R Square	0.970193689	
Standard Error	0.581494657	
Observations	10	

Based on R-Square, F and t-statistics
we would accept this model as a
statistically significant result

ANOVA

	df	SS	MS	F	Significance F
Regression	1	99.39491171	99.39491171	293.9494796	1.36187E-07
Residual	8	2.705088285	0.338136036		
Total	9	102.1			

	Coefficients	Standard Error	t Stat	P-value	Lower 95%	Upper 95%
Intercept	2.452271286	0.439699368	5.577154446	0.000524041	1.438322725	3.466219847
x Log(Build)	10.43898793	0.608866455	17.14495493	1.36187E-07	9.034939369	11.8430365

Table 6.5   Prediction Interval Around the Linear Regression of a Transformed Logarithmic Function

			Regression Coefficients			Pred					
			Intercept	Log(Build)		Interval					
			2.452	10.439		90%					
Obs	y Cum Design Queries	Production Build Number	Unity	x Log(Build)	ŷ Regression Output	t	Standard Error	Z Matrix	Half Prediction Interval	LPL	UPL
1	3	1	1	0.0000	2.452	1.860	0.581	0.572	1.356	1.097	3.808
2	5	2	1	0.3010	5.595	1.860	0.581	0.238	1.203	4.392	6.798
3	8	3	1	0.4771	7.433	1.860	0.581	0.135	1.152	6.281	8.585
4	8	4	1	0.6021	8.737	1.860	0.581	0.103	1.136	7.601	9.873
5	9	5	1	0.6990	9.749	1.860	0.581	0.102	1.135	8.614	10.884
6	11	6	1	0.7782	10.575	1.860	0.581	0.116	1.143	9.433	11.718
7	11	7	1	0.8451	11.274	1.860	0.581	0.139	1.154	10.120	12.428
8	12	8	1	0.9031	11.880	1.860	0.581	0.167	1.168	10.712	13.048
9	13	9	1	0.9542	12.414	1.860	0.581	0.198	1.183	11.230	13.597
10	13	10	1	1.0000	12.891	1.860	0.581	0.230	1.199	11.692	14.090

be determined in a similar manner.) When plotted in Figure 6.2 we can observe that the Prediction Interval is symmetrical around the Regression Line. The upper graph is the linearised version in logarithmic space and the lower graph is the same data but plotted in linear space. The potential scatter of points around the Regression Line again follows a Student's t-Distribution, which as we discussed previously in Volume II Chapter 6, tends towards a Normal Distribution as the number of regression points increase.

Visually, it may appear to be three parallel lines but this is just an optical illusion; it is slightly hourglass shaped.

Note that both vertical scales are the same, hence why we did not need to reverse the transformation.

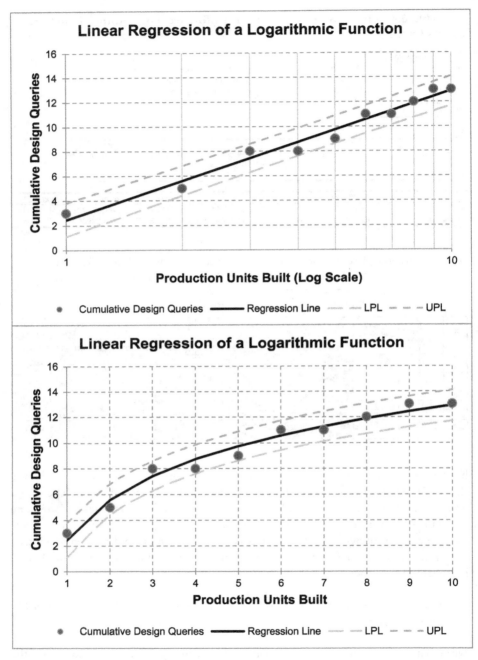

Figure 6.2 Prediction Interval Around the Linear Regression of a Transformed Logarithmic Function

Transforming Nonlinear Regression | 291

> ## Caveat augur
>
> Before we get carried away by the excitement of all this, recall our wise warning from Chapter 5.
>
> **Never take the log of a date**
>
> Just because we can, doesn't mean that we should! It implies that we know when time began.

Let's turn our attention now to the second of these standard linear transformations – the Exponential Functions.

6.1.2 Simple Linear Regression with an Exponential Function

Let's consider whether the annual cost of purchasing the raw material and a number of bought out components to manufacture a standard domestic air conditioning unit is broadly in line with the general rate of inflation of 3.3%. First we must review the underlying rate of escalation. Figure 6.3 confirms that we appear to have an Exponential Function. (Even though the example depicts unique sequential values of the independent variable this is not a requirement. There can be missing values or duplicated values in the data.)

Taking a simplistic Geometric Mean of the nine annual rates of inflation is equivalent simply to taking the ninth root of the ratio of the last year's cost divided by the base year's cost (see Volume II Chapter 2 on Geometric Means) and would confirm that the average rate of inflation was 3.3% per annum. However, as estimators, always keen

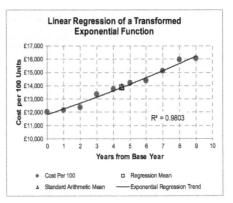

Figure 6.3 Linear Regression of a Transformed Exponential Function

292 | Transforming Nonlinear Regression

to look at things from an alternative perspective (*it's in the blood*) we can see that had we just looked at years −07 through to −13, we would have concluded that the average rate was much higher, being just a shade under 4.4%, Is this an indicator that someone is cashing in on global warming?). Who's to say that any two points are any better or more reliable than any other pair? Surely it is better to give them all equal billing and to run a regression through the entire data set at our disposal?

Table 6.6 shows us the result of performing a regression of the Log of the Cost per 100 units (y-value) with the Relative Year (i.e. number of years since our defined Base Year (x-value). However, the coefficients are both logarithmic values and need to be returned to normal linear values for us to fully understand them. (Recall Chapter 5, Section 5.2.3). In this case as we have used the default Common Logs (base 10), we only have to raise ten to the power of the appropriate coefficient to get a value that is really meaningful (see Table 6.7.)

Table 6.6 Linear Regression of a Transformed Exponential Function

SUMMARY OUTPUT

Regression Statistics	
Multiple R	0.990086834
R Square	0.980271939
Adjusted R Square	0.977805931
Standard Error	0.006969417
Observations	10

Based on R-Square, F and t-statistics we would accept this model as a statistically significant result

ANOVA

	df	SS	MS	F	Significance F
Regression	1	0.019308344	0.019308344	397.5137425	4.17496E-08
Residual	8	0.000388582	4.85728E-05		
Total	9	0.019696926			

	Coefficients	Standard Error	t Stat	P-value	Lower 95%	Upper 95%
Intercept	4.072875408	0.004096301	994.2812119	1.17255E-21	4.06342932	4.082321495
x Relative Year No	0.015298383	0.000767308	19.93774668	4.17496E-08	0.013528969	0.017067798

Table 6.7 Reversing the Logarithmic Transformation of the Regression Coefficients

Output Coefficient Relating to:	Output Coefficient Log Value (Rounded)	$10^{Coefficient\ Log\ Value}$	Description of Untransformed Coefficient
Intercept	4.0729	£11,827	Theoretical Base Year cost
Relative Year No	0.0153	1.036	Mean escalatory factor per annum => 3.6% mean escalation per annum

This statement immediately appears to contradict itself by referring to a *'Theoretical Value'* for the Base Year Cost, prompting the question *'Why do I need that when I have the actual cost?'* However, think of it instead as a level of cost that is more representative of the underlying pattern or trend. The *'actual'* cost may be artificially high or low in the context of what happens after (which we know) or before (which we either don't know or have excluded.) If we had defined −12 to be the Base Year then we would probably not be having this debate as the simple graphical trend appears to go very close to that point.

If we are still uncomfortable with that rationale, then an alternative strategy would be to create an index based on the Base Year's actual cost index being 1 (equivalent to the logarithmic Origin); this is simply a normalisation step of dividing all annual costs by the Base Year Cost. We can then take the Log of the Index and force a Regression through the origin. This would give us the Best Fit rate of escalation from that fixed start point. As we can see from Figure 6.4, the result (equivalent to 3.35% average escalation) is much closer to the ninth root of the Geometric Mean with which we began this little exploration. (In this case then, we might conclude why bother doing the regression?)

However, sticking with the original regression, we can create the regression line and reverse the transformation as in Table 6.8 by raising ten (the logarithmic base we used) to the power of the corresponding log-based output column we want (e.g. Linear UPL = $10^{\text{Log UPL}}$)

Figure 6.5 presents the data graphically. In logarithmic scale (upper graph) the Lower and Upper Prediction Limits are equidistant from the Regression Line. However, once the reverse transformation has taken place, as shown in Table 6.8, the Prediction Interval

Year	Material & Bought Out Cost per 100	Material & Bought Out Cost Index	y Log(Index)	x Relative Year No
−05	£11,999	1.0000	0.0000	0
−06	£12,145	1.0122	0.0053	1
−07	£12,345	1.0288	0.0123	2
−08	£13,335	1.1113	0.0458	3
−09	£13,750	1.1459	0.0592	4
−10	£14,210	1.1843	0.0734	5
−11	£14,380	1.1984	0.0786	6
−12	£15,125	1.2605	0.1006	7
−13	£15,950	1.3293	0.1236	8
−14	£16,071	1.3394	0.1269	9
Arithmetic Mean		1.1610	0.0626	4.5
$10^{\text{Arithmetic Mean}}$	(Anti-log) >	1.155		

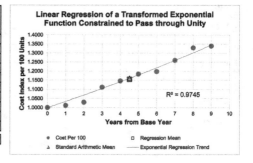

Figure 6.4 Linear Regression of a Transformed Exponential Function Constrained Through Unity

Table 6.8 Prediction Interval Around the Linear Regression of a Transformed Exponential Function

			Regression Coefficients			Pred Interval						Linear Values	
			Intercept	Rel Year								ŷ	
			4.073	0.015	Log Value	90%			Logarithmic Values			Regression	
Year	Material & Bought Out Cost per 100	y Log(Cost)	x Unity	Relative Year No	ŷ Regression Output	t	Standard Error	Z Matrix	Half Prediction Interval	LPL	UPL	LPL	Output	UPL
−05	£11,999	4.0791	1	0	4.073	1.860	0.007	0.345	0.015	4.058	4.088	£11,425	£11,827	£12,244
−06	£12,145	4.0844	1	1	4.088	1.860	0.007	0.248	0.014	4.074	4.103	£11,849	£12,251	£12,666
−07	£12,345	4.0915	1	2	4.103	1.860	0.007	0.176	0.014	4.089	4.118	£12,286	£12,690	£13,108
−08	£13,335	4.1250	1	3	4.119	1.860	0.007	0.127	0.014	4.105	4.133	£12,735	£13,145	£13,568
−09	£13,750	4.1383	1	4	4.134	1.860	0.007	0.103	0.014	4.120	4.148	£13,196	£13,617	£14,050
−10	£14,210	4.1526	1	5	4.149	1.860	0.007	0.103	0.014	4.136	4.163	£13,670	£14,105	£14,554
−11	£14,380	4.1578	1	6	4.165	1.860	0.007	0.127	0.014	4.151	4.178	£14,155	£14,611	£15,081
−12	£15,125	4.1797	1	7	4.180	1.860	0.007	0.176	0.014	4.166	4.194	£14,652	£15,134	£15,632
−13	£15,950	4.2028	1	8	4.195	1.860	0.007	0.248	0.014	4.181	4.210	£15,163	£15,677	£16,209
−14	£16,071	4.2060	1	9	4.211	1.860	0.007	0.345	0.015	4.196	4.226	£15,687	£16,239	£16,811

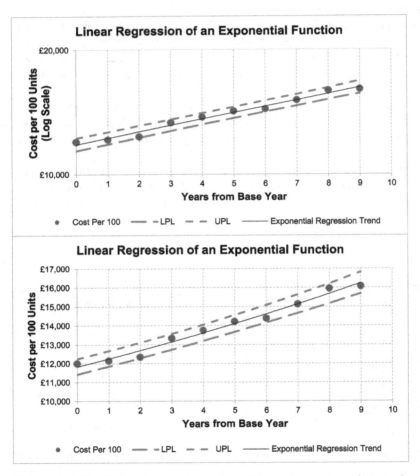

Figure 6.5 Prediction Interval Around the Linear Regression of a Transformed Exponential Function

becomes positively skewed due to the compressing and stretching properties of logarithms (see Chapter 5, Section 5.1.2). We may also recall the following:

- Points are scattered around a Linear Regression Line in a Student's t-Distribution (Chapter 5) or approximately Normally Distributed if we have more than 30 data points.
- By implication, the nature of the linear transformation of the nonlinear Exponential Function suggests that the scatter of the transformed points in logarithmic space will also be approximately Normally Distributed.
- If the logarithmic values are approximately Normally Distributed, then the equivalent linear values can be assumed to be approximately Lognormally Distributed (Volume II Chapter 4.) This leads us to conclude that the scatter of points is positively skewed. This also leads us to conclude that the Regression Line represents the 50% Confidence Level for the predicted dependent variable and not the Mean, because the Mean of the Normal Distribution translates as the Median of the Lognormal Distribution (covered in Volume II Chapter 4), so data is more closely packed under the line than over it. (In linear space, they are the same value, so it doesn't matter which we cite.)

Other options in Microsoft Excel

In Chapter 4 we discussed the Microsoft Excel Function **TREND** which allows us to create a forecast quickly based on the underlying Simple Linear Regression trend. For Exponential Functions, there is a similar function within Excel that allows us to generate a growth forecast. The full syntax of the function is:

- **GROWTH(*known_ys, known_xs, new_xs, const*)**
  - o Where **known_ys** and **known_xs** refer to the location of the y and x range of values; the location must be a set of contiguous cells (i.e. adjacent to each other) in either a row or a column format, but not mixed. Excel assumes (logically) that the values are paired together in the order that they are presented.
  - o The range **new_xs** refers to the location of a set of values for which we want to create the Line of Best Fit values; this could be one or more of the **known_xs** or one or more new values, or both. In theory, this value accepts a range input which can be enabled by Ctrl+Shift+Enter but it works just as well and more straightforwardly with a single cell that we can then copy across a range. For a single *new x* cell we can just press the Enter key as we would for any normal calculation.
  - o The final parameter, **const** allows us to force the regression through the origin rather than the Mean of the data. As a general rule, we should have a good reason for doing this; to force it through the origin set **const** to FALSE; otherwise set it to TRUE. The implication of the Log Intercept being zero is that in real linear space the constant is 1.

296 | Transforming Nonlinear Regression

> ## Caveat augur
>
> As a general rule, array formulae or array functions such as **GROWTH** are discouraged as they do not function correctly if the user forgets to enable them with Ctrl+Shift+Enter instead of the usual Enter key.
>
> Unfortunately, a failure to use Ctrl+Shift+Enter may not necessarily return an error so the user may not be aware of their mistake. To avoid this, we can restrict the function's range parameter *new_xs* to a single cell, which can then be copied to adjacent cells. In this way **GROWTH** will work with the normal Enter key.

We have another option to perform an Exponential Regression in Microsoft Excel using the more advanced function **LOGEST**, equivalent and very similar to **LINEST** that we discussed in Chapter 4. (*Yes, I agree, perhaps Microsoft could have helped us by being more logical in naming the function as* EXPEST *to align our thinking to Exponential Functions, but perhaps that sounded too much like a dead rodent or insect. Anyway, they didn't so let's stop waffling about it and move on.*)

The full syntax of **LOGEST** is:

- **LOGEST(*known_ys, known_xs, const, stats*)** where:
  - o   *known_ys* and *known_xs* refer to the location of the y and x range of values; the location must be a set of contiguous cells (i.e. adjacent to each other) in either a row or a column format, but not mixed. Excel assumes (logically) that the values are paired together in the order that they are presented. This relates directly to the untransformed y-variable data – we don't have to take log values first.
  - o   The parameter *const* allows us to force the Regression through the origin rather than the Mean of the data. As a general rule, we should have a good reason for doing this; to force it through the origin set *const* to FALSE; otherwise set it to TRUE.
  - o   The final parameter *stats* is asking us to specify whether we want Excel to calculate the supporting statistics (TRUE) or not (FALSE). As a general rule we should always say 'TRUE' to this, otherwise how else will we know whether the **LOGEST** results are any good?

As with **LINEST**, **LOGEST** is an array function (which we normally discourage from using . . . see Volume I Chapter 3) but which we can interrogate more easily in combination with the **INDEX** function:

INDEX(LOGEST(known_ys, known_xs, Const, Stats), *row_num, column_num*)

This gives us the outputs as summarised in Table 6.9.

Transforming Nonlinear Regression | 297

Table 6.9 LOGEST Output Data for a Simple Exponential Regression

		column_num	
		1	2
row_num	1	Exponential Base (which is raised to the power of x)	Intercept
	2	Standard Error for the Natural Log of the Slope	Standard Error for the Natural Log of the Intercept
	3	Coefficient of Determination R-Square	Standard Error for y Estimate
	4	F-statistic for the Regression Model	Number of Degrees of Freedom in the Residuals
	5	Sum of Squares for the Regression Model	Sum of Squares for the Residual Error

Table 6.10 A Comparison on LOGEST Output with LINEST Output on the Transformed Data

INDEX		LINEST of LOG data			LINEST of LN data			LOGEST of data		
row	col	Value	Description	10^{value}	Value	Description	e^{value}	Value	Description	e^{value}
1	1	0.0153	< Log(m) m >	1.036	0.0352	< LN(m) m >	1.036	1.036	< m	N/A
1	2	4.0729	< Log(c) c >	11827.022	9.3781	< LN(c) c >	11827.022	11827.022	< c	N/A
2	1	0.0008	< SE Log(m) SE m >	1.002	0.0018	< SE LN(m) SE m >	1.002	0.0018	< SE LN(m) SE m >	1.002
2	2	0.0041	< SE Log(c) SE c >	1.009	0.0094	< SE LN(c) SE c >	1.009	0.0094	< SE LN(c) SE c >	1.009
3	1	0.9803	< R-Square	N/A	0.9803	< R-Square	N/A	0.9803	< R-Square	N/A
3	2	0.0070	< SE Log(y) SE y >	1.016	0.0160	< SE LN(y) SE y >	1.016	0.0160	< SE LN(y) SE y >	1.016
4	1	397.514	< F-Statistic	N/A	397.514	< F-Statistic	N/A	397.514	< F-Statistic	N/A
4	2	8	< Res df	N/A	8	< Res df	N/A	8	< Res df	N/A
5	1	0.0193	< Reg SS	See note c	0.1024	< Reg SS	See note c	0.1024	< Reg SS	See note c
5	2	0.0004	< Res SS		0.0021	< Res SS		0.0021	< Res SS	

In terms of how this differs from using the transformation option with **LINEST** we can repeat the example from Figure 6.3 and Table 6.6. Table 6.10 compares the results.

Notes:

a. Some of the data are identical in all three versions
b. The reversal of the appropriate logarithmic transformation shows that Common Logs and Natural Logs give equivalent results. To convert between the two, we just have to multiply by the Natural Log of 10 i.e.
c. To convert between the Sum of Squares (SS) for Common Logs and Natural Logs, we just need to multiply by the square of the Natural Log of 10, i.e.

The only real difference between the middle and right-hand options is that with **LOGEST** we don't have to convert back the Regression Coefficients to real numbers (*in both senses of the term – don't worry about it, mathematicians have a strange sense of humour.*)

6.1.3 Simple Linear Regression with a Power Function

Now let's consider Power Function. Along with the linear relationship group, this group probably forms the most common relationships we will find and use as estimators, Chilton's Law, and its Cost-Weight equivalents in other industries are Power Functions. The classic Learning Curve which we will discuss in Volume IV is another example.

Let's consider such a Learning Curve in Figure 6.6. As we will come to expect after Volume IV, we have what appears to be a good fit to a Power Trendline in Microsoft Excel. This can be corroborated if we first transform the data into a linear format by taking the logarithm of both the dependent and independent variables (i.e. the y and x variables) as illustrated in Figure 6.6. (Even though the example depicts unique sequential values of the independent variable this is not a requirement. There can be missing values or duplicated values in the data, and the data need not be organised sequentially.) If we run our Linear Regression on our logarithmic values, it will give us the Regression Output Summary in Table 6.11.

Build Number	Assembly Time (Hrs)	y Log(Time)	x Log(Build)
1	1,024	3.0103	0.0000
2	875	2.9420	0.3010
3	654	2.8156	0.4771
4	612	2.7868	0.6021
5	525	2.7202	0.6990
6	567	2.7536	0.7782
7	421	2.6243	0.8451
8	444	2.6474	0.9031
9	451	2.6542	0.9542
10	408	2.6107	1.0000
Arithmetic Mean	5.50	598.10	2.7565	0.6560
10^Arithmetic Mean		(Anti-log)>	570.81	4.53

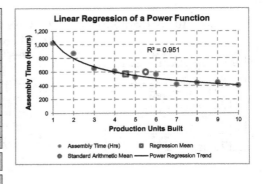

Figure 6.6 Linear Regression of a Transformed Power Function

Table 6.11 Linear Regression of a Transformed Power Function
SUMMARY OUTPUT

Regression Statistics	
Multiple R	0.975200411
R Square	0.951015842
Adjusted R Square	0.944892822
Standard Error	0.03185154
Observations	10

Based on R-Square, F and t-statistics we would accept this model as a statistically significant result

ANOVA

	df	SS	MS	F	Significance F
Regression	1	0.157573422	0.157573422	155.3181085	1.6061E-06
Residual	8	0.008116165	0.001014521		
Total	9	0.165689586			

	Coefficients	Standard Error	t Stat	P-value	Lower 95%	Upper 95%
Intercept	3.029138415	0.024084662	125.7704367	1.78566E-14	2.973599085	3.084677744
x Log(Build)	-0.415640446	0.033350838	-12.46266859	1.6061E-06	-0.492547617	-0.338733275

Transforming Nonlinear Regression | 299

Following that we can generate the Prediction Interval in the transformed linear space before reversing the transformation to get it all back into a nice, friendly real space format by raising 10 (if using Common Logs) to the power of the log data in the appropriate columns as illustrated in Table 6.12.

In terms of calculating the key parameters to allow us to use the basic formula in real space, and not have to convert to and fro in Log space, we only need to convert the intercept into a theoretical value at value Build Unit 1. (For a logarithmic intercept to equal zero, it implies that the corresponding linear value must be 1, as by definition any logarithmic base raised to the power of zero equals 1.)

Caveat augur

Just as we counselled for logarithmic functions, recalling our wise warning from Chapter 5.

Never take the Log of a date

Just because we can, doesn't mean that we should!

6.1.4 Reversing the transformation of Logarithmic, Exponential and Power Functions

Although we have just considered the reversal of these transformations in the respective sections, wouldn't you like a one-stop-shop option for them all (see Table 6.13)? (*Despite what you may think at times, I am trying to empathise with you!*)

However, if we have a linear model that ticks all our 'goodness of fit' criteria, but we also have a valid linearised model of a different functional form, how do we compare the two and decide which is the 'best of the best'? We cannot simply compare the statistics reported by Microsoft Excel at a face value because in effect the transformation process

Table 6.12 Prediction Interval Around the Linear Regression of a Transformed Power Function

			Regression Coefficients			Pred Interval						Linear Values		
			Intercept	Log(Build)		90%								
		Log Value	3.029	-0.416	Log Value				Logarithmic Values					
Build Number	Assembly Time (Hrs)	y Log(Time)	Unity	x Log(Build)	ŷ Regression Output	t	Standard Error	Z Matrix	Half Prediction Interval	LPL	UPL	LPL	ŷ Regression Output	UPL
1	1,024	3.0103	1	0.0000	3.029	1.860	0.032	0.572	0.074	2.955	3.103	901.3	1,069.4	1,268.8
2	875	2.9420	1	0.3010	2.904	1.860	0.032	0.238	0.066	2.838	2.970	688.8	801.7	933.1
3	654	2.8156	1	0.4771	2.831	1.860	0.032	0.135	0.063	2.768	2.894	585.8	677.4	783.3
4	612	2.7868	1	0.6021	2.779	1.860	0.032	0.103	0.062	2.717	2.841	520.8	601.0	693.6
5	525	2.7202	1	0.6990	2.739	1.860	0.032	0.102	0.062	2.676	2.801	474.7	547.8	632.1
6	567	2.7536	1	0.7782	2.706	1.860	0.032	0.116	0.063	2.643	2.768	439.7	507.8	586.5
7	421	2.6243	1	0.8451	2.676	1.860	0.032	0.139	0.063	2.615	2.741	411.8	476.3	550.9
8	444	2.6474	1	0.9031	2.654	1.860	0.032	0.167	0.064	2.590	2.718	388.9	450.6	522.1
9	451	2.6542	1	0.9542	2.633	1.860	0.032	0.198	0.065	2.568	2.697	369.6	429.1	498.1
10	408	2.6107	1	1.0000	2.613	1.860	0.032	0.230	0.066	2.548	2.679	353.0	410.7	477.7

300 | Transforming Nonlinear Regression

Table 6.13 Reversing the Transformation of the Standard Functions

	Regression Line	Intercept Coefficient	Slope Coefficient
Regression Value	$Y = MX + C$	C	M
Logarithmic Function	No need to reverse the transformation: $c^y = mx$ $$y = \frac{\log_b x}{\log_b c} + \frac{\log_b m}{\log_b c}$$	$$C = \frac{\log_b m}{\log_b c}$$ $m = b^{MC}$	$$M = \frac{1}{\log_b c}$$ $c = b^{-M}$
Exponential Function	Raise the logarithmic base (e.g. b=10) to the power of the Regression Line's calculated logarithmic values $y = cm^x$	$C = \log_b c$ Raise the logarithmic base (e.g. b=10) to the power of the Intercept Coefficient $c = b^C$	$M = \log_b M$ Raise the logarithmic base (e.g. b=10) to the power of the Slope Coefficient $m = b^M$
Power Function	Raise the logarithmic base (e.g. b=10) to the power of the Regression Line's calculated logarithmic values $y = cx^m$	$C = \log_b c$ Raise the logarithmic base (e.g. b=10) to the power of the Intercept Coefficient $c = b^C$	$M = m$ No need to make a reverse transformation

has distorted the measurement scale. (*We wouldn't compare a physical measurement in inches with one made in centimetres, would we?*) We will address this in Section 6.4 after we have discussed the delights of Multi-Linear Regression.

6.2 Multiple Linear Regression of a multi-linear transformation

This is where things can begin to get a little more complicated. (*OK, I was trying to be gentle with you. So, let me re-phrase that . . . things will get a lot more complicated.*)

Just as we could extend the concept of linear regression into multi-linear regression, so too can we extend the principle to linearised transformations of multiple nonlinear regression . . . at least to a point. We have to stick with the same basic function type throughout or limit their combination to two mutually exclusive groups:

- Logarithmic Functions can only be used in conjunction with basic Linear Functions (and *vice versa*)
- Power Functions can only be used in conjunction with Exponential Functions and *vice versa*

Figure 6.7 illustrates the combinations as the horizontal rows of the diagram. The reason is simply that the left-hand side of the equation is the dependent variable that we are trying to estimate and we can only do one of two things but not both; we either leave it in a linear form or we transform it to a logarithm. (*It boils down to a choice of LILI – log it or leave it*). What we choose to do on the right-hand side with the multiple independent variables or drivers is unrestricted. We can have any combination of linear and logarithmic-based x variables, but the function types will be restricted to a choice of two depending on our LILI decision about y.

For LILI, horizontal integration works; vertical integration doesn't.

However, there are some 'rules' that we should or have to respect:

- **Never take the log of a date** – always keep it linear (*yes, I know I'm repeating myself – I prefer to think of it as positive reinforcement, although some would call it nagging!*)
- **Never take the log if the variable could legitimately be zero or negative** – always keep it linear

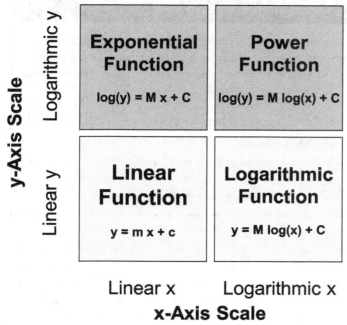

Figure 6.7 Natural Pairings of Function Types for Linear Regression

302 | Transforming Nonlinear Regression

- **Never take the log of a Categorical Variable,** e.g. a switch – always keep it linear because TRUE and FALSE take the notional values of 1 and 0 respectively *(and giving them notional values of 1 and 2 just distorts the idea of a switch)*
- In general, avoid using a variable in both linear and logarithmic forms as they will be highly correlated – work out which gives the best contribution to the behaviour of the dependent variable. *(Well spotted, this is more of a guideline than a rule – there can be exceptions that work.)*

6.2.1 Multi-linear Regression using linear and linearised Logarithmic Functions

In Chapter 4 we saw that a Multiple Linear Regression was a combination of a number of independent linear variables (x's) each contributing an effect on the dependent variable (y) through its slope coefficient, plus a single constant. (Note that 'plus' does not exclude 'minus' as it is generally taken that 'minus' means 'plus a negative value'.)

We can extend this such that the linear combinations are all logarithmic transformations, or a combination of linear variables and logarithmic transformations.

For the Formula-philes: Multiple linearised Logarithmic Regression

Consider a range of n drivers $x_1, x_2, x_3 \ldots x_n$ all of which have a logarithmic relationship with the value of y, the entity which we wish to estimate. Let β_i represent the linear contribution (slope) of the corresponding $\log_b x_i$ variable, and β_0 is the Intercept Constant of the model.

Multi-linear model
$$y = \beta_0 + \beta_1 \log_b x_1 + \beta_2 \log_b x_2 + \beta_3 \log_b x_3 + \ldots + \beta_n \log_b x_n \quad (1)$$

Applying the Additive and Multiplicative
Properties of Logs to (1):
$$y = \log_b \left(b^{\beta_0} x_1^{\beta_1} x_2^{\beta_2} x_3^{\beta_3} \ldots x_n^{\beta_n} \right) \quad (2)$$

Taking the anti-log of (2)
to the base b:
$$b^y = b^{\beta_0} x_1^{\beta_1} x_2^{\beta_2} x_3^{\beta_3} \ldots x_n^{\beta_n}$$

A logical question we may ask ourselves is *'What if all the logarithmic bases are different?'* the answer to which is that it doesn't matter – it is purely cosmetic and can be resolved easily through a basic data normalisation step. It is a fundamental property of Logarithms that the Log of a number to a base can be expressed as a Log to another base by applying a constant factor which is the Log of the first base to the log of the second base.

It must be said that such multiple log combination models are fairly rare but not altogether impossible.

Transforming Nonlinear Regression | 303

For the Formula-philes: Normalisation between log bases

Consider four constants a, b, c and d:

Suppose the relationship between a, b and c is:

$$a = \log_b c \qquad (1)$$

Raising b to the power of each side of the equation of (1):

$$b^a = b^{\log_b c} \qquad (2)$$

Using the definition of a log of a value as that power to which the base must be raised to get the value, (2) becomes:

$$b^a = c \qquad (3)$$

Taking the log of (2) to the base d:

$$a \log_d b = \log_d c \qquad (4)$$

Substituting a from (1) in (3):

$$\log_b c \log_d b = \log_d c \qquad (5)$$

Simplifying (5):

$$\log_b c = \frac{\log_d c}{\log_d b}$$

Another rare relationship is one that combines linear and logarithmic functions. For simplicity here we will consider only one of each function type: Linear and Logarithmic.

Don't forget that the linear value may be a categorical value to depict a 'level shift' between two conditions. (There could be more than two categories but this does impose certain limitations, although these can be overcome using multiple 'binary on/off switches' – see the discussion in Chapter 5.

For the Formula-philes: Combined linear and logarithmic function Regression

Consider two drivers x_1 and x_2 the first of which has a linear relationship with the value of y, and the second of which has a logarithmic relationship with the value of y; y being the entity which we wish to estimate. Let β_i represent the linear contribution (slope) of the corresponding x_i or $\log_b x_i$ variable, and β_0 be the Intercept Constant of the model.

Multi-linear model:

$$y = \beta_0 + \beta_1 x_1 + \beta_2 \log_b x_2 \qquad (1)$$

Applying additive and multiplicative properties of Logs to (1):

$$y = \log_b \left(b^{\beta_0 + \beta_1 x_1} x_2^{\beta_2} \right) \qquad (2)$$

Taking the Anti-log of (2) to the base b:

$$b^y = b^{\beta_0 + \beta_1 x_1} x_2^{\beta_2} \qquad (3)$$

Simplifying (3) with constants $k_0 = b^{\beta_0}$ and $k_1 = b^{\beta_1}$:

$$b^y = k_0 k_1^{x_1} x_2^{\beta_2}$$

Multiple logarithmic transformations

However, as we discussed in Chapter 4 Section 4.7, multicollinearity between regression variables is not necessarily a reason to reject a model out of hand. Consider Table 6.14 which adds an extra variable (the total number of product variants in production) to our earlier example in Figure 6.1. If we were to calculate the correlation between the two predictor variables then we would find that it was very high at 0.91. However, how many of us made the mistake (*as I did initially*) of calculating the correlation between the Production Build Number and the Number of Variants in Production and getting 0.94? As illustrated in Figure 6.8 the correct calculation is done on the Logarithmic values as the predicator

Table 6.14 Example of a Multi-Linear Regression Using Two Logarithmic Function Transformations

Obs	y Cum Design Queries	Production Build Number	Cumulative Production Variants	x1 Log(Build)	x2 Log(Variants)
1	3	1	1	0.0000	0.0000
2	5	2	1	0.3010	0.0000
3	8	3	2	0.4771	0.3010
4	8	4	2	0.6021	0.3010
5	9	5	2	0.6990	0.3010
6	11	6	3	0.7782	0.4771
7	11	7	3	0.8451	0.4771
8	12	8	4	0.9031	0.6021
9	13	9	4	0.9542	0.6021
10	13	10	4	1.0000	0.6021
	Correlation with y			0.9867	0.9834
	Collinearity between x1, x2			0.955	

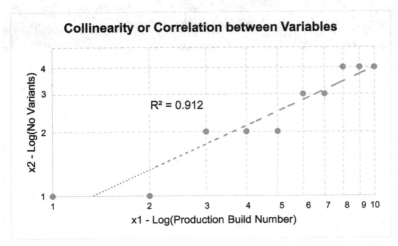

Figure 6.8 Multicollinearity Between Two Predicator Variables

Transforming Nonlinear Regression | 305

variables here are the Log values, not the linear values. As you can see the results are not too different because of the high degree of correlation between values and their Logs (*but we might as well do it correctly if we're doing it.*) Similarly, we should also check that each of the predictor variables (the log values in this case) are correlated well with the dependent variable, the Number of Design Queries raised; the values are 0.987 and 0.983 (*we would have been surprised if they had been significantly different because of the inherent multicollinearity.*)

We can see the results of this Multi-linear Regression of two Logarithmic Function transformations in Table 6.15. All the usual statistical measures give us a positive indicator that the model is valid. If we compare this to our previous result using only a single Logarithmic Function transformation on the Production Build Number in Table 6.14, then we will see:

- R-Square has increased (but we would expect this because Regression will always try its level best to fit the data it has (*doesn't everyone love a trier?*)
- More importantly, the Adjusted R-Square has also increased, closing the gap on the unadjusted R-Square. This is telling us that the predictive power of the model has improved with the addition of the extra variable.
- The F-Statistic, and all the t-Statistics are significant

So that's it then, we have a valid model . . . or do we? The tendency here may be to 'cut and run', but as responsible estimators (*no, that is not an oxymoron, thank you*) we have that little issue of multicollinearity to consider. What does it mean here? Well, for one thing we cannot make wild changes in one predicator without considering the impact on the other. For instance we cannot logically have a greater number of variants in production

Table 6.15 Multi-Linear Regression Output with Two Logarithmic Function Transformations

SUMMARY OUTPUT

Regression Statistics	
Multiple R	0.996381565
R Square	0.992776222
Adjusted R Square	0.990712286
Standard Error	0.324598146
Observations	10

Based on R-Square, Adjusted R-Square, F and t-statistics, we would accept this model as a statistically significant result

ANOVA

	df	SS	MS	F	Significance F
Regression	2	101.3624523	50.68122615	481.0110388	3.20387E-08
Residual	7	0.737547695	0.105363956		
Total	9	102.1			

	Coefficients	Standard Error	t Stat	P-value	Lower 95%	Upper 95%
Intercept	3.033240237	0.279854643	10.83862752	1.25518E-05	2.371489161	3.694991313
X Variable 1	5.712909454	1.145261639	4.988300716	0.001586056	3.004796008	8.4210229
X Variable 2	6.87653327	1.591306121	4.321313905	0.003474593	3.113692224	10.63937432

306 | Transforming Nonlinear Regression

at any one time than we have units of Production built (*that would be a real oxymoron!*) There will potentially be some cross-over between the contributions from each predictor variable, so any sensitivity analysis we do should bear that in mind. Ideally, we would want to see and play with some more data (*I use 'play' deliberately because I find this kind of thing to be fun. OK, perhaps I do need therapy.*)

In the lower third of Table 6.16 (*colloquially speaking, I could have said 'lower half of the table' but that would have just been a poor estimate*), we can examine what our model would give us based on alternative assumptions on the number of variants in production after we have built 20 units in total. As we can see, there is a consistency in the pattern.

If we refer to the simple model we created in Section 6.1.1, we would have generated a 90% Prediction Interval of between 14.685 and 17.383 (or 15 to 17 because we can't have a fraction of a query.) Our current (more sophisticated) model would suggest that the number of queries could be anything between 13.804 and 17.766 (interpreted as 14 to 18). The results are not inconsistent.

So, which model should we have used, and was it all worth it? In terms of this specific dataset, then we can justifiably claim that either model would suffice in terms of the results they give. As we have said several times already, estimators need to use their judgement to make informed decisions; this would be one such judgement call. One factor we need to consider is whether we can reasonably predict the future values of both variables; if we can't they're are not much use as drivers!

Combined linear and logarithmic transformations

Consider the cost of producing tankers (the ship freighter variety, not the road haulage truck type). Suppose we have data on two basic types of tankers (chemical and petrochemical), summarised in Table 6.17. If we consider the ship's deadweight to be a reasonable indicator of its Rough Order of Magnitude (RoM or ROM) cost

Table 6.16 Multi-Linear Regression Sensitivity Analysis Using Prediction Intervals

Obs	y Cum Design Queries	Production Build Number	Cumulative Production Variants	Unity	x1 Log(Build)	x2 Log(Variants)	ŷ Regression Output	t	Standard Error	Z Matrix	Half Prediction Interval	LPL	UPL
					Intercept 3.033	Log(Build) 5.713	Log(Variants) 6.877				90%		
1	3	1	1	1	0.0000	0.0000	3.033	1.895	0.325	0.743	0.812	2.221	3.845
2	5	2	1	1	0.3010	0.0000	4.753	1.895	0.325	0.598	0.777	3.976	5.530
3	8	3	2	1	0.4771	0.3010	7.829	1.895	0.325	0.215	0.678	7.151	8.507
4	8	4	2	1	0.6021	0.3010	8.543	1.895	0.325	0.122	0.652	7.891	9.194
5	9	5	2	1	0.6990	0.3010	9.096	1.895	0.325	0.318	0.706	8.390	9.803
6	11	6	3	1	0.7782	0.4771	10.760	1.895	0.325	0.134	0.655	10.105	11.414
7	11	7	3	1	0.8451	0.4771	11.142	1.895	0.325	0.148	0.659	10.483	11.801
8	12	8	4	1	0.9031	0.6021	12.333	1.895	0.325	0.271	0.693	11.639	13.026
9	13	9	4	1	0.9542	0.6021	12.625	1.895	0.325	0.220	0.679	11.946	13.304
10	13	10	4	1	1.0000	0.6021	12.886	1.895	0.325	0.230	0.682	12.204	13.568
	Alternative Predictions	20	4	1	1.3010	0.6021	14.606	1.895	0.325	1.592	0.990	13.616	15.596
		20	5	1	1.3010	0.6990	15.272	1.895	0.325	0.851	0.837	14.436	16.109
		20	6	1	1.3010	0.7782	15.817	1.895	0.325	0.580	0.773	15.044	16.590
		20	7	1	1.3010	0.8451	16.277	1.895	0.325	0.596	0.775	15.503	17.052
		20	8	1	1.3010	0.9031	16.676	1.895	0.325	0.766	0.817	15.859	17.493

Table 6.17 Tanker Costs and Deadweights

Petrochemical tankers			Chemical tankers		
Tanker No	Deadweight (k Tonnes)	Cost $m	Tanker No	Deadweight (k Tonnes)	Cost $m
P1	5.8	15	C1	4.71	16
P2	6.5	11.5	C2	7	17.4
P3	8	12	C3	7	14
P4	9	26	C4	7.5	15.5
P5	16.5	25	C5	8.4	26.5
P6	20	33.75	C6	9	25.6
P7	36.2	46	C7	10.8	23.5
P8	37	40.5	C8	12	27
P9	37	43	C9	12	30.5
P10	37.4	41.5	C10	13	25
P11	46	44	C11	13	22
P12	46.6	43	C12	13	25
P13	50	46	C13	13	23.35
P14	50	48	C14	14	25
P15	50	46	C15	16.4	33
P16	50.3	47	C16	17	29
P17	50.3	47	C17	17	31
P18	50.4	45.6	C18	18.5	36.9
P19	51	47	C19	18.5	36.9
P20	51	45.2	C20	20	34
P21	51	48	C21	25	35
P22	51.8	45	C22	25.5	35
P23	52	47.5	C23	36.2	46
P24	52	46.5	C24	37	41.75
P25	52	46.5	C25	37	41.75
P26	74	50.75	C26	44	50
P27	74	58	C27	45	50
P28	100	65	C28	45	51
P29	110	60	C29	45	50
P30	114	60.5			
P31	115	68			
P32	115	69.5			

308 | Transforming Nonlinear Regression

> Paraphrasing the Encyclopaedia Britannica (1998) Deadweight tonnage measures the weight of the total contents on board a ship including people (crew and passengers), cargo, and consumables (fuel, food, and water ... but excluding boiler water.)

(i.e. a high-level cost driver) then we can produce the logarithmic trendlines shown in Figure 6.9.

In Table 6.17, the list to the left relates to petrochemical tankers, and the list to the right refers to those that are chemical tankers (*Perhaps in hindsight, the use of the word 'list' was an unfortunate choice here in relation to ships! Please note that if you see that a tanker that lists to the left or right, this is not a reliable means of determining whether it is a chemical or petrochemical tanker.*)

We could produce two separate estimating relationships for the different kinds of tankers using a Simple Linear Regression of the Logarithmic Function transformation (which is what is indicated by the two trendlines in Figure 6.9.) From a visual perspective the two lines appear to have a constant difference (approximately). We can improve this visual perspective by plotting the deadweight on a logarithmic scale. Perhaps then we can develop a single cost estimating relationship that covers the two with a differential factor value between the two types. We can do this using a binary true or false

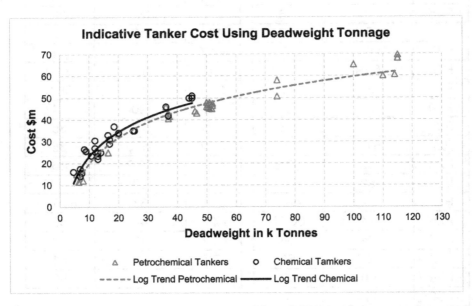

Figure 6.9 Indicative Tanker Costs Using Deadweight Tonnage

'switch' in relation to the type of tanker, in which takes a linear value of 0 (petrochemical) or 1 (chemical).

If we consider the distribution of potential values for the slope parameters, there is a high degree of overlap between the two individual Regression result slope parameters, as illustrated in Figure 6.10 thus supporting the consideration of a joint model.

Figure 6.10 also illustrates that if we run the combined Regression with effectively double the number of data points in comparison with either of the two individual models, not only does it focus on a value between the two individual values, it compresses the range of likely values (e.g. the 95% Confidence Interval is narrower.) Note that it doesn't matter whether we draw this as a Student's t-Distribution or a Normal Distribution the graphs would look to be the same. In Volume II Chapter 5 we demonstrated that a Student's t-Distribution converges towards a Normal Distribution for sample sizes of 30 and above.

Table 6.18 gives the results of a combined regression using Linear and Transformed Logarithmic Function parameters.

I can probably guess what you are thinking ... What does this all mean, especially that negative intercept?

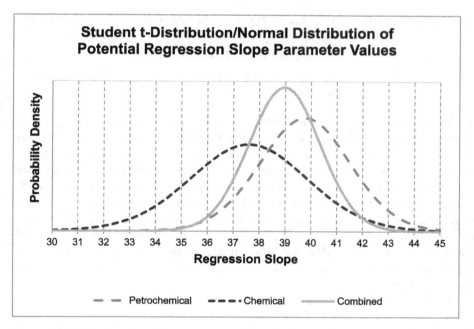

Figure 6.10 Regression Slope Parameter – Student's t-Distribution or Normal Distribution

310 | Transforming Nonlinear Regression

Table 6.18 Combined Tanker Regression Model with Linear and Logarithmic Function Parameters

SUMMARY OUTPUT

Regression Statistics	
Multiple R	0.972640067
R Square	0.9460287
Adjusted R Square	0.944167621
Standard Error	3.350836607
Observations	61

Based on R-Square, Adjusted R-Square, F and t-statistics we would accept this model as a statistically significant result

ANOVA

	df	SS	MS	F	Significance F
Regression	2	11415.00035	5707.500173	508.3226137	1.70893E-37
Residual	58	651.2301462	11.22810597		
Total	60	12066.23049			

	Coefficients	Standard Error	t Stat	P-value	Lower 95%	Upper 95%
Intercept	-18.64891764	2.305367565	-8.089346759	4.33915E-11	-23.2636122	-14.03422308
Log(DWT)	38.97838206	1.379093121	28.26377818	1.31327E-35	36.21782724	41.73893687
Tanker Type	2.544463678	1.013981373	2.509379113	0.014909127	0.514759435	4.57416792

Cost in $m	equals	38.978 $m multiplied by the Log of the Ship's Deadweight in Tonnes
	plus	2.544 $m only if it's a Chemical Tanker
	less	18.649 $m to compensate for unknown variables and random factors

In this particular model we can never have a ship that can carry nothing, i.e. its deadweight cannot be zero, so we can never get a situation when we have a ship that costs -18.649 $m, but could we have one that is theoretically free? The hypothetical answer implied by this regression is 'yes' when we have a petrochemical tanker with a deadweight of around 3 tonnes. Small Coastal petrochemical tankers with a deadweight of 1.5 to 2 tonnes do exist (*but they were not built for free*), so our model is not perfect for all cases (*and why would we expect it to be with just two parameters?*) The 90% Confidence Interval we have on the Intercept alone ranges from -23.26 to -14.03 $m but that is not going to resolve our little dilemma in this case; even considering the higher end of the Prediction Interval, we will still be looking at negative cost ranges for the smallest tankers.

In short, we have four options:

1. We can reject the model as unfit for purpose, *stick our bottom lip out and sulk about life not being fair to us as estimators*
2. We can reject the intercept on the grounds of it being inappropriate to be negative (*in effect, we would be invoking a Backward Stepwise Procedure on a point of logic*)

Transforming Nonlinear Regression | 311

3. Include other data points or other potential cost drivers (*which unfortunately we don't have to hand otherwise we would have used them in the first place*)
4. We can accept the model as it stands and limit the scope to which we can apply it

The preferred solution is not to reject the model as unfit for purpose but to limit its scope of use. As well as the very small tankers, the model has not considered the ultra large tankers, the largest of which was the Seawise Giant (now scrapped) which had a deadweight of nearly 565 kilotonnes, far bigger than anything we have used in the model!

However, what if we were to take option 2 and reject the negative intercept? We would then get the results in Table 6.19. What is this now telling us?

| Cost in $m | equals | 28.197 $m multiplied by the Log of the ship's deadweight in tonnes |
| | less | 2.897 $m only if it's a petrochemical tanker |

We now get a valid result for the lower end around a more realistic minimum deadweight of 1.5 tonnes, giving us a value of just over 2 $m. However, the trade-off for this lower end improvement to the model is a worsening of the fit at the upper end . . . a bit like a see-saw; if we push down on one side it pops up on the other . . . and vice versa, which is the case here – raising the intercept, lowers the slope (. . . *now we can consider sulking again!*)

What's more, it may or may not have escaped our attention that the model's 'switch parameter' has flipped the other way. In this case the default position is now to calculate the cost of a chemical tanker and subtract a value if we want a petrochemical tanker, whereas before the default was to calculate the cost of a petrochemical one and to add a cost penalty for a chemical one.

Table 6.19 Combined Tanker Regression Model Constrained Through the Origin

SUMMARY OUTPUT

Regression Statistics	
Multiple R	0.993150431
R Square	0.986347778
Adjusted R Square	0.969167232
Standard Error	4.846752676
Observations	61

Based on R-Square, Adjusted R-Square, F and t-statistics we would accept this model as a statistically significant result

ANOVA

	df	SS	MS	F	Significance F
Regression	2	100133.7428	50066.87141	2131.320374	5.11111E-55
Residual	59	1385.969678	23.4910115		
Total	61	101519.7125			

	Coefficients	Standard Error	t Stat	P-value	Lower 95%	Upper 95%
Intercept	0	#N/A	#N/A	#N/A	#N/A	#N/A
Log(DWT)	28.1969659	0.512541792	55.0139839	2.10668E-52	27.17137214	29.22255966
Tanker Type	-2.897484877	1.097374976	-2.640378121	0.010580559	-5.093327131	-0.701642622

312 | Transforming Nonlinear Regression

6.2.2 Multi-Linear Regression using linearised Exponential and Power Functions

Let's move our attention now to the second group of Multi-Linear Regressions, the ones that use linearised Exponential and Power Functions. Remember from Sections 6.1.2 and 6.1.3, these types of Regressions require us to use the logarithm of our dependent y variable instead of y itself, and as a consequence we have to be able to reverse the transformation again in order to interpret the results.

Multiple exponential transformations

Most practical examples of Exponential Functions that we will experience as estimators are likely to be time based. Suppose we have sales volumes that are growing exponentially over time. Also, over a period of time we may also notice or expect that the unit cost of production will escalate through inflation. Both of these are examples of Exponential Functions. If we put them together then the total cost of production will also be an Exponential Function. Unfortunately, we should not/cannot perform a multiple exponential regression on such data as the independent variable is the same. If we do stick it all in a model and run a regression we will soon find that one of the two coefficients will be practically indistinguishable from zero, and the one that remains will be the product of the two 'separate' exponential rates.

For the Formula-philes: Multiple linearised Exponential Function Regression

Consider a range of n drivers $x_1, x_2, x_3 \ldots x_n$ all of which have an exponential relationship with the value of y, the entity which we wish to estimate. Let β_i represent the linear contribution (slope) of the corresponding x_i variable, and β_0 is the Intercept Constant of the model.

Multi-linear model:

$$\log_b y = \beta_0 + \beta_1 x_1 + \beta_2 x_2 + \beta_3 x_3 + \ldots + \beta_n x_n \quad (1)$$

Taking the Anti-log of (1) to the base b:

$$y = b^{\beta_0 + \beta_1 x_1 + \beta_2 x_2 + \beta_3 x_3 + \ldots + \beta_n x_n} \quad (2)$$

Applying the additive property of powers to (2)

$$y = b^{\beta_0} b^{\beta_1 x_1} b^{\beta_2 x_2} b^{\beta_3 x_3} \ldots b^{\beta_n x_n} \quad (3)$$

Let ρ_i be a series of constants such that $\rho_i = b^{\beta_i}$, (3) becomes:

$$y = \rho_0 \rho_1^{x_1} \rho_2^{x_2} \rho_3^{x_3} \ldots \rho_n^{x_n}$$

For the Formula-phobes: Mind your p's and q's and remember the ones that count

Suppose that the cost of production is increasing by **p%** per annum. Suppose also that the size of the customer queue drives an increase in production volumes of **q%** per annum.

Let's say that **p% = 2%** and **q% = 4%**. The total cost of production will increase at a rate of **6.08%** per annum (not 6% or 8%)

The total cost of production will be growing at a rate of **1+p%** relative to the previous year. The size of the customer queue will be growing at a rate of **1+q%** relative to the previous year. The total cost of production will be increasing at a rate of per year. So, in our example, **1.02 × 1.04 = 1.0608**

Remember: It's the 'one' that makes the difference!

A more practical example of a Regression involving multiple Exponential Functions might be where we have an underlying rate of inflation but the level of cost is dependent on the relative proportion of two different products. The second variable here is in effect a switch – but in this case, it is more like a dimmer switch than a traditional binary on/off switch.

In the example in Table 6.20 we have the Annual Cost of Quality Failures which includes any underlying rate of inflation, and any growth or shrinkage per year. In the imaginary organisation in question, quality failures are categorised into two Cat A and Cat B. It is suspected that the cost of a Cat B failure is greater than that of a Cat A. We know the relative proportions of the two categories of quality failures.

Table 6.20 Using a "Dimmer Switch" as a Variable Category

Cost of Failures	y Log(Cost)	x1 Relative Year No	x2 %Cat A	Ref Only %Cat B
£ 12,600	4.1004	0	90%	10%
£ 16,000	4.2041	1	37%	63%
£ 14,700	4.1673	2	62%	38%
£ 18,300	4.2625	3	25%	75%
£ 14,800	4.1703	4	85%	15%
£ 19,300	4.2856	5	29%	71%
£ 18,100	4.2577	6	48%	52%
£ 18,400	4.2648	7	57%	43%
£ 18,400	4.2648	8	69%	31%
£ 20,500	4.3118	9	45%	55%

314 | Transforming Nonlinear Regression

Table 6.21 Multi-Linear Regression of Multiple Transformed Exponential Functions (1)

SUMMARY OUTPUT

Regression Statistics	
Multiple R	0.996749843
R Square	0.99351025
Adjusted R Square	0.991656036
Standard Error	0.006011049
Observations	10

Based on R-Square, Adjusted R-Square, F and t-statistics we would accept this model as a statistically significant result

ANOVA

	df	SS	MS	F	Significance F
Regression	2	0.038720687	0.019360343	535.812033	2.2019E-08
Residual	7	0.000252929	3.61327E-05		
Total	9	0.038973616			

	Coefficients	Standard Error	t Stat	P-value	Lower 95%	Upper 95%
Intercept	4.253370643	0.006529282	651.4300684	5.30491E-18	4.237931345	4.268809942
x1 Relative Year No	0.015643327	0.000670406	23.33410067	6.73869E-08	0.014058068	0.017228586
x2 %Cat A	-0.17340221	0.009156948	-18.93668162	2.84788E-07	-0.195054951	-0.151749468

Table 6.22 Multi-Linear Regression of Multiple Transformed Exponential Functions (2)

SUMMARY OUTPUT

Regression Statistics	
Multiple R	0.996749843
R Square	0.99351025
Adjusted R Square	0.991656036
Standard Error	0.006011049
Observations	10

Based on R-Square, Adjusted R-Square, F and t-statistics we would accept this model as a statistically significant result

ANOVA

	df	SS	MS	F	Significance F
Regression	2	0.038720687	0.019360343	535.812033	2.2019E-08
Residual	7	0.000252929	3.61327E-05		
Total	9	0.038973616			

	Coefficients	Standard Error	t Stat	P-value	Lower 95%	Upper 95%
Intercept	4.079968433	0.005091421	801.3418212	1.24462E-18	4.067929136	4.092007731
x1 Relative Year No	0.015643327	0.000670406	23.33410067	6.73869E-08	0.014058068	0.017228586
x2 %Cat B	0.17340221	0.009156948	18.93668162	2.84788E-07	0.151749468	0.195054951

If we run a multiple Linear Regression on the log of the Cost of Quality Failures against the Relative Year number and the Relative Proportion of the two Categories of Failure we will get the results in Table 6.21. We only have to use one of the two variables for the Relative Proportion of Failure types – either Cat A or Cat B, because they are wholly dependent on each other; in fact, they are perfectly negatively correlated. If we hadn't spotted this already then the Regression Result would have highlighted one of them as being superfluous.

If we had chosen to run with the %Cat B as our second dependent variable, we would have got the answer in Table 6.22.

One thing becomes obvious when we compare the two results. They are similar but not identical. The R-Square, F-Statistic, Coefficient of x_1 and its associated p-value are all the same, but the sign of the x_2 Coefficient has changed, but the value of the intercept and its t-Statistic and p-value have changed! What's more, the difference in the two intercepts equals the value of the Coefficient of our 'dimmer switch' – *you're probably thinking that can't be just a coincidence, can it? You're right, it's not.*

The results are depicted graphically in Figure 6.11. Note that the fixed difference is only fixed in Log Space; in real terms it is subject to inflationary/deflationary drivers, hence the divergence shown between the nominal upper and lower limits. (*We say 'nominal' because there are still Prediction Intervals around both these nominal limits.*)

For the Formula-phobes: A dimmer glimmer

The model is implying that there is a fixed cost difference between the two types of failures at fixed economic and temporal conditions.

We can use either Category A or Category B as the baseline and calculate the cost assuming that all Failures are that one Category and add or subtract the difference to take account of the relative proportion of the other Category.

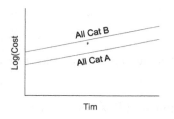

The time-dimension allows us to model net inflation (or deflation) and growth or shrinkage in overall quality failures.

The dimmer switch lets us model the costs between the two extremes of being all one or the other!

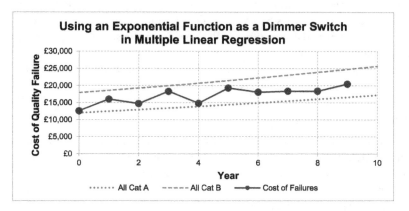

Figure 6.11 Using an Exponential Function as a "Dimmer Switch" in Multiple Linear Regression

316 | Transforming Nonlinear Regression

> ## For the Formula-philes: The intercept is linked to the Dimmer Switch parameter
>
> Consider our two versions of the same model which forecasts the Cost of Quality Failure as an Exponential function of the Proportion of one of the two failure categories, and the rate of change per year, regardless of whether it is due to inflationary elements, or growth or shrinkage in the volume of quality failures.
>
> Multi-linear transformed model:
>
> $$\log(Cost) = Intercept_A + Coeff_A \times A\% + Rate \times Year \quad (1)$$
>
> Alternative multi-linear transformed model:
>
> $$\log(Cost) = Intercept_B + Coeff_B \times B\% + Rate \times Year \quad (2)$$
>
> Subtracting (2) from (1) and re-arranging:
>
> $$Intercept_B = Intercept_A + Coeff_A \times A\% - Coeff_B \times B\% \quad (3)$$
>
> Substituting in (3) the empirical result that $Coeff_B = -Coeff_A$:
>
> $$Intercept_B = Intercept_A + Coeff_A \times (A\% + B\%) \quad (4)$$
>
> But, by definition A% + B% =1 $\quad Intercept_B = Intercept_A + Coeff_A$
>
> ... which substantiates the empirical result we found by regression.

Incidentally, whilst we introduced the idea of binary on/off switches for linear functions and dimmer switches for Exponential Functions, the two concepts extend across both – we can have dimmer switches between 0 and 1 with linear functions and binary on/off switches with Exponential Functions; it all depends on the specific requirement or case for which we are trying to model our data.

(*Please note that whilst no prior electrical knowledge or experience is required to fit either kind of switch in our models, we should still be mindful not to get them cross-wired!*)

Multiple power transformations

Possibly the most frequently occurring family of curves used by estimators other than linear ones are multiple Power Functions where each independent driver or predictor variable has a power function relationship with the dependent variable for which we are trying to establish an estimating relationship. (*Yes, I know you can all come up with lots of relationships that aren't, but think of all the ones that are!*)

Let's bring this to life and look at an example of the hours required to manufacture a product in steady state production. Suppose we have data from other similar products

Transforming Nonlinear Regression | 317

For the Formula-philes: Multiple linearised Power Function Regression

Consider a range of n drivers $x_1, x_2, x_3 \ldots x_n$ all of which have a power relationship with the value of y, the entity which we wish to estimate. Let β_i represent the linear contribution (slope) of the corresponding $\log_b x_i$ variable, and β_0 is the Intercept Constant of the model.

Multi-linear model:

$$\log_b y = \beta_0 + \beta_1 \log_b x_1 + \beta_1 \log_b x_2 + \ldots + \beta_n \log_b x_n \quad (1)$$

Using the Multiplicative Property of Logs with (1):

$$\log_b y = \beta_0 + \log_b x_1^{\beta_1} + \log_b x_2^{\beta_2} + \ldots + \log_b x_n^{\beta_n} \quad (2)$$

Using the Additive Property of Logs with (2):

$$\log_b y = \log_b \left(b^{\beta_0} x_1^{\beta_1} x_2^{\beta_2} \ldots x_n^{\beta_n} \right) \quad (3)$$

Let γ_0 be a constant such that $\gamma_0 = b^{\beta_0}$, (3) becomes :

$$\log_b y = \log_b \left(\gamma_0 x_1^{\beta_1} x_2^{\beta_2} \ldots x_n^{\beta_n} \right) \quad (4)$$

Taking the anti-log to the base b of (4): $y = \gamma_0 x_1^{\beta_1} x_2^{\beta_2} \ldots x_n^{\beta_n}$

(basic functionality, and manufacturing processes, facilities and technologies) where we know:

- The average hours at a steady state rate (i.e. that which we want to predict for a new product)
- The finished weight of the products
- The number of units we made to get to that steady state

The rationale for using the latter two as predictor variables or cost drivers is that:

- Weight is a commonly used indicator of relative cost.
- The number of units prior to steady state to give an indication of the total learning to date.

Consider the data in Table 6.23, showing the transformation of the dependent y-variable and both the independent x-variables into logarithmic values. If we then perform a Multi-Linear Regression of the log data in Microsoft Excel we get the result in Table 6.24.

318 | Transforming Nonlinear Regression

Table 6.23 Multi-Linear Regression of Multiple Transformed Power Functions

Product ID	Ave Hours at Steady State	Finished Weight (kg)	Units Prior to Steady State	y Log(Hours)	x1 Log(Weight)	x2 Log(Prior Qty)
A	4635	990	25	3.666	2.996	1.398
B	2275	783	88	3.357	2.894	1.944
C	5011	512	6	3.700	2.709	0.778
D	3523	862	36	3.547	2.936	1.556
E	1954	669	90	3.291	2.825	1.954
F	5499	808	12	3.740	2.907	1.079
G	2404	565	42	3.381	2.752	1.623
H	1783	558	96	3.251	2.747	1.982
I	2263	673	54	3.355	2.828	1.732

Table 6.24 Multi-Linear Regression of Multiple Transformed Power Functions

SUMMARY OUTPUT

Regression Statistics	
Multiple R	0.995864937
R Square	0.991746973
Adjusted R Square	0.988995963
Standard Error	0.019759967
Observations	9

Based on R-Square, F and t-statistics we would accept this model as a statistically significant result

ANOVA

	df	SS	MS	F	Significance F
Regression	2	0.281521305	0.140760652	360.5029725	5.62134E-07
Residual	6	0.002342738	0.000390456		
Total	8	0.283864042			

	Coefficients	Standard Error	t Stat	P-value	Lower 95%	Upper 95%
Intercept	1.957171036	0.206736806	9.466969514	7.90866E-05	1.451304296	2.463037776
x1 Log(Weight)	0.765475024	0.072421066	10.56978402	4.2192E-05	0.58826706	0.942682988
x2 Log(Prior Qty)	-0.42126619	0.016808143	-25.06321965	2.6561E-07	-0.462394235	-0.380138145

[In terms of transforming the data back into real numbers (*there goes that 'ad hoc' play on words joke again*) we only need to concern ourselves with the intercept. As we used Common Logs to transform our data into a multi-linear compatible format, we need to raise ten (i.e. the base for Common Logs) to the power of the Intercept Coefficient to get the real constant.

In this case, it is $10^{1.957} = 90.609$ hours,

Q: What does this number represent?

A: It's the theoretical value we get if we have zero for both the x-variables; zero is always the log of 'one'

When Log(x_1) = 0   =>   a weight of 1 kg
When Log(x_2) = 0   =>   a Build Number of 1

All of which implies that the untransformed value for the intercept represents the 'Hours per kilogram for the First Unit'

We don't need to transform the coefficients of the two x-variables as these are the power indices which we can use either directly as powers or indirectly as log transform multipliers:

$$\text{Hours} = 90.609 \times \text{Weight}^{0.765} \times \text{PriorQty}^{-0.421}$$

(Don't worry about the negative power; this is characteristic of a Learning Curve as we will discover to our lasting delight in Volume IV. No? Oh well, it must be just me then who gets excited about learning curves. Never mind, I'll continue with the medication as the doctor prescribed)

Just for the fun of it (and the practice), let's compute the Prediction Interval for our model. While we're at it, we can also do a little bit of sensitivity analysis. Tables 6.25 and 6.26 illustrates this. The first table shows the results of the calculation for the 90% Prediction Interval in Log Space using the Z-matrix introduced in Chapter 4. The second table converts the Log Interval back into linear-based numbers to which we can more easily relate. To reverse the log transformation, we simply raise the number ten to the power of the Regression Output and to repeat this for each of the Lower and Upper Prediction Limits (LPL and UPL)

In this specific case there is quite a wide spread of uncertainty around the Regression Output Median. If we examine the linear value for LPL and UPL we will notice that the predicted data is positively skewed, consistent with a Lognormal Distribution i.e. UPL – Median > Median – LPL for all values. The sensitivity analysis performed in the lower section of Table 6.26 illustrates that each predictor variable individually contributes significantly to the dependent variable.

Table 6.25 90% Prediction Interval for a Regression Based on Multiple Power Functions (1)

Product ID	Ave Hours at Steady State	Finished Weight (kg)	Units Prior to Steady State	Log Value y Log(Hours)	Unity	Regression Coeffs		Log Value ŷ Regression Output	Pred Interval 90%	Logarithmic Values				
						Intercept 1.957 x1 Log(Weight) 0.765	Log(Prior Qty) -0.421 x2 Log(Prior Qty)		t	Standard Error	Z Matrix	Half Prediction Interval	LPL	UPL
A	4635	990	25	3.666	1	2.996	1.398	3.661	1.943	0.020	0.446	0.046	3.615	3.708
B	2275	783	88	3.357	1	2.894	1.944	3.353	1.943	0.020	0.247	0.043	3.310	3.396
C	5011	512	6	3.700	1	2.709	0.778	3.703	1.943	0.020	0.773	0.051	3.652	3.754
D	3523	862	36	3.547	1	2.936	1.556	3.549	1.943	0.020	0.224	0.042	3.506	3.591
E	1954	669	90	3.291	1	2.825	1.954	3.297	1.943	0.020	0.229	0.043	3.254	3.339
F	5499	808	12	3.740	1	2.907	1.079	3.728	1.943	0.020	0.341	0.044	3.684	3.773
G	2404	565	42	3.381	1	2.752	1.623	3.380	1.943	0.020	0.228	0.043	3.337	3.423
H	1783	558	96	3.251	1	2.747	1.982	3.225	1.943	0.020	0.376	0.045	3.180	3.270
I	2263	673	54	3.355	1	2.828	1.732	3.392	1.943	0.020	0.136	0.041	3.351	3.433
1		300	10		1	2.477	1.000	3.432	1.943	0.020	2.096	0.068	3.365	3.500
2		300	150		1	2.477	2.176	2.937	1.943	0.020	2.243	0.069	2.867	3.006
3		1200	10		1	3.079	1.000	3.893	1.943	0.020	1.114	0.056	3.837	3.949
4		1200	150		1	3.079	2.176	3.397	1.943	0.020	1.096	0.056	3.342	3.453

320 | Transforming Nonlinear Regression

Table 6.26 90% Prediction Interval for a Regression Based on Multiple Power Functions (2)

Product ID	Ave Hours at Steady State	Finished Weight (kg)	Units Prior to Steady State	Log Values				Linear Values		
				$\hat{y}$ Regression Output	Half Prediction Interval	LPL	UPL	$\hat{y}$ Regression Output	LPL	UPL
A	4635	990	25	3.661	0.046	3.615	3.708	4585	4123	5099
B	2275	783	88	3.353	0.043	3.310	3.396	2255	2043	2489
C	5011	512	6	3.703	0.051	3.652	3.754	5049	4489	5680
D	3523	862	36	3.549	0.042	3.506	3.591	3537	3207	3900
E	1954	669	90	3.297	0.043	3.254	3.339	1980	1795	2184
F	5499	808	12	3.728	0.044	3.684	3.773	5347	4827	5923
G	2404	565	42	3.380	0.043	3.337	3.423	2399	2175	2646
H	1783	558	96	3.225	0.045	3.180	3.270	1677	1512	1860
I	2263	673	54	3.392	0.041	3.351	3.433	2467	2245	2711
1		300	10	3.432	0.068	3.365	3.500	2704	2315	3160
2		300	150	2.937	0.069	2.867	3.006	864	737	1013
3		1200	10	3.893	0.056	3.837	3.949	7815	6872	8887
4		1200	150	3.397	0.056	3.342	3.453	2497	2197	2838

Before we proceed any further, time for that 'timely reminder':

Never take the log of a date.

So, if we think that date is a good predicator variable, then we should not be considering it as part of a multiple power function model. Instead we should look at combining it as an exponential function with other exponential functions or with variables that are truly power functions.

That brings us nicely to the last of our 'standard' groups.

Combined exponential and power transformations

This is another commonly occurring type of model in estimating; it is one which combines Exponential and Power Functions.

For the Formula-philes: Combined Exponential and Power Function Regression

Consider two drivers x_1 and x_2 the first of which has an exponential relationship with the value of y, and the second of which has a power relationship with the value of y; y being the entity which we wish to estimate. Let β_i represent the linear contribution (slope) of the corresponding x_i or $\log_b x_i$ variable, and β_0 be the Intercept Constant of the model.

Multi-linear model:

$$\log_b y = \beta_0 + \beta_1 x_1 + \beta_2 \log_b x_2 \quad (1)$$

Applying Additive and Multiplicative Properties of Logs to (1):

$$\log_b y = \log_b \left(b^{\beta_0 + \beta_1 x_1} x_2^{\beta_2} \right) \quad (2)$$

Transforming Nonlinear Regression | 321

Taking the anti-log of (2) to the base b: $\qquad y = b^{\beta_0 + \beta_1 x_1} x_2^{\beta_2}$ $\qquad$ (3)

Simplifying (3) with constants $k_0 = b^{\beta_0}$ and $k_1 = b^{\beta_1}$: $\qquad y = k_0 k_1^{x_1} x_2^{\beta_2}$

Before we get onto the general case, let's consider the two distributions. At first glance we may think that we might be able to use this model form to calibrate data that we suspect comes from a Gamma Distribution or Chi-Squared Distribution (compare the formula-phile section for this with the respective Probability Density Functions (PDF) in Volume II Chapter 5). However, this would require us to set values of x_1 to equal those of x_2. Here, following a multi-linear transformation x_1 would be used in its pure linear form and x_2 would be taking its logarithmic value. Due to the high degree of multicollinearity between a number and its log (to whatever base), coupled with an annoying tendency to overcompensate for missing values and exaggerate rare events, this type of Regression model may return what appears to be statistically significant coefficients, but these may not bear any resemblance to the real Gamma parameters (Recall the last 'rule' or guideline before this section, and refer to the discussion on multicollinearity in the context of Stepwise Regression in Chapter 4 Section 4.7). We may have to reject a model on the basis of a low R-Square. We will revisit an option for solving this little riddle in Section 6.3, where we will also revisit and discuss the overcompensation and exaggeration issues.

However, in a more general context, as we discussed in Section 6.1.3, Power Functions are very commonly found especially in cost-size relationships. Frequently the Exponential Function will be used to model the effects of a date-related variable, or to operate as either a binary on/off switch or a graduated dimmer switch, as we discussed in previous sections.

The date-related variable could be used as an indicator of the underlying level of technology employed, or as an economic index such as a measure of inflation. Here, it doesn't matter if we use the absolute date or a relative date, (i.e. the elapsed time since some meaningful start date or baseline) as the only difference it will make will be in the value of the intercept.

The binary switch could relate to the difference between two variants or versions. (We will be exploring the use of the binary switch in multi-variant learning curves in Volume IV Chapter 6, *if we can wait that long*.)

Building on the example we have just developed for multiple power functions, we can add an example of a date-related Exponential Function in order to assess the impact of inflation on the cost of labour hours. In Table 6.27 we have replaced the average labour hours with the average labour cost and introduced the year at which the nominally steady state production was reached as an indicator of economic inflation. In this particular case we have decided to use a relative date/year based on the elapsed time since 2000 (which happens to be our first record, although we could equally use any earlier or later date – exponential functions can cope with negative x-values, unlike Logarithmic and Power Functions.)

In the Regression Output data in Table 6.28, which passes all our R-Square, F and t-Statistic tests for significance, we have two coefficients that we need to transform back

322 | Transforming Nonlinear Regression

Table 6.27 Combined Exponential and Power Function Regression Input Data

Product ID	Ave Cost on reaching Steady State	Finished Weight (kg)	Units Prior to Steady State	Year Steady State Commenced	y Log(Cost)	x1 Log(Weight)	x2 Log(Prior Qty)	x3 Relative Year
A	€ 234,207	990	25	2000	5.370	2.996	1.398	0
B	€ 125,239	783	88	2003	5.098	2.894	1.944	3
C	€ 275,856	512	6	2003	5.441	2.709	0.778	3
D	€ 207,364	862	36	2005	5.317	2.936	1.556	5
E	€ 124,939	669	90	2008	5.097	2.825	1.954	8
F	€ 351,606	808	12	2008	5.546	2.907	1.079	8
G	€ 161,188	565	42	2010	5.207	2.752	1.623	10
H	€ 129,125	558	96	2013	5.111	2.747	1.982	13
I	€ 174,885	673	54	2015	5.243	2.828	1.732	15

Table 6.28 Combined Exponential and Power Function Regression Output Data

SUMMARY OUTPUT

Regression Statistics	
Multiple R	0.995139395
R Square	0.990302415
Adjusted R Square	0.984483863
Standard Error	0.020041711
Observations	9

Based on R-Square, F and t-statistics we would accept this model as a statistically significant result

ANOVA

	df	SS	MS	F	Significance F
Regression	3	0.205089685	0.068363228	170.1974221	1.88009E-05
Residual	5	0.002008351	0.00040167		
Total	8	0.207098036			

	Coefficients	Standard Error	t Stat	P-value	Lower 95%	Upper 95%
Intercept	3.726377711	0.249575557	14.93086005	2.4391E-05	3.084823318	4.367932104
X Variable 1	0.743252086	0.087790838	8.466169192	0.000377613	0.517578553	0.968925619
X Variable 2	-0.415821085	0.019551273	-21.26823582	4.26005E-06	-0.466079233	-0.365562938
X Variable 3	0.010925718	0.001878032	5.817640607	0.002118636	0.006098082	0.015753354

to be able to appreciate those values: the intercept and the Coefficient of the Relative Year. The latter is the average annual rate of inflation, which as a power ten, yields: $10^{0.0109} = 1.025$, or 2.5% average annual inflation.

By adding the additional time-based variable, and changing the dependent variable from hours to weight, we inevitably get slightly different contributions in terms of the weight and prior quantity drivers than we did in the previous example in Table 6.24. (*It's one of those little idiosyncrasies of life that estimators find difficult to explain to non-believers!*)

The procedure for developing a prediction interval around this Regression model is the same as for the other multi-linear transformation Regression models we have discussed.

Transforming Nonlinear Regression | 323

6.3 Stepwise Regression and multi-linear transformations

What if we don't know which drivers are really important and whether we should transform any in relationship to the dependent variable and how? How do we decide which model is better! There is no single answer to that conundrum, only a set of guidelines.

Just as with normal, 'uncomplicated' Multi-linear Regression (*although to some that might seem like an oxymoron*) Stepwise Regression is a very simple procedure that we can use to sort out the wheat from the chaff in determining whether the relationship between our dependent variable and a combination of Linear and Logarithmic Functions, or a combination Exponential and Power Functions give us a statistically valid result. In reality though, it can be more like speed dating for estimators! We may hit lucky and find the perfect relationship quickly, or we may feel very awkward moving from one to another, feeling less and less confident that we're doing the right thing!

Don't worry; you will not be alone in thinking this. Leaving the analogy behind, the problem arises because a set of values and their log values are highly correlated linearly, (and perfectly correlated from a rank perspective, by definition.) This can result in some relationships between dependent and independent variables compensating for the absence of other independent variables.

So, where do we start? Just for the fun of it (and to bring this procedure to life), why don't we pretend that we haven't created the model in Section 6.2.2 yet? With a bit of luck, if we step through the procedure, we should end up with the model we created. (*If we don't, I'll be going back to rework the example, so you'll never know any different! Unethical? Not at all; we all agreed that estimating is an iterative process.*)

The following procedure is a variation on the Stepwise Procedure using Backwards Elimination described in Chapter 4. This is not just a question of a personal preference for Backward Elimination over Forward Selection, we shall see shortly that it is probably more reliable and more robust in this instance. (We'll briefly discuss a possible Forward Selection option afterwards.)

6.3.1 Stepwise Regression by Backward Elimination with linear transformations

Even though it's a Backward Procedure (*hmm, perhaps I could have phrased that better!*), we're going to begin by looking for the most significant single independent variable or variables (if it's a close call) in our list of candidate variables, just as we would do in a Forward Selection Procedure. (*If you're not following all this 'to-ing' and 'fro-ing' then you might want to read Chapter 4 Section 4.7 first if you haven't done so already; it may save you some angst.*) Let's break the procedure into three sections:

- Stepping up   Getting started (Functional Identification)
- Stepping back   Variable selection by Backward Elimination
- Stepping sideways   Functional validation and trial substitution

324 | Transforming Nonlinear Regression

Stepping Up

1. For each 'candidate' predictor variable calculate the R-Square with the dependent variable using Microsoft Excel's **RSQ** function, or if we are feeling brave and chipper, using **INDEX** with parameters 3 and 1 in conjunction with **LINEST**. (*Yes, you are probably wise at this stage to keep it simple and use the first one.*)

   o Do this for each type of functional relationship (Linear, Logarithmic, Exponential and Power) except where the variable is date related in which case we don't need to consider the logarithmic or power options. (*Doing all this may sound a little tedious but it will be worth it later.*)

   o Table 6.29 does this in the two groups of Linear-Logarithmic Function pairings, and Exponential-Power Function pairings for our example. Note that Relative Year is still a date or time so we should not consider taking its log value.

2. Determine which variable and function type return the highest R-Square

   o This will determine whether we should be building a model with linear values for our dependent y-variable, or its log values. In other words, 'Do we log it

Table 6.29 Step 1 – Calculate R-Square for All Candidate Variables and their Logarithms

			Linear — Logarithmic Function Pairings				
		Linear y	Linear	Logarithmic	Linear	Logarithmic	Linear
			x1	Log(x1)	x2	Log(x2)	x3
Product ID	Year Steady State Commenced	Ave Cost on reaching Steady State	Finished Weight (kg)	Log(Weight)	Units Prior to Steady State	Log(Prior Qty)	Relative Year
A	2000	€ 234,207	990	2.996	25	1.398	0
B	2003	€ 125,239	783	2.894	88	1.944	3
C	2003	€ 275,856	512	2.709	6	0.778	3
D	2005	€ 207,364	862	2.936	36	1.556	5
E	2008	€ 124,939	669	2.825	90	1.954	8
F	2008	€ 351,606	808	2.907	12	1.079	8
G	2010	€ 161,188	565	2.752	42	1.623	10
H	2013	€ 129,125	558	2.747	96	1.982	13
I	2015	€ 174,885	673	2.828	54	1.732	15
		R-Square	0.07	0.05	0.79	0.81	0.09

			Exponential – Power Function Pairings				
		log(y)	Exponential	Power	Exponential	Power	Exponential
			x1	Log(x1)	x2	Log(x2)	x3
Product ID	Year Steady State commenced	y Log(Cost)	Finished Weight (kg)	Log(Weight)	Units Prior to Steady State	Log(Prior Qty)	Relative Year
A	2000	5.370	990	2.996	25	1.398	0
B	2003	5.098	783	2.894	88	1.944	3
C	2003	5.441	512	2.709	6	0.778	3
D	2005	5.317	862	2.936	36	1.556	5
E	2008	5.097	669	2.825	90	1.954	8
F	2008	5.546	808	2.907	12	1.079	8
G	2010	5.207	565	2.752	42	1.623	10
H	2013	5.111	558	2.747	96	1.982	13
I	2015	5.243	673	2.828	54	1.732	15
		R-Square	0.08	0.06	0.88	0.85	0.10

Transforming Nonlinear Regression | 325

Table 6.30 Step 2 – Rank the R-Squares for all Candidate Variables and their Logarithms

R-Square (y, x)				R-Square (y, log x)			
Linear Function				Logarithmic Function			
Ave Cost on Reaching Steady State		y	Rank	Ave Cost	y	Rank	Best Rank
Weight (kg)	x1	0.07	8	Log(x1)	0.05	10	8
Prior Units	x2	0.79	4	Log(x2)	0.81	3	3
Relative Year	x3	0.09	6	Log(x3)	#N/A	#N/A	6
						Total	17

R-Square (log y, x)				R-Square (log y, log x)			
Exponential Function				Power Function			
Ave Cost on Reaching Steady State		Log(y)	Rank	Ave Cost	Log(y)	Rank	Best Rank
Weight (kg)	x1	0.08	7	Log(x1)	0.06	9	7
Prior Units	x2	0.88	1	Log(x2)	0.85	2	1
Relative Year	x3	0.10	5	Log(x3)	#N/A	#N/A	5
						Total	13

or leave it?', or LILI for short. (*Hmm, we could call this step 'Pick a LILI' . . . OK, perhaps not.*)

o   This decision will reduce the options we consider going forward to being either a combination of Linear and Logarithmic Functions, or Exponential and Power Functions.

o   Table 6.30 takes the calculated R-Square data from Table 6.29 and summarises the ranking.

3.   Based on the function type of the highest R-Square, we can keep all the others in that same pairing group (i.e. either Linear + Logarithmic group, or Exponential + Power group), and discard all the other group of pairings from step 1. (*It may seem like wasted effort but we had to do it to get this far so it wasn't a nugatory step.*)

o   If the first and second ranked R-Square values are very close in value and come from opposite groups, then we may wish to base our selection on the sum of the best ranks in the two groups. (*I know I'm giving you another decision to make – think of it as character building through empowerment!*)

o   In our example this would suggest that the Exponential-Power Function pairings offer the best option for a valid model on both counts. So we will reject the Linear-Logarithmic Function pairings.

Stepping Back

Before we continue we have another choice to make; we can now either select the highest R-Square from each pairing, setting aside the others, or we can just take the attitude of '*let's throw them all into the mix and let the Regression algorithm sort them out*'. Perversely (*you might think*) we are going to do the latter, but ensure that we only have one of each of a pair by the time we have finished, otherwise we run the risk of multicollinearity spoiling our party. We will discuss the rationale for this at the end as the example may help explain what is going on.

4.   Taking our 'starter variable' from Step 3, we can now create a regression model with all the candidate predictor variables and their two functional types (i.e. each variable and its

326 | Transforming Nonlinear Regression

logarithmic value) unless the variable is date-related in which case we should ignore the logarithmic option as invalid. This is illustrated in Table 6.31 (the lower half of Table 6.29).

o Note that the only element that is fixed in the model at present is that we are using the logarithmic transformation of the Cost (y-variable)

o The results of the first regression are shown in Table 6.32.

o What do our 'goodness of fit' measures tell us?

- ✔ **R-Square:** Very high value (*giving us that feel good factor*)
- ✔ **Adjusted R-Square:** Very high value (*maintaining that feel good factor that the additional variables are adding value to the model*)

Table 6.31 Step 4 – Initial Regression Input Variables

			Exponential – Power Function Pairings				
			Exponential	Power	Exponential	Power	Exponential
		log(y)	x1	Log(x1)	x2	Log(x2)	x3
Product ID	Year Steady State Commenced	y Log(Cost)	Finished Weight (kg)	Log(Weight)	Units Prior to Steady State	Log(Prior Qty)	Relative Year
A	2000	5.370	990	2.996	25	1.398	0
B	2003	5.098	783	2.894	88	1.944	3
C	2003	5.441	512	2.709	6	0.778	3
D	2005	5.317	862	2.936	36	1.556	5
E	2008	5.097	669	2.825	90	1.954	8
F	2008	5.546	808	2.907	12	1.079	8
G	2010	5.207	565	2.752	42	1.623	10
H	2013	5.111	558	2.747	96	1.982	13
I	2015	5.243	673	2.828	54	1.732	15

Table 6.32 Step 4 – Initial Regression Output Report – Rejecting 'Log(x1 Weight)'

SUMMARY OUTPUT

Regression Statistics	
Multiple R	0.996026525
R Square	0.992068838
Adjusted R Square	0.978850234
Standard Error	0.023398918
Observations	9

ANOVA

	df	SS	MS	F	Significance F
Regression	5	0.205455508	0.041091102	75.05095373	0.002381099
Residual	3	0.001642528	0.000547509		
Total	8	0.207098036			

	Coefficients	Standard Error	t Stat	P-value	Lower 95%	Upper 95%
Intercept	5.187799867	2.95104376	1.757954232	0.177000167	-4.203738444	14.57933818
x1 Weight	0.00042406	0.000787102	0.53876068	0.627463175	-0.00208085	0.002928969
Log(x1 Weight)	0.137364598	1.234795323	0.111244832	0.918447408	-3.792305216	4.067034411
x2 Prior Units	0.000729933	0.000956236	0.763339343	0.500799934	-0.002313238	0.003773103
Log(x2 Prior Units)	-0.472800284	0.081150903	-5.826186325	0.010071036	-0.731058677	-0.214541891
x3 Relative Year	0.012531294	0.003000853	4.175910562	0.025009885	0.00298124	0.022081348

Transforming Nonlinear Regression | 327

✔ **F-Statistic:** Very high value (*suggesting that there is virtually no chance of a fluke relationship —around a quarter of one percent*)

✘ **t-Statistics:** 91% chance that the contribution associated with the $Log(x_1$ Weight) variable is really zero, (*and an early indication that others may follow!*)

5. If we then follow our usual iterative procedure in evaluating the key statistics and rejecting the parameters that cannot be supported statistically (e.g. the parameter p-values > 5% say), we will get the results in Tables 6.33 and 6.34.

    o   Table 6.33: Reject variable x_2 Prior Units as there is a 39.5% probability that the true contribution is zero. All other p-values are still excellent. It is worth drawing our attention to the fact that in the previous iteration the p-value of variable x_1 Weight was also looking a bit dodgy, but with the removal of its *alter ego* $Log(x_1$ Weight) this 'threat' appears to have disappeared. This is due to the high correlation between a variable and its log values, i.e. inherent multicollinearity.

    o   Table 6.34: This shows the end state of our usual Backward Stepwise procedure where no further variables are rejected on statistical grounds: our key statistics of R-Square, Adjusted R-Square, the significance of F and all the parameter t-Statistics' p-values all pass muster.

    o   From a sensibility perspective all the parameter values (Coefficients) also pass the 'Does it make sense?' test. The negative values associated with the $Log(x_2$ Prior

Table 6.33 Step 5 – Regression Output Report Iteration 2 Rejecting 'x2 Prior Units'

SUMMARY OUTPUT

Regression Statistics	
Multiple R	0.996010101
R Square	0.99203612
Adjusted R Square	0.984072241
Standard Error	0.020305811
Observations	9

ANOVA

	df	SS	MS	F	Significance F
Regression	4	0.205448732	0.051362183	124.566942	0.00018926
Residual	4	0.001649304	0.000412326		
Total	8	0.207098036			

	Coefficients	Standard Error	t Stat	P-value	Lower 95%	Upper 95%
Intercept	5.516020269	0.052108326	105.8567929	4.77549E-08	5.371344362	5.660696175
x1 Weight	0.000511085	7.54397E-05	6.774744478	0.002477368	0.000301631	0.000720539
x2 Prior Units	0.0007593	0.000797582	0.952003069	0.395004771	-0.001455141	0.002973742
Log(x2 Prior Units)	-0.474320663	0.069417643	-6.832854637	0.002399608	-0.667054937	-0.281586388
x3 Relative Year	0.012702923	0.002233645	5.687083314	0.004720476	0.006501331	0.018904514

328 | Transforming Nonlinear Regression

Table 6.34 Step 5 – Regression Output Report Iteration 3 – All Variables Significant

SUMMARY OUTPUT

Regression Statistics	
Multiple R	0.995103856
R Square	0.990231685
Adjusted R Square	0.984370696
Standard Error	0.020114666
Observations	9

ANOVA

	df	SS	MS	F	Significance F
Regression	3	0.205075037	0.068358346	168.9529993	1.91451E-05
Residual	5	0.002022999	0.0004046		
Total	8	0.207098036			

	Coefficients	Standard Error	t Stat	P-value	Lower 95%	Upper 95%
Intercept	5.497233572	0.047773114	115.0696087	9.40046E-10	5.374428872	5.620038272
x1 Weight	0.000462271	5.48149E-05	8.433316655	0.000384605	0.000321365	0.000603177
Log(x2 Prior Units)	-0.410936237	0.019459526	-21.11748473	4.41281E-06	-0.460958541	-0.360913934
x3 Relative Year	0.011671423	0.001934864	6.032168057	0.001802463	0.006697698	0.016645149

Units) is in line with what we would expect from a learning curve perspective (as we will explore in Volume IV.) The negative merely indicates that as the number of Prior Units increases, we would expect the cost to reduce. In terms of the other parameters, their positive values indicate that as the Weight increases and the time increases, so will cost.

Yes, you've picked up the nuance of the 'usual Backward Stepwise procedure', that this is not necessarily the end state here.

Stepping sideways or a step to the side

6. Now that we have what we think is a statistically sound model (Table 6.34), we can challenge some of our basic assumptions about the function types we have chosen in terms of their inherent multicollinearity. To do this we need to create a Correlation Matrix for our independent variables using both their linear and logarithmic values

   o Identify those variables where the Correlation between one variable and another is similar when we use linear or log values – we'll use these in the next step

   o Also, identify those variables where the correlation between a variable's values and its log values is particularly higher than others (they will all be high unless they cover an extensive range of values, in which case they will still not be small, but they could have been larger.)

   o Table 6.35 illustrates the case for our example. Here we have very high correlation between weight values and their log values (*I bet you were*

Transforming Nonlinear Regression | 329

Table 6.35 Step 6 – Regression Independent Variable Correlation Matrix

Regression Variable Correlation Matrix	x1 Weight	x2 Prior Units	x3 Relative Year	Log(x1 Weight)	Log(x2 Prior Units)
x1 Weight	**1**	-0.205	-0.533	0.996	-0.012
x2 Prior Units	-0.205	**1**	0.378	-0.159	0.934
x3 Relative Year	-0.533	0.378	**1**	-0.484	0.410
Log(x1 Weight)	0.996	-0.159	-0.484	**1**	0.038
Log(x2 Prior Units)	-0.012	0.934	0.410	0.038	**1**

expecting me to make a joke about the weight of logs, weren't you?). We also have very high correlation between the number of Prior Units and their log values, but not quite as high. We will also note that the correlation between weight and any other variable is similar to that for the log of weight and that other variable.

> The reason for the difference is down to the relative range of values being logged. Ranges closer to (but not including zero or negative numbers) will have lower correlations with their Log values than those further away. In this case the weight values are significantly higher than the number of Prior Units.

Note also that the Correlation Matrix is symmetrical about the Top-Left to Bottom-Right diagonal.

7. Taking the model we were provisionally happy with from step 5, we can try interchanging a variable and its logarithmic equivalent one at a time, commencing with the highest match from Step 6, to see if we can improve the model's key statistics. In this case we have three options we can try:
   o Substituting $Log(x_1$ Weight) for $(x_1$ Weight)
   o Substituting $(x_2$ Prior Units) for $Log(x_2$ Prior Units)
   o Substituting both the above

Table 6.36 illustrates that the first of these creates an improvement in R-Square, Adjusted R-Square and the F-Statistic in comparison with Table 6.34
... which is the result we had in Section 6.2.2. (*Don't you just love it when a plan comes together?*)

If we were to review the other two options then we would find that they would dramatically reduce the F-Statistic to around 14.7 and 24.8 respectively, which while being quite respectable in their own right, don't hold a torch to the 170 level in Table 6.36. However, both these options would suggest that neither x_1 Weight or $Log(x_1$ Weight) were adding anything statistically significant to the model, with a probability of some 50% or more that their contributions are zero.

330 | Transforming Nonlinear Regression

Table 6.36 Step 7 – Regression Output Report Iteration 4 – All Variables Significant

SUMMARY OUTPUT

Regression Statistics	
Multiple R	0.995139395
R Square	0.990302415
Adjusted R Square	0.984483863
Standard Error	0.020041711
Observations	9

ANOVA

	df	SS	MS	F	Significance F
Regression	3	0.205089685	0.068363228	170.1974221	1.88009E-05
Residual	5	0.002008351	0.00040167		
Total	8	0.207098036			

	Coefficients	Standard Error	t Stat	P-value	Lower 95%	Upper 95%
Intercept	3.726377711	0.249575557	14.93086005	2.4391E-05	3.084823318	4.367932104
Log(x1 Weight)	0.743252086	0.087790838	8.466169192	0.000377613	0.517578553	0.968925619
Log(x2 Prior Units)	-0.415821085	0.019551273	-21.26823582	4.26005E-06	-0.466079233	-0.365562938
x3 Relative Year	0.010925718	0.001878032	5.817640607	0.002118636	0.006098082	0.015753354

6.3.2 Stepwise Regression by Forward Selection with linear transformations

We saw in Chapter 4 that there were two recognised procedures for Stepwise Regression. In the case of true Multi-linear Regression then the choice is down to personal preference, although we may find that Backward Elimination may be more reliable than Forward Selection at finding the best solution. In the previous section we have just explored Backward Elimination in the context of multiple linear transformations. Let's briefly turn our attention to the alternative.

In step 3 of the stepping up phase of the Stepwise Procedure for Backward Elimination, we determined that the function types we should consider for our example are the Exponential-Power Function pairing. We also ascertained previously in step 2, the R-Square values of each variable with the dependent variable, and also that for its log equivalent. We also ranked them in ascending order of R-Square.

If we take the variable with the highest ranked R-Square, x_2 Prior Units, and run a simple Linear Regression we will get the result if Table 6.37. This will give us a reference point from which to measure any improvement.

We will now add the next most significant variable with the caveat that it is not the '*Logarithmic Alter Ego*' of the first variable (as that would guarantee us a multicollinearity issue to resolve!). Referring back to Table 6.30 we would then pick variable x_3 Relative Year as we should not use the second highest ranked variable, Log(x_2 Prior Units), in this case as our first variable, x_2 Prior Units, outranks it. Table 6.38 presents the results of this second regression.

Whilst R-Square, Adjusted R-Square and F-Statistic are all acceptable, but markedly less than Iteration 1, we would instinctively reject this model on the basis that the contribution of x_3 Relative Year being zero has a probability of some 80%. However, with everything we've discussed about multicollinearity of variable values with their log equivalents, it might be worth trying to substitute x_2 Prior Units with Log(x_2 Prior

Transforming Nonlinear Regression | 331

Table 6.37 Forward Regression Output Report Iteration 1 – Best Single Variable

SUMMARY OUTPUT

Regression Statistics	
Multiple R	0.939530982
R Square	0.882718466
Adjusted R Square	0.865963961
Standard Error	0.058905221
Observations	9

ANOVA

	df	SS	MS	F	Significance F
Regression	1	0.182809261	0.182809261	52.68544049	0.00016869
Residual	7	0.024288775	0.003469825		
Total	8	0.207098036			

	Coefficients	Standard Error	t Stat	P-value	Lower 95%	Upper 95%
Intercept	5.489376336	0.036059478	152.2311632	1.39265E-13	5.404109219	5.574643453
x2 Prior Units	-0.004400405	0.000606244	-7.258473702	0.00016869	-0.005833944	-0.002966866

Table 6.38 Forward Regression Output Report Iteration 2

SUMMARY OUTPUT

Regression Statistics	
Multiple R	0.940253624
R Square	0.884076877
Adjusted R Square	0.845435836
Standard Error	0.06325537
Observations	9

ANOVA

	df	SS	MS	F	Significance F
Regression	2	0.183090585	0.091545293	22.87922001	0.001557795
Residual	6	0.024007451	0.004001242		
Total	8	0.207098036			

	Coefficients	Standard Error	t Stat	P-value	Lower 95%	Upper 95%
Intercept	5.483533981	0.044552229	123.0810256	1.93957E-11	5.374518605	5.592549357
x2 Prior Units	-0.004470824	0.000703099	-6.358738696	0.000709685	-0.006191245	-0.002750402
x3 Relative Year	0.001295371	0.004885263	0.265158872	0.799761637	-0.010658436	0.013249178

Units). In this case only it 'improves' the significance of x_3 Relative Year by some 10% probability, so in this case our instinct would have been right. (In a more marginal situation, this may not have been the case.)

Our next step then is to reject the variable x3 Relative Year and to add the next most significant variable, which is x1 Weight. The results are shown in Table 6.39.

Although showing an improvement on the previous two iterations, we would reject this model on the basis that there was a 50% probability that the contribution of x1 Weight was zero. Again our instinct may be to reject the x1 Weight variable out of hand, but perhaps we should try our little side-step manoeuvre and substitute each variable with its log equivalent one at a time, and together, to see if we can get an improvement that is statistically sound overall.

Of these the 'best of the bunch' is the substitution of both variables with their log equivalents. This gives the model results depicted in Table 6.40, which shows that the

332 | Transforming Nonlinear Regression

Table 6.39 Forward Regression Output Report Iteration 4

SUMMARY OUTPUT

Regression Statistics	
Multiple R	0.943583895
R Square	0.890350567
Adjusted R Square	0.853800757
Standard Error	0.061519891
Observations	9

ANOVA

	df	SS	MS	F	Significance F
Regression	2	0.184389854	0.092194927	24.35992272	0.001318315
Residual	6	0.022708182	0.003784697		
Total	8	0.207098036			

	Coefficients	Standard Error	t Stat	P-value	Lower 95%	Upper 95%
Intercept	5.420734918	0.112695228	48.10083808	5.41297E-09	5.144979629	5.696490207
x1 Weight	9.02226E-05	0.000139611	0.646240975	0.54203659	-0.000251394	0.000431839
x2 Prior Units	-0.004314561	0.000646938	-6.669204284	0.000550007	-0.005897562	-0.002731561

Table 6.40 Forward Regression Output Report Iteration 5

SUMMARY OUTPUT

Regression Statistics	
Multiple R	0.961592206
R Square	0.924659572
Adjusted R Square	0.899546095
Standard Error	0.05099486
Observations	9

ANOVA

	df	SS	MS	F	Significance F
Regression	2	0.191495181	0.095747591	36.81925857	0.000427646
Residual	6	0.015602855	0.002600476		
Total	8	0.207098036			

	Coefficients	Standard Error	t Stat	P-value	Lower 95%	Upper 95%
Intercept	4.513811168	0.533528961	8.460292686	0.00014898	3.20831283	5.819309507
Log(x1 Weight)	0.463533677	0.186898196	2.480139922	0.047796279	0.006210266	0.920857088
Log(x2 Prior Units)	-0.360134135	0.043377043	-8.302413225	0.000165512	-0.466273935	-0.253994335

variable $Log(x_1\ Weight)$ just creeps in under the radar with only a 4.78% chance that the true contribution is zero.

For completion the other two options would be rejected based on the contribution of the x_1 Weight or $Log(x_1\ Weight)$ variables being zero is 50% or greater.
Whilst this gives us an acceptable model, a quick comparison with the results of our first iteration using a single variable in Table 6.37 is that it is inferior to the simpler model:

R-Square:	Increases with the addition of the second variable
Adjusted R-Square:	Decreases with the addition of the second variable (*suggesting that the Regression is force fitting the best fit!*)
F-Statistic:	Reduces (*suggesting that the simple regression is the better fit*)

Transforming Nonlinear Regression | 333

In this particular instance, a Stepwise Regression by Forward Selection would lead us to a Simple Linear Regression based on the Prior Units whereas the Backward Elimination procedure gives us a more robust explanation of the observed data, which just goes to show that sometimes it pays us to take step back to take in the whole picture rather than push forward to get a quick result!

6.4 Is the Best Fit really the better fit?

Suppose that we have diligently prepared our alternative nonlinear models so that they pass all the health checks of R-Square (or Adjusted R-Square), F Significance, t-parameter significances, sensibility and homoscedasticity, how do we then decide which of our Best Fit relationships is the best of the best.

Caveat augur

Recall that logs change our perspective, but they don't change reality (Chapter 5).

Mind you don't trip over the log!

The answer to that lies in the nature of the alternative nonlinear functions in question. In some instances, this requires us to transform the model data back into linear space and to make the comparison from a linear perspective. (*We wouldn't compare a physical measurement taken in metric units with one taken in imperial units without taking account of the difference in scale, would we?*) The key is in what is happening to the dependent y-variable as summarised in Table 6.41:

For the Formula-philes: Calculating the Standard Error of the Estimate (SEE)

Consider a series of values $y_1 \ldots y_n$ with $Y_1 \ldots Y_n$ being the associated logarithmic values to a known base. Let $\widehat{Y_1} \ldots \widehat{Y_n}$ be the predicted values of a linearised Exponential, Power or combined Exponential/Power Function Regression. Let $\hat{y}_1 \ldots \hat{y}_n$ be the equivalent linear predicted values of the regression.

The Standard Error of the Estimate (SEE or SE) is the square root of the Mean Square Error (MSE) with v independent variables given by:

$$SEE = \sqrt{\frac{\sum_{i=1}^{n}(y_i - \hat{y}_i)}{n - 1 - v}}$$

334 | Transforming Nonlinear Regression

Table 6.41 Statistic to Use as the Basis of Comparing Alternative Best Fit models

Comparing	Linear or Logarithmic Combinations (Linear y)	Exponential or Power Combinations (Logarithmic y)
Linear or Logarithmic Combinations (Linear y)	**Use Adjusted R-Square** **Bigger is better**	Convert the exponential or power values back to Linear Space and compare the equivalent **Standard Error of the Estimate (SEE or SE)** **Smaller is better**
Exponential or Power Combinations (Logarithmic y)	Convert the exponential or power values back to Linear Space and compare the equivalent **Standard Error of the Estimate (SEE or SE)** **Smaller is better**	**Use Adjusted R-Square** **Bigger is better**

When we are conducting a Stepwise Regression routine we can rely on the Adjusted R-Square to guide us through the variable addition or elimination routine, as we will have already decided on the basic functional form for that 'branch' of the Stepwise Decision Tree. The only time we need to concern ourselves with the Standard Error of the Estimate (SEE) in Stepwise Regression is when we are choosing the basic functional form at the start i.e. whether we will be transforming the dependent y-variable or keeping it linear.

Let's look at a basic example of two alternative models using the same data; one a linear model and the other a power function version.

In Figure 6.12 we have the Regression Results and Input data for a Linear Cost Model with respect to the weight of nine similar products. They all pass our criteria for R-Square, F-Statistic Significance and t-Statistic P-values (see Chapter 4). In Figure 6.13 we have an alternative power function model, which also scores well against those same statistics. In fact the power model version, which utilises the Log(Cost) in relation to the Log(Weight), scores better against all these measures. It would be very easy to assume that this model is the better fit to our data. (*I think that some of us have already guessed that it is not going to be, haven't we?*) Based on Table 6.41, as we are comparing across Log and Linear Space we need to consider which is the better model using the Standard Error in Linear Space.

Transforming Nonlinear Regression 335

INPUT DATA >

	y	x1
	Cost £	Weight Kg
	5123	10
	6527	21
	6388	34
	9253	42
	7722	59
	9182	63
	8348	71
	10702	85
	11092	97

SUMMARY OUTPUT

Regression Statistics	
Multiple R	0.907072211
R Square	0.822779996
Adjusted R Square	0.797462853
Standard Error	906.6035088
Observations	9

ANOVA

	df	SS	MS	F	Significance F
Regression	1	26711840.55	26711840.55	32.49892701	0.000734755
Residual	7	5753509.455	821929.9221		
Total	8	32465350			

	Coefficients	Standard Error	t Stat	P-value	Lower 95%	Upper 95%
Intercept	4896.171288	662.8970207	7.386020958	0.000151211	3328.668916	6463.673659
Weight Kg	62.80385562	11.0167069	5.700783017	0.000734755	36.7534833	88.85422794

Figure 6.12 Linear Cost-Weight Regression Model

INPUT DATA >

			y	x1
Cost £	Weight Kg		Log Cost	Log Weight
5123	10		3.7095	1.0000
6527	21		3.8147	1.3222
6388	34		3.8054	1.5315
9253	42		3.9663	1.6232
7722	59		3.8877	1.7709
9182	63		3.9629	1.7993
8348	71		3.9216	1.8513
10702	85		4.0295	1.9294
11092	97		4.0450	1.9868

SUMMARY OUTPUT

Regression Statistics	
Multiple R	0.911100819
R Square	0.830104702
Adjusted R Square	0.805833945
Standard Error	0.049035414
Observations	9

ANOVA

	df	SS	MS	F	Significance F
Regression	1	0.082237376	0.082237376	34.20184657	0.000631733
Residual	7	0.016831303	0.002404472		
Total	8	0.099068679			

	Coefficients	Standard Error	t Stat	P-value	Lower 95%	Upper 95%
Intercept	3.381360755	0.090973005	37.16883662	2.65273E-09	3.166243781	3.596477728
Log Weight	0.317954346	0.054367578	5.848234483	0.000631733	0.189395452	0.44651324

Figure 6.13 Power Log(Cost)-Log(Weight) Regression Model

In Figure 6.14 we show how Microsoft Excel calculates the Regression Standard Error from the Analysis of Variance Table (ANOVA) for the Linear Regression model in Figure 6.12

1. The Model's error term is calculated by taking the difference between the observed Cost and the Model Cost

336 | Transforming Nonlinear Regression

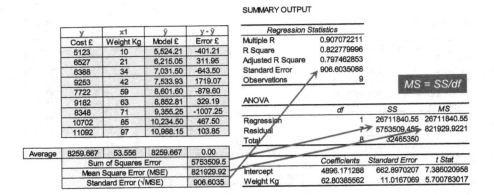

Figure 6.14 Derivation of Regression Standard Error in Microsoft Excel

2. The Sum of Squares Error (SSE) is calculated using the Excel function **SUMSQ(*range*)** where range is the range of cells calculated in Step 1
3. The Mean Square Error (MSE) is calculated by dividing the SSE from Step 2 by the Residual Degrees of Freedom (df). (In this case, 7 df.)
4. The Standard Error of the Estimate (SEE, or SE) is the Square Root of the Mean Square Error (MSE) from Step 3

In Table 6.42 we have calculated the Regression Model from the output in Figure 6.13 in Log Space:

1. The linear values of the Regression Model are the Anti-Log of the Regression Log Values. As the model was run with Common Logs (base 10), these are calculated by taking 10 to the power of the log values
2. The model's linear error term is calculated by taking the difference between the observed Cost (linear not log value) and the Model's Linear value calculated at Step 1
3. The Sum of Squares Error (SSE) is calculated using the Excel function **SUMSQ(*range*)** where range is the range of Linear Error values calculated in Step 2
4. The Mean Square Error (MSE) is calculated by dividing the SSE from Step 3 by the Residual Degrees of Freedom (df). (In this case, 7 df.)
5. The Standard Error of the Estimate (SEE, or SE) is the Square Root of the Mean Square Error (MSE) from Step 4, giving us a value of 951.6472 hours

When we look back to Figure 6.12 or Figure 6.14 at the standard error for our linear model, we had a value of 906.6035.

Transforming Nonlinear Regression | 337

Table 6.42 Derivation of the Standard Error in a Linear Space for a Power Function Regression

			Slope	Intercept			
		Regression Coefficients	0.3179543	3.3813608			
		y	x1	$\hat{y}$	y - $\hat{y}$	10^$\hat{y}$	Cost - 10^$\hat{y}$
Cost £	Weight Kg	Log Cost	Log Weight	Model Log £	Error £	Model £	Linear Error
5123	10	3.7095	1.0000	3.6993	0.0102	5004	119.03
6527	21	3.8147	1.3222	3.8018	0.0129	6335	191.72
6388	34	3.8054	1.5315	3.8683	-0.0629	7384	-996.16
9253	42	3.9663	1.6232	3.8975	0.0688	7897	1355.68
7722	59	3.8877	1.7709	3.9444	-0.0567	8799	-1076.54
9182	63	3.9629	1.7993	3.9535	0.0095	8984	198.02
8348	71	3.9216	1.8513	3.9700	-0.0484	9332	-984.04
10702	85	4.0295	1.9294	3.9948	0.0346	9882	820.39
11092	97	4.0450	1.9868	4.0131	0.0319	10305	786.63
Average	8259.667	53.556	3.9047	1.6461	3.9047	0.0000	8213.5875	46.0792
				Sum of Squares Error	0.0168313		6339427.29
				Mean Square Error	0.0024045		905632.47
				Standard Error	0.0490		951.6472

From a standard error perspective, the linear model is the better model despite the other measurements pointing us in the wrong direction!

6.5 Regression of Transformed Generalised Nonlinear Functions

Before we get too deep into the details of this technique, let's do a bit to manage our expectations just so we don't get carried away in the euphoria of thinking we have a potential solution that will allow us to transform any Generalised Nonlinear Function into a linear format. There is, therefore, a big 'but' associated with this technique (. . . and I'm going to resist the temptation to ask whether my 'but' looks big in this. . . With apologies and respect to Arabella Weir, 1997: "Does my bum look big in this?") So, unconventionally we will begin this section with a health warning.

Caveat augur

This technique is not a panacea for all our nonlinear ills.

It should be considered to be one that gives us an 'indication only' of a potential model that can be used for interpolation or 'close' extrapolation i.e. slightly to the left or right of the data range we have.

There is a real risk of over fitting the data to a function type i.e. getting a better R-Square for one Generalised Function type than we might have been happy to

(Continued)

338 | Transforming Nonlinear Regression

> accept for another. The technique does have a tendency to follow the line of least resistance when seeking out those parameters that maximises the opportunity to make a linear transformation.
>
> In theory our confidence in its output should improve with more data from a wider range. However, this is not always the case.

Where we have small data samples, the variation in the data around the 'best fit' trendline may be such that the 3-in-1 Solver technique 'points us' towards a different functional form than the one we may intuitively expect (as we saw in Chapter 5.) If we have reasonable grounds based on a sound rationale to expect that there is a model of a particular form then there is probably no need to 'force fit' the other two generalised forms as a matter of routine. Sometimes simple a '*shabby chic*' piece of furniture is better than a more refined one with an obvious flaw. It doesn't pretend to be better than it really is, and it is still functional. Where the result is 'close' (in terms of R-Square) and using additional data is not an option, then we may be justified in trusting our intuition, using Solver merely as a tool to calibrate the data relationship we choose.

So, let's have a closer look at the three-stage process involved:

   i.   Identify the functional type and determine the offset adjustment
  ii.   Perform the appropriate linear transformation
 iii.   Perform the Linear Regression

Recalling the three Generalised Forms of these functions from Chapter 5, we have: The technique works on the basis that if we knew the value of the constants a and/or b then we could use these to normalise our data into a form where we could then apply a standard transformation into a linear form using logarithms. In other words, this is a two-stage normalisation process of adjustment followed by transformation prior to performing a linear regression.

For the Formula-philes: Generalised transformable function types

Function type	Standard form	Generalised form	Transform to Standard form by letting
Logarithmic Functions	$m^y = cx$	$m^y = c(x+a)$	$X = x + a$
Exponential Functions	$y = cm^x$	$y = cm^x - b$	$Y = y + b$

| **Power Functions** | $y = cx^m$ | $y = c(x+a)^m - b$ | $X = x + a$ |
| | | | $Y = y + b$ |

The Standard forms are those where the Constants a and b in the Generalised forms are equal to zero.

In theory we can establish a model in which we can determine the Best Fit parameters for the Offset constant, and the slope and intercept of the transformed data. In practice, however, it can sometimes be difficult to get the model to converge appropriately, because as we saw in Chapter 5, it is not always clear which form of Generalised Function our data might be, as some offset adjustments may look like more than one Generalised form. In the event that an adjustment gives us a perfect standard function, we would get an R-Square of 1. So, inherently we should be looking for the offset constant(s) that give us the highest R-Square by minimising the variation around a straight line or curve, but that does not guarantee it being right! It all depends on the data scale and pattern of scatter in our sample.

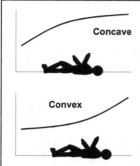

For concave relationships, the R-Square for Generalised Logarithmic and Power Functions will tend to 'move' in the same way as the offset parameter changes, especially where data ranges are quite compact.

For convex relationships, Generalised Exponential, Logarithmic and Power Functions will also tend to 'move' in the same way, thus exhibiting some form of collective correlation. Note that Exponential Functions are always Convex as we demonstrated in Chapter 5 Section 5.2.3 even when they have negative power, i.e. sloping down.

Nevertheless, with that caveat in mind, this technique can be helpful and will allow us to search for the best fit option across the three generalised forms; but we shouldn't place too much credence on the precise values.

If we have no inkling of what an appropriate offset value or values might be, we can choose either to 'Guess and Iterate' at random, use the Random-Start Bisection Method (see Chapter 5) or we can use Microsoft Excel's **Solver** algorithm to determine the 'best solution'. (Solver comes as a free Add-in to Microsoft Excel. Once loaded it will always appear in the Analysis section of the Data ribbon.)

Solver is probably the best option, as it is a very flexible tool that allows us to define a formula with random parameters and then seeks to find the best fit value to some

340 | Transforming Nonlinear Regression

criterion we have defined, subject to a number of optional constraints that again we can choose to impose. **Solver** does this by an iterative routine, such as:

- The Coefficient of Determination, R-Square is be maximised
- The parameters' values must not create conditions where we would want to take the Log of a negative number

From Chapter 5 we discussed using **Solver** to find the values of the offset parameters that maximises the R-Square (or minimises the Sum of Squares Error) for all three Generalised forms at once. If we have that 'inkling' of what an appropriate value of the offset may be then we can use that as the starting point. Whichever technique we choose the layout for both options can be the same, as illustrated in Table 6.43; the references to Solver here clearly refer to the second option but in the case of the first option we could replace the label 'Solver Variable' with something like 'Current Guess'!

As we always advocate drawing a picture of our data as a sensibility check, Figure 6.15 illustrates the three starting positions. In this case we have also decided to use the argument that we should only use a common x offset parameter for Logarithmic and Power Functions and a common y offset for Exponential and Power Functions. As the data is clearly concave in nature, we can disregard the Generalised Exponential Function as a viable option before we start with Solver; it curves the wrong way.

Suppose that in this particular case we want to restrict the offset in x to integer values. We can specify that in the Solver Constraints. Running our Solver 3-in-1 model to maximise the R-Squares, we get the result in Table 6.44 and Figure 6.16, suggesting that

Table 6.43 Using Solver to Maximise the R-Square of Three Generalised Functions – Set-Up

		Generalised Logarithmic Function		Generalised Exponential Function		Generalised Power Function				3-in-1 Solvers
		Offset a		Offset b		Offset a		Offset b		
Solver Variable >		0		0		0		0		
Minimum Offset >		-0.999		-0.999		-0.999		-0.999		
Maximum Offset >		54		26		54		26		
Raw Input Data		**Transformed Data**		**Transformed Data**		**Transformed Data**				
x	y	X = x+a	Log(x+a)	Y=y+b	Log(y+b)	X = x+a	Log(x+a)	Y=y+b	Log(y+b)	
1	1	1.000	0.0000	1.000	0.0000	1.000	0.0000	1.000	0.0000	
2	2	2.000	0.3010	2.000	0.3010	2.000	0.3010	2.000	0.3010	
3	3	3.000	0.4771	3.000	0.4771	3.000	0.4771	3.000	0.4771	
4	4	4.000	0.6021	4.000	0.6021	4.000	0.6021	4.000	0.6021	
5	5	5.000	0.6990	5.000	0.6990	5.000	0.6990	5.000	0.6990	
7	7	7.000	0.8451	7.000	0.8451	7.000	0.8451	7.000	0.8451	
12	9	12.000	1.0792	9.000	0.9542	12.000	1.0792	9.000	0.9542	
17	11	17.000	1.2304	11.000	1.0414	17.000	1.2304	11.000	1.0414	
22	12	22.000	1.3424	12.000	1.0792	22.000	1.3424	12.000	1.0792	
27	13	27.000	1.4314	13.000	1.1139	27.000	1.4314	13.000	1.1139	
		R^2 with y		**R^2 with x**		**R^2 based on Log Transforms**				**R^2 Sum**
R^2 to Maximise >		0.9696		0.7189		↳	0.9705	↵		2.659

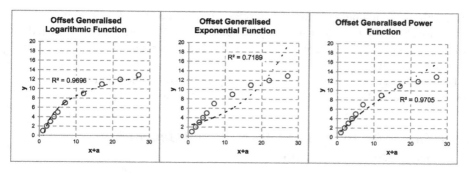

Figure 6.15 3-in-1 Solver Graphical Sensibility Check (Before)

Table 6.44 Using Solver to Maximise the R-Square of Three Generalised Functions – Result

		Generalised Logarithmic Function		Generalised Exponential Function		Generalised Power Function				3-in-1 Solvers
		Offset a		Offset b		Offset a		Offset b		
Solver Variable >		3		26	< Maximum	2		26	< Maximum	
Minimum Offset >		-0.999		-0.999		-0.999		-0.999		
Maximum Offset >		54		26		54		26		
Raw Input Data		Transformed Data		Transformed Data		Transformed Data				
x	y	X = x+a	Log(x+a)	Y=y+b	Log(y+b)	X = x+a	Log(x+a)	Y=y+b	Log(y+b)	
1	1	4.000	0.6021	27.000	1.4314	3.000	0.4771	27.000	1.4314	
2	2	5.000	0.6990	28.000	1.4472	4.000	0.6021	28.000	1.4472	
3	3	6.000	0.7782	29.000	1.4624	5.000	0.6990	29.000	1.4624	
4	4	7.000	0.8451	30.000	1.4771	6.000	0.7782	30.000	1.4771	
5	5	8.000	0.9031	31.000	1.4914	7.000	0.8451	31.000	1.4914	
7	7	10.000	1.0000	33.000	1.5185	9.000	0.9542	33.000	1.5185	
12	9	15.000	1.1761	35.000	1.5441	14.000	1.1461	35.000	1.5441	
17	11	20.000	1.3010	37.000	1.5682	19.000	1.2788	37.000	1.5682	
22	12	25.000	1.3979	38.000	1.5798	24.000	1.3802	38.000	1.5798	
27	13	30.000	1.4771	39.000	1.5911	29.000	1.4624	39.000	1.5911	
		R^2 with y		R^2 with x		R^2 based on Log Transforms				R^2 Sum
R^2 to Maximise >		0.9961		0.9036		0.9947				2.894

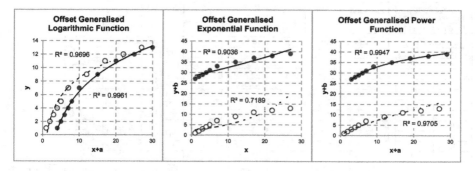

Figure 6.16 3-in-1 Solver Graphical Sensibility Check (After)

342 | Transforming Nonlinear Regression

a Logarithmic Trendline with an x-value offset by +3 is our Best Fit. We should really now rationalise why this offset is appropriate from a TRACEability perspective rather than just accepting it. (In Figure 6.16, the hollow data points with the dotted trendline is the original dataset.)

Let's look at some examples of each type of generalised function and how (or if) we can then complete the regression . . . *but be warned, this is going to be a bit of a rollercoaster ride!*

6.5.1 Linear Regression of a Transformed Generalised Logarithmic Function

Let's consider the case of some design alterations being introduced into a complex product to make it easier to produce. Intuitively we would probably expect that the number of new alterations would reduce over time as we produced more units; this makes it sound unlikely to be linear function. We can use our 3-in-1 **Solver** technique to determine whether the data we have is one of the three remaining standard or generalised function types.

In Table 6.45 and Figure 6.17 we show results of our 3-in-1 Solver. As the data is concave in nature, it cannot be an exponential relationship. For the time being we have limited the x and y offsets to integer values as all our data values are integer in nature; we will revisit this shortly. The best result in terms of R-Square is for a generalised Power Function, but only at the expense of a large positive y-offset, for which there is no obvious logical reason we might make to support it. This is a good example of how **Solver** will do its level best to give us the best numerical result but which is not

Table 6.45 Using 3-in-1 Solver to Identify a Generalised Logarithmic Function

		Generalised Logarithmic Function		Generalised Exponential Function		Generalised Power Function				3-in-1 Solvers
		Offset a		Offset b		Offset a		Offset b		
Solver Variable >		-2	< Minimum	31	< Maximum	-2	< Minimum	31	< Maximum	
Minimum Offset >		-2		0		-2		0		
Maximum Offset >		12		31		12		31		
Raw Input Data		**Transformed Data**		**Transformed Data**		**Transformed Data**				
x	y	X = x+a	Log(x+a)	Y=y+b	Log(y+b)	X = x+a	Log(x+a)	Y=y+b	Log(y+b)	
3	11	1	0.0000	42	1.6232	1	0.0000	42	1.6232	
4	16	2	0.3010	47	1.6721	2	0.3010	47	1.6721	
5	19	3	0.4771	50	1.6990	3	0.4771	50	1.6990	
6	23	4	0.6021	54	1.7324	4	0.6021	54	1.7324	
7	25	5	0.6990	56	1.7482	5	0.6990	56	1.7482	
8	27	6	0.7782	58	1.7634	6	0.7782	58	1.7634	
9	28	7	0.8451	59	1.7709	7	0.8451	59	1.7709	
10	29	8	0.9031	60	1.7782	8	0.9031	60	1.7782	
11	29	9	0.9542	60	1.7782	9	0.9542	60	1.7782	
12	31	10	1.0000	62	1.7924	10	1.0000	62	1.7924	
			R^2 with y		R^2 with x		R^2 based on Log Transforms			R^2 Sum
R^2 to Maximise >			0.9905		0.8734			0.9911		2.855

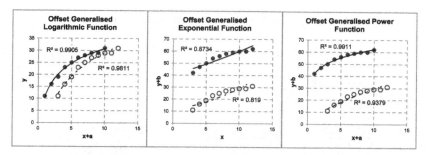

Figure 6.17 3-in-1 Solver Graphical Sensibility Check for a Generalised Logarithmic Function

necessarily the best logical result. (*Recall what we said about 'overfitting'.*) In this case if we were to increase our maximum allowable y-offset, **Solver** would pick that, so on grounds of credibility (logic), we are probably doing the right thing in rejecting this particular option.

On the other hand, we have a viable result for a generalised Logarithmic Function using an x-offset of –2. The question we must ask ourselves is '*Can we justify this value?*' after all it has hit the minimum limit we allowed. We started off on the premise that we didn't want to use non-integer values (but we could review this decision based on the discussion that follows) and we could not have a value of –3 or less as this would encourage **Solver** to take the logarithm of negative numbers, (*and that would just upset it.*) However, consider the following rationale:

> Let us assume that the first unit produced is considered to be the baseline standard from which all design alterations to improve manufacturability are measured. By that definition then there will be no alterations for unit number one. Suppose that the design process takes longer than the manufacturing or production process, then there will be a lag between the improvement opportunity being identified and the solution being embodied in the product as an alteration. If this lag is equivalent to an integer number of production cycles then we will have an integer lag in the relationship.
>
> In effect the logarithmic relationship of design alterations is with the number of units built when the improvement opportunity is identified, not embodied as such. So it is quite feasible (and we could argue highly likely) that any such lag is not an integer relationship as we first suspected; not all alterations will necessarily take the same time to resolve, so we are really looking at the average lag, which does not have to be integer. With this in mind, if we were to re-run our 3-in-1 **Solver** allowing non-integer values it would find the result –1.722 as shown in Table 6.46. However, you have probably already spotted that the best result (in terms of maximum R-Square) is generated for a Generalised Power Function with zero y-offset

344 | Transforming Nonlinear Regression

Table 6.46 Using 3-in-1 Solver to Identify a Generalised Logarithmic Function

		Generalised Logarithmic Function		Generalised Exponential Function		Generalised Power Function				3 in 1 Solvers
		Offset a		Offset b		Offset a		Offset b		
Solver Variable >		-1.722		31	< Maximum	-2.999	< Minimum	-10.936		
Minimum Offset >		-2.999		-10.999		-2.999		-10.999		
Maximum Offset >		12		31		12		31		
Raw Input Data		**Transformed Data**		**Transformed Data**		**Transformed Data**				
x	y	X = x+a	Log(x+a)	Y=y+b	Log(y+b)	X = x+a	Log(x+a)	Y=y+b	Log(y+b)	
3	11	1.278	0.1065	42.000	1.6232	0.001	-3.0000	0.064	-1.1946	
4	16	2.278	0.3575	47.000	1.6721	1.001	0.0004	5.064	0.7045	
5	19	3.278	0.5156	50.000	1.6990	2.001	0.3012	8.064	0.9065	
6	23	4.278	0.6312	54.000	1.7324	3.001	0.4773	12.064	1.0815	
7	25	5.278	0.7225	56.000	1.7482	4.001	0.6022	14.064	1.1481	
8	27	6.278	0.7978	58.000	1.7634	5.001	0.6991	16.064	1.2059	
9	28	7.278	0.8620	59.000	1.7709	6.001	0.7782	17.064	1.2321	
10	29	8.278	0.9179	60.000	1.7782	7.001	0.8452	18.064	1.2568	
11	29	9.278	0.9675	60.000	1.7782	8.001	0.9031	18.064	1.2568	
12	31	10.278	1.0119	62.000	1.7924	9.001	0.9543	20.064	1.3024	
		R^2 with y		R^2 with x		R^2 based on Log Transforms				R^2 Sum
R^2 to Maximise >		0.9915		0.8734			0.9982			2.863

and -2.55 x-offset. (*What do they say about the best laid plans of mice and men? Have some more cheese!*) Let's look at this a little more before we accept or reject either the generalised Logarithmic or Generalised Power Function option.

We would hope that as we approached the 'true' relationship that the R-Square would improve progressively, only deteriorating as we went past the optimum. We would also expect this then to be true if we were approaching it from the left or right in terms of an x-offset, or from above or below in terms of a y-offset. The issue we have here is that for a Power Function we have both dimensions working at the same time. A good test for stability is to try alternative starting values or 'seeds' to run the Solver. If we get the same Solver solution using say minimum or maximum (or zero if this is between the two) then the result would appear to be stable. For a Power Function we should also try 'crossing' the starting points Min x with Max y and vice versa. In this particular example, we will find that the Generalised Power Function is not particularly stable in the values it determines. If there's one thing that an estimator likes it is stability or repeatability. (It's the R in TRACEability)

So yet again we have a judgement call to make. As we will see from Figure 6.18 the impact on making the wrong call can be quite dramatic, and potentially too big a difference to cover by a simple three-point estimate. Ideally, if we could we would get more data and refresh the exercise but on the assumption that if there had been more available we would have used it in the first place, we should make a note to revisit this at a future iteration of the estimate in line with good practice (see Volume I Chapter 2.) Based on the inherent instability in the Generalised Power Function in this instance, we will make an executive decision to accept the Generalised Logarithmic Function.

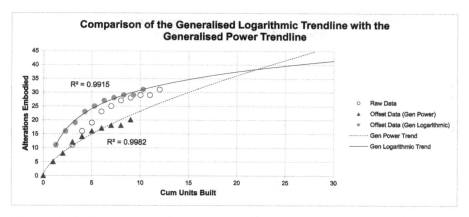

Figure 6.18 Comparison of Generalised Logarithmic Trendline with a Generalised Power Trendline

(*You're right, we could have just discussed a simple example which was unequivocally a Generalised Logarithmic Function, but that would have avoided this particular slice of reality. Life as an estimator is full of 'decision calls', usually down to the data (or lack of it) that we have available to us.*)

In further defence of accepting the Generalised Logarithmic Function in preference to what appears to be a better fit solution (more precise) offered by the Generalised Power Function, we must not lose sight of the fact that precision does not mean that it is any more correct. Remember that **Solver** will give us the best result it can within the limitations of the 'Reduced Gradient Method' algorithm it uses and the constraints with which we feed it. Note that to get this 'superior' R-Square for the Generalised Power Function, **Solver** has had to buffer up to the minimum x-offset we allowed (that is possibly too much of a coincidence for some of us.)

Putting this argument to one side some of us may be wondering why are we bothering with all this anyway when we had such a good R-Square with the raw data in the first place, and we are only improving it by a small amount – are we being guilty of being overly precise? Let's just reflect on two graphs:

- In earlier Figure 6.17 (left hand plot), showing the original data with the hollow markers, we have the data below the dotted trendline of a standard Logarithmic Function at the 'edges', but above it in the centre, whereas the adjusted data with the solid markers is a better (but not a perfect fit). The middle plot for an Exponential Function returns an acceptable R-Square in excess of 0.8, but the trendline is not representative of the data, bending in the wrong direction (*if you squint a little you will see that it is not quite a straight line.*)

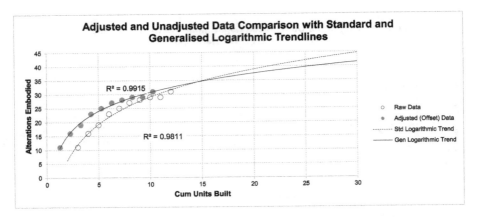

Figure 6.19 Extrapolating Standard and Generalised Logarithmic Trendlines

- In Figure 6.19 we show the impact of extrapolating both the Standard and Generalised Logarithmic Trendlines. Quite clearly going forward we will get completely different estimate, even after we reverse the impact of the offset adjustment

Even if we were to elect to use the Standard Logarithmic Function as the basis of the model, we could still use the Generalised Logarithmic Function solution we have found to test the sensitivity of our estimate going forward.

The last stage of our procedure is to perform the Linear Regression of the transformed Logarithmic data using the Cumulative Alterations Embodied as the y-variable, and the Log of the Number of Cumulative Units Produced less our Offset of -1.722 as our x-variable. We get the results shown in Table 6.47, giving the relationship:

$$\text{Alterations Embodied} = 22.376 \times \log_{10}(\text{Build No} - 1.722) + 8.382$$

Or, more pragmatically, we should take the Number of Alterations rounded to the nearest integer on the basis that the answer must be an integer value. (If we leave it as a real number, then we are implying it to be the average value for that number of build units.)

Going back to the issue of sensitivity analysis, even the Prediction Intervals around the two Regression Lines (Standard Logarithmic and Generalised Logarithmic) will give diverging, and eventually non-overlapping results as illustrated in Figure 6.20. This underlines the fact that for extrapolation close to the data range analysed, the range estimates will give reasonably similar answers, but the further away from this comfort zone we go, then the more important it is that we select the most appropriate functional form of the relationship. Widening the Prediction Intervals will delay the effect, but is not a long term solution.

Transforming Nonlinear Regression | 347

Table 6.47 Using 3-in-1 Solver to Identify a Generalised Logarithmic Function

SUMMARY OUTPUT

Regression Statistics	
Multiple R	0.995746233
R Square	0.991510561
Adjusted R Square	0.990449381
Standard Error	0.638019279
Observations	10

Based on R-Square, F and t-statistics we would accept this model as a statistically significant result

ANOVA

	df	SS	MS	F	Significance F
Regression	1	380.3434512	380.3434512	934.3473078	1.42513E-09
Residual	8	3.256548806	0.407068601		
Total	9	383.6			

	Coefficients	Standard Error	t Stat	P-value	Lower 95%	Upper 95%
Intercept	8.382226218	0.543247035	15.42986095	3.09482E-07	7.129496308	9.634956128
Log(x-1.722)	22.37560529	0.732016083	30.56709518	1.42513E-09	20.68757317	24.0636374

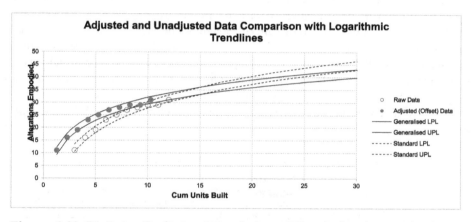

Figure 6.20 Diverging Prediction Intervals Around Standard and Generalised Logarithmic Functions

Some of us may still be feeling a little uncomfortable with the decision to reject the Generalised Power Function option. Suppose we did our Regression on the Generalised Power Function instead and generated the 90% Prediction Interval around it. In Figure 6.20 we can compare this with the equivalent Prediction Interval for the Generalised Logarithmic Function Regression from Table 6.47. The latter interval is narrower than that for the Power Function. It's all down to the compression property of logarithms we discussed in Chapter 5. In the case of a Power Function we have two compressions at

348 | Transforming Nonlinear Regression

work, in the Logarithmic Function, we only have one. (*Do you feel a little more comfortable with the decision now?*)

6.5.2 Linear Regression of a Transformed Generalised Exponential Function

Now let's consider another example, one in which our 3-in-1 Solver points us to a Generalised Exponential Function. In this case let's consider a temperature cooling problem. Some of us may recall vaguely, if we did physics lessons at school, that this sounds like Newton's Law of Cooling and therefore we know that it is basically an offset exponential in form, but let's assume that we don't know that. (*There will be many cases where we won't have that 'gut feeling' or prior knowledge of what it might be.*)

In our example the temperature of an object removed from a source of heat is allowed to cool. Initially the temperature is checked every five minutes, and then latterly at ten-minute intervals. There may be some inaccuracies in the temperature taken and the precise time that they are taken relative to the start point. Table 6.48 shows the data we have collected and the initial set-up of our 3-in-1 Solver Template. We have chosen 'seed' starter values for our offsets that are either zero or one, based on either those values (zeroes) that return a standard form of the function, or simply those that avoid creating a logarithm of less than or equal to zero. (In practice we should be able to prime Solver with any values; it may just have to work a little harder to find an optimum solution.)

Running Solver returns the results in Table 6.49 which are summarised pictorially as a 'before and after' view in Figure 6.21, in which the solid 'blobs' are the 'after' event. One

Table 6.48 3-in-1 Solver Set-Up for a Generalised Exponential Function Example

		Generalised Logarithmic Function		Generalised Exponential Function		Generalised Power Function				3-in-1 Solvers
		Offset a		Offset b		Offset a		Offset b		
Solver Variable >		1.0000		0.0000		1		0.0000		
Minimum Offset >		0.0001		-27.30		0.0001		-27.3		
Maximum Offset >		60		30		60		30		
Minutes	**Temp °C**	**Transformed Data**		**Transformed Data**		**Transformed Data**				
x	y	X = x+a	Log(x+a)	Y=y+b	Log(y+b)	X = x+a	Log(x+a)	Y=y+b	Log(y+b)	
0	90.2	1.00	0.0000	90.200	1.9552	1.00	0.0000	90.200	1.9552	
5	77.8	6.00	0.7782	77.800	1.8910	6.00	0.7782	77.800	1.8910	
10	67.3	11.00	1.0414	67.300	1.8280	11.00	1.0414	67.300	1.8280	
15	59.2	16.00	1.2041	59.200	1.7723	16.00	1.2041	59.200	1.7723	
20	52.0	21.00	1.3222	52.000	1.7160	21.00	1.3222	52.000	1.7160	
25	46.7	26.00	1.4150	46.700	1.6693	26.00	1.4150	46.700	1.6693	
30	41.6	31.00	1.4914	41.600	1.6191	31.00	1.4914	41.600	1.6191	
40	35.1	41.00	1.6128	35.100	1.5453	41.00	1.6128	35.100	1.5453	
50	30.3	51.00	1.7076	30.300	1.4814	51.00	1.7076	30.300	1.4814	
60	27.4	61.00	1.7853	27.400	1.4378	61.00	1.7853	27.400	1.4378	
				R^2 with y				R^2 based on Log Transforms		Sum
R^2 to Maximise >				0.9262	R^2 with x	0.9733	↳	0.8383	↵	2.738

Table 6.49 3-in-1 Solver Results for the Generalised Exponential Function Example

		Generalised Logarithmic Function		Generalised Exponential Function		Generalised Power Function				3-in-1 Solvers
		Offset a		Offset b		Offset a		Offset b		
Solver Variable >		8.0369		-21.2538		60	< Maximum	-12.4090		
Minimum Offset >		0.0001		-27.30		0.0001		-27.3		
Maximum Offset >		60		30		60		30		
Minutes	Temp °C	Transformed Data		Transformed Data		Transformed Data				
x	y	X = x+a	Log(x+a)	Y=y+b	Log(y+b)	X = x+a	Log(x+a)	Y=y+b	Log(y+b)	
0	90.2	8.04	0.9051	68.9462	1.8385	60.00	1.7782	77.7910	1.8909	
5	77.8	13.04	1.1152	56.5462	1.7524	65.00	1.8129	65.3910	1.8155	
10	67.3	18.04	1.2562	46.0462	1.6632	70.00	1.8451	54.8910	1.7395	
15	59.2	23.04	1.3624	37.9462	1.5792	75.00	1.8751	46.7910	1.6702	
20	52.0	28.04	1.4477	30.7462	1.4878	80.00	1.9031	39.5910	1.5976	
25	46.7	33.04	1.5190	25.4462	1.4056	85.00	1.9294	34.2910	1.5352	
30	41.6	38.04	1.5802	20.3462	1.3085	90.00	1.9542	29.1910	1.4652	
40	35.1	48.04	1.6816	13.8462	1.1413	100.00	2.0000	22.6910	1.3559	
50	30.3	58.04	1.7637	9.0462	0.9565	110.00	2.0414	17.8910	1.2526	
60	27.4	68.04	1.8327	6.1462	0.7886	120.00	2.0792	14.8910	1.1758	
		R^2 with y		R^2 with x		R^2 based on Log Transforms				Sum
R^2 to Maximise >		0.9965		0.9999		↳	0.9994	↵		2.996

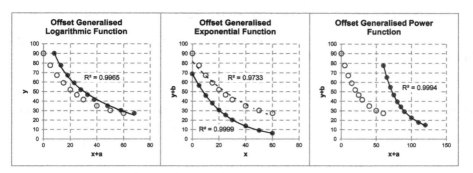

Figure 6.21 3-in-1 Solver Graphical Sensibility Check for a Generalised Exponential Function

remarkable thing here is that Solver can return values that yield very good R-Squares for all three functional types. In order to select the most appropriate one we have some apply some sound judgement:

- The Generalised Power Function is achieved by adjusting the time variable by the maximum offset we chose that in all honesty was chosen at random as equal to the maximum values of the recorded times. We have already learnt to be highly suspicious of Solver settling on a random maximum (or minimum). Let's reject this one.

350 | Transforming Nonlinear Regression

- As for the Logarithmic Function unlike the Power Function, it has not defaulted to a maximum allowed value but has instead picked an offset value of just over 8 minutes. If we reflect on this, we can probably reject it purely on the grounds of why should we adjust the start point by eight minutes (or thereabouts)? There is no logical reason to do this; it's another case of '**Just because we can, doesn't mean we should!**' *Why would we have to wait 8 minutes for it to start cooling?*
- The logic for accepting the Generalised Exponential Function negative offset of around 21.25° C is that it reflects the ambient temperature of the environment in which the cooling is taking place. The reason it is a negative offset is that cooling is a relative relationship that is based on the difference between the start and end values of the object and environment respectively, so the variable in question is a relative temperature drop not an absolute one.

We can now use this offset value and functional form to remodel our data and perform a Linear Regression based on the Log of the temperature drop relative to the time taken in minutes. We illustrate the results in Table 6.50. A quick check of the key statistical measures indicates that this is a valid result. Expressing the result formulaically with parameter values rounded to three decimal places, we have:

$$\log_{10}(\text{Temp}°\text{C} - 21.254) = 1.8399 - 0.01756 \times \text{Minutes from start}$$

Reversing the Log Transformation (taking the antilog of both sides of the equation), in which $69.170 = 10^{1.8399}$ and $0.960 = 10^{-0.01756}$, and re-arranging:

$$\text{Temp}°\text{C} = 21.254 + 69.170 \times 0.960^{\text{Minutes from Start}}$$

Hopefully, we will all be cool with this result. (*You just knew that was coming, didn't you? That's the trouble with being an estimator, everything becomes so predictable.*)

Table 6.50 Solver Regression Results for the Generalised Exponential Function Example

SUMMARY OUTPUT

Regression Statistics	
Multiple R	0.999951288
R Square	0.999902579
Adjusted R Square	0.999890401
Standard Error	0.003611045
Observations	10

Based on R-Square, F and t-statistics we would accept this model as a statistically significant result

ANOVA

	df	SS	MS	F	Significance F
Regression	1	1.070680506	1.070680506	82109.62793	2.46314E-17
Residual	8	0.000104317	1.30396E-05		
Total	9	1.070784824			

	Coefficients	Standard Error	t Stat	P-value	Lower 95%	Upper 95%
Intercept	1.839921532	0.001935389	950.672806	1.67863E-21	1.835458517	1.844384547
x - Minutes	-0.017559366	6.1279E-05	-286.547776	2.46314E-17	-0.017700675	-0.017418056

6.5.3 Linear Regression of a Transformed Generalised Power Function

In both the two previous examples we saw that it is possible to create conditions where a Generalised Power Function can give us a very good theoretical fit to data that reasonably might be expected to be a Generalised Logarithmic or Generalised Exponential Function. This is particularly so with small data samples such as the ones we are dealing with here. This can happen the other way around too, where we might reasonably expect a Generalised Power Function the 3-in-1 Solver technique may give us a 'best' solution that is basically Logarithmic or Exponential in form. Consequently, for small data samples use of the 3-in-1 Solver technique should be considered only where we have no preconception of the function type with which we are dealing, as this technique can be a little on the fragile side with some datasets.

We have an additional problem when we expect data to be a Generalised Power Function. The two offset parameters will tend to compete with each other like sparring siblings, and it is not always easy to predict which will win in any particular situation. Let's look at these issues using a variation to the 3-in-1 Solver that we can employ, when we have an expectation of a Generalised Power Function based on a relationship such as a Learning Curve. We will consider three cases:

1. One that examines the optimum values of the x-offset alone
2. One that only considers the y-offset
3. And finally one in which the values of both offsets are free to roam independently of each other.

This may give us a different insight into how our model behaves, and what is really important in driving its shape. As we will see in Volume IV, three alternative Learning Curve models take the form of Generalised Power Functions, each of which mirror the situations described above, and all of which are variations on a basic Unit Learning Curve model which is a standard Power Function.

Generalised Power Function – Example 1

Let's consider Example 1 in Table 6.51 in which we believe that the recurring hours to assemble a complex product on a production line has benefited from the assembly of 6 prior developments units. There is also a view that there may be some test activities for which there is an opinion that the activity time will be fixed, i.e. no learning. These assumptions allow us to test three alternative learning curve models in which we have either an x-offset, y-offset or both:

> In Volume IV Chapter 2 we will discuss these models, known as 'Stanford-B', 'De Jong' and 'S-Curve' Unit Learning respectively. Where the offsets are all zero, the three models are all equivalent to the 'standard' Crawford Unit Learning Curve.

352 | Transforming Nonlinear Regression

Table 6.51 3-in-1 Solver Set-up for a Generalised Power Function – Example 1

			Generalised Power Function with x-offset only		Generalised Power Function with y-offset only		Generalised Power Function with both x and y-offsets				3 in 1 Solvers
			Offset a		Offset b		Offset a		Offset b		
	Solver Variable >		0 < Minimum		0.00 < Maximum		0 < Minimum		0.00 < Maximum		
	Minimum Offset >		0		-400		0		-400		
	Maximum Offset >		6		0		6		0		
Build	Hours	Transformed Data	Transformed Data		Transformed Data		Transformed Data				
x	y	Log(x)	Log(y)	X = x+a	Log(x+a)	Y=y+b	Log(y+b)	X = x+a	Log(x+a)	Y=y+b	Log(y+b)
1	6021	0.0000	3.7797	1	0.0000	6021	3.7797	1	0.0000	6021	3.7797
2	5447	0.3010	3.7362	2	0.3010	5447	3.7362	2	0.3010	5447	3.7362
3	5315	0.4771	3.7255	3	0.4771	5315	3.7255	3	0.4771	5315	3.7255
4	4762	0.6021	3.6778	4	0.6021	4762	3.6778	4	0.6021	4762	3.6778
5	4927	0.6990	3.6926	5	0.6990	4927	3.6926	5	0.6990	4927	3.6926
6	4296	0.7782	3.6331	6	0.7782	4296	3.6331	6	0.7782	4296	3.6331
7	4429	0.8451	3.6463	7	0.8451	4429	3.6463	7	0.8451	4429	3.6463
8	4111	0.9031	3.6139	8	0.9031	4111	3.6139	8	0.9031	4111	3.6139
9	3997	0.9542	3.6017	9	0.9542	3997	3.6017	9	0.9542	3997	3.6017
10	4027	1.0000	3.6050	10	1.0000	4027	3.6050	10	1.0000	4027	3.6050
R^2 (Raw Data) >		0.9361		R^2 with log(y)		R^2 with log(x)		R^2 based on Log Transforms of x & y			Sum
	R^2 to Maximise >			0.9361		0.9361			0.9361		2.808

- From the definition of the problem it seems reasonable to assume that the minimum x-offset would be zero (i.e. no benefit carried over from the development units) and the maximum x-offset would be 6 (i.e. full benefit carried over.)
- There may be fixed negative y-offset to represent the 'unlearnable' activities. Here we have set the minimum to be a nominal negative value of -400, (*we can always change it later*.) The maximum in this case would be zero to signify that everything is subject to learning.
- The third case is simply an amalgamation of these other two

Table 6.52 summaries the result of the 3-in-1 Solver. In this case we have restricted the x-offset to being integer values only, so that we can more easily demonstrate a particular issue in the next example. (We will be discussing the concept of non-integer build numbers in the section on Equivalent Unit Learning in Volume IV Chapter 5.)

In this particular case, Solver has determined also that the best x-offset is 3 units to the right but also that the y-offset adjustment is zero hours, i.e. everything is learnable.

Using an x-offset adjustment of 3 units we can convert our data into a Standard Power Function and apply the logarithmic transformation of both the dependent hours and the offset build number in order to perform a Linear Regression. In this case we get the following results from Table 6.53:

$$\text{Log}_{10} \text{ Hours} = 3.9913 - 0.3551 \times \log_{10} (\text{Build No} + 3)$$

Or, reversing the log transformation, in which $9802.2 = 10^{3.9913}$, we get:

$$\text{Hours} = 9802.2 \times (\text{Build No.} + 3)^{-0.3551}$$

Table 6.52 3-in-1 Solver Result for a Generalised Power Function – Example 1

				Generalised Power Function with x-offset only		Generalised Power Function with y-offset only		Generalised Power Function with both x and y-offsets				3 in 1 Solvers
				Offset a		Offset b		Offset a		Offset b		
		Solver Variable >		3		0.00 < Maximum		3		0.00 < Maximum		
		Minimum Offset >		0		-400		0		-400		
		Maximum Offset >		6		0		6		0		

Build	Hours	Transformed Data		Transformed Data		Transformed Data		Transformed Data			
x	y	Log(x)	Log(y)	X = x+a	Log(x+a)	Y=y+b	Log(y+b)	X = x+a	Log(x+a)	Y=y+b	Log(y+b)
1	6021	0.0000	3.7797	4	0.6021	6021	3.7797	4	0.6021	6021	3.7797
2	5447	0.3010	3.7362	5	0.6990	5447	3.7362	5	0.6990	5447	3.7362
3	5315	0.4771	3.7255	6	0.7782	5315	3.7255	6	0.7782	5315	3.7255
4	4762	0.6021	3.6778	7	0.8451	4762	3.6778	7	0.8451	4762	3.6778
5	4927	0.6990	3.6926	8	0.9031	4927	3.6926	8	0.9031	4927	3.6926
6	4296	0.7782	3.6331	9	0.9542	4296	3.6331	9	0.9542	4296	3.6331
7	4429	0.8451	3.6463	10	1.0000	4429	3.6463	10	1.0000	4429	3.6463
8	4111	0.9031	3.6139	11	1.0414	4111	3.6139	11	1.0414	4111	3.6139
9	3997	0.9542	3.6017	12	1.0792	3997	3.6017	12	1.0792	3997	3.6017
10	4027	1.0000	3.6050	13	1.1139	4027	3.6050	13	1.1139	4027	3.6050

R^2 (Raw Data) >	0.9361	R^2 with log(y)	0.9563	R^2 with log(x)	0.9361	R^2 based on Log Transforms of x & y	↳ 0.9563 ↵	Sum	2.849
R^2 to Maximise >									

Table 6.53 3-in-1 Solver Regression Result for a Generalised Power Function – Example 1

SUMMARY OUTPUT

Regression Statistics	
Multiple R	0.9778871
R Square	0.956263181
Adjusted R Square	0.950796078
Standard Error	0.013633167
Observations	10

Based on R-Square, F and t-statistics we would accept this model as a statistically significant result

ANOVA

	df	SS	MS	F	Significance F
Regression	1	0.032509759	0.032509759	174.9122498	1.01857E-06
Residual	8	0.001486906	0.000185863		
Total	9	0.033996665			

	Coefficients	Standard Error	t Stat	P-value	Lower 95%	Upper 95%
Intercept	3.991323462	0.024588041	162.3278321	2.3206E-15	3.934623337	4.048023587
Log(x+3)	-0.355085887	0.026848702	-13.22543949	1.01857E-06	-0.416999106	-0.293172669

Generalised Power Function – Example 2

In our second example, depicted in Table 6.54, we have taken the results from Example 1, adjusted the Build Number by the 3 Units offset, applied a labour costing rate of $100 per hour and added a fixed cost of $300k for material.

It does not seem unreasonable to expect that our 3-in-1 Solver would return a y-offset estimate of around $300k. It doesn't; in fact, it fails miserably!

Before we throw the technique into the non-recyclable bin, let's try to understand what is going on here, (*after all as we agreed our expectations were not really that unreasonable, were they?*)

In order to understand what's going on here, let's recall Figures 5.24 and 5.26 from Chapter 5, on Generalised Power Functions; the top-right hand quadrant of the two graphs are reproduced in Figure 6.22.

Table 6.54 3-in-1 Solver Set-Up for a Generalised Power Function – Example 2

				Generalised Power Function with x-offset only		Generalised Power Function with y-offset only		Generalised Power Function with both x and y-offsets				3-in-1 Solvers
				Offset a		Offset b		Offset a		Offset b		
		Solver Variable >		0	< Minimum	0.00	< Maximum	0	< Minimum	0.00	< Maximum	
		Minimum Offset >		0		-500		0		-500		
		Maximum Offset >		6		0		6		0		
		Transformed Data		Transformed Data		Transformed Data		Transformed Data				
Build	$k			X = x+a	Log(x+a)	Y=y+b	Log(y+b)	X = x+a	Log(x+a)	Y=y+b	Log(y+b)	
x	y	Log(x)	Log(y)									
4	902.1	0.6021	2.9553	4	0.6021	902	2.9553	4	0.6021	902	2.9553	
5	844.7	0.6990	2.9267	5	0.6990	845	2.9267	5	0.6990	845	2.9267	
6	831.5	0.7782	2.9199	6	0.7782	832	2.9199	6	0.7782	832	2.9199	
7	776.2	0.8451	2.8900	7	0.8451	776	2.8900	7	0.8451	776	2.8900	
8	792.7	0.9031	2.8991	8	0.9031	793	2.8991	8	0.9031	793	2.8991	
9	729.6	0.9542	2.8631	9	0.9542	730	2.8631	9	0.9542	730	2.8631	
10	742.9	1.0000	2.8709	10	1.0000	743	2.8709	10	1.0000	743	2.8709	
11	711.1	1.0414	2.8519	11	1.0414	711	2.8519	11	1.0414	711	2.8519	
12	699.7	1.0792	2.8449	12	1.0792	700	2.8449	12	1.0792	700	2.8449	
13	702.7	1.1139	2.8468	13	1.1139	703	2.8468	13	1.1139	703	2.8468	
R^2 (Raw Data) >		0.9568		R^2 with log(y)		R^2 with log(x)		R^2 based on Log Transforms of x & y				Sum
		R^2 to Maximise >			0.9568		0.9568			0.9568		2.870

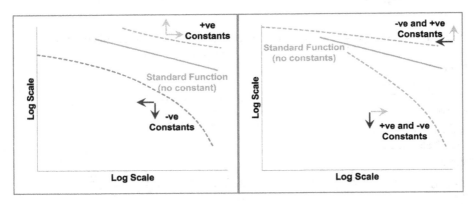

Figure 6.22 Generalised Increasing Power Functions with Unlike-Signed Offset Constants

In our Example 2 we have an expectation that our raw data is like the upper line. From this we want to offset it downwards to remove the unlearnable material cost.

If our data had been a perfect relationship, then Solver would have done its magic, but given that our data contains all the random vagaries of real life, then it will be scattered around the lines. It would be a tall order to expect Excel to tell the difference between the random scatter around what is nearly a straight line and the transformed straight line.

It also explains why in Example 1 where we were expecting to detect an unlearnable element of the task, such as a test of fixed duration that we failed to do so. It was another case of the Unlike-signed Offset Constants competing with one another.

Transforming Nonlinear Regression | 355

Generalised Power Function – Example 3

In this particular we would have exactly the same issue as we did for Example 2, so we know now that Solver will not be able to deal with it as it stands. In fact it further confuses – if we were to run it with the original unadjusted Build Number plus the Material Cost (see Table 6.55) as Solver is trying to compensate for the introduction of a constant in the dependent variable (Cost) by changing the offset in the Build Number it had previously determined from Example 1 – it's fitting numbers because it can! So, we should be asking ourselves whether it matters or not.

If we were to take the Solver result for Example 3 somewhat blindly, we could calculate the supporting Linear Regression shown in Table 6.56.

Table 6.55 3-in-1 Solver Result for a Generalised Power Function – Example 3

			Generalised Power Function with x-offset only		Generalised Power Function with y-offset only		Generalised Power Function with both x and y-offsets				3-in-1 Solvers
			Offset a		Offset b		Offset a		Offset b		
	Solver Variable >		2		0.00	< Maximum	2		0.00	< Maximum	
	Minimum Offset >		0		-500		0		-500		
	Maximum Offset >		6		0		6		0		
Build	$ k	Transformed Data	Transformed Data		Transformed Data		Transformed Data				
x	y	Log(x)	Log(y)	X = x+a	Log(x+a)	Y=y+b	Log(y+b)	X = x+a	Log(x+a)	Y=y+b	Log(y+b)
1	902.1	0.0000	2.9553	3	0.4771	902	2.9553	3	0.4771	902	2.9553
2	844.7	0.3010	2.9267	4	0.6021	845	2.9267	4	0.6021	845	2.9267
3	831.5	0.4771	2.9199	5	0.6990	832	2.9199	5	0.6990	832	2.9199
4	776.2	0.6021	2.8900	6	0.7782	776	2.8900	6	0.7782	776	2.8900
5	792.7	0.6990	2.8991	7	0.8451	793	2.8991	7	0.8451	793	2.8991
6	729.6	0.7782	2.8631	8	0.9031	730	2.8631	8	0.9031	730	2.8631
7	742.9	0.8451	2.8709	9	0.9542	743	2.8709	9	0.9542	743	2.8709
8	711.1	0.9031	2.8519	10	1.0000	711	2.8519	10	1.0000	711	2.8519
9	699.7	0.9542	2.8449	11	1.0414	700	2.8449	11	1.0414	700	2.8449
10	702.7	1.0000	2.8468	12	1.0792	703	2.8468	12	1.0792	703	2.8468
R^2 (Raw Data) >		0.9428		R^2 with log(y)		R^2 with log(x)		R^2 based on Log Transforms of x & y			Sum
	R^2 to Maximise >		0.9576		0.9576		0.9428	↳	0.9576	↵	2.858

Table 6.56 3-in-1 Solver Regression Result for a Generalised Power Function – Example 3

SUMMARY OUTPUT

Regression Statistics	
Multiple R	0.97856378
R Square	0.957587072
Adjusted R Square	0.952285456
Standard Error	0.008276304
Observations	10

Based on R-Square, F and t-statistics we would accept this model as a statistically significant result

ANOVA

	df	SS	MS	F	Significance F
Regression	1	0.012372084	0.012372084	180.6217354	9.00234E-07
Residual	8	0.000547978	6.84972E-05		
Total	9	0.012920062			

	Coefficients	Standard Error	t Stat	P-value	Lower 95%	Upper 95%
Intercept	3.043858897	0.011971949	254.2492412	6.41132E-17	3.016251534	3.071466261
Log(x+2)	-0.187373415	0.013941932	-13.4395586	9.00234E-07	-0.219523568	-0.155223261

356 | Transforming Nonlinear Regression

In truth, we would have expected to make an adjustment of +3 to the Build Number and − $300k base on how we had articulated the data. Table 6.57 shows the Regression results for this theoretical model, and in Table 6.58 and Figure 6.23 we compare a forward projection of both sets of results up to the 100th production unit. In the short term the two models are reasonably comparable but by the time we get to the 100th unit,

Table 6.57 Regression Result Expected for Offset Generalised Power Function − Example 3

SUMMARY OUTPUT

Regression Statistics	
Multiple R	0.9778871
R Square	0.956263181
Adjusted R Square	0.950796078
Standard Error	0.013633167
Observations	10

Based on R-Square, F and t-statistics we would accept this model as a statistically significant result

ANOVA

	df	SS	MS	F	Significance F
Regression	1	0.032509759	0.032509759	174.9122498	1.01857E-06
Residual	8	0.001486906	0.000185863		
Total	9	0.033996665			

	Coefficients	Standard Error	t Stat	P-value	Lower 95%	Upper 95%
Intercept	2.991323462	0.024588041	121.657655	2.32947E-14	2.934623337	3.048023587
Log(x+a)	-0.355085887	0.026848702	-13.22543949	1.01857E-06	-0.416999106	-0.293172669

Table 6.58 Comparison of Forward Projection of Alternative Regression Results for Example 3

		Solver Based Result					Expected Result				
Offset a		2					3				
Offset b		0					-300				
Intercept					3.0439						2.9913
Slope					-0.1874						-0.3551
$ k	Build				$ k						$ k
y	x	y+b	x+a	Log(y+b)	Log(x+a)	10^9 - b	y+b	x+a	Log(y+b)	Log(x+a)	10^9 - b
902	1	902	3	2.9553	0.4771	900.45	602	4	2.7797	0.6021	899.16
845	2	845	4	2.9267	0.6021	853.20	545	5	2.7362	0.6990	853.51
832	3	832	5	2.9199	0.6990	818.26	532	6	2.7255	0.7782	818.82
776	4	776	6	2.8900	0.7782	790.78	476	7	2.6778	0.8451	791.18
793	5	793	7	2.8991	0.8451	768.26	493	8	2.6926	0.9031	768.43
730	6	730	8	2.8631	0.9031	749.28	430	9	2.6331	0.9542	749.25
743	7	743	9	2.8709	0.9542	732.92	443	10	2.6463	1.0000	732.75
711	8	711	10	2.8519	1.0000	718.60	411	11	2.6139	1.0414	718.35
700	9	700	11	2.8449	1.0414	705.88	400	12	2.6017	1.0792	705.62
703	10	703	12	2.8468	1.0792	694.46	403	13	2.6050	1.1139	694.26
	20		22		1.3424	619.90		23		1.3617	621.95
	30		32		1.5051	577.87		33		1.5185	583.22
	40		42		1.6232	549.17		43		1.6335	557.81
	50		52		1.7160	527.63		53		1.7243	539.36
	60		62		1.7924	510.52		63		1.7993	525.11
	70		72		1.8573	496.41		73		1.8633	513.64
	80		82		1.9138	484.46		83		1.9191	504.12
	90		92		1.9638	474.13		93		1.9685	496.04
	100		102		2.0086	465.05		103		2.0128	489.06

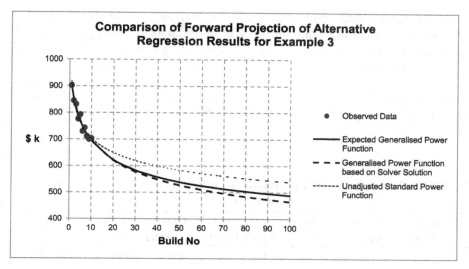

Figure 6.23 Comparison of Forward Projection of Alternative Regression Results for Example 3

they are some 5% different. We have also shown a Standard Power Function Trendline in Figure 6.23 for completeness.

So, it would appear that using Solver to normalise a Generalised Function into a Standard form can be a little challenging – a bit like having a teenager in the house; so long as we nurture it, we will get something positive, but the will be occasions when it becomes downright uncooperative. The next section reviews the warning signs and when and when not to try it.

6.5.4 Generalised Function transformations: Avoiding the pitfalls and tripwires

So what can do with our 'temperamental teenager' technique? There are a few options we can consider:

- Rip this section out of the book on the basis that it is so unreliable that we will never use it. (*A little bit extreme perhaps?*)
- Keep it happy by feeding it with more data more. (*Probably not realistic on the basis that if we had more data why wouldn't we have used it in the first place?*)
- Be more selective in our choice of offset parameter limits, especially if we can get an independent alternative view of their values (including those sourced through expert judgement)

358 | Transforming Nonlinear Regression

The last one is probably the realistic option, but even then we can often expect that the offset values will tend towards one or other of the limits suggested. Just as with all our techniques, none of them are tools that work for all cases. *(Remember that a hammer and chisel are very useful tools to have ... but not if we only want to saw wood!)*

In Table 6.59 we have a summary of instances where the first impression may be that our data could be a Standard Logarithmic, Exponential or Power Function, but due to the naturally scattering of data around such 'Best Fits', they could equally be considered

Table 6.59 Hot Spots where Generalised Functions can be Misinterpreted as Standard Functions

Alternative Generalised Function Type	Direction of x-offset	Direction of y-offset	Function of First Impression	Slope
Generalised Exponential	N/A	+ve y	Standard Logarithmic	Decreasing
			Standard Exponential	Decreasing
				Increasing
		-ve y	Standard Logarithmic	Decreasing
Generalised Logarithmic	+ve x	N/A	Standard Logarithmic	Decreasing
				Increasing
			Standard Exponential	Decreasing
	-ve x	N/A	Standard Exponential	Decreasing
			Standard Power	Increasing
Generalised Power	+ve x	+ve y	Standard Logarithmic	Increasing
			Standard Exponential	Increasing
			Standard Power	Decreasing
				Increasing
		-ve y	Standard Logarithmic	Decreasing
				Increasing
			Standard Exponential	Decreasing
				Increasing
	-ve x	+ve y	Standard Logarithmic	Decreasing
			Standard Power	Decreasing
				Increasing
		-ve y	Standard Logarithmic	Decreasing
			Standard Exponential	Decreasing

Transforming Nonlinear Regression | 359

to be a Generalised form of the same or another type. (Refer back to the diagrams we drew in Chapter 5 Sections 5.3.1 to 5.3.3.)

Note: The Offsets here are those that are used to define a Generalised Function in relation to a Standard Function of the equivalent type. In order to use these to normalise a Generalised Form back into a Standard FORM to allow us to then transform the data prior to Linear Regression we must reverse the sign of the offset in adjusting the data.

The frequency with which the Generalised Power Function occurs in Table 6.59 gives us some insight into why we found the Learning Curve examples in the previous section so difficult to converge as we expected.

If we are interpolating within our data, or extrapolating to a position in the close neighbourhood (i.e. slightly to the left or right of our data) then we are probably better going for the simplest model we can justify. However, where we want (or need – *we rarely do this sort of thing just for fun*) to extrapolate to more distant positions, we should try to justify whether a more sophisticated model is more appropriate. In that case we may have no choice other than to estimate the offset adjustments of a Generalised Logarithmic, Exponential Or Power Function by expert judgement or reasoned argument. It is essential that such judgements are recorded in a Basis of Estimate to maintain TRACEability at all times.

6.6 Pseudo Multi-linear Regression of Polynomial Functions

Some of us may have noticed that there is a Trendline option in Microsoft Excel's graphics facility for a polynomial function. You may know also that higher order polynomials can be used to fit many curved relationships.

If this is the case, then as a general rule of thumb, the advice must be . . . **DON'T DO IT!** Just because you can, doesn't mean you should! We can all exceed the speed limit (if we can drive), but that doesn't make it right!

Why do we say 'don't do it'? Simply because when we extrapolate the trendline, there may be an unexpected and unpleasant surprise around the next bend! For instance, if we revisit the example we used in Section 6.1.1 for Design Queries, we considered this to be a Logarithmic Function and consequently fitted a Logarithmic Trendline, reproduced in the left-hand graph of Figure 6.24. If instead we had fitted a Cubic Trendline (i.e. a Polynomial of Order 3), then we would get the result in the right-hand graph of Figure 6.24.

Visually, the Cubic appears to be as good a fit as the Logarithmic Trendline; in fact the Cubic returns the higher R-Square, but in Figure 6.25, look what happens when we extrapolate both versions forward by ten units. Which one do you think is more credible?

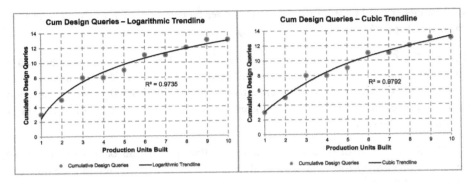

Figure 6.24 Using a Cubic Trendline Instead of a Logarithmic Trendline

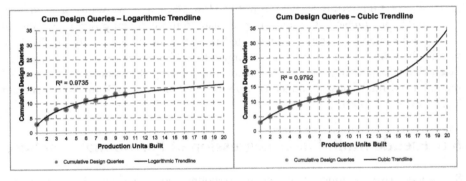

Figure 6.25 Extrapolating a Cubic Trendline in Comparison with a Logarithmic Trendline

If instead we had used a Quartic Trendline (Polynomial of Order 4) we would have got the even more bizarre extrapolation in Figure 6.26. Yet again the R-Square may delude us into thinking that bigger is better, whereas it appears that this is not the case.

However, as with all rules of thumb, there are some exceptions where it is acceptable to use a polynomial relationship in some form; for example:

- Where there is a physical relationship that ties the variables together, such as the volume of a fluid that flows through a pipe is a quadratic function dependent on the pipe diameter
- The Cumulative Value of a straight line is always a Quadratic Equation i.e. a Polynomial of Order 2 that passes through the origin (see Chapter 2 on Exploiting Linear Properties)
- There may be cases where we have a logical model that is the product of two or three drivers that are highly correlated, but even then a stronger case could probably

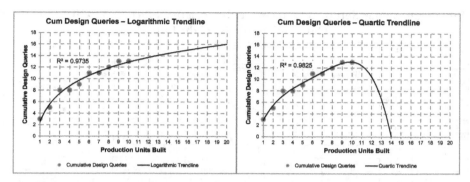

Figure 6.26 Extrapolating a Quartic Trendline in Comparison with a Logarithmic Trendline

be made for a Generalised or Standard Power Function relationship, as justification of a cubic term would not necessarily justify lower order terms as well, which would be implicitly implied with a true Polynomial Regression.

6.6.1 Offset Quadratic Regression of the Cumulative of a straight line

Let's consider the cumulative output from a production line where the quarterly output is perceived to be increasing fairly linearly but for which we do not have a complete set of records, as depicted by Table 6.60. From the properties of a straight line (as discussed in Chapter 2), the cumulative of this increasing output is a quadratic equation or a polynomial of order 2 through the origin. Using this property we can determine a forecast for the nominal weekly output by fitting a Quadratic Equation to the Cumulative data using a Pseudo-Multi-linear Regression to find the Best Fit Curve. However, we must first find when the production line started and offset our model so that the first week of production output is Week 1. Week 0 is then the last week of no production output. If we know the answer to this, then we can use this to determine the relative start point, otherwise we will have to derive it, for which Microsoft Excel's **Solver** can again be used. The procedure is as follows:

1. Enter an initial guess at an offset adjustment constant that will be used to calculate the Relative Time rather than the current Week Number Label. Zero is a good place to start (i.e. no offset) shown at the top of Table 6.60 in Cell E2
2. Create a value for the Relative Time (T + Offset)
3. The independent predictor variable is the Relative Time, denoted as T+Offset (Column E, Cells E4:E8), but we will also create a pseudo predictor variable from

362 | Transforming Nonlinear Regression

Table 6.60 Solver Setup to Determine the Relative Start Time Offset for the Cumulative of a Straight Line

◢	A	B	C	D	E	F	G	H
1					Offset		Quadratic	
2			Solver Variable >		0.00	Relative	Model through	
3	Period Ending		Cumulative Output from Start	Time (Weeks)	Relative Time (T+Offset)	Time Squared $(T+Offset)^2$	the Origin using LINEST Coeffs	Error Term (Actual – Model)
4	Y1	Q2	20	26	26	676	17.94	2.06
5		Q3	50	39	39	1521	53.59	-3.59
6	Y2	Q3	375	91	91	8281	374.04	0.96
7		Q4	500	104	104	10816	498.61	1.39
8	Y3	Q2	800	130	130	16900	801.12	-1.12
10					Sum of Errors Constrained to Zero >			-0.30
11					Sum of Square Errors to Minimise >			21.23
13		LINEST Function Coefficients			2nd Index Parameter			
14					2	1		
15		1st Index Parameter		1	-0.6781	0.0526		

the square of the Relative Time, denoted here as $(T+Offset)^2$ in Column F (Cells F4:F8).

Note that we have not expressed this as an independent variable because that would be incorrect. Potentially, this could give us a multicollinearity problem as there is a very high correlation between the Relative Time and its square, even though the two variables constitute a nonlinear relationship.

4. In the lower section of Table 6.60 we have created a table of **LINEST** function output values for the Coefficients, based on the starting value of the Solver seed variable in Cell E2. Note that these values will change when we run Solver ... the Excel Formula for these Coefficients are:

Cell E14 = **INDEX(LINEST(C4:C8,E4:F8,FALSE,TRUE),D15,E14)**
Cell F14 = **INDEX(LINEST(C4:C8,E4:F8,FALSE,TRUE),D15,F14)**

Note that the FALSE parameter in **LINEST** sets the intercept to zero. Note also that the order of the output Coefficients is the reverse of their natural input sequence (as explained in Chapter 3).

5. In Column G (Cells G4:G8) we have calculated the Model Quadratic Equation based on the Coefficients determined in Step 4 (Cells E14:F14). The product can be calculated using the **SUMPRODUCT** function in Excel, for instance:

G4 = **SUMPRODUCT($E$15:$F$15,E4:F4)**

Transforming Nonlinear Regression | 363

6. In Column H, we calculate the difference between Column C and Column G, the Observed Value minus the Modelled Value. In Cell H10 we have shown the current Sum of the Error terms; this will be used as a Solver Constraint and be set to zero (unbiased)
7. In Cell H11 we have calculated the Sum of Squares of the Error Terms above it using the Excel Function **SUMSQ(H4:H8)**
8. We can now open the Solver dialogue box and set the Sum of Squares Error as the objective to minimise (Cell H11) by varying the Offset value in Cell E2, subject to the constraint that the Sum of the Error terms in Cell H10 is zero. Ideally as a matter of convenience we might want to ask Solver to constrain the offset variable to integer values, but this appears to be a step too far for **Solver** as it returns a message saying that it cannot converge in this instance.
9. Finally, just check that the tick box under the Constraints window to 'Make Unconstrained Variables Non-negative' is clear before clicking 'Solve' to get the result in Table 6.61.
10. However, if we were to give the Solver Seed Variable (Cell E2) a different value, say –20, we would get the answer in Table 6.62. A quick glance at the calculated Coefficients in Cells E15:F15 tells us that this is the same solution with a horizontal shift in time. Between the two points, the Cumulative Curve goes negative! Perhaps somewhat counter-intuitively the answer we should use is the second one with the smallest (most negative) offset. This gives us a theoretical origin (or Cumulative Start Point) of 13.96 weeks, as opposed to one at –1.76 weeks. This will prevent us from having a Cumulative Curve that dips below zero. We will discuss this further at the end of this section.

Table 6.61 Solver Solution for the Relative Start Time Offset for the Cumulative of a Straight Line

	A	B	C	D	E	F	G	H
1					Offset		Quadratic Model through the Origin using LINEST Coeffs	
2			**Solver Variable >**		1.76	Relative		
3	Period Ending		Cumulative Output from Start	Time (Weeks)	Relative Time (T+Offset)	Time Squared $(T+Offset)^2$		Error Term (Actual – Model)
4	Y1	Q2	20	26	27.76	771	17.51	2.49
5		Q3	50	39	40.76	1661	53.47	-3.47
6	Y2	Q3	375	91	92.76	8604	374.33	0.67
7		Q4	500	104	105.76	11185	498.81	1.19
8	Y3	Q2	800	130	131.76	17361	800.88	-0.88
10					Sum of Errors Constrained to Zero >			0.00
11					Sum of Square Errors to Minimise >			20.84
13	**LINEST Function Coefficients**				2nd Index Parameter			
14					2	1		
15	1st Index Parameter			1	-0.8232	0.0524		

364 | Transforming Nonlinear Regression

Table 6.62 Solver Solution for the Relative Start Time Offset for the Cumulative of a Straight Line

	A	B	C	D	E	F	G	H
1					Offset		Quadratic	
2			Solver Variable >		-13.96	Relative	Model through	
3	Period Ending		Cumulative Output from Start	Time (Weeks)	Relative Time $(T+Offset)$	Time Squared $(T+Offset)^2$	the Origin using LINEST Coeffs	Error Term (Actual − Model)
4	Y1	Q2	20	26	12.04	145	17.51	2.49
5		Q3	50	39	25.04	627	53.47	-3.47
6	Y2	Q3	375	91	77.04	5935	374.33	0.67
7		Q4	500	104	90.04	8107	498.81	1.19
8	Y3	Q2	800	130	116.04	13465	800.87	-0.87
10					Sum of Errors Constrained to Zero >			0.00
11					Sum of Square Errors to Minimise >			20.84
13		LINEST Function Coefficients			2nd Index Parameter			
14					2	1		
15		1st Index Parameter		1	0.8237	0.0524		

Table 6.63 Solver Solution for the Relative Start Time Offset for the Cumulative of a Straight Line

	A	B	C	D	E	F	G	H
1					Offset		Quadratic	
2			Solver Variable >		-14.00	Relative	Model through	
3	Period Ending		Cumulative Output from Start	Time (Weeks)	Relative Time $(T+Offset)$	Time Squared $(T+Offset)^2$	the Origin using LINEST Coeffs	Error Term (Actual − Model)
4	Y1	Q2	20	26	12	144	17.43	2.57
5		Q3	50	39	25	625	53.33	-3.33
6	Y2	Q3	375	91	77	5929	373.98	1.02
7		Q4	500	104	90	8100	498.40	1.60
8	Y3	Q2	800	130	116	13456	800.36	-0.36
10					Sum of Errors Constrained to Zero >			1.51
11					Sum of Square Errors to Minimise >			21.43
13		LINEST Function Coefficients			2nd Index Parameter			
14					2	1		
15		1st Index Parameter		1	0.8291	0.0524		

11. From a purely practical perspective we may want to round the result to the nearest integer (Solver won't in this instance) giving us the position in Table 6.63. We then have to decide whether the Sum of the Errors is close enough to zero for us to ignore.

12. Using the Cumulative Output (C4:C8) with the Relative Time (E4:E8) and the Square of the Relative Time (F4:F8) we can then run a Multi-linear Regression to get the results in Table 6.63, having first elected to force the intercept constant to be zero.

Transforming Nonlinear Regression | 365

In Chapter 2, Section 2.2, we discovered that the Coefficient of the Square term in the quadratic was half the rate of the underlying straight line:

$$\text{Cumulative Output} = \frac{m}{2}n^2 + \left(\frac{m}{2} + c\right)n$$

where m is the slope and c the intercept of the straight line
and n is a series of integers

So, in this case, using the coefficients from Table 6.64 and with $0.1047 = 2 \times 0.0524$ (using unrounded values), the underlying straight line is:

$$\text{Output} = 0.1047 \times (\text{Time}{-}14) + (0.8291 - 0.0524)$$

$$\text{Output} = 0.1047 \times (\text{Time} - 14) + 0.7767$$

Incidentally, what if we had forgotten to force the regression through the origin? In this particular case we would have got the results in Table 6.65 which would have us reject the model in favour of the Null Hypothesis that the Intercept Coefficient should be zero. Based on the previous discussion of Chapter 2, this is what we might have expected.

Using this result we can then create a weekly forecast going forward (or backward) by disaggregating the Cumulative Modelled Value as depicted in Table 6.66 or by deriving the weekly straight line equivalent as we derived in Chapter 2:

Table 6.64 Quadratic Polynomial Function Regression Output Through the Origin

SUMMARY OUTPUT

Regression Statistics	
Multiple R	0.999989915
R Square	0.999979831
Adjusted R Square	0.666639774
Standard Error	2.635994576
Observations	5

ANOVA

	df	SS	MS	F	Significance F
Regression	2	1033504.155	516752.0773	74369.21656	1.34462E-05
Residual	3	20.84540222	6.948467406		
Total	5	1033525			

	Coefficients	Standard Error	t Stat	P-value	Lower 95%	Upper 95%
Intercept	0	#N/A	#N/A	#N/A	#N/A	#N/A
X1 (T+Offset)	0.829130243	0.076412819	10.85066943	0.001674861	0.585950548	1.072309938
X2 (T+Offset)^2	0.052369816	0.000764536	68.49878284	6.85629E-06	0.04993672	0.054802912

Table 6.65 Quadratic Polynomial Function Regression Output Rejecting the Intercept

SUMMARY OUTPUT

Regression Statistics	
Multiple R	0.999975449
R Square	0.999950898
Adjusted R Square	0.999901797
Standard Error	3.228360322
Observations	5

Based on the t-statistic for the intercept we would accept the Null Hypothesis that the intercept is statistically indistinguishable from zero

ANOVA

	df	SS	MS	F	Significance F
Regression	2	424499.1554	212249.5777	20364.92583	4.91016E-05
Residual	2	20.84462074	10.42231037		
Total	4	424520			

	Coefficients	Standard Error	t Stat	P-value	Lower 95%	Upper 95%
Intercept	0.036266648	4.18824414	0.008659153	0.993877169	-17.98429343	18.05682673
X1 (T+Offset)	0.827809849	0.178912891	4.626887671	0.043673982	0.058009811	1.597609888
X2 (T+Offset)^2	0.052379115	0.001424797	36.76251272	0.000739108	0.046248709	0.058509521

Table 6.66 Forecast Data from the Output of a Quadratic Polynomial Function Regression

				Regression Coefficients			Intercept
		Cumulative Data		0.8291	0.0524		0
		Weekly Data >		0.1047	↵		0.7768

		Offset in Weeks >		-14	Relative Time Squared (T+Offset)^2	Quadratic Model through Origin using Regression Coefficients	Weekly Output from Cumulative Increments	Weekly Output by implied straight lLine
Period ending		Cumulative Output from Start	Time (Week Number)	Relative Time (T+Offset)				
Y1	Q2		14	0	0	0.00		
		20	26	12	144	17.49		
	Q3	50	39	25	625	53.46		
	Q4		52	38	1444	107.13		
Y2	Q1		65	51	2601	178.50		
	Q2		78	64	4096	267.57		
	Q3	375	91	77	5929	374.34		
	Q4	500	104	90	8100	498.82		
Y3	Q1		117	103	10609	640.99		
	Q2	800	130	116	13456	800.87		
	Q3		131	117	13689	813.90	13.03	13.03
			132	118	13924	827.03	13.14	13.14
			133	119	14161	840.28	13.24	13.24
			134	120	14400	853.62	13.35	13.35
			135	121	14641	867.07	13.45	13.45
			136	122	14884	880.63	13.55	13.55
			137	123	15129	894.29	13.66	13.66
			138	124	15376	908.05	13.76	13.76
			139	125	15625	921.92	13.87	13.87
			140	126	15876	935.89	13.97	13.97
			141	127	16129	949.97	14.08	14.08
			142	128	16384	964.16	14.18	14.18
			143	129	16641	978.44	14.29	14.29

Linear slope = Twice the Cumulative Squared Term's Coefficient

Linear intercept = Difference between the Two Quadratic Coefficients

We promised to come back to the question of multiple solutions that Solver can give us:
   With any polynomial or order n, there are potentially n solutions. In the case of a Quadratic Function there are therefore two possible answers. So how do you know which one to choose? The answer is by applying common sense; the glib answer is 'the one that's right is on the right!' This glib response implies that we should draw a graph . . .

For the Formula-phobes: Make the right choice with Cumulative quadratics

If we think we are looking at the Cumulative of an increasing straight line, plot the data and fit a Polynomial Trendline of Order 2. Project the Trendline backwards until it intersects the horizontal axis, and goes below, it then turns and then re-appears above the axis. Where it crosses the axis, it provides the two solutions to our problem. In this case we must take the larger value of the two on the right. If we take the one on the left we will be left with negative cumulative values

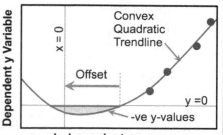

The right one is the right one if it's going up on the right!

On the other hand, if we have an underlying decreasing straight line and fit a Polynomial Trendline of order 2, then everything gets turned upside down and around. When it comes to choosing an offset value to force the Trendline through the origin, this means that we must choose the smaller value on the left this time rather than the larger on one the right.

So, if it's going up on the left then we're left with the left because we can't really be left with the right!

368 | Transforming Nonlinear Regression

Alternatively, we can go straight to the solution using the General Solution of a Quadratic Equation. (*Oh, fond memories again of maths lessons at school!*) Using the example in this section we can draw the scatter graph of Cumulative values against time and fit an unconstrained Polynomial Trendline of Order 2. If we select the option to display the equation on the plot we get Figure 6.27. (In this particular case it barely goes beneath the y=0 axis.)

We can solve this using the standard equation for the General Solution to a Quadratic Equation as discussed in Chapter 2. (*Sorry, was that a shiver or shudder from getting a flashback to school maths lessons?*)

For the Formula-philes: Determining the Quadratic Offset Solutions

From the example in Figure 6.27 we can extract the unconstrained Quadratic Regression Equation

Example Trendline:

$$y = 0.052x^2 - 0.6388x - 1.2868 \qquad (1)$$

General Solution of a Quadratic Equation of the form ...

$$ax^2 + bx + c = 0 \qquad (2)$$

... is given by:

$$x = \frac{-b \pm \sqrt{b^2 - 4ac}}{2a} \qquad (3)$$

Substituting the values of a, b and c from Equation (1) into (3) we get to 4 decimal places:

$$x = \frac{0.6388 \pm \sqrt{0.4081 + 0.2697}}{0.1048} \qquad (4)$$

Simplifying:

$$x = -1.7602 \text{ or } x = 13.9511$$

To use it as an offset in this case we need to deduct the greater of the two from the time value to get the Relative Time. Note that the figure is very similar to the approximate value we determined using the Microsoft Excel **Solver** algorithm in Table 6.62.

6.6.2 Example of a questionable Cubic Regression of three linear variables

Consider the data on a number of ships of comparable type and function, gleaned from an internet search, as summarised in Table 6.67. (Yes, these are actual ships, but the data is definitely tertiary in nature by definition of its source [see Volume

Transforming Nonlinear Regression | 369

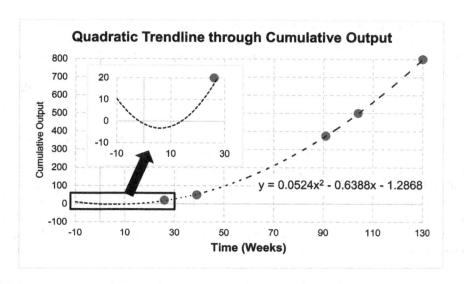

Figure 6.27 Unconstrained Quadratic Trendline Example

Table 6.67 Ship Dimensions and Gross Tonnage Based on an Internet Search

Ship No	Ship Name	Type	Gross Tonnage	Length (metres)	Beam (metres)	Depth (metres)	L x B x D		Type	Description	Qty
1	Svalbard	OPV	6375	103.7	19.1	10.8	21391		FPV	Fisheries Patrol Vessel	4
2	River Class	OPV	2109	79.5	13.6	9.6	10380		FRV	Fisheries Research Vessel	5
3	Galatea	THV	3569	84.2	16.5	7.2	10003		ILV	Irish Lights Vessel	1
4	Roisin	OPV	1784	78.0	14.0	6.8	7426		NLV	Northern Light Vessel	2
5	Norna	FPV	1190	71.4	11.6	7.4	6129		OPV	Offshore Patrol Vessel	4
6	Island	OPV	1017	59.5	10.98	6.81	4449		RRS	Royal Research Ship	3
7	Pourquoi Pas	FRV	7854	107.6	20				SRS	Sonar Research Ship	1
8	James Clark Ross	RRS	5732	99.04	18.85				THV	Trinity House Vessel	4
9	James Cook	RRS	5368	89.2	18.6						
10	Alliance	SRS	3180	93	15.2						
11	Endeavour	FRV	2983	74	15.8						
12	Mermaid	THV	2820	82	14.6						
13	Thalassa	FRV	2803	73.65	14.9	Not known					
14	Granuaile	ILV	2625	79.7	16.1						
15	Scotia	FRV	2619	68.6	15						
16	Patricia	THV	2541	86.3	13.94						
17	Jura	FPV	2181	84	13						
18	Pharos	NLV	1986	79.58	14						
19	Charles Darwin	RRS	1936	69.4	14.4						
20	Corystes	FRV	1289	52.5	12.8						
21	Vigilant	FPV	1190	71.4	11.6						
22	Sulisker	FPV	1177	71.3	11.6						
23	Pole Star	NLV	1174	51.52	12						
24	Alert	THV	302	39.3	8.0						
1 to 6	Correlation with Gross Tonnage			94.3%	96.4%	74.2%	96.7%				
1 to 24	Correlation with Gross Tonnage			84.5%	93.6%						

I Chapter 2] – but for the purposes of the example we will assume that they are factually correct.)

Gross tonnage is often used as a primary cost driver to give an indication of the likely cost of building a ship. Suppose we want to derive a parametric relationship for the Gross Tonnage of ships of these or similar types based on other factors as a precursor to determining the potential cost. Gross Tonnage is in fact a volume measure (*even though it might sound like weight*) so it seems reasonable to suggest the gross tonnage is related to the Product of the length, beam (breadth) and depth of the ships. Unfortunately we can see that we only have a full set of data for six ships (*not really what you would call a statistically significant sample size.*) However, for the ones that we do have, we can plot the relationship in Figure 6.28 and perhaps take some encouragement that it is not completely random!

It also seems reasonable to assume that there will be some basic relationship between the relative dimensions of length, beam and depth. (We wouldn't expect a ship to be wider than it was long, would we, or be so deep as to cause excessive drag in the water?)

If we look at the relationship between the three linear dimensions Length L, Beam B and Depth D, we can see from Figure 6.29 that there is a reasonably strong linear relationship between Length and Beam but also implicitly it cannot be extrapolated backwards to the dashed line which marks the equality relationship of length = beam. (*If this were to be the case then we would be looking at something like a Stone Age coracle, which hardly compares with the other ships we are comparing.*)

The evidence for a linear relationship between depth and length or depth and beam is much less compelling (Figure 6.30), there being only six data points and much lower Coefficients of Determination (R^2). However, if for the time being we assume that the relationship is broadly linear for the range of values we are likely to consider, then we can express each of length, beam and depth as linear functions of each other. From

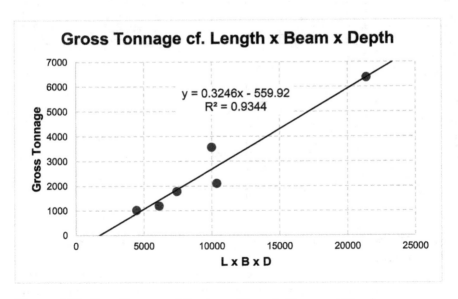

Figure 6.28 Gross Tonnage cf. Product of Length, Beam and Depth

Transforming Nonlinear Regression | 371

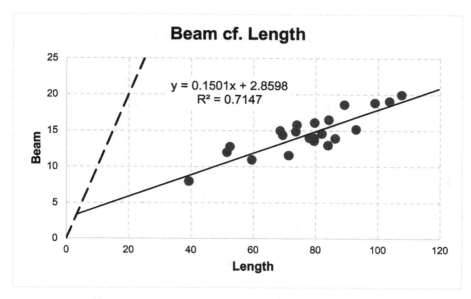

Figure 6.29 Ship Length cf. Beam

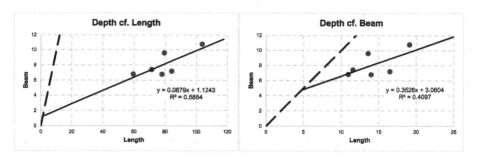

Figure 6.30 Ship Depth cf. Length and Beam

this we can derive a theoretical cubic relationship for the product of all three through the origin for any one of length, breadth or depth. Previously we had suggested that there was a potential correlation between the gross tonnage and the product of these three dimensions. (*If we think of it as a giant rectangular packing case in which we can fit the ship, then the volume of the packing case must be an indicator of the useful capacity or volume of the ship.*)

On that pretext we can now, with some feeling of self-justification, test whether these relationships hold good where we only have the length or breadth of the ship as an indicator of gross tonnage. Based on our data from Table 6.67, we show the Gross

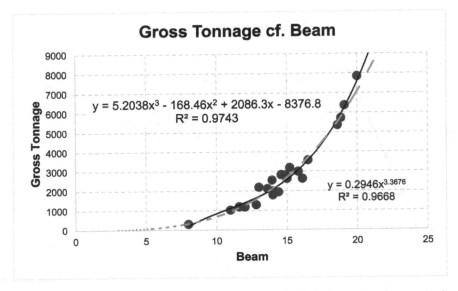

Figure 6.31 Gross Tonnage as a Cubic Function of Ship's Beam (Pre-Regression)

Tonnage plotted against the Beam in Figure 6.31, through which we have fitted a cubic polynomial relationship. The result highlights two key points:

i. There is a very impressive R-Square of some 0.9743, with a fairly tight degree of scatter around the trendline
ii. There is a distinct joggle in the trendline between the lowest value and the origin. With different data (possibly just one extra point) this could actually be an S-bend, which intuitively would not feel right

We have also shown a Power Trendline through the data. Perhaps not unsurprisingly the power exponent is a little over 3, suggesting that the rationale of a cubic function was not too unreasonable.

For the Formula-philes: Justification of a cubic parametric relationship

Consider a number of ships with broadly comparable functionality. Denote the length as L, Beam as B and Depth as D.

Suppose L is approximately a linear function of B with slope a and intercept b: $\qquad L = aB + b \qquad$ (1)

Suppose D is also approximately a linear function of B with slope c and intercept d:
$$D = cB + d \quad (2)$$

From (1) and (2) the product of L, B and D can be expressed in relation to a single variable L (or B or D):
$$LBD = B(aB + b)(cB + d) \quad (3)$$

Expanding (3):
$$LBD = acB^3 + (ad + bc)B^2 + bdB$$

.... which is a cubic polynomial through the origin

Note: This justification of a cubic polynomial through the Origin assumes that our proposition that the three dimensions are truly linearly related. This is probably an over-simplification and for that reason we should expect some compensatory constant term when we compare gross tonnage with the cubic function.

We could easily have used the length as the single predicator variable. Figure 6.32 repeats the exercise to give us gross tonnage in relation to the beam. Fitting a cubic polynomial through the origin yields a reasonable R-Square of 0.87, however, we should be a little concerned about the degree of scatter in the middle ground in comparison to the edges.

Note that this relationship has been predicated based on a logical relationship between multiplicative drivers with some (although not a lot) of supporting evidence; we haven't

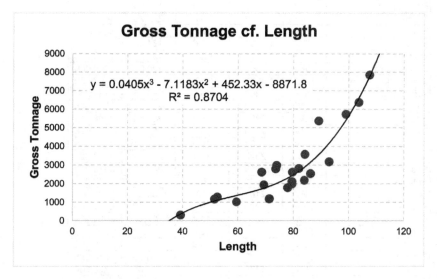

Figure 6.32 Gross Tonnage as a Cubic Function of Ship's Length (Pre-Regression)

374 | Transforming Nonlinear Regression

done it '*just because it works*'. Neither can we necessarily imply that the relationship can be carried over to ships of other types such as oil tankers or warships.

Having established that the relationship has some merit, we can now perform a full regression analysis and create a Prediction Interval around it. Table 6.68 illustrates the set-up in which B, B^2 and B^3 are all used as predicator variables for gross tonnage. Based on this data (but resisting the temptation to force the regression through the origin for the reason stated) we get the result in Table 6.69. Two things should jump out at us (*sorry, didn't mean to startle you!*)

i. The calculated Values of the Coefficients of B, B^2 and B^3 are identical to those generated by the Polynomial Trendline using the Chart Utility in Excel in Figure 6.31 (*Don't you love it when everything stacks up?*)

ii. The p-value associated with the Intercept is 9.9% and we could therefore accept the Null Hypothesis that the value is insignificantly different zero (*it really depends on whether we are using the 5% or 10% significance level as our acceptance/rejection threshold*) ... which we should have determined before we started.

Table 6.68 Regression Input Data for Cubic Polynomial Function Regression

Ship Number	Ship Name	Type	Gross Tonnage	B Beam	B^2	B^3
1	Svalbard	OPV	6375	19.1	364.81	6967.871
2	River Class	OPV	2109	13.6	184.96	2515.456
3	Galatea	THV	3569	16.5	272.25	4492.125
4	Roisin	OPV	1784	14.0	196.00	2744.000
5	Norna	FPV	1190	11.6	134.56	1560.896
6	Island	OPV	1017	10.98	120.56	1323.753
7	Pourquoi Pas	FRV	7854	20	400.00	8000.000
8	James Clark Ross	RRS	5732	18.85	355.32	6697.829
9	James Cook	RRS	5368	18.6	345.96	6434.856
10	Alliance	SRS	3180	15.2	231.04	3511.808
11	Endeavour	FRV	2983	15.8	249.64	3944.312
12	Mermaid	THV	2820	14.6	213.16	3112.136
13	Thalassa	FRV	2803	14.9	222.01	3307.949
14	Granuaile	ILV	2625	16.1	259.21	4173.281
15	Scotia	FRV	2619	15	225.00	3375.000
16	Patricia	THV	2541	13.94	194.32	2708.871
17	Jura	FPV	2181	13	169.00	2197.000
18	Pharos	NLV	1986	14	196.00	2744.000
19	Charles Darwin	RRS	1936	14.4	207.36	2985.984
20	Corystes	FRV	1289	12.8	163.84	2097.152
21	Vigilant	FPV	1190	11.6	134.56	1560.896
22	Sulisker	FPV	1177	11.6	134.56	1560.896
23	Pole Star	NLV	1174	12	144.00	1728.000
24	Alert	THV	302	8.0	64.00	512.000

Transforming Nonlinear Regression | 375

Table 6.69 Regression Output for Gross Tonnage as a Cubic Polynomial Function of a Ship's beam (1)

SUMMARY OUTPUT

Regression Statistics	
Multiple R	0.98707943
R Square	0.974325802
Adjusted R Square	0.970474672
Standard Error	320.0628224
Observations	24

Based on the Significance of the t-statistic, we would accept the null hypothesis that the intercept is statistically indistinguishable from zero

ANOVA

	df	SS	MS	F	Significance F
Regression	3	77751319.13	25917106.38	252.9973953	4.55049E-16
Residual	20	2048804.206	102440.2103		
Total	23	79800123.33			

	Coefficients	Standard Error	t Stat	P-value	Lower 95%	Upper 95%
Intercept	-8376.785357	4841.224574	-1.73030299	0.098978341	-18475.40286	1721.832144
B Beam	2086.285968	1098.735711	1.898806007	0.072113341	-205.6365628	4378.208499
B^2	-168.4638176	80.56923222	-2.090919982	0.049507848	-336.528291	-0.399344191
B^3	5.203818746	1.90685546	2.729005347	0.012929339	1.226187957	9.181449534

Table 6.70 Regression Output for Gross Tonnage as a Cubic Polynomial Function of a Ship's Beam (2)

SUMMARY OUTPUT

Regression Statistics	
Multiple R	0.995463789
R Square	0.990948155
Adjusted R Square	0.942467026
Standard Error	334.9132733
Observations	24

Based on the Significance of the t-statistic, we would accept the null hypothesis that the coefficient of the B variable is statistically indistinguishable from zero

ANOVA

	df	SS	MS	F	Significance F
Regression	3	257868219.1	85956073.03	766.3229754	8.39188E-21
Residual	21	2355504.913	112166.9006		
Total	24	260223724			

	Coefficients	Standard Error	t Stat	P-value	Lower 95%	Upper 95%
Intercept	0	#N/A	#N/A	#N/A	#N/A	#N/A
B Beam	196.8027342	127.1526869	1.547767011	0.136617368	-67.62575394	461.2312223
B^2	-31.8017236	16.65413131	-1.909539621	0.069949072	-66.43588565	2.832438451
B^3	2.024942944	0.53445482	3.788800975	0.001075278	0.913483301	3.136402587

Let's be strict with ourselves and reject the intercept, i.e. set it to zero. If we re-run our regression, forcing it through the origin, we get the revised result in Table 6.70. This time we are faced with having to reject the contribution from the B-variable term as the t-statistic is telling us that there is a13.7% chance that it could be zero.

We can now re-run the regression using only B^2 and B^3 as predicator variables, giving us the result in Table 6.71. This time all our standard measures for goodness of fit are positive leading us to accept the model.

376 | Transforming Nonlinear Regression

Table 6.71 Regression Output for Gross Tonnage as a Cubic Polynomial Function of a Ship's Beam (3)

SUMMARY OUTPUT

Regression Statistics	
Multiple R	0.994945005
R Square	0.989915562
Adjusted R Square	0.944002633
Standard Error	345.3726579
Observations	24

Based on R-Square, F and t-statistics
we would accept this model as a
statistically significant result

ANOVA

	df	SS	MS	F	Significance F
Regression	2	257599514	128799757	1079.789594	6.73439E-22
Residual	22	2624210.002	119282.2728		
Total	24	260223724			

	Coefficients	Standard Error	t Stat	P-value	Lower 95%	Upper 95%
Intercept	0	#N/A	#N/A	#N/A	#N/A	#N/A
B^2	-6.186055735	1.916796232	-3.227289178	0.003874539	-10.16124782	-2.210863653
B^3	1.215977353	0.115119365	10.56275246	4.41734E-10	0.977234402	1.454720303

Finally, based on this latter regression and using the technique described in Chapter 4, we can now generate a Prediction Interval around the regression curve to give us an uncertainty range as depicted in Figure 6.33. (Clearly we can ignore the values that give us a negative value in one sense, yet it does serve to remind us that perhaps this model is not quite as robust as at first it looks. Also, the Prediction Interval around a zero value is nonsensical – how can we have a volume if one of the dimensions is zero?)

If we were to try fitting a Power Trendline for gross tonnage as a power of the beam we would find that we get a very similar answer in terms of the Median Regression value, as shown in Table 6.72 and Figure 6.34, but with very different Prediction Intervals. The question we must ask ourselves are:

- Is it worth all the extra effort to derive a polynomial relationship?
- Which Prediction Interval looks more intuitively correct, the power function or the cubic? (At least we don't get nonsensical or negative values with the former. (*That's a bit of a hint to where I am on this!*)

Even if we decide that the Polynomial Regression has some merits then we cannot rely on the Chart Utility values alone. Whilst they are calculated correctly in the background by Excel, they do not express any measures of goodness of fit. Consequently, the graph trendline is good for a quick sensibility check, but not for determining whether it is a statistically significant result or not.

As we said at the beginning of this section in relation to Polynomial Regressions, as a general rule of thumb, unless we have a convincing case otherwise, the advice must

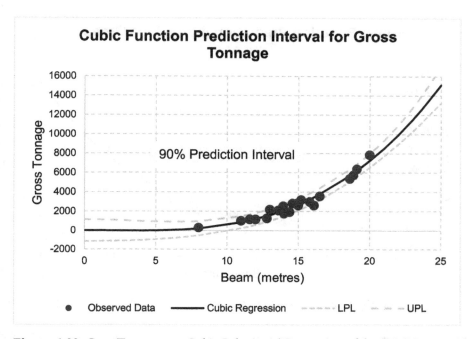

Figure 6.33 Gross Tonnage as a Cubic Polynomial Regression of the Ship's Beam with Prediction Interval

Table 6.72 Regression Output for Gross Tonnage as a Power Function of a Ships' Beam

SUMMARY OUTPUT

Regression Statistics	
Multiple R	0.983257244
R Square	0.966794808
Adjusted R Square	0.965285481
Standard Error	0.057383597
Observations	24

ANOVA

	df	SS	MS	F	Significance F
Regression	1	2.109242602	2.109242602	640.5469935	9.24218E-18
Residual	22	0.072443299	0.003292877		
Total	23	2.181685901			

	Coefficients	Standard Error	t Stat	P-value	Lower 95%	Upper 95%
Intercept	-0.530772994	0.153528785	-3.457156218	0.002242967	-0.849172207	-0.212373782
Log B	3.367630193	0.133060422	25.30902988	9.24218E-18	3.091679768	3.643580619

remain ... **DON'T DO IT!** Just because you can, doesn't mean you should! Polynomial models are the model of last resort.

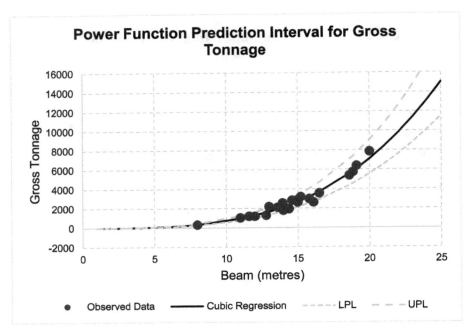

Figure 6.34 Gross Tonnage as Power Function Regression of the ship's beam with Prediction Interval

6.7 Chapter review

Well this chapter may have turned out to be a little more mind bending than first expected, but hopefully we have straightened out a few things as we went along.

In this chapter, we combined the techniques of the last two chapters, allowing us to take Logarithmic, Exponential or Power Function relationships, transform them into linear functions, from which we can then exploit the power of Linear Regression to find the 'Best Fit' relationship through the data, and also be able to test its credibility. In doing this we needed to recognise that the best fit line passes through the Arithmetic Mean of any logarithmic data, which is the Geometric Mean of the untransformed data; this is the trade-off for being able to use a simpler model.

To reverse the transformation for the Regression result to get 'real values' again, we concluded that we only had to do this for exponential and Power Function Regressions, and not the Logarithmic Function type because the predicted or dependent variable was still in linear space for the latter.

Our journey then extended the principle of Multi-linear Regression to these transformable functions, recognising along the way that there are logical restrictions on the combinations we can use, and this is determined by what we decide to do for the primary driver and its functional relationship with the dependent y-variable. Valid combinations for Multi-linear Transformed Regressions are:

- Any number of Linear Functions and Logarithmic Functions
- Any number of Exponential Functions and Power Functions

We cannot cross mix the two groups!

The same messages, limitations and techniques can be applied to generalised versions of the standard function types which include a constant term that prevent us from performing a simple logarithmic transformation. To deal with these we can use Microsoft Excel's Solver to determine the best value for the constant which we can then use to offset the observed values i.e. normalise the data into a form where we can apply a simple logarithmic transformation. However, as we saw these can sometimes be notoriously difficult to converge to an acceptable or expected result due to their sheer flexibility coupled with their ability to mimic other standard or generalised functional forms.

Although we can adapt multi-linear regression to allow us to perform polynomial Regression, this is **not recommended** unless we know (or strongly suspect) that there is a true polynomial relationship at work. One such example is the cumulative value of a straight line, which is always a quadratic function, i.e. polynomial of order 2. It may be possible to justify its use where we can demonstrate a high correlation between two drivers and the dependent value we are trying to estimate, but these may not be valid with low values close to zero as prediction intervals may extend below zero.

References

Dolan, A (2006) 'The obsessive disorder that haunts my life' Daily Mail, 3rd April, [online] Available from: http://www.dailymail.co.uk/tvshowbiz/article-381802/The-obsessive-disorder-haunts-life.html [Accessed 12–01–2017].

The Editors of Encyclopaedia Britannica (1998) 'Tonnage', Chicago, Encyclopaedia Britannica [online] Available from: https://www.britannica.com/technology/tonnage#ref265291 [Accessed 22/11/2016].

Weir, A (1997) *Does my bum look big in this?*, London, Hodder & Stoughton.

7 Least Squares Nonlinear Curve Fitting without the logs

What if we have what looks from the historical data to be a good nonlinear relationship between the variable we are trying to estimate and one or more drivers, but the relationship doesn't fit well into one or more of our transformable functions? Do we:

i.   Throw in the proverbial towel, and make a 'No bid' recommendation because it's too difficult and we don't want to make a mistake?
ii. Think '*What the heck; it's close enough!*' and proceed with an assumption of a Generalised form of the nearest transformable function, estimate the missing constant, or just put up with the potential error?
iii. Shrug our shoulders, and say '*Oh well! We can't win the all!*', and go off for a coffee (*other beverages are available*), and compile a list of experts whose judgement we trust?
iv. Use a Polynomial Regression and only worry if we get bizarre predictions?
v.   Consider another nonlinear function, and find the 'Best Fit Curve' using the Least Squares technique from first principles?

A word (or two) from the wise?

"*It is common sense to take a method and try it. If it fails, admit it frankly and try another; but above all, try something.*"

Franklin D Roosevelt
1882–1945
American president

Apart from the first option which goes against the wise advice of Roosevelt, all of these might be valid options, although we should really avoid option iv **(just because we can it doesn't mean we should!)** However, in this chapter, we will be looking at option (v): how we can take the fundamental principles that underpin Regression, and apply them to those stubborn curves that don't transform easily.

Least Squares Nonlinear Curve Fitting without the logs | 381

7.1 Curve Fitting by Least Squares . . . without the logarithms

In Volume II we explored the concepts of how data can be described in terms of representative values (Volume II Chapter 2 – Measures of Central Tendency), the degree and manner of how the data is scattered around these representative values (Volume II Chapter 3 – Measures of Dispersion and Shape), and examples of some very specific patterns of data scatter (Volume II Chapter 4 – Probability Distributions.)

Sometimes we may want to describe our data in terms of a probability distribution but not really know which is appropriate. We can apply the principle of Least Squares Errors to our sample data and measure whether any particular distribution is a better fit than another. To help us do this we can enlist the support of Microsoft Excel's **Solver** facility.

Even though here we will be discussing the fitting of data to Cumulative Distribution Functions (CDFs), we can try applying this Solver technique to any non-transformable curve. What we cannot do is give you a guarantee that it will always work; that depends heavily on how typical our random sample is of the curve we are trying to fit and how large our data sample is.

Initially we are going to consider the distribution of a Discrete Random Variable, before applying the principle to Continuous Random Variables.

7.1.1 Fitting data to Discrete Probability Distributions

Let's consider the simplest of all Discrete Probability Distributions, the Uniform or Rectangular Distribution. For simplicity, we will consider a standard six-sided die (i.e. one that is 'true' and not weighted towards any particular value, the values of the faces being the integers 1 to 6.)

Suppose we threw the die 30 times and got each integer face-up, exactly five times each; what could we determine from that, (*apart from how lucky or fluky we were?*) We could infer that we had a one-in-six chance of throwing any number (*which in this case we already knew*), or we could infer that the cumulative probability of throwing less than or equal to each successive integer from 1 to 6 is a straight line. (*Again, something we already knew.*)

This straight line could be projected backwards to zero, where clearly we had no results, giving us the simple picture in Figure 7.1 (left-hand graph). Realistically, we would probably get a different number of scores for each integer, as illustrated by the cumulative graph on the right-hand side. With a greater number of observations, the scatter around the theoretical cumulative probability line would be, more than likely, reduced, but with fewer observations it could be significantly more scattered.

If we recall the concept of Quartiles from Volume II Chapter 3 we could determine the five Quartile start and end values using the **QUARTILE.INC*(array, quart)*** function in

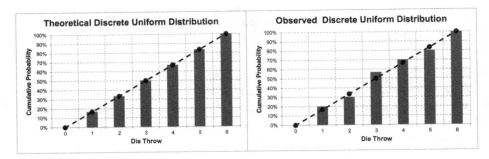

Figure 7.1 Theoretical and Observed Discrete Uniform Distribution

Table 7.1 Casting Doubts? Anomalous Quartile Values of a Die

Quartile Parameter	Quartile Value Returned	Associated Confidence Level with Quartile	Uniform Distribution CDF for Quartile Value Returned
0	1	0%	17%
1	2.25	25%	> 33% and < 50%
2	3.5	50%	> 50% and < 67%
3	4.75	75%	> 67% and < 83$_\%$
4	6	100%	100%

Quartile Array based on the successive integers 1, 2, 3, 4, 5, 6

Microsoft Excel," where quart takes the integer values 0 to 4. If the array is taken to be the six values on the faces on the die then we will get the Quartile values in Table 7.1.

No prizes for spotting that this in blatantly inconsistent with the CDF from Figure 7.2. So what went wrong? (*Answer: We did! Our logic was incomplete* — recall the telegraph poles and spaces analogy from Volume II Chapter 4.)

> ## For the Formula-phobes: Resolving doubts cast over quartile values of a die
>
> Consider the telegraph poles and spaces analogy again. If we had seven poles and six spaces this would define the Sextiles synonymous with a die.
>
>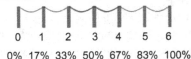
>
> The sextiles then return the same values as the CDF of the Discrete Uniform Distribution.
>
> This also works for other quantiles. All we needed to do to correct our original mistake was add the last value on the left that has zero probability of occurring, as an additional leading telegraph pole.

> We can take this as a rule of thumb to adjust any quartile, decile, percentile or any other quantile to align with the empirical Cumulative Probability.

If instead we include the last value for which we have zero probability of getting – in this case the number 0, we can recalculate our quartiles based on the integers 0 to 6, summarised in Table 7.2.

For the Formula-philes: Calculating quantiles for a discrete range

Consider a range of discrete consecutive numbers I_i arranged in ascending order of value from I_1 to I_n. Let $Q(r, n)\%$ represent the Confidence Level of the rth quantile based on n intervals, such that r can be any integer from 0 to n

$Q(r,n)\%$ is defined by:
$$Q(r,n)\% = 100\left(\frac{r-1}{n-1}\right)$$

To adjust the Quantile to give the correct Confidence Level for the discrete range of values add a preceding term I_0 with zero probability of occurring. The Quantile is then recalculated based on n+1 intervals and r can be any integer from 0 to n+1

$Q(r,n+1)\%$ becomes
$$Q(r,n+1)\% = 100\left(\frac{r-1}{n}\right)$$

The fictitious first term (when $r = 1$) is then at the 0% Confidence Level and the last term (when $r = n+1$) occurs at the 100% Confidence Level, and is in line with the Uniform Distribution's CDF.

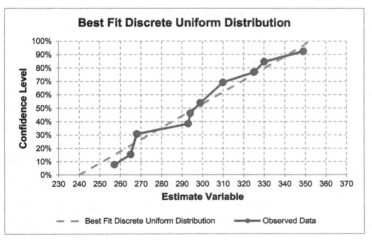

Figure 7.2 Theoretical and Observed Cumulative Discrete Uniform Distribution

384 | Least Squares Nonlinear Curve Fitting without the logs

Table 7.2 Resolving Doubts? Quartile Values of a Die Re-Cast

Quartile Parameter	Quartile Value Returned	Associated Confidence Level with Quartile	Uniform Distribution CDF for Quartile Value Returned
0	0	0%	0%
1	1.5	25%	> 17% and < 33%
2	3	50%	50%
3	4.5	75%	> 67% and < 83%
4	6	100%	100%
Quartile Array based on the successive integers 0, 1, 2, 3, 4, 5, 6			

This adjustment is very similar in principle to Bessel's Correction Factor for Sample Bias (Volume II Chapter 3), and that which we discussed previously in Volume II Chapter 6 on Q-Q plots.

Suppose we have a traffic census in which the volume of traffic is being assessed at a busy road junction. At the busiest time of day between 8 and 9 a.m. over a three-day period, the 'number of vehicles per quarter of an hour' have been recorded in Table 7.3. We suspect that the data is uniformly distributed but have to estimate the range. We can adapt the technique described and justified for the die problem to help us. With the die, we knew what the lower and upper bounds were. In this case we don't; we only know what we have observed, and the actual minimum and maximum may fall outside of those that we have observed, i.e. they have not occurred during the sampling period. In deriving the empirical cumulative probabilities or confidence levels we will assume that there is one extra data point (n+1). This will allow us to assume that the last observed point is not at the 100% Confidence Level; furthermore, it is unbiased in the sense that the Confidence Level of our last observed point is as far from 100% as the first is from 0%. The symmetry of this logic can be further justified if we were to arrange our data in descending order; neither the first or last points are ever at the minimum or maximum confidence level.

The model setup to use with Microsoft Excel's Solver is illustrated in Table 7.3.

The data is arranged in ascending order of observed values. In terms of our 'Adjusted Observed Cum %', this is simply the Running Cumulative Total Number of Observations divided by the Total Number of Observations plus one.

In setting up our theoretical Discrete Uniform Distribution model we will assume that there is one more value than the difference between the assumed maximum and minimum (*the telegraph poles and spaces analogy again*) just as we did for the die example. The Theoretical Confidence Level % is the difference between the Observed Values and the assumed minimum divided by the Range (Maximum − Minimum).

Least Squares Nonlinear Curve Fitting without the logs | 385

Table 7.3 Solver Set-Up for Fitting a Discrete Uniform Distribution to Observed Data

Solver Variables for Uniform Distribution	Min >	256	i.e. @ 100%
	Max >	350	i.e. @ 0%
No of Discrete Values (Max - Min + 1) >		95	

	Vehicles per 1/4 Hour	Occurrences	Cumulative Occurrences	Adjusted Observed Cum %	Modelled Uniform Distribution	Delta % (Observed – Model)
Model Min >	256				0%	
	257	1	1	7.7%	1.1%	6.6%
	265	1	2	15.4%	9.5%	5.9%
	268	2	4	30.8%	12.6%	18.1%
	293	1	5	38.5%	38.9%	-0.5%
	294	1	6	46.2%	40.0%	6.2%
	299	1	7	53.8%	45.3%	8.6%
	310	2	9	69.2%	56.8%	12.4%
	325	1	10	76.9%	72.6%	4.3%
	330	1	11	84.6%	77.9%	6.7%
	349	1	12	92.3%	97.9%	-5.6%
Model Max >	350				100%	

Sum		12				62.8%
Median	296.50					6.4%
Mean	299.00					6.3%
				Sum of Squares Error (SSE) >		0.0768

The Solver Variables (minimum and maximum) can be chosen at random; here we have taken them to be one less and one more than the observed minimum and maximum. The Solver objective is to minimise the Sum of Squares Error (SSE) subject to a number of constraints:

- The Theoretical Minimum is an integer less than or equal to the observed Minimum
- The Theoretical Maximum is an integer less than or equal to the observed Maximum
- The Minimum is less than or equal to the maximum minus 1
- The Median Error is to be zero.

Note that the last constraint here is different from that used for Regression in which we assume that the Sum of Errors is zero, and therefore the Mean Error is zero too. We may recall from Volume II Chapter 2 that there are benefits from using the Median over the Arithmetic Mean with small sample sizes in that it is more 'robust' being less susceptible to extreme values. If we are uncomfortable in using the Median instead of the Mean we can exercise our personal judgement and use the Mean.

Table 7.4 and Figure 7.2 illustrate the Solver solution derived using the Zero Median Error constraint.

386 | Least Squares Nonlinear Curve Fitting without the logs

Table 7.4 Solver Results for Fitting a Discrete Uniform Distribution to Observed Data

	Solver Variables for Uniform Distribution		Min >	240	i.e. @ 100%
			Max >	352	i.e. @ 0%
	No of Discrete Values (Max - Min + 1) >			113	

	Vehicles per 1/4 Hour	Occurrences	Cumulative Occurrences	Adjusted Observed Cum %	Modelled Uniform Distribution	Delta % (Observed − Model)
Model Min >	240				0%	
	257	1	1	7.7%	15.0%	-7.4%
	265	1	2	15.4%	22.1%	-6.7%
	268	2	4	30.8%	24.8%	6.0%
	293	1	5	38.5%	46.9%	-8.4%
	294	1	6	46.2%	47.8%	-1.6%
	299	1	7	53.8%	52.2%	1.6%
	310	2	9	69.2%	61.9%	7.3%
	325	1	10	76.9%	75.2%	1.7%
	330	1	11	84.6%	79.6%	5.0%
	349	1	12	92.3%	96.5%	-4.2%
Model Max >	352				100%	
Sum		12				-6.7%
Median	296.50					0.0%
Mean	299.00					-0.7%
				Sum of Squares Error (SSE) >		0.0310

In some instances, however, we may know that we will have a discrete distribution but not know the nature of the distribution. For instance, the sports minded amongst us will know that we score or count goals, points, shots, fish caught, etc., in integer values. Let's 'put' in a golf example (*sorry!*)

Are elite golfers normal?

We can use this approach to try and address that provocative question of whether elite golfers can ever be described as just being "normal" golfers! To answer this, let's consider a number of such elite professional and top-notch amateurs golfers whose scores were recorded at a major tournament in 2014 on a par 72, 18-hole course.

> For the non-golfers amongst us (myself included) the 'tee' is where the golfer starts to play a particular hole and the 'green' is the location of the hole and flag (pin) where the grass is cut very short. The 'fairway' is a section of reasonably short grass between the tee and the green from where most golfers are expected to play their second and third shots etc towards the green unless they are expected to reach the green in one shot. This contrasts with the 'rough' which is where they don't want to be playing any shot as it is characterised by long uneven areas of grass. I mention this just in case you have no idea of golfing terms, but want to understand the areas or risk and uncertainty.

Least Squares Nonlinear Curve Fitting without the logs | 387

For those unfamiliar with scoring in golf:
Each hole has a 'standard' number of shots that a competent golfer could be expected to need to complete the hole. This is based on the estimated number of approach shots required to reach the green from the tee plus two putts on the green. This estimate is called the 'par' for the hole. Typically, a hole might have a par of 3, 4, or 5. Anything above 5 would be very exceptional.

The number of approach shots is determined by the length and complexity of the fairway leading to the green (e.g. location of hazards, dog-legged or straight approach etc.) In this example, the sum of the individual par values of all the 18 holes is 72.

(*Notice how I worked a bottom-up detailed parametric estimate into the example using metrics and complexity factors?*)

We know that there must be at least one shot per hole and typically we would expect an absolute minimum around 36 based on approach shots, and even then, that assumes that the last approach shot on each hole goes straight in the hole, and how fluky would that be if it happened eighteen times on the bounce? Realistically we can expect that the number of cumulative scores for 18 holes against the par of 72 will be distributed around it. Suppose we want to assess the probability of getting any particular score? Is the distribution symmetrical or skewed?

The distribution of scores on any hole is likely to be positively skewed as our ability to score more than par for the hole is greater than our ability to score less than par (both in skill terms and mathematically), as it is bounded by 1 as an absolute minimum, but unbounded as a maximum. The phenomenon of central tendency (*otherwise known as 'swings and roundabouts'*) suggests that we can probably expect a more symmetrical result from a number of players and that the best overall score for the round will be greater than the sum of the best scores for each hole.

Although the Normal Distribution is a Continuous Distribution we have already commented in Volume II Chapters 4 that it can be used as an approximation to certain Discrete Distributions. Let's see if it works here.

Table 7.5 summarises the scores for each of two rounds of 18 holes by all 155 professional and amateur golfers at that Major Golf Tournament in 2014. The data reflects their first and second round scores before the dreaded 'cut' is made, at which point only the lowest scoring 'half' continue to play a third and fourth round. In this particular case, because we have a large sample size, we will revert to the more usual case of setting a constraint that the Mean Error is zero. (If we were to set the Median Error to zero we would get a very slightly different, but not inconsistent answer.) For the eagle-eyed amongst us (*no golf pun intended, but I'll take it*) we do have one, and potentially two outliers in the rounds of 90 and 84, using Tukey's Fences (Volume II Chapter 7) (the Upper Outer Fence sits at 91 and the Upper Inner Fence at 83.5 based on first and third quartiles of 71 and 76 respectively). This might encourage us to take the Median again rather than the Mean to mitigate against the extreme values.

388 | Least Squares Nonlinear Curve Fitting without the logs

Table 7.5 Solver Results for Fitting a Normal Distribution to Discrete Scores at Golf (Two Rounds)

	Solver Variables for Normal Distribution		Mean α >	73.06
			Std Dev σ >	3.24
			6σ Range >	19.46

	Round Score	Rounds at that Score	Running Total ≤ Round Score	Adjusted Observed Cum %	Modelled Normal Distribution	Delta % (Observed - Model)
Model Optimistic μ-6σ >	54				0.0%	
Mid-way Smoothing Point >	60				0.0%	
	65	1	1	0.3%	0.6%	-0.3%
	66	3	4	1.3%	1.5%	-0.2%
	67	2	6	1.9%	3.1%	-1.2%
	68	10	16	5.1%	5.9%	-0.8%
	69	16	32	10.3%	10.5%	-0.2%
	70	23	55	17.7%	17.3%	0.4%
	71	31	86	27.7%	26.3%	1.4%
Observed Data	72	31	117	37.6%	37.2%	0.4%
	73	40	157	50.5%	49.3%	1.2%
	74	42	199	64.0%	61.4%	2.6%
	75	31	230	74.0%	72.5%	1.4%
	76	23	253	81.4%	81.8%	-0.4%
	77	23	276	88.7%	88.8%	0.0%
	78	11	287	92.3%	93.6%	-1.3%
	79	13	300	96.5%	96.6%	-0.2%
	80	4	304	97.7%	98.4%	-0.6%
	81	3	307	98.7%	99.3%	-0.6%
	82	1	308	99.0%	99.7%	-0.7%
	84	1	309	99.4%	100.0%	-0.6%
	90	1	310	99.7%	100.0%	-0.3%
Mid-way Smoothing Point >	92				100.0%	
Model Optimistic μ+6σ >	93				100.0%	
Sum			310			0.0%
Median	72.00					-0.3%
Mean	71.87					0.0%
Std Dev	2.50			Sum of Squares Error (SSE) >		0.0018

In setting up the Model in Table 7.5 we have again calculated our 'Adjusted Observed Cum %' as the Running Cumulative Total Number of Observations divided by the Total Number of Observations plus one.

The Theoretical Distribution is calculated using Microsoft Excel's **NORM.DIST** function with a Mean and standard deviation equal to our Solver Parameters at the top, and the Cumulative Value parameter set to TRUE. The x-values here are the round scores.

The Solver starting parameters can be chosen at random, but it would be sensible to start with values that are similar to the observed Mean or Median of around 72 and a standard deviation of around 2.5.

Taking the results of the Solver from Table 7.5, despite the inclusion of the potential outliers, the scores do appear to be Normally Distributed. Figures 7.3 and 7.4 illustrate this.

If we were to consider these elite golfers in two groups, those who qualified for the third and fourth rounds of the tournament and those who didn't, we can see whether they are still Normally Distributed.

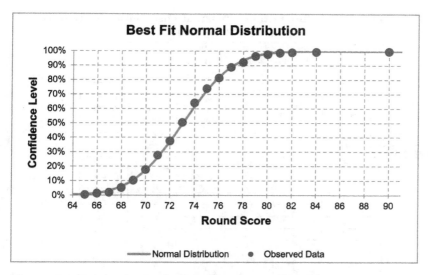

Figure 7.3 Best Fit Normal Distribution to Golf Tournament Round Scores (Two Rounds)

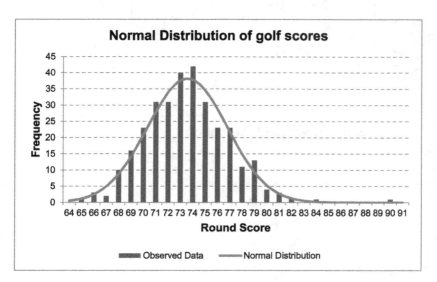

Figure 7.4 Normal Distribution of Golf Scores (Two Rounds – All Competitors)

- Table 7.6 and Figures 7.5 and 7.6 analyses the score of the ongoing qualifiers across all four rounds. Again, there is strong support that their scores are Normally Distributed.

Table 7.6 Solver Results for Fitting a Normal Distribution to Discrete Scores at Golf (Top half)

		Solver Variables for Normal Distribution	Mean μ >	70.83
			Std Dev σ >	2.73
			6σ Range >	16.40

	Round Score	Rounds at that Score	Running Total ≤ Round Score	Adjusted Observed Cum %	Modelled Normal Distribution	Delta % (Observed - Model)
Model Optimistic μ-6σ >	54				0.0%	
Mid-way Smoothing Point >	60				0.0%	
Observed Data	65	5	5	1.7%	1.6%	0.1%
	66	5	10	3.5%	3.9%	-0.4%
	67	12	22	7.6%	8.0%	-0.4%
	68	25	47	16.3%	15.0%	1.2%
	69	28	75	26.0%	25.1%	0.8%
	70	31	106	36.7%	38.1%	-1.4%
	71	46	152	52.6%	52.5%	0.1%
	72	35	187	64.7%	66.6%	-1.9%
	73	40	227	78.5%	78.6%	-0.1%
	74	29	256	88.6%	87.7%	0.9%
	75	18	274	94.8%	93.6%	1.2%
	76	8	282	97.6%	97.1%	0.5%
	77	3	285	98.6%	98.8%	-0.2%
	78	2	287	99.3%	99.6%	-0.3%
	79	1	288	99.7%	99.9%	-0.2%
Mid-way Smoothing Point >	83				100.0%	
Model Optimistic μ+6σ >	87				100.0%	

Sum		288				0.0%
Median	71.00					-0.1%
Mean	71.22					0.0%
Std Dev	2.63		Sum of Squares Error (SSE) >			0.0010

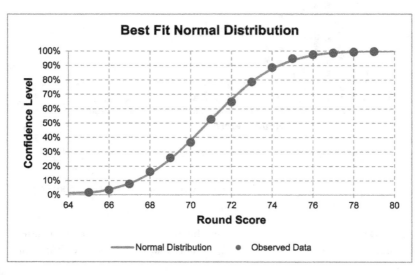

Figure 7.5 Best Fit Normal Distribution to Golf Tournament Round Scores (Top Half)

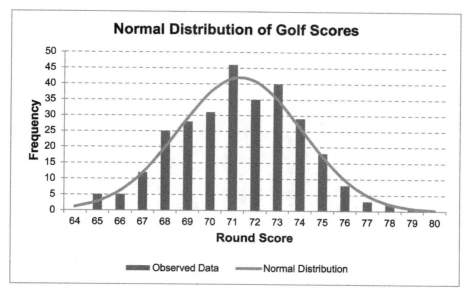

Figure 7.6 Normal Distribution of Golf Scores (Four Rounds – Top Competitors)

- Figure 7.7 shows the results of a similar analysis based on the scores of those failing to make the cut (i.e. not qualifying to continue in to the third round).

Finally, just out of interest, we can compare the scores of those eliminated with those who qualify to continue into the third round. For this we have used just the scores from the first two rounds (to normalise playing conditions, etc.). Both groups appear to be distributed Normally as illustrated in Figure 7.8 with a difference of some 4 shots between their Mean scores over two rounds. Also, there is a wider range of scores for the 'bottom half' being some 2.5 shots across the six-sigma spread of scores, thus indicating that those eliminated are less consistent with a bigger standard deviation. The overlap between the two distributions also illustrate that even elite golfers have good days and bad days.

So, in conclusion to our provocative question, we can surmise that, statistically speaking, elite golfers are indeed just normal golfers, and if we can parody *Animal Farm* (1945) by George Orwell, '*All golfers are normal but some are more normal than others*'!

7.1.2 Fitting data to Continuous Probability Distributions

If we were to draw a sample of values at random from a truly Continuous Distribution (not just an approximation where in reality only discrete integer values will occur) then in many instances, we will not draw the minimum or maximum values (where they exist)

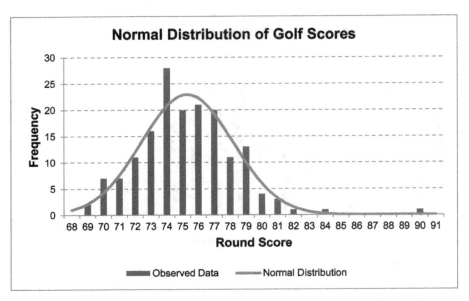

Figure 7.7 Normal Distribution of Golf Scores (Two Rounds – Eliminated Competitors)

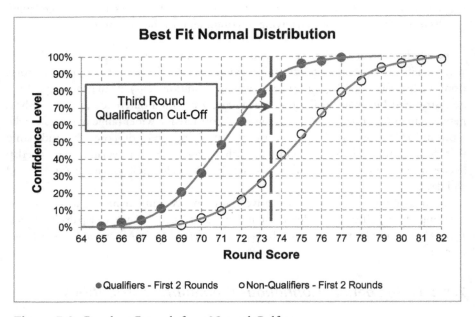

Figure 7.8 Random Rounds from Normal Golfers

Least Squares Nonlinear Curve Fitting without the logs | 393

or the very extreme values where the distribution tends towards infinity in either or both directions. As a consequence, we need to use the same, or at least a similar, method of compensating for values outside of the range we have observed. We can use the same argument that we used for discrete distributions at the start of the preceding section, and assume that there is always one more point in our sample than there actually is! A fuller explanation of why this works and is valid can be found in Volume II Chapter 4.

> An alternative adjustment can be achieved by subtracting a half from the running cumulative total and dividing by the true sample size. This also equalises the difference between 0% and the first observation Confidence Level, and the last observation and 100%. For simplicity's sake, we will stick with the same adjustment that we used for Discrete Distributions.

Let's consider a typical problem where we have a range of values for which we want to establish the most appropriate form of Probability Distribution that describes the observed data. Suppose we have a number of component repairs the time for which vary. Table 7.7 summarises the observed values and sets up a Solver Model to create the best fit Continuous Uniform Distribution. (Afterwards we will compare the data against a number of other continuous distributions.)

The procedure to set up the model is very similar to that for Discrete Distributions:

1. Arrange the data in ascending order of value
2. The 'Adjusted Observations Cum %' are calculated by dividing the Running Total number of observations divided by the Total Observations plus one
3. The Modelled Uniform Distribution is calculated by determining the difference between the Observed Value and the assumed Minimum Value and dividing by the Range (Max − Min)
4. The Delta % expresses the difference in the Observed and Theoretical Confidence Values
5. The Solver objective is to minimise the Sum of Squares Error (Delta %) by varying the Minimum and Maximum values of the Theoretical Distribution, subject to the constraint that the Median Delta % is zero. We have reverted to the median here because of the small batch size.

The results of the Solver algorithm are shown in Table 7.8 and Figure 7.9.

Whilst this is the best fit Continuous Uniform Distribution to the observed data, it is clearly not a particularly good fit as the data is arced across the theoretical distribution, which in this case is a straight line. Ideally, the slope of a trendline through errors would be flat, indicating a potential random scatter. However, we should never say 'never' in a case of distribution fitting like this with a small sample size as this could be just a case of sampling error.

Table 7.7 Solver Set-Up for Fitting a Continuous Uniform Distribution to Observed Data

	Solver Variables for Uniform Distribution		Min >	4.00
			Max >	13.00
			Range >	9.00

	Repair Time (Hrs)	Qty at that Value	Running Total ≤ Repair Time	Adjusted Observed Cum %	Modelled Uniform Distribution	Delta % (Observed – Model)
Model Min >	4.00				0%	
	4.61	1	1	9.1%	6.8%	2.3%
	4.88	1	2	18.2%	9.8%	8.4%
	5.23	1	3	27.3%	13.7%	13.6%
	6.03	1	4	36.4%	22.6%	13.8%
	6.83	1	5	45.5%	31.4%	14.0%
	8.04	1	6	54.5%	44.9%	9.7%
	9.26	1	7	63.6%	58.4%	5.2%
	10.33	1	8	72.7%	70.3%	2.4%
	11.04	1	9	81.8%	78.2%	3.6%
	12.60	1	10	90.9%	95.6%	-4.6%
Model Max >	13.00				100%	
Sum		10				68.3%
Median	7.44					6.8%
Mean	7.89					6.8%
				Sum of Squares Error (SSE) >		0.0809

Table 7.8 Solver Result for Fitting a Continuous Uniform Distribution to Observed Data

	Solver Variables for Uniform Distribution		Min >	2.73
			Max >	12.93
			Range >	10.20

	Repair Time (Hrs)	Qty at that Value	Running Total ≤ Repair Time	Adjusted Observed Cum %	Modelled Uniform Distribution	Delta % (Observed – Model)
Model Min >	2.73				0%	
	4.61	1	1	9.1%	18.4%	-9.3%
	4.88	1	2	18.2%	21.1%	-2.9%
	5.23	1	3	27.3%	24.5%	2.8%
	6.03	1	4	36.4%	32.3%	4.0%
	6.83	1	5	45.5%	40.2%	5.3%
	8.04	1	6	54.5%	52.0%	2.5%
	9.26	1	7	63.6%	64.0%	-0.4%
	10.33	1	8	72.7%	74.5%	-1.8%
	11.04	1	9	81.8%	81.5%	0.4%
	12.60	1	10	90.9%	96.7%	-5.8%
Model Max >	12.93				100%	
Sum		10				-5.3%
Median	7.44					0.0%
Mean	7.89					-0.5%
				Sum of Squares Error (SSE) >		0.0191

Least Squares Nonlinear Curve Fitting without the logs | 395

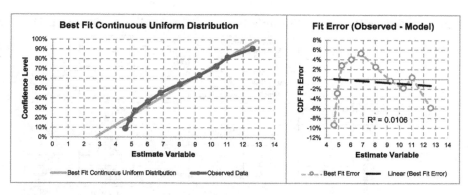

Figure 7.9 Solver Result for Fitting a Continuous Uniform Distribution to Observed Data

However, let's compare other potential distributions, starting with the Normal Distribution. The Model set-up is identical to that for a Continuous Uniform Distribution with the exception that the Solver Parameters are the Model Distribution Mean and standard deviation, and that the Model Normal Distribution can be written using the Excel function **NORM.DIST(*Repair Time, Mean, Std Dev, TRUE*)**. The Solver starting parameters can be based on the Observed Mean or Median and sample standard deviation, giving us the result in Table 7.9 and Figure 7.10.

Again, we would probably infer that the Normal Distribution is not a particularly good fit at the lower end, but is better 'at the top end' than the Continuous Uniform Distribution. Let's now try our flexible friend the Beta Distribution which can be modelled with four parameters using the Microsoft Excel function as **BETA.DIST(*Repair Time, alpha, beta, TRUE, Start, End*)**. Table 7.10 and Figure 7.11 illustrate the output from such a Solver algorithm. We do have to set a couple of additional constraints, however. We may recall from Volume II Chapter 4 (*unless we found we had a much more interesting social life*) that alpha and beta should both be greater than one in normal circumstances, so we can set a minimum value of 1 for both these parameters.

In the output, we may have noticed that the Best Fit Beta Distribution has a parameter value of alpha = 1. Volume II Chapter 4 informs us that this is in fact a right-angled triangular distribution with the Mode at the minimum value. (If it had given a beta parameter of 1, then this would have signified a right-angled triangular distribution with the Mode at the maximum value. Both parameters equalling one would have signified a uniform distribution.)

Visually, from Figure 7.11, this is a very encouraging result, giving what appears to be a very good result. This is not unusual as the Beta Distribution is very flexible.

Based on this outcome we can now turn our attention to fitting a Triangular Distribution instead, the outcome of which is shown in Table 7.11 and Figure 7.12.

Least Squares Nonlinear Curve Fitting without the logs

Table 7.9 Solver Result for Fitting a Normal Distribution to Observed Data

		Solver Variables for Normal Distribution		Mean m >	7.77
				Std Dev s >	3.61
				6σ Range >	21.66
	Repair Time (Hrs)	Qty at that Value	Running Total ≤ Repair Time	Adjusted Observed Cum %	Modelled Normal Distribution	Delta % (Observed – Model)
Model Optimistic μ-3σ >	-3.06				0.1%	
Mid-way Smoothing Point >	0.78				2.6%	
Observed Data	4.61	1	1	9.1%	19.1%	-10.0%
	4.88	1	2	18.2%	21.2%	-3.0%
	5.23	1	3	27.3%	24.1%	3.2%
	6.03	1	4	36.4%	31.5%	4.9%
	6.83	1	5	45.5%	39.7%	5.7%
	8.04	1	6	54.5%	53.0%	1.6%
	9.26	1	7	63.6%	66.0%	-2.4%
	10.33	1	8	72.7%	76.1%	-3.4%
	11.04	1	9	81.8%	81.7%	0.1%
	12.60	1	10	90.9%	91.0%	0.0%
Mid-way Smoothing Point >	15.60				98.5%	
Model Optimistic μ+3σ >	18.60				99.9%	
Sum		10				-3.3%
Median	7.44					0.0%
Mean	7.89					-0.3%
Std Dev	2.82			Sum of Squares Error (SSE) >		0.0195

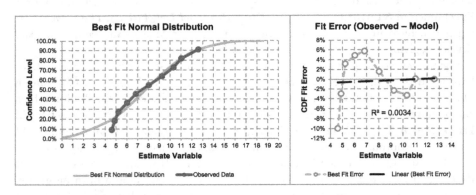

Figure 7.10 Solver Result for Fitting a Normal Distribution to Observed Data

We may recall from Volume II Chapter 4 that Microsoft Excel does not have an in-built function for the Triangular Distribution, and we have to 'code' one manually as a conditional calculation:

- If the Repair Time ≤ Mode, then

    Confidence Level=(Repair Time-Min)2/((Mode-Min) × Range)

- If the Repair Time > Mode then

    Confidence Level=(1-(Max-Repair Time)^2)/((Max-Mode) × Range)

Least Squares Nonlinear Curve Fitting without the logs | 397

Table 7.10 Solver Result for Fitting a Beta Distribution to Observed Data

		Solver Variables for Beta Distribution	alpha	1.00
			beta	1.97
			Start	3.53
			End	16.73

	Repair Time (Hrs)	Qty at that Value	Running Total ≤ Repair Time	Adjusted Observed Cum %	Modelled Beta Distribution	Delta % (Observed − Model)
Model Minimum >	3.53				0%	
Mid-way Smoothing Point >	4.07				7.9%	
Observed Data	4.61	1	1	9.1%	15.4%	-6.3%
	4.88	1	2	18.2%	19.1%	-0.9%
	5.23	1	3	27.3%	23.7%	3.5%
	6.03	1	4	36.4%	33.8%	2.5%
	6.83	1	5	45.5%	43.2%	2.3%
	8.04	1	6	54.5%	56.1%	-1.5%
	9.26	1	7	63.6%	67.4%	-3.7%
	10.33	1	8	72.7%	75.9%	-3.2%
	11.04	1	9	81.8%	80.9%	0.9%
	12.60	1	10	90.9%	89.8%	1.1%
Mid-way Smoothing Point >	14.66				97.4%	
Model Maximum >	16.73				100%	
Sum		10				-5.4%
Median	7.44					0.0%
Mean	7.89					-0.5%
				Sum of Squares Error (SSE) >		0.0094

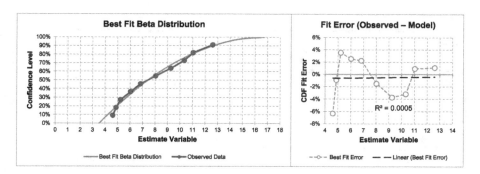

Figure 7.11 Solver Result for Fitting a Beta Distribution to Observed Data

Again, Solver has found a right-angled Triangular Distribution as the Best Fit. However, you will note that it is not identical to the special case of the Beta Distribution. Why not? It's a reasonable question to ask.

The values are very similar but not exact, and the only thing we can put it down to is that the Beta Distribution function in Microsoft Excel is a close approximation

Table 7.11 Solver Result for Fitting a Triangular Distribution to Observed Data

				Solver Variables for Triangular Distribution	Min	3.54
					Mode	3.54
					Max	16.86
					Range	13.32

		Repair Time (Hrs)	Qty at that Value	Running Total ≤ Repair Time	Adjusted Observed Cum %	Modelled Triangular Distribution	Delta % (Observed − Model)
Model Minimum >		3.54				0%	
Mid-way Smoothing Point >		4.08				7.9%	
	Observed Data	4.61	1	1	9.1%	15.4%	-6.3%
		4.88	1	2	18.2%	19.1%	-0.9%
		5.23	1	3	27.3%	23.8%	3.5%
		6.03	1	4	36.4%	33.9%	2.5%
		6.83	1	5	45.5%	43.3%	2.2%
		8.04	1	6	54.5%	56.1%	-1.6%
		9.26	1	7	63.6%	67.4%	-3.8%
		10.33	1	8	72.7%	76.0%	-3.2%
		11.04	1	9	81.8%	80.9%	0.9%
		12.60	1	10	90.9%	89.8%	1.1%
Mid-way Smoothing Point >		14.73				97.4%	
Model Maximum >		16.86				100%	

Sum		10		-5.6%
Median	7.44			0.0%
Mean	7.89			-0.6%
			Sum of Squares Error (SSE) >	0.0094

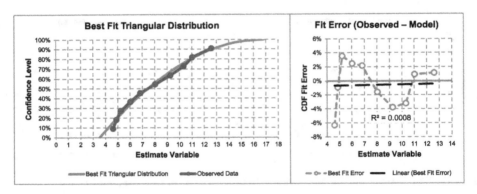

Figure 7.12 Solver Result for Fitting a Triangular Distribution to Observed Data

to the true distribution values. We cannot lay the 'blame' at Solver's door in getting stuck at a local minimum close to the absolute minimum as this would go away if we typed the results of one model into the other and vice versa. It doesn't!

We will have to console ourselves with the thought that they are both accurate enough, as in reality they will be both precisely incorrect in an absolute sense. (*One more or less data point will give us a slightly different answer.*)

Least Squares Nonlinear Curve Fitting without the logs | 399

There are other distributions we could try, but most of us will have lives to lead, and will be wanting to get on with them. Nevertheless, the principles of the curve fitting are similar.

Despite this, overall, we may conclude that the right-angled Triangular Distribution is a reasonable representation of the sample distribution in question.

7.1.3 Revisiting the Gamma Distribution Regression

In Section 7.2.2 we concluded that it was a little too difficult to use a Multi-Linear Regression to fit a Gamma Distribution PDF due to collinearity and other issues, and I rashly promised to get back to you on that particular conundrum. (*That time has come, but please do try to curtail the excitement level; it's unbecoming!*)

In the last section we could have tried an offset Gamma Distribution to our data, and this would have worked, giving us the solution in Table 7.12 and Figure 7.13 (although not as convincing as a right-angled Triangular Distribution.) If we look at the error distribution in Figure 7.13 it is does not appear to be homoscedastic, leaving quite a sinusoidal looking error pattern around the fitted curve. However, this example does conveniently illustrate the difficulties of using a transformed Multi-Linear Regression with small sample sizes. Here, we have a frequency value of one for each observed value. If we were to take the logarithmic transformation of this frequency it would always be zero, regardless of the observed Repair Time and its logarithm.

Table 7.12 Solver Result for Fitting a Gamma Distribution to Observed Data

	Repair Time (Hrs)	Qty at that Value	Running Total ≤ Repair Time	Adjusted Observed Cum %	Modelled Beta Distribution	Delta % (Observed – Model)
		Solver Variables for Gamma Distribution	alpha	1.17		
			beta	4.05		
			Start	3.73		
Model Minimum >	3.73				0.0%	
Quarter-way Smoothing Point >	3.95				2.9%	
Mid-way Smoothing Point >	4.17				6.4%	
	4.61	1	1	9.1%	13.7%	-4.6%
	4.88	1	2	18.2%	18.1%	0.1%
	5.23	1	3	27.3%	23.6%	3.6%
	6.03	1	4	36.4%	35.3%	1.0%
	6.83	1	5	45.5%	45.5%	-0.1%
	8.04	1	6	54.5%	58.3%	-3.8%
	9.26	1	7	63.6%	68.3%	-4.7%
	10.33	1	8	72.7%	75.2%	-2.4%
	11.04	1	9	81.8%	78.9%	2.9%
	12.60	1	10	90.9%	85.3%	5.6%
Quarter-way Smoothing Point >	17.82				95.7%	
Mid-way Smoothing Point >	23.04				98.8%	
Model Pessimistic >	33.49				99.9%	
Sum		10				-2.3%
Median	7.44					0.0%
Mean	7.89					-0.2%
			Sum of Squares Error (SSE) >			0.0117

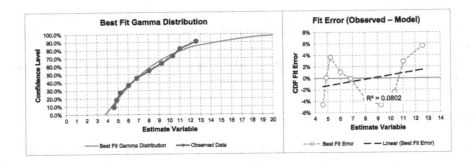

Figure 7.13 Solver Result for Fitting a Gamma Distribution to Observed Data

In short the regression would force every coefficient to be zero in order to achieve the result that 0 = 0 + 0 +0 ! *(Which is neat, but now falls into the category of 'useless'!)*

However, it is not always this catastrophic. Instead let's consider a large sample of data (100 points) collected at a busy set of traffic lights which are set to change after a set time interval. The data we have collected is shown in Figure 7.14 and Table 7.13. Incidentally, this data appears to be a very close approximation to the cumulative discrete integer values of a Continuous Gamma Distribution; it will help us to understand a little better what is happening when we try to fit discrete data to Continuous Distributions. However, a quick peak at the error terms in Table 7.13 suggests that the scatter is not homoscedastic with the right-hand tail consisting exclusively of negative errors.

Incidentally, in this model we have had to use a Weighted Sum of Errors and a Weighted Sum of Squares of Errors to take account of the frequency in which the error occurs. In Microsoft Excel, we can do this as follows:

Weighted Sum of Errors = **SUMPRODUCT**(*Frequency, Delta%*)
Weighted Sum of Squares Error = **SUMPRODUCT**(*Frequency, Delta%, Delta%*)

... where **Frequency** is the Frequency of Occurrence range of values, and **Delta%** is the Delta % (Model − Observed) range of values

In this case we have used Microsoft Solver to calculate the Least Squares Error around the Gamma CDF to determine potential parameters for α and β of 2.44 and 2.39 respectively. (*What was that? It sounds all Greek to you?*) Let's see if we can replicate the result using a Transformed Nonlinear Regression of the PDF data (i.e. the frequency per queue length rather than the cumulative data).

From Volume II Chapter 4 and here in Section 7.2.2, the Linear Transformation gives us the relationship that the logarithm of the Queue Length Frequency is a function of two independent variables (Queue Length and the logarithm of the Queue Length) and a constant term.

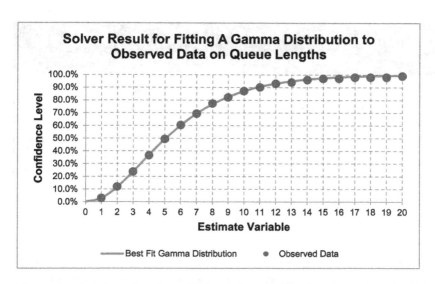

Figure 7.14 Solver Result for Fitting a Gamma Distribution to Observed Data on Queue Lengths

Table 7.13 Solver Results for Fitting a Gamma Distribution to Observed Data on Queue Lengths

	Solver Variables for Gamma Distribution		alpha	2.4436		
			beta	2.3873		
			Start	0.00		

	Cars Queuing at Traffic Lights	Frequency of Occurrence	Running Total ≤ Queue Length	Adjusted Observed Cum %	Modelled Beta Distribution	Delta % (Observed − Model)
Model Minimum >	0				0.0%	
	1	3	3	3.0%	2.8%	0.1%
	2	9	12	11.9%	11.7%	0.2%
	3	12	24	23.8%	23.8%	-0.1%
	4	13	37	36.6%	36.9%	-0.3%
	5	13	50	49.5%	49.3%	0.2%
	6	11	61	60.4%	60.2%	0.2%
	7	9	70	69.3%	69.3%	0.0%
	8	8	78	77.2%	76.7%	0.5%
Observed Data	9	5	83	82.2%	82.6%	-0.4%
	10	5	88	87.1%	87.1%	0.0%
	11	3	91	90.1%	90.5%	-0.4%
	12	3	94	93.1%	93.1%	0.0%
	13	1	95	94.1%	95.0%	-0.9%
	14	2	97	96.0%	96.4%	-0.4%
	15	1	98	97.0%	97.4%	-0.4%
	16	0	98	97.0%	98.2%	-1.1%
	17	1	99	98.0%	98.7%	-0.7%
	18	0	99	98.0%	99.1%	-1.1%
	19	0	99	98.0%	99.3%	-1.3%
	20	1	100	99.0%	99.5%	-0.5%
Quarter-way Smoothing Point >	21.05				99.7%	
Mid-way Smoothing Point >	22.11				99.8%	
Model Pessimistic >	24.22				99.9%	

Sum		100		Weighted Sum (Tgt = 0)	0.0%
Median	5.50				
Mean	6.24			Weighted Mean (Tgt = 0)	0.0%
				Weighted Sum of Squares Error (SSE) >	0.0008

For the Formula-philes: Multi-linear transformation of a Gamma Distribution PDF

Consider a Gamma Distribution PDF of dependent variable y with respect to independent variable x with shape and scale parameters α and β

Gamma Distribution PDF is:
$$y = \frac{1}{\beta^{\alpha}\Gamma(\alpha)} x^{\alpha-1} e^{-\frac{x}{\beta}} \tag{1}$$

Taking Natural Logs of (1):
$$\ln y = \ln\left(\frac{1}{\beta^{\alpha}\Gamma(\alpha)}\right) + (\alpha - 1)\ln x - \frac{x}{\beta}$$

This immediately gives us a problem (see Table 7.14) that Queue Lengths of 16, 18 and 19 vehicles have not been observed so we must reject the zero values from our analysis to avoid taking on Mission Impossible, i.e. the log of zero. We have also had to make another adjustment to the data because we are trying to fit a Continuous Distribution to Discrete values – we're offsetting the Queue Length by one half to the left. (*This may seem quite random but is entirely logical.*)

For the Formula-phobes: Offsetting Discrete Movements in a Continuous Distribution

When we are using a Continuous Probability Distribution to model a Discrete Variable, the CDF is measuring the cumulative probability between consecutive integer values.

On the other hand, the PDF is measuring the relative density at each discrete value. To equate this to a tangible probability we are implying that it is the area of the trapezium or rectangle that straddles the integer value. In theory, this is always half a unit too great, so we have to adjust it downwards.

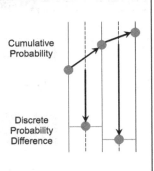

On running the Regression without these zero values we get the statistically significant result in Table 7.15.

From the earlier formula-phile summary of the Gamma Distribution we can translate the Regression Coefficients to get the Gamma's Shape and Scale Parameters as summarised in Table 7.16.

Least Squares Nonlinear Curve Fitting without the logs | 403

Table 7.14 Regression Input Data Preparation highlighting Terms to be Omitted

Queue Length	Frequency	Ln (Frequency)	Unity	Ln(Queue)	OffsetQueue	Note
1	3	1.0986	1	-0.6931	0.50	
2	9	2.1972	1	0.4055	1.50	
3	12	2.4849	1	0.9163	2.50	
4	13	2.5649	1	1.2528	3.50	
5	13	2.5649	1	1.5041	4.50	
6	11	2.3979	1	1.7047	5.50	
7	9	2.1972	1	1.8718	6.50	
8	8	2.0794	1	2.0149	7.50	
9	5	1.6094	1	2.1401	8.50	
10	5	1.6094	1	2.2513	9.50	
11	3	1.0986	1	2.3514	10.50	
12	3	1.0986	1	2.4423	11.50	
13	1	0	1	2.5257	12.50	
14	2	0.6931	1	2.6027	13.50	
15	1	0	1	2.6741	14.50	
16	0	#NUM!	2	2.7408	15.50	Omit
17	1	0	1	2.8034	16.50	
18	0	#NUM!	2	2.8622	17.50	Omit
19	0	#NUM!	3	2.9178	18.50	Omit
20	1	0	1	2.9704	19.50	

Table 7.15 Regression Output Data for Queue Length Data Modelled as a Gamma Function

SUMMARY OUTPUT

Regression Statistics	
Multiple R	0.933729178
R Square	0.871850177
Adjusted R Square	0.853543059
Standard Error	0.3744644
Observations	17

Based on R-Square, F and t-statistics we would accept this model as a statistically significant result.

ANOVA

	df	SS	MS	F	Significance F
Regression	2	13.35589379	6.677946895	47.62356353	5.67579E-07
Residual	14	1.963130215	0.140223587		
Total	16	15.31902401			

	Coefficients	Standard Error	t Stat	P-value	Lower 95%	Upper 95%
Intercept	2.188145439	0.206448532	10.59898766	4.52323E-08	1.745357377	2.630933502
Ln(Queue)	1.000699222	0.220765338	4.532863858	0.000468788	0.527204664	1.47419378
Queue	-0.304811702	0.03861967	-7.89265428	1.6021E-06	-0.387642656	-0.221980749

However, we would probably agree that as estimates of the parameter values these Regression Results are disappointingly different to those values derived using the Least Squares Error technique on the CDF in Table 7.13.

The problem is that Regression is assuming (quite rightly in one sense) that Queue Lengths of 16, 18 and 19 are just missing values. However, the adjacent, somewhat

404 | Least Squares Nonlinear Curve Fitting without the logs

Table 7.16 Interpretation of Regression Coefficients as Gamma Function Parameters

Coefficient of:	Coefficient Value	Equal to:	Parameter Values
Ln(Queue)	1.00699222	$\alpha - 1$	$\alpha = 2.00699222$
Queue	-0.304811702	$-\dfrac{1}{\beta}$	$\beta = 3.28071394$
Intercept	2.188145439	$\ln\left(\dfrac{1}{\beta^{\alpha}\Gamma(\alpha)}\right)$	$\ln\left(\dfrac{1}{\beta^{\alpha}\Gamma(\alpha)}\right) = 96.1004559$

proportionately inflated unity values at 17 and 20 are artificially driving a higher degree of positive skewness than is rightly due – the Regression is 'tilted' a little in their favour.

In order to clarify what is happening here, let's be a little pliable ourselves in how we use the data; let's re-run both techniques using only the observed data for Queue Lengths of 1.15, thus ensuring a contiguous range (i.e. no missing values) in both techniques.

In Table 7.17 we have still used the full 100 observations to derive the Adjusted Observed Cum%, but have only used the first 15 queue lengths on which to minimise the Least Squares Error. This gives us a revised Gamma Distribution with shape and scale parameters 2.44 and 2.39. (They will have changed slightly in the third decimal place because we have excluded the tail-end errors.)

Table 7.18 shows the regression results for this reduced set of observations. The coefficients are then mapped to the Gamma Shape and Scale Parameters in Table 7.19, with values of 2.41 and 2.40, which are similar to those generated by the CDF Solver technique.

The implication we can draw from this is that with very large sample sizes we can probably expect to get very compatible results using the two techniques, but with small sample sizes the Regression technique can be adversely affected by disproportionate or unrepresentative random events.

The reason why the two techniques do not give identical results is down to differences in how they treat the random errors or residuals; the Cumulative Technique inherently compensates for random error and hence is minimising a smaller set of values. We can observe this by comparing the minimised Solver SSE of 0.0007 in Table 7.17 with the Residual Sum of Squares of 0.6213 in Table 7.18. Also, the log transformation then forces the line of best fit through the Geometric Mean, not the Arithmetic Mean.

Finally, we may recall from Section 7.1 that any Linear Regression that involves a logarithmic transformation passes through the Arithmetic Mean of the transformed Log

Table 7.17 Solver Results for Fitting a Gamma Distribution Using Restricted Queue Length Data

Solver Variables for Gamma Distribution		alpha	2.4444
		beta	2.3857
		Start	0.00

	Cars Queuing at Traffic Lights	Frequency of Occurrence	Running Total ≤ Queue Length	Adjusted Observed Cum %	Modelled Beta Distribution	Delta % (Observed – Model)
Model Minimum >	0				0.0%	
	1	3	3	3.0%	2.8%	0.1%
	2	9	12	11.9%	11.7%	0.2%
	3	12	24	23.8%	23.8%	-0.1%
	4	13	37	36.6%	36.9%	-0.3%
	5	13	50	49.5%	49.3%	0.2%
	6	11	61	60.4%	60.2%	0.2%
	7	9	70	69.3%	69.4%	-0.1%
Observed Data	8	8	78	77.2%	76.8%	0.5%
	9	5	83	82.2%	82.6%	-0.4%
	10	5	88	87.1%	87.1%	0.0%
	11	3	91	90.1%	90.5%	-0.4%
	12	3	94	93.1%	93.1%	0.0%
	13	1	95	94.1%	95.0%	-0.9%
	14	2	97	96.0%	96.4%	-0.4%
	15	1	98	97.0%	97.4%	-0.4%
	16	0	98	97.0%	98.2%	
	17	1	99	98.0%	98.7%	
	18	0	99	98.0%	99.1%	Excluded
	19	0	99	98.0%	99.4%	
	20	1	100	99.0%	99.5%	
Quarter-way Smoothing Point >	21.05				99.7%	
Mid-way Smoothing Point >	22.10				99.8%	
Model Pessimistic >	24.21				99.9%	

Sum		100		Weighted Sum (Tgt = 0)	0.0%
Median	5.50				
Mean	6.24			Weighted Mean (Tgt = 0)	0.0%
				Weighted Sum of Squares Error (SSE) >	0.0007

Table 7.18 Revised Regression Output Data for Queue Length Data Modelled as a Gamma Function

SUMMARY OUTPUT

Regression Statistics	
Multiple R	0.971123354
R Square	0.943080569
Adjusted R Square	0.933593997
Standard Error	0.227543768
Observations	15

Based on R-Square, F and t-statistics we would accept this model as a statistically significant result.

ANOVA

	df	SS	MS	F	Significance F
Regression	2	10.29436085	5.147180423	99.4121575	3.40066E-08
Residual	12	0.621313999	0.051776167		
Total	14	10.91567484			

	Coefficients	Standard Error	t Stat	P-value	Lower 95%	Upper 95%
Intercept	2.27222829	0.126795818	17.92037251	5.00558E-10	1.995963936	2.548492645
Ln(Queue)	1.407865717	0.157764541	8.923841235	1.20765E-06	1.064126311	1.751605124
Queue	-0.417276198	0.03305062	-12.62536683	2.74024E-08	-0.489287313	-0.345265084

406 | Least Squares Nonlinear Curve Fitting without the logs

Table 7.19 Interpretation of the Revised Regression Coefficients as Gamma Function Parameters

Coefficient of:	Coefficient Value	Equal to:	Parameter Values
Ln(Queue)	1.407865717	$\alpha - 1$	$\alpha = 2.407865717$
Queue	-0.417276198	$-\dfrac{1}{\beta}$	$\beta = 2.39649423$
Intercept	2.27222829	$\ln\left(\dfrac{1}{\beta^{\alpha}\Gamma(\alpha)}\right)$	$\ln\left(\dfrac{1}{\beta^{\alpha}\Gamma(\alpha)}\right) = 99.3580893$

data which is the Geometric Mean of the raw data. If we want to force the regression through the Arithmetic Mean of the raw data then we can always set up a model to do that and use Solver to find the slope and intercept. That said, we would then have to re-create all the measures for goodness of fit long-hand in order to verify that the best fit is indeed a good fit. We must ask ourselves in the context of being estimators '*Is it worth the difference it will make. Anything we do will only be approximately right regardless.*'

7.2 Chapter review

Wow, perhaps not the easiest technique, but if followed through logically one step at a time it is possible to fit nonlinear, untransformable curves to our data, exploiting the principles of Least Squares Regression, and the flexibility of Microsoft Excel's Solver.

On a note of caution, we may find that our data may not meet the desired attribute of homoscedasticity, but if we think that the fit is good, and has good predictive potential (perhaps bounded), and we have no other realistic alternative, then the technique is a good fallback.

Reference

Orwell, G (1945) *Animal Farm*, London, Secker and Warburg.

8 The ups and downs of Time Series Analysis

Despite the musing attributed to Albert Einstein, time is an important consideration for estimators. Sometimes it will be on our side and other times, it will seem to be against us – especially when we have a deadline to meet. It is certainly something we need to consider when it comes to issues of data normalisation, as we discussed in Volume I Chapter 6.

> ### A word (or two) from the wise?
>
> *"The only reason for time is so that everything doesn't happen at once."*
> Often attributed to **Albert Einstein**
> (1879–1955)
> Theoretical physicist

There are many instances where the time of year can be seen as a driver of cost or demand on a service or system. There are other occasions where it may be seen to be a useful predicator of some other value rather than be a true driver in itself. Energy consumption is a classic example in those countries where heating and lighting costs traditionally rise in the winter months in comparison with summer when they fall. Such seasonal variations become largely predictable.

Employees who are fortunate enough to receive annual bonus payments on top of their salary are often likely to receive them in relation to their organisation's accounting calendar, which may be geared to the calendar year or to the tax year. Aficionados of economic statistics published by governments are likely to see 'spikes' in unadjusted indices on average earnings as illustrated in Figure 8.1; these spikes correspond to the end of the UK Tax Year in March *(Well, 5th April if you want to be picky!)* and the Gregorian calendar year in December. The dip in summer months can be rationalised by reductions in overtime attributable to higher levels of holidays taken. Following the global financial crisis and ensuing recession, we can see that the UK average earnings entered a new phase or cycle with a slowdown in the underlying increasing trend (Source Data from the Office of National Statistics, 2009).

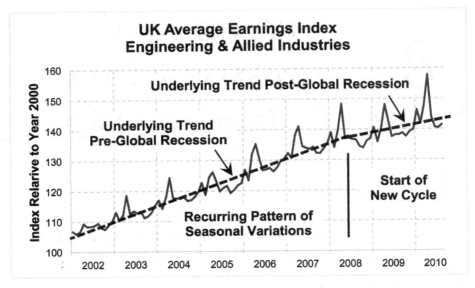

Figure 8.1 Example of a Time Series with Seasonal Variation and Beginnings of a Cyclical Impact
© Crown Copyright 2010. Adapted from data from the Office for National Statistics licensed under the Open Government Licence v.3.0.

8.1 The bits and bats ... and buts of a Time Series

A Time Series can be considered to be a series of values of a variable obtained at successive, often equal, time intervals. The implication of this definition is that the time is not necessarily constant but it does follow a repeating and predictable pattern. For instance, it is not unusual for statement of accounts, project performance etc. to be published on a monthly basis. The twelve calendar months are not equal but they are predictable (*even with the irritation of leap years.*) Some finance departments often prefer to report in multiples of completed weeks within the month. As a consequence, a rolling three-month period or quarter would always comprise of 13 weeks, divided into two months of four weeks and one month of five weeks. Therefore, in a nominally steady-state environment, every third month we might reasonably expect the reported costs to be some 25% higher than in the preceding two months.

More importantly, where we have a repeating pattern of behaviour, in which the values observed vary from the underlying trend in a manner that depends on their relative position in the repeating pattern, then we can analyse this behaviour and use it to forecast future values. This procedure by which we deconstruct and dissect the time series data into the elements that define its repeating pattern and underlying trend is called Time Series Analysis.

The ups and downs of Time Series Analysis | 409

Definition 8.1   Time Series Analysis

Time Series Analysis is the procedure whereby a series of values obtained at successive time intervals is separated into its constituent elements that describe and calibrate a repeating pattern of behaviour over time in relation to an underlying trend.

We would usually be doing this analysis with the intent of predicting a future set of values.

Note: Time here does not have to be taken as a calendar date, and can be used to infer an elapsed time relative to the start or end point of some natural cycle like a Project Start Date.

A typical Time Series can be defined in terms of four component elements, T, S, C and R:

1 T, Underlying Trend: The basic underlying movement in the data as a whole
2 S, Seasonal Variation: The level of fluctuation up or down from the underlying trend, the timing of which can be attributed to the relative position in a repeating pattern of a fixed duration such as a year
3 C, Cyclical Variation: The level of undulation up or down from the underlying trend, the timing of which can be attributed to the relative position in a longer repeating cycle of fixed duration such as a year
4 R, Residual Variation: The unexplained difference between the observed data and the theoretical value implied by the underlying trend and adjustments accorded by the Seasonal and Cyclical Variation Factors

In many practical circumstances (*apart from a few occasions, some might say rare, unless it was their personal experience otherwise*) the third element (cyclical variation) is often not present, or is ignored as being relatively minor. The Cyclical Variation is often used to explain a more gentle or general 'sway' in the data, synonymous perhaps with more global economic conditions such as prosperity, recession, depression and recovery. However, the term can equally be used to reflect the different phases of a project life cycle such as conceptualisation, development, implementation, use, and retirement.

For many practical purposes, only the underlying trend, seasonal variation and residual variation need to be considered. For simplicity, we will adopt that approach here during our discussions, but before those of us who work in industries where cyclical variation is important think it's time to go off in a melodramatic huff, the principles we will discuss here for seasonality apply equally to cyclicality.

An example of a Time Series that many of us may relate to is that of domestic energy consumption such as gas. Figure 8.2 illustrates the three components of T, S and R in relation to the gas consumed in a household (*mine as it happens*) over a four-year period for heating, hot water and cooking. (*Yes, I use these techniques at home too; being an estimator can be a lifestyle choice, not just a career option!*) In this case, we can see that there are four seasonal elements, which we can nominally refer to as Winter (Jan–Mar), Spring (Apr–Jun), Summer (Jul–Sep) and Autumn (Oct–Dec). Here we are using date as a predicator variable not a cost driver in the absolute sense as the values refer to cumulative units billed at the end of the time period or season stated.

Any residual variation is that which cannot be explained by the underlying trend or seasonal variation. This could be due to one or more of a number of factors, including:

- Variation in annual weather conditions, ambient temperatures, etc.
- Some records of units consumed are estimated values by the gas supply company where meter readings were not available
- Differences in the dates between the nominally quarterly meter readings or record dates (i.e. there may not be a perfect repeating pattern of days between reading)

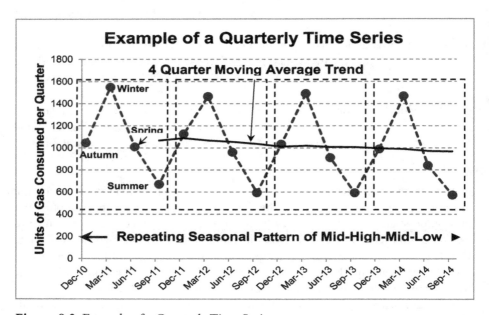

Figure 8.2 Example of a Quarterly Time Series

The ups and downs of Time Series Analysis | 411

In analysing this data into its constituent elements, we have to decide what type of model we are going to assume.

8.1.1 Conducting a Time Series Analysis

The procedure to perform a Time Series Analysis is basically the same regardless of the technique and model adopted. We can summarise it in six basic steps:

1. Determine the number of unique seasons
2. Determine the type of model (*You didn't expect there to be only one, did you?*)
3. Determine the underlying trend (i.e. increasing, decreasing or steady state; linear or non-linear, etc.)
4. Determine the seasonal variation values (using either a Classical Decomposition or Multi-Linear Regression technique, or Microsoft Excel Solver)
5. Apply the seasonal variations to the underlying trend in the model
6. Create a forward forecast by extrapolating the pattern (*on the assumption that we're not just doing this for the fun of it*).

In many cases the number of seasons we should be considering will be implied through common sense, and may be dictated by the natural cycle of data over a rolling period of time, e.g.

- 12 months in a year
- 4 quarters in a year
- 3 months in a quarter
- 52 weeks in a year
- 7 days in a week

In other cases, we may have to detect it using a numerical technique. By far the simplest technique (and the only ones we will consider here) is the simple Moving Average or Moving Median (see Chapter 3) in which we would choose the Moving Average Interval or base that gives us the smoothest trend.

8.2 Alternative Time Series Models

As with virtually all estimating and forecasting problems, there is more than one method (approach and technique) we can use, not least of which is the fundamental choice of model that expresses how the dynamics of the underlying trend and seasonal variations interact. There are two fundamental Time Series Models in common usage; their names describe how their constituent elements interact:

412 | The ups and downs of Time Series Analysis

i.   Additive/Subtractive Model
ii. Multiplicative Model

8.2.1 Additive/Subtractive Time Series Model

This is the simpler of the two models and is probably the more commonly used as a consequence, often as a matter of convenience rather than necessarily an ignorance of the alternative:

> **Definition 8.2 Additive/Subtractive Time Series Model**
>
> The Predicted Value is a function of the forecast value attributable to the underlying Trend plus or minus adjustments for its relative Seasonal and Cyclical positions in time.

For the Formula-philes: Additive/Subtractive Seasonal Variation with linear trend

For an Additive/Subtractive Time Series with Trend, T, Seasonal Variation, S, Cyclical Variation, C and Residual Variation, R
The Projected Model value, P can be
expressed as
$$P = T + S + C$$
The Observed or Actual Value,
A can be expressed as
$$A = T + S + C + R$$

Let's look at this in relation to our example of units of gas consumed by quarter. (*Note that we are not doing this in relation to the cost of gas in order to eliminate the vagaries of gas pricing models – that would open up a whole new can of worms.*) In Figure 8.3 we have uncluttered our previous picture by removing the actual data so that we can concentrate on the principles of the model. (*Why let the facts get in the way of a good story?*) Here we can see that our model reflects that winter consumption is raised, whilst in summer it is reduced relative to both spring and autumn. In essence we could draw four parallel straight lines through our data to represent the trend for each of the four seasons. This gives us our Additive/Subtractive Model in which to move between the lines we simply add or subtract a constant. The underlying trend is simply the weighted average of these

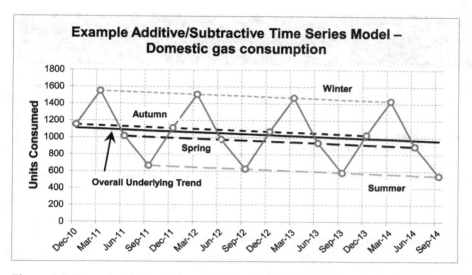

Figure 8.3 Example Additive/Subtractive Time Series Model – Domestic Gas Consumption

four parallel seasonal trends, but in fact any reference line parallel to the four seasons will suffice. (We will return to this particular observation later.)

8.2.2 Multiplicative Time Series Model

As an alternative, where the trend is observed to be more exponential than linear in nature, we might want to consider a Multiplicative Model instead:

> **Definition 8.3 Multiplicative Time Series Model**
>
> The Predicted Value is a function of the forecast value attributable to the underlying trend multiplied by appropriate seasonal and cyclical factors.

Let's consider it in relation to our gas consumption data. In Figure 8.4 we have replaced the linear trend with an exponential one but still looking at the units of gas consumed by quarter. Over the four-year period we can draw four other exponential lines, one for each season. This gives us our Multiplicative Model whereby to move between the lines we simply factor by a constant greater than or less than one. In this particular case the effect of the exponential reduction is very subtle, being

414 | The ups and downs of Time Series Analysis

> ### For the Formula-philes: Multiplicative Seasonal Variation with linear trend
>
> For a Multiplicative Time Series with Trend, T, Seasonal Variation, S, Cyclical Variation, C and Residual Variation, R
> The Projected Model value, P, can be expressed as
>
> $$P = T \times S \times C$$
>
> The Observed or Actual value, A, can be expressed as
>
> $$A = T \times S \times C + R$$

at a rate of 1% per quarter. A visual comparison between Figures 8.3 and 8.4 shows that the four seasons in the former are funnelling in rather than running in parallel. (This is more obvious when we look at the outer seasonal values of summer and winter values.)

The choice of model we use should be influenced by the nature of the data plus a dollop of common sense. In this particular case, whilst short term extrapolation could be performed using the simpler Additive/Subtractive Model, for longer term predictions we ought to be considering the Multiplicative Model as the implication of a sustained linear reduction would be that in future we could be consuming negative units of gas, which is clearly nonsensical (*but quite appealing, it must be said.*) If we had been

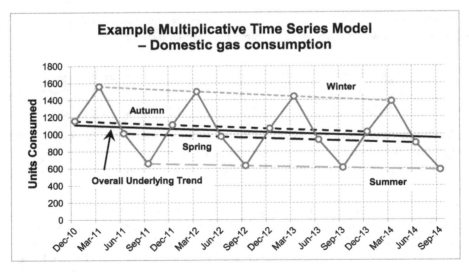

Figure 8.4 Example Multiplicative Time Series Model – Domestic Gas Consumption

The ups and downs of Time Series Analysis | 415

considering electricity net consumption, then this may have been acceptable where a user is generating electricity via renewable methods such as solar panels and wind turbines.

In order to fulfil our analysis, there are three basic techniques we are going to consider:

- Classical Decomposition Technique (CDT)
- Multi-Linear Regression (MLR)
- Microsoft Excel Solver

The three techniques can all be used in conjunction with either Additive/Subtractive or Multiplicative Time Series Models, and with linear or exponential trends (and to some extent meandering ones as well.) Initially we will consider the Classical Decomposition of a Time Series into its component parts.

8.3 Classical Decomposition: Determining the underlying trend

On the face of it, we might think that this is the easy bit, but in truth it is the most difficult part of this technique or procedure, but it is key to our getting the right seasonal adjustments.

Based on our discussion in Chapters 2, 3, 4 and 6 we might think that we have a number of choices to determine an underlying linear trend:

i.   Linear Regression (Chapter 4)
ii.   Offset Moving Average or Median (Chapter 3)
iii. Offset Cumulative Average (Chapter 3)
iv.   Disaggregated Cumulative Quadratic (Chapter 2)

Or, an underlying exponential trend:

i.   Transformed Exponential Regression (Chapter 6)
ii.   Offset Geometric Mean (Chapter 3)
iii. Offset Cumulative Geometric Mean (Chapter 2)

Also, we might think that these techniques would be reasonably consistent with each other, within each trend type

Disappointingly, we would be wrong.

Let's start by throwing the proverbial cat in amongst the pigeons and dispel a myth or two. Regrettably, there are numerous textbooks that advise using Linear Regression to determine the underlying linear trend, and whilst this may seem an intuitive approach, it is fundamentally flawed, as we will discover. It may work satisfactorily in some circumstances but there are others where it is not an appropriate.

(Note: No real pigeons have been harmed in the introduction to this technique.)

In order to demonstrate the issues with which we will have to contend, let's begin by considering a perfect Additive/Subtractive Time Series i.e. one in which all our observed data points sit on a predetermined model and where the Residual Variations are all zero, as is the case with Figure 8.5 which has been constructed from the data in Table 8.1:

- Decreasing linear trend of 10 units per period and an intercept of 710
- Four Recurring Seasonal Variations of +100, +300, -100, -300 relative to the underlying linear trend

Let's bear this model in mind over the next few sections and discussion points.

8.3.1 See-Saw... Regression flaw?

If, instead of constructing the data in this manner, we had just stumbled across the Observed Data in the right-hand column, our natural inclination might be to calculate the underlying linear trend using Linear Regression (or at least plotted the data in Microsoft Excel

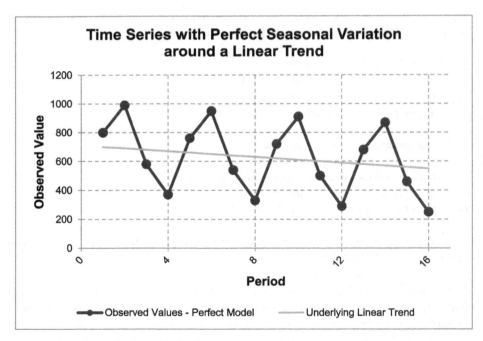

Figure 8.5 Time Series with Perfect Seasonal Variation Around a Linear Trend

The ups and downs of Time Series Analysis | 417

Table 8.1 Time Series with Perfect Seasonal Variation Around a Linear Trend

Data Points	Sequential Period No	Recurring Season No	True Underlying Trend	True Seasonal Variation	Observed Value (Perfect Model)
1	1	1	700	100	800
2	2	2	690	300	990
3	3	3	680	-100	580
4	4	4	670	-300	370
5	5	1	660	100	760
6	6	2	650	300	950
7	7	3	640	-100	540
8	8	4	630	-300	330
9	9	1	620	100	720
10	10	2	610	300	910
11	11	3	600	-100	500
12	12	4	590	-300	290
13	13	1	580	100	680
14	14	2	570	300	870
15	15	3	560	-100	460
16	16	4	550	-300	250

and created a Linear Trendline using the graphical utility.) This would give us the graph in Figure 8.6, with a somewhat steeper line than we started with in Figure 8.5.

On closer inspection, we will notice that for the first two seasons, the seasonal data is diverging from the Regression Lines, whereas the third and fourth seasons are converging towards it. In short, the Regression Line is not truly representative of the pattern in the data!

Before we do anything rash like throwing this book in the bin or start thinking about a career change, let's try to understand what is going on here.

If we dropped the first period on the left with a positive seasonal variation and re-calculated the linear trend we would get a slightly steeper slope. If instead we added an extra period on the right, following the same pattern, this would also have a positive seasonal variation which would then cause the underlying best fit straight line to be shallower in slope. If we did both, in effect rolling the whole data pattern forward by one period then we might expect that we would get a compensatory effect. Well, it does to a degree, but not totally; Figure 8.7 illustrates what we get.

The reason for the change is that we have substituted a value close to and above the trendline on the left with one on the right and below, creating a tilting effect on the trendline. The position of the data average has also moved down the trendline to the right by one unit. The Sum of Squares Error is then recalculated giving us a different albeit

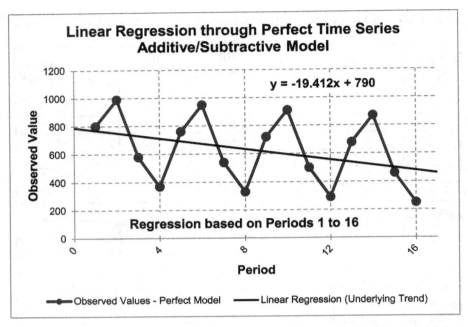

Figure 8.6 Linear Regression Through Perfect Time Series Additive/Subtractive Model

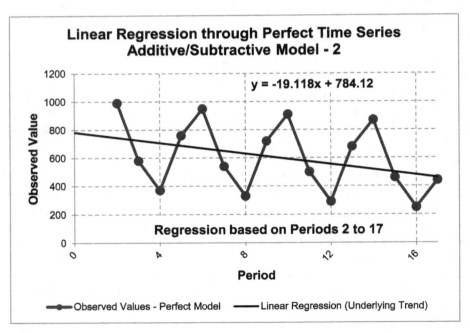

Figure 8.7 Linear Regression Through Perfect Time Series Additive/Subtractive Model – 2

The ups and downs of Time Series Analysis | 419

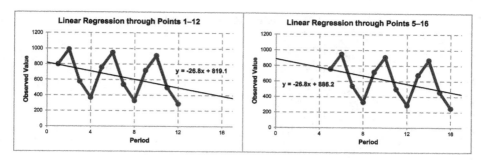

Figure 8.8 Time Series Linear Regression Trends are Unreliable – 1

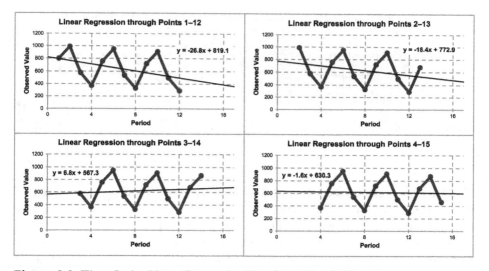

Figure 8.9 Time Series Linear Regression Trends are Unreliable – 2

similar result ... and that's all using this 'Perfect World' Simulated Time Series data – we can't even blame it on a poor set of actuals!

However, as this is a perfect set of data from a Time Series point of view, we should be able to draw the same conclusions for trend and seasonal variations if we were to drop or add a complete set of four seasons ... or so we might think. (*Or would we, now that the seeds of doubt have been sown?*)

Figure 8.8 illustrates that the problem does not go away if we roll the regression by an entire pattern of four seasons and compare the results of a Regression of Data Points 1–12 with one for Data Points 5–16 from Table 8.1. This time we get the same slope but a different intercept, but both are different than we had for Data Points 1–16! (*Don't you just hate it when things you thought you could rely on, let you down?*)

It gets worse if we roll the data one period at a time, as illustrated in Figure 8.9 ... a veritable see-saw of trendlines!

420 | The ups and downs of Time Series Analysis

Table 8.2 Range of Regression Results for Slope and Intercept

Data Points	Intercept	Slope
1 to 16	790.0	-19.4
1 to 12	819.1	-26.8
2 to 13	772.9	-18.4
3 to 14	567.3	6.8
4 to 15	630.3	-1.6
Average	697.4	-10.0
5 to 16	886.2	-26.8

Table 8.3 Linear Regression Lines of Best Fit for Each Season

First season		Second season		Third season		Fourth season			
Data Points	Observed Value (Perfect Model)	Data Points	Observed Value (Perfect Model)	Data Points	Observed Value (Perfect Model)	Data Points	Observed Value (Perfect Model)		
1	800	2	990	3	580	4	370		
5	760	6	950	7	540	8	330		
9	720	10	910	11	500	12	290		
13	680	14	870	15	460	16	250		
									Average
Slope	-10		-10		-10		-10		-10
Intercept	810		1010		610		410		710

Table 8.2 summarises the values calculated for slope and intercept. At least if we take the average of the results for Data Points 1–12, 2–13, 3–14 and 4–15, we get the correct average for the slope ... but not the intercept, which should be 710.

However, if we consider the linear trend on each seasonal element, and compare them we get a more satisfying result. Using every fourth data point beginning with the first, second, third and fourth data point from Table 8.1, we can get Lines of Best Fit for each season. The results are shown in Table 8.3. The slope of each line in this perfect world scenario is the same and matches that of the true underlying trend that we used to construct the data initially. The intercepts, however, are all different but the average of the four intercepts equals that of the true underlying trend. (*Don't you love it when a plan finally comes together, even if we have to go into somewhat of a recovery mode?*)

Later we'll look at whether this technique works with real data and not just hypothetical 'perfect world' data.

8.3.2 Moving Average Seasonal Smoothing

If instead we simply took a Moving Average of our data, then in this particular case we would get a steady state linear trend as depicted in Figure 8.10 and Table 8.4. As we

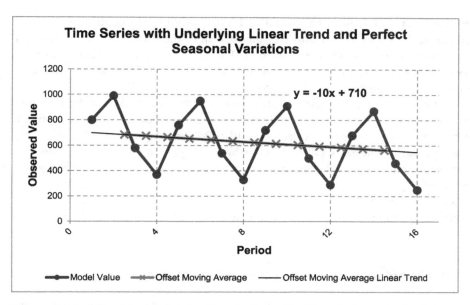

Figure 8.10 Offset Moving Average Trend of a Perfect Time Series

Table 8.4 Offset Moving Average Trend of a Perfect Time Series

Data Points	Sequential Period Number	Recurring Season Number	True Underlying Trend	True Seasonal Variation	Observed Value (Perfect Model)	Moving Average Offset Qtr	Offset Moving Average
1	1	1	700	100	800		
2	2	2	690	300	990		
3	3	3	680	-100	580		
4	4	4	670	-300	370	2.5	685
5	5	1	660	100	760	3.5	675
6	6	2	650	300	950	4.5	665
7	7	3	640	-100	540	5.5	655
8	8	4	630	-300	330	6.5	645
9	9	1	620	100	720	7.5	635
10	10	2	610	300	910	8.5	625
11	11	3	600	-100	500	9.5	615
12	12	4	590	-300	290	10.5	605
13	13	1	580	100	680	11.5	595
14	14	2	570	300	870	12.5	585
15	15	3	560	-100	460	13.5	575
16	16	4	550	-300	250	14.5	565

discussed in Chapter 3, a Simple Moving Average always lags the true location of the data by (n-1)/2 periods where n is the Moving Average Base. In this case, based on a

422 | The ups and downs of Time Series Analysis

Moving Average of 4, the offset lag will be 1.5 periods. (Incidentally, the easiest way of calculating the Offset Period is to take the Median of the Range of Period Numbers using Microsoft Excel's **MEDIAN** function. (We can also use the **AVERAGE** function as we are dealing with consecutive equally spaced values. They will return the same value because Integers are uniformly distributed.)

In this case, it doesn't matter whether we roll the data backwards or forwards, or take a contiguous subset of the data we will get a consistent underlying trend. Furthermore, if we were to project the Offset Moving Average trend back to the Intercept, we would get a value of 710, which conveniently matches the value we assumed in setting up this hypothetical case. (*Don't you love it when a plan finally does come together?*)

Sorry to spoil the party atmosphere, but before we celebrate too much, we need to review whether this will work with real data. *Hold that thought, we'll get back to it later.*

8.3.3 Cumulative Average Seasonal Smoothing

In Chapter 3 we explored how we can use the Cumulative Average as a Smoothing technique that returns the Line of Best Fit slope that is half that of the underlying line of Best Fit through the raw 'unaveraged' data. To overcome this, we can look to plot the data at the midpoint or median of the range, just as we did for Moving Averages, except this time the range is cumulative. Table 8.5 illustrates the results.

Table 8.5 Cumulative Average Trend of a Perfect Time Series

Data Points	Sequential Period Number	Recurring Season Number	True Underlying Trend	True Seasonal Variation	Observed Value (Perfect Model)	Cumulative Average Offset Qtr	Cumulative Average
1	1	1	700	100	800		800
2	2	2	690	300	990		895
3	3	3	680	-100	580		790
4	4	4	670	-300	370	2.5	685
5	5	1	660	100	760	3.0	700
6	6	2	650	300	950	3.5	742
7	7	3	640	-100	540	4.0	713
8	8	4	630	-300	330	4.5	665
9	9	1	620	100	720	5.0	671
10	10	2	610	300	910	5.5	695
11	11	3	600	-100	500	6.0	677
12	12	4	590	-300	290	6.5	645
13	13	1	580	100	680	7.0	648
14	14	2	570	300	870	7.5	664
15	15	3	560	-100	460	8.0	650
16	16	4	550	-300	250	8.5	625

In Figure 8.11 (left hand graph) we can compare the Cumulative Average with our 'Perfect World' unadjusted Seasonal Data. The initial values where the Cumulative Average is still looking to find its 'basic steady state trend', is somewhat erratic, but this is also due to the absence of a full indicative pattern of seasons. If we only look at values from where we have completed the first full set of seasons (right hand graph of Figure 8.11), we get a much tighter scatter of values around the Best Fit Straight Line through them. However, the values of the slope and intercept are still influenced by the dampened seasonal effects, as illustrated by the regular pattern of the scatter around the line.

Unfortunately, this still doesn't match our true underlying trend line that we used to construct this data in the first place. Before we get too dejected, let's look at the Cumulative Average for completed seasonal sets only (e.g. after 4, 8, 12 and 16 weeks) as shown in the left-hand graph of Figure 8.12. Here we have a perfect linear trend for the Cumulative Average, but as we know from Chapter 2, the slope of a Cumulative Average Line is half that of the true line and that the Intercept

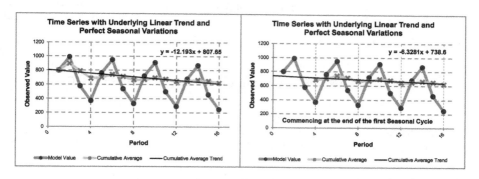

Figure 8.11 Cumulative Average Smoothing of a Perfect Time Series – 1

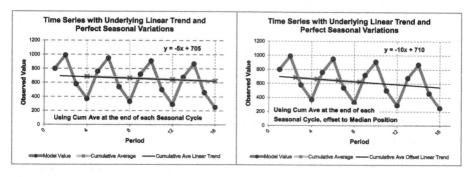

Figure 8.12 Cumulative Average Smoothing of a Perfect Time Series – 2

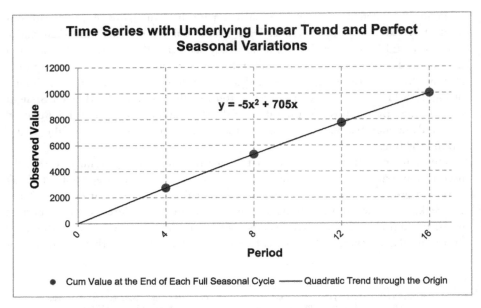

Figure 8.13 Cumulative Smoothing of a Perfect Time Series

also needs to be adjusted by subtracting the numerical value of the Cumulative Average Slope:

Cumulative Average: $y = -5x + 705$
True underlying trend: $y = -10x + 710$

We can get this by offsetting the Cumulative Average value to the left to the mid-point of the time period to which it relates as shown in the right-hand graph of Figure 8.12. A linear trend through this offset data gives us the true underlying trendline from which we started. (*another dose of mathe-magic perhaps?*)

Perhaps it may not come as a surprise then that if we were to plot every fourth Cumulative value and fit a Quadratic Trendline (Polynomial of Order 2) through the data, we would get a coefficient for the squared term equivalent to half the true underlying linear trend (as discussed in Chapter 2.) Also, the intercept of the true trend is equal to the coefficient for the non-squared term less the coefficient of the squared term, as illustrated in Figure 8.13.

8.3.4 What happens when our world is not perfect? Do any of these trends work?

Let's now return to our gas consumption example and see whether any of the Average of the Four Seasonal Linear Regression trends, Offset Moving Average or Offset Cumulative Average Techniques work satisfactorily for us.

In Table 8.6 we have extracted the data from Figure 8.2 and arranged it into the four seasons. (*Sounds like a cue for some music – for those partial to a bit of Vivaldi or even Frankie Valli.*) The four seasonal trends can be calculated using the **SLOPE** and **INTERCEPT** functions in Microsoft Excel using the relevant 'Observed Values' and 'Data Points' as the *known_y's* and *known_x's* respectively. Figure 8.14 depicts the results of the overall average.

In Table 8.7 we have calculated the Offset Moving Average and the Offset Annual Cumulative Average values for our data. This includes the equivalent Offset Quarter Number for both the Moving Average and Cumulative Average from which we can calculate the respective slope and intercept of the underlying trend.

Table 8.6 Underlying Linear Trend as the Average of the Individual Seasonal Trends

First season		Second season		Third season		Fourth season		
Data Points	Observed Value	Data Points	Observed Value	Data Points	Observed Value	Data Points	Observed Value	
1	1046	2	1549	3	1009	4	670	
5	1127	6	1464	7	961	8	594	
9	1035	10	1491	11	912	12	592	
13	990	14	1468	15	840	16	572	
								Average
Slope	-6.5		-5.4		-13.9		-7.4	-8.3
Intercept	1095		1536.2		1055.6		681	1091.95

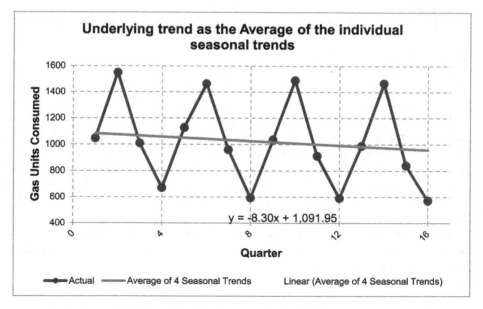

Figure 8.14 Underlying Trend as the Average of the Individual Seasonal Trends

426 | The ups and downs of Time Series Analysis

Table 8.7 Example of Domestic Gas Consumption – Underlying Linear Trend

Date	Quarter Number (Underlying Trend)	Recurring Quarter (Season)	Actual Gas Units Consumed	Cumulative Average	Moving Average Offset Qtr	Moving Average	Cumulative Average Offset Qtr	Last Season Cum Ave
Dec-10	1	1	1046	1046				
Mar-11	2	2	1549	1298				
Jun-11	3	3	1009	1201				
Sep-11	4	4	670	1069	2.5	1069	2.5	1069
Dec-11	5	1	1127	1080	3.5	1089		
Mar-12	6	2	1464	1144	4.5	1068		
Jun-12	7	3	961	1118	5.5	1056		
Sep-12	8	4	594	1053	6.5	1037	4.5	1053
Dec-12	9	1	1035	1051	7.5	1014		
Mar-13	10	2	1491	1095	8.5	1020		
Jun-13	11	3	912	1078	9.5	1008		
Sep-13	12	4	592	1038	10.5	1008	6.5	1038
Dec-13	13	1	990	1034	11.5	996		
Mar-14	14	2	1468	1065	12.5	991		
Jun-14	15	3	840	1050	13.5	973		
Sep-14	16	4	572	1020	14.5	968	8.5	1020
				Slope		-9.54		-8.03
				Intercept		1103.62		1088.76

- We can calculate the Offset value by taking either the **MEDIAN** or **AVERAGE** of the equivalent Moving Average range or overall Cumulative Average.
- In effect the Moving Average Offset is always 1.5 quarters earlier than the last quarter in the range (Last Quarter − First Quarter)/2
- The Cumulative Average Offset is always half the most recent quarter plus a half

Figures 8.15 and 8.16 depict the results graphically.

If we use all three models for the underlying trend then we can compare what (if any) difference they make. We have done this in Table 8.8 using them to predict our likely consumption over the next two years. Even though we haven't yet looked at the seasonal variation, so long as we take a complete set of one or more years, the sum of the underlying trend values should equal the sum of the seasonal values. Clearly in this case we have a slight difference between one technique (Offset Moving Averages) and the other two techniques, which are relatively similar in value. (*What do they say about 'best laid plans of mice and men'? Pass me some more cheese, please.*) We will also note that a Simple Linear Regression would give us a very bullish (some might say foolhardy) reduction.

Whilst we will probably agree to reject the Simple Linear Regression option in favour of the other for reasons we have already discussed, we are probably now torn between the others thinking:

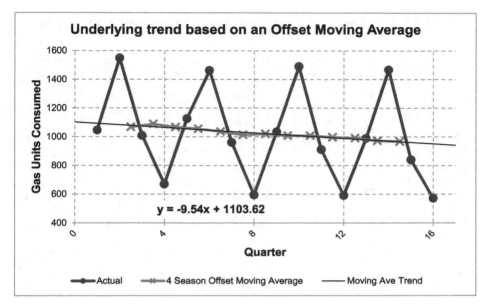

Figure 8.15 Underlying Trend Based on an Offset Moving Average

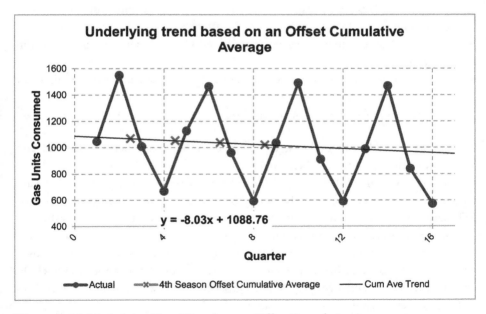

Figure 8.16 Underlying Trend Based on an Offset Cumulative Average

428 | The ups and downs of Time Series Analysis

Table 8.8 Two-Year Forward Forecast Based on Alternative Underlying Linear Trends

Underlying Trend Slope				-8.30	-9.54	-8.03	*-18.93*
Underlying Trend Intercept				1091.95	1103.62	1088.76	*1180.90*
Date	Quarter Number (Underlying Trend)	Actual Average Gas Units		Average of 4 Seasonal Trends	Based on Offset Moving Average	Based on Offset Cum Average	*Based on Simple Linear Regression*
Dec-10 to Sep 14		1020.00		1021.40	1022.52	1020.55	*1020.00*
Dec-14	17			951	941	952	*859*
Mar-15	18			943	932	944	*840*
Jun-15	19			934	922	936	*821*
Sep-15	20			926	913	928	*802*
Dec-15	21			918	903	920	*783*
Mar-16	22			909	894	912	*764*
Jun-16	23			901	884	904	*746*
Sep-16	24			893	875	896	*727*
2015-2016 Total				7374	7264	7394	*6343*

- Surely, it must be better to take either the Average of the four individual seasonal trends or the Offset Cumulative Average as we have similar values using two different techniques?
- Surely, the Offset Moving Average is more reliable as it uses more data points in determining the trend?
- But then again, all three techniques are using the same data points, but in different ways
- Hey, instead let's just give a range estimate!

Perhaps now would be a good time to look at the residual data between the data and each of the underlying trends (Table 8.9).

From this we can see that whilst a Simple Linear Regression gives us the Least Sum of Squares Error (residuals) and a Residuals' Sum of zero as we would expect, the distribution of the annual sums of those errors is the maximum of the four values shown.

Setting this aside and looking at our three proposed techniques we can conclude that in this particular case the Offset Cumulative Average Technique gives us the lowest absolute Sum of Residuals or Errors, but also it gives us the worst Sum of Squares value! On the other hand, the Offset Moving Average gives the largest absolute Sum of Residuals but the smallest equivalent Sum of Squares!

(*Was that "argh!" a cry of pain, or just frustration, or perhaps annoyance?*)

The ups and downs of Time Series Analysis | 429

Table 8.9 Residual Variation of Alternative Underlying Linear Trends Relative to Actual Values

| | Model slope | -8.30 | -9.54 | -8.03 | -18.93 | Residuals | | | |
| | Model intercept | 1091.95 | 1103.62 | 1088.76 | 1180.90 | | | | |
Date	Quarter Number (Underlying Trend)	Actual Gas Units Consumed	Based on Average of 4 Seasonal Trends	Based on Offset Moving Average	Based on Offset Cum Average	Based on Simple Linear Regression	Actual – Average of 4 Seasonal Trends	Actual – Offset Moving Average	Actual – Offset Cum Average	Actual – Simple Linear Regression
Dec-10	1	1046	1084	1094	1081	1162	-38	-48	-35	-116
Mar-11	2	1549	1075	1085	1073	1143	474	464	476	406
Jun-11	3	1009	1067	1075	1065	1124	-58	-66	-56	-115
Sep-11	4	670	1059	1065	1057	1105	-389	-395	-387	-435
Dec-11	5	1127	1050	1056	1049	1086	77	71	78	41
Mar-12	6	1464	1042	1046	1041	1067	422	418	423	397
Jun-12	7	961	1034	1037	1033	1048	-73	-76	-72	-87
Sep-12	8	594	1026	1027	1025	1029	-432	-433	-431	-435
Dec-12	9	1035	1017	1018	1017	1011	18	17	18	24
Mar-13	10	1491	1009	1008	1009	992	482	483	482	499
Jun-13	11	912	1001	999	1000	973	-89	-87	-88	-61
Sep-13	12	592	992	989	992	954	-400	-397	-400	-362
Dec-13	13	990	984	980	984	935	6	10	6	55
Mar-14	14	1468	976	970	976	916	492	498	492	552
Jun-14	15	840	967	961	968	897	-127	-121	-128	-57
Sep-14	16	572	959	951	960	878	-387	-379	-388	-306
Average			1020.0	1021.4	1022.5	1020.6	1020.0			

Sum of Quarters 1–4	-11	-45	-1	-260
Sum of Quarters 5–8	-6	-20	0	-85
Sum of Quarters 9–12	11	16	12	101
Sum of Quarters 13–16	-16	9	-20	244
Sum of Residuals/Errors	-22	-40	-9	0
Sum of Squares Error	1,564,918	1,556,541	1,566,905	1,526,472
Residual Slope	-10.63	-9.39	-10.90	0.00
Residual Intercept	88.95	77.28	92.14	0.00

Perhaps we should just recognise that we have three alternative ways to determine the underlying trend, none of them necessarily any better or worse than another. In the 'Perfect World' situation they all gave the same result, but with actual data with real random variation or errors, they all treat these random variations slightly differently.

Finally, before we move on, some of us may have spotted that in Table 8.9 the average of each linear forecast is slightly different than the average of the observed values, which is why the sum of the residuals is not zero, unlike a Simple Linear Regression through the data. This is simply because the number of data points used in each technique differs in effect! For example, in respect of the Moving Average technique the first data point is only included in the first Moving Average but dropped for the second. Similarly, the second data point is only used in the first two Moving Averages and dropped for the third. The third point is used three times, then dropped but for data points 4 through to 13, each point is used four times, after which the pattern reverses and the last data point is only used in the last Moving Average.

A similar explanation can be made for the Cumulative Average Technique in which each point is used once less than each previous point.

430 | The ups and downs of Time Series Analysis

8.3.5 Exponential trends and seasonal funnels

Having calculated the Offset Moving Average or Offset Cumulative Average, we may come to the conclusion that the underlying trend is not a linear function. Logically over a period of time many estimating relationships or situations in which Time Series are used, an overall decreasing trend may be better described by an Exponential Function. For instance, in the case of domestic gas consumption, if we were to continue the linear trend used in Section 8.3.4 then we would eventually be consuming a negative quantity. (*I like the idea of the gas supplier paying me, but it's not going to happen, is it?*) We cannot necessarily imply the same about an increasing trend.

If we believe that the seasonal variations are still additive/subtractive around the exponential trend then we can still use the Offset Moving Average or Offset Cumulative Average Technique. However, if we observe that the seasonal variations are creating a funnelling effect around the underlying trend, then we should look to modify our technique to compensate accordingly where we see:

- A decreasing trend with the seasonal variations converging on both sides
- An increasing trend with the seasonal variations diverging on both sides

In these cases we really should be using a Moving Geometric Mean and a Cumulative Geometric Mean as these are the logarithmic equivalent to the Moving Average and Cumulative Average (see Chapter 3). In Table 8.10 we have created another 'Perfect World' Time Series where the underlying trend is a 5% reduction per period with a theoretical start value at time zero of 750 (intercept). In this case we have applied seasonality factors to increase or decrease the expected values in different seasons as opposed to additive/subtractive differences. In this case the product of the four seasonality factors is 1. Figure 8.17 illustrates the funnelling effect around the Moving Geometric Mean.

If we were to calculate the Best Fit exponential trend through the Moving Geometric Mean we would get an intercept of 750 and an exponent of $e^{-0.05129}$ which equals 0.95, which is identical to the parameters of the hypothetical model from which we started. (*Whoopee!*)

Other than reading it from the graph we could derive these using a combination of the **INDEX** and **LOGEST** functions in Excel (see Chapter 6):

For the Intercept: **INDEX(LOGEST(*y-range, x-range*, TRUE, TRUE), 1, 2)** = 750
For the Exponent: **INDEX(LOGEST(*y-range, x-range*, TRUE, TRUE), 1, 1)** = 0.95

... where the **x-range** and **y-range** refer to the two right-hand columns of Table 8.10.

Similarly, if we wanted to use the Moving Cumulative Geometric Mean of the Observed Data plotted with an Offset to the left equivalent to the Median of the Period Range included, we would get the results in Table 8.11 and Figure 8.18. Note: we only use the Cumulative Geometric Means based on completed sets of Seasons (4, 8, 12, 16), similar to the approach we took with the Moving Cumulative Average.

Table 8.10 Time Series with Perfect Seasonality Factors Around an Exponential Trend

Data Points	Sequential Period Number	Recurring Season Number	True Underlying Trend	True Seasonal Variation Factor	Observed Value (Perfect Model)	Moving Geometric Mean Offset Period Number	Moving Geometric Mean
1	1	1	712.5	125%	890.6		
2	2	2	676.9	150%	1015.3		
3	3	3	643.0	80%	514.4		
4	4	4	610.9	67%	407.3	2.5	660
5	5	1	580.3	125%	725.4	3.5	627
6	6	2	551.3	150%	827.0	4.5	595
7	7	3	523.8	80%	419.0	5.5	566
8	8	4	497.6	67%	331.7	6.5	537
9	9	1	472.7	125%	590.9	7.5	510
10	10	2	449.1	150%	673.6	8.5	485
11	11	3	426.6	80%	341.3	9.5	461
12	12	4	405.3	67%	270.2	10.5	438
13	13	1	385.0	125%	481.3	11.5	416
14	14	2	365.8	150%	548.6	12.5	395
15	15	3	347.5	80%	278.0	13.5	375
16	16	4	330.1	67%	220.1	14.5	356

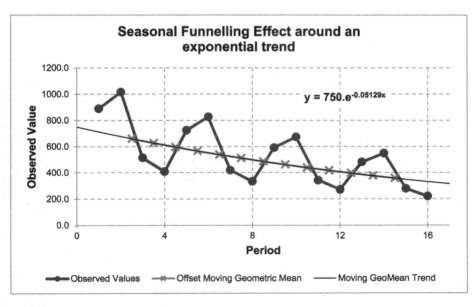

Figure 8.17 Seasonal Funnelling Effect Around an Exponential Trend

432 | The ups and downs of Time Series Analysis

This would mean that the only one of our three techniques that we haven't tried with the exponential trend is the Mean of the four individual seasonal trends. Using our Perfect World data, we get the results in Table 8.12, in which we get identical values for the four exponents, but different values for the intercepts. To get the underlying trend intercept we simply take the Geometric Mean of the four individual ones. Now we have a full set of three techniques that give us the theoretical model data with which we started. (*I can feel a nice warm glow of satisfaction coming on – some may call it smugness, which is not a feeling we should be encouraging!*) In this particular case, because the four seasonal exponents are the same, their Average and Geometric Mean will be the same also. However, as we will see where we have non-identical exponents we need only take their simple average and not the Geometric Mean as their values are unchanged between Linear and Logarithmic Space.

Now you are probably thinking that '*all's well and good*' when we're messing around with Perfect World data, but what about reality? How do any of these techniques stand up if we apply them to our gas consumption example?

Table 8.13 computes the Moving Geometric Mean for a rolling group of 4 Seasonal Quarters using Microsoft Excel's function **GEOMEAN(*range*)** and the Cumulative Geometric Mean at the end of every twelve-month period (on completion of every set of four seasons.) In Figures 8.19 and 8.20 these Geometric Means are plotted with an offset to the left to counter the natural lag, equivalent to the Median position.

Table 8.11 Determining the Exponential Trend Using an Offset Cumulative Geometric Mean

Data Points	Sequential Period Number	Recurring Season Number	True Underlying Trend	True Seasonal Variation Factor	Observed Value (Perfect Model)	Equivalent Sequential Period Number	Cumulative Geometric Mean
1	1	1	712.5	125%	890.6		891
2	2	2	676.9	150%	1015.3		951
3	3	3	643.0	80%	514.4		775
4	4	4	610.9	67%	407.3	2.5	660
5	5	1	580.3	125%	725.4		672
6	6	2	551.3	150%	827.0		696
7	7	3	523.8	80%	419.0		647
8	8	4	497.6	67%	331.7	4.5	595
9	9	1	472.7	125%	590.9		595
10	10	2	449.1	150%	673.6		602
11	11	3	426.6	80%	341.3		572
12	12	4	405.3	67%	270.2	6.5	537
13	13	1	385.0	125%	481.3		533
14	14	2	365.8	150%	548.6		534
15	15	3	347.5	80%	278.0		511
16	16	4	330.1	67%	220.1	8.5	485

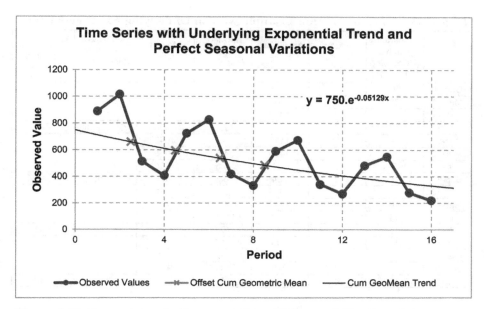

Figure 8.18 Determining the Exponential Trend Using an Offset Cumulative Geometric Mean (1)

Table 8.12 Underlying Exponential Trend as a Function of the Individual Seasonal Trends

	First season		Second season		Third season		Fourth season			
	Data Points	Observed Value	Data Points	Observed Value	Data Points	Observed Value	Data Points	Observed Value		
	1	890.6	2	1015.3	3	514.4	4	407.3		
	5	725.4	6	827.0	7	419.0	8	331.7		
	9	590.9	10	673.6	11	341.3	12	270.2		
	13	481.3	14	548.6	15	278.0	16	220.1		
									Average	GeoMean
Exponent	0.95		0.95		0.95		0.95		0.95	
Intercept	937.5		1125		600		500			750

The two results are different but of a similar order of magnitude. Before we compare how similar they are when projecting forward, perhaps we should really try the third technique (. . .but only if you want to! Sure? Well, OK then . . .)

In Table 8.14, we have extracted the data for each corresponding quarter and calculated the Intercept and Exponent (Slope) for each season in turn using Microsoft Excel's **INDEX** and **LOGEST** functions:

For the intercept: **INDEX(LOGEST**(*y-range*, *x-range*, TRUE, TRUE), 1, 2)
For the exponent: **INDEX(LOGEST**(*y-range*, *x-range*, TRUE, TRUE), 1, 1)

. . . where the *x-range* and *y-range* refer to the corresponding data points and Observed Values respectively.

Table 8.13 Gas Consumption Example with an Underlying Exponential Trend

Date	Quarter Number (Underlying Trend)	Recurring Quarter (Season)	Actual Gas Units Consumed	Moving Geometric Mean Offset Qtr	Moving Geometric Mean	Cumulative Geometric Mean Offset Qtr	Cumulative Geometric Mean
Dec-10	1	1	1046				
Mar-11	2	2	1549				
Jun-11	3	3	1009				
Sep-11	4	4	670	2.5	1023	2.5	1023
Dec-11	5	1	1127	3.5	1042		
Mar-12	6	2	1464	4.5	1028		
Jun-12	7	3	961	5.5	1015		
Sep-12	8	4	594	6.5	985	4.5	1018
Dec-12	9	1	1035	7.5	964		
Mar-13	10	2	1491	8.5	969		
Jun-13	11	3	912	9.5	956		
Sep-13	12	4	592	10.5	955	6.5	993
Dec-13	13	1	990	11.5	945		
Mar-14	14	2	1468	12.5	941		
Jun-14	15	3	840	13.5	922		
Sep-14	16	4	572	14.5	914	8.5	973

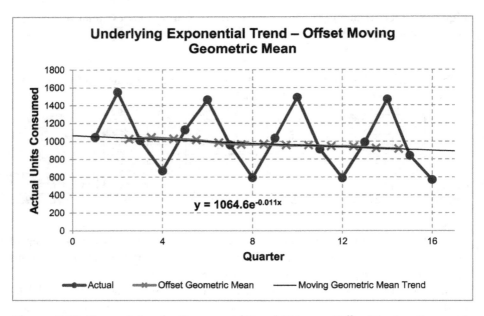

Figure 8.19 Determining the Exponential Trend Using an Offset Moving Geometric Mean

We have used the Arithmetic Mean (Simple Average) of the four individual exponents to get the net overall exponent of the underlying trend, but the Geometric Mean of the four intercept values for the equivalent overall result.

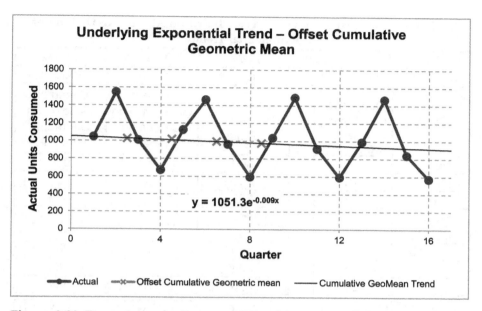

Figure 8.20 Determining the Exponential Trend Using an Offset Cumulative Geometric Mean (2)

Table 8.14 Underlying Exponential Trend of Gas Consumed in Relation to the Individual Seasonal Trends

First season		Second season		Third season		Fourth season	
Data Points	Observed Value	Data Points	Observed Value	Data Points	Observed Value	Data Points	Observed Value
1	1046	2	1549	3	1009	4	670
5	1127	6	1464	7	961	8	594
9	1035	10	1491	11	912	12	592
13	990	14	1468	15	840	16	572

							Average	GeoMean
Exponent	0.9938		0.9964		0.9851		0.9881	0.9908
Intercept	1095.28		1535.88		1063.10		682.76	1051.19

So how do the three techniques compare with forward predictions?

In comparison with the previous two techniques, this technique yields another specifically different but not inconsistent result. So, how does it compare creating a two year forward forecast of our gas usage? Unfortunately, the answer is 'not very well' from the results in Table 8.15. The three forecasts are reasonably comparable amongst themselves but are significantly lower than the equivalent linear forecast in Table 8.8, which is perhaps a little unexpected given the fact that exponentials have that 'tailing off' effect.

Despite our reservations about the reducing linear forecast eventually becoming negative, and that the exponential trend always being positive, we have to wait another

436 | The ups and downs of Time Series Analysis

Table 8.15 Two-Year Forward Forecast Based on Alternative Underlying Exponential Trends

Underlying Trend Exponent		0.9908	0.9895	0.9912
Underlying Trend Intercept		1051.19	1064.61	1051.26
Date	Quarter No (Underlying Trend)	Average of 4 Seasonal Trends	Based on Offset Moving Geometric Mean	Based on Offset Cum Geometric Mean
Dec-14	17	899	889	905
Mar-15	18	891	880	897
Jun-15	19	883	871	889
Sep-15	20	875	862	882
Dec-15	21	867	853	874
Mar-16	22	859	844	866
Jun-16	23	851	835	859
Sep-16	24	843	826	851
2015-2016 Total		6966	6858	7023

thirteen years (*'Unlucky' was the cry!*) for the two forecasts to become comparable on an annual basis. When we review the sum of the errors between the fitted trends and the observed values, then in this particular case the linear model is the better representation, but the two sets are comparable when we look at the Sum of Squares Error! (Compare Table 8.16 with the equivalent Linear error results in Table 8.9.)

8.3.6 Meandering trends

What if the underlying trend is neither linear nor exponential? (*You do ask some difficult questions! I can tell you are an estimator!*)

If we have sufficient data to cover a number of potential cyclical variations in addition to the seasonal variations, we can choose to model the underlying trend with both types of variation in mind (in which case move on to the next section.) Alternatively, if we have insufficient data to support an analysis of cyclical variations, we can always look to 'splice' two linear or exponential trends together.

The easiest way to splice the two trends together is probably to run an Offset Moving Average or Offset Moving Geometric Mean and to detect at which point the net trend appears to change direction, as highlighted by Figure 8.21.

Rather than simply observe and fit the change in direction of the underlying trend, it is better to try to articulate and hypothesise (*not just guess*) why the trend has changed. For instance, in terms of gas consumption this may reflect the culminations of a domestic economy drive, the completion of a home insulation project, or the completion of a

The ups and downs of Time Series Analysis | 437

Table 8.16 Residual Variation of Alternative Underlying Exponential Trends Relative to Actual Values

| | Model Exponent | | 0.9908 | 0.9895 | 0.9912 | Residuals | | |
| | Model Intercept | | 1051.19 | 1064.61 | 1051.26 | | | |
Date	Quarter Number (Underlying Trend)	Actual Gas Units Consumed	Average of 4 Seasonal Trends	Based on Offset Moving Geometric Mean	Based on Offset Cum Geometric Mean	Actual – Average of 4 Seasonal Trends	Actual – Offset Moving Geometric Mean	Actual – Offset Cum Geometric Mean
Dec-10	1	1046	1042	1053	1042	4	-7	4
Mar-11	2	1549	1032	1042	1033	517	507	516
Jun-11	3	1009	1023	1031	1024	-14	-22	-15
Sep-11	4	670	1013	1020	1015	-343	-350	-345
Dec-11	5	1127	1004	1010	1006	123	117	121
Mar-12	6	1464	995	999	997	469	465	467
Jun-12	7	961	986	989	988	-25	-28	-27
Sep-12	8	594	977	978	980	-383	-384	-386
Dec-12	9	1035	968	968	971	67	67	64
Mar-13	10	1491	959	958	963	532	533	528
Jun-13	11	912	950	948	954	-38	-36	-42
Sep-13	12	592	941	938	946	-349	-346	-354
Dec-13	13	990	933	928	938	57	62	52
Mar-14	14	1468	924	918	929	544	550	539
Jun-14	15	840	916	908	921	-76	-68	-81
Sep-14	16	572	907	899	913	-335	-327	-341

		Sum of Residuals	752	733	699
		Sum of Squares Residuals	1,596,037	1,585,805	1,593,720

natural downsizing in the number of people living at home. (*I always suspected that it was our offspring who kept turning the thermostat up – doesn't seem to happen anymore – strange that! It's nothing to do with the natural rotation of the earth as my offspring would have had me believe.*)

Note: The Cumulative Average Seasonal Smoothing technique will not work as well as the Moving Average Seasonal Smoothing technique as it is slower to respond to changes and inherently carries the first trend into the second, creating a damping effect.

A viable alternative is to take the average of the individual trends so long as we impose the same equivalent point in time where the trends change.

We may discover that our data does not appear to have any specific underlying trend, but that the seasons appear to be acting completely independently of each other, but that there is regular pattern to each. In this case, see Section 8.6.3 instead. (*I know what you were thinking but let's face it, making a rash decision like changing your career choice is a bit reactionary.*)

8.4 Determining the seasonal variations by Classical Decomposition

Having determined the underlying trend, we now need to determine each seasonal fluctuation or variation around that trend. As the mechanics of the calculations are slightly different, we will look at the simpler Additive/Subtractive type of model first.

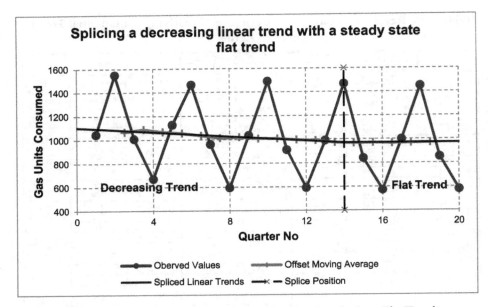

Figure 8.21 Splicing a Decreasing Linear Trend with a Steady State Flat Trend

8.4.1 The Additive/Subtractive Model

The technique assumes that we have already determined the underlying trend.
The procedure is as follows:

1. Flag each season with a separate identifier
2. If we have calculated the underlying trend by more than one technique, choose one (*there is nothing stopping us from repeating the procedure for the other version of the trend, other than a compelling need that we should try to get out more!*)
3. Determine the value indicated by the underlying trend for each period of time
4. Calculate the difference between the Observed Value (Actual) and the underlying trend
5. Calculate the Average Variation for each season
6. Apply the Average Variation to the underlying trend for each season

(*Yes, it really is that simple!*) Table 8.17 illustrates the procedure using our domestic gas consumption example assuming the underlying linear trend based on the Offset Moving Average Technique illustrated previously in Table 8.9.

In this case we have flagged the seasons as 1 through to 4, and in this example for each season we have four values. The average difference between each seasonal observation and the underlying trend can be calculated as:

The ups and downs of Time Series Analysis | 439

1. Autumn: (-48+71+17+10)/4 = 13
2. Winter: (464+418+483+498)/4 = 466
3. Spring: (-66-76-87-121)/4 = -87
4. Summer: (-395-433-397-379)/4 = - 401

It just goes to show that the summers are never as good as the winters are bad!

Table 8.17 Calculation of Additive/Subtractive Seasonal Variations Using Classical Decomposition

Date	Quarter Number (Underlying Trend)	Recurring Quarter (Season)	Actual Gas Units Consumed	Moving Average Offset Qtr	4 Qtr Moving Average	Underlying Linear Trend	Variation from Underlying Trend				Seasonally Adjusted Trend	Residual Variation
							Q1	Q2	Q3	Q4		
Dec-10	1	1	1046			1094	-48				1107	-61
Mar-11	2	2	1549			1085		464			1550	-1
Jun-11	3	3	1009			1075			-66		988	21
Sep-11	4	4	670	2.5	1069	1065				-395	664	6
Dec-11	5	1	1127	3.5	1089	1056	71				1069	58
Mar-12	6	2	1464	4.5	1068	1048		418			1512	-48
Jun-12	7	3	961	5.5	1056	1037			-76		950	11
Sep-12	8	4	594	6.5	1037	1027				-433	626	-32
Dec-12	9	1	1035	7.5	1014	1018	17				1030	5
Mar-13	10	2	1491	8.5	1020	1008		483			1474	17
Jun-13	11	3	912	9.5	1008	999			-87		911	1
Sep-13	12	4	592	10.5	1008	989				-397	588	4
Dec-13	13	1	990	11.5	996	980	10				992	-2
Mar-14	14	2	1468	12.5	991	970		498			1436	32
Jun-14	15	3	840	13.5	973	961			-121		873	-33
Sep-14	16	4	572	14.5	968	951				-379	550	22

	Average	1020				1023	13	466	-87	-401	1020	0
	Slope					-9.54	Seasonal Adjustments					
	Intercept					1103.62						

Table 8.18 Completed Time Series Model with Additive/Subtractive Seasonal Variations

Date	Quarter Number (Underlying Trend)	Recurring Quarter (Season)	Actual Gas Units Consumed	Underlying Linear Trend	Average Seasonal Variation				Seasonally Adjusted Trend	Residual Variation	% Residual Variation
					Q1	Q2	Q3	Q4			
Dec-10	1	1	1046	1094	13				1107	-61	-5.5%
Mar-11	2	2	1549	1085		466			1550	-1	-0.1%
Jun-11	3	3	1009	1075			-87		988	21	2.2%
Sep-11	4	4	670	1065				-401	664	6	0.9%
Dec-11	5	1	1127	1056	13				1069	58	5.5%
Mar-12	6	2	1464	1046		466			1512	-48	-3.2%
Jun-12	7	3	961	1037			-87		950	11	1.2%
Sep-12	8	4	594	1027				-401	626	-32	-5.1%
Dec-12	9	1	1035	1018	13				1030	5	0.4%
Mar-13	10	2	1491	1008		466			1474	17	1.2%
Jun-13	11	3	912	999			-87		911	1	0.1%
Sep-13	12	4	592	989				-401	588	4	0.7%
Dec-13	13	1	990	980	13				992	-2	-0.2%
Mar-14	14	2	1468	970		466			1436	32	2.2%
Jun-14	15	3	840	961			-87		873	-33	-3.8%
Sep-14	16	4	572	951				-401	550	22	4.0%
Dec-14	17	1		941	13				954		
Mar-15	18	2		932		466			1398		
Jun-15	19	3		922			-87		835		
Sep-15	20	4		913				-401	512		
Dec-15	21	1		903	13				916		
Mar-16	22	2		894		466			1359		
Jun-16	23	3		884			-87		797		
Sep-16	24	4		875				-401	473		

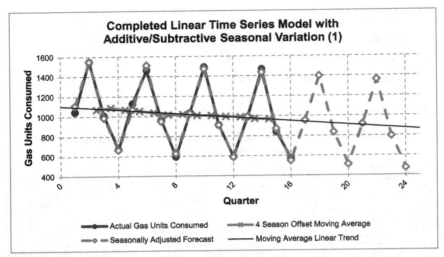

Figure 8.22 Completed Linear Time Series Model with Additive/Subtractive Seasonal Variation (1)

We can now use this data to complete our Time Series Model. In Table 8.18 we have added the appropriate average seasonal variation to the underlying trend and calculated the residual variation based on the difference between actual values and the seasonally adjusted trend data. We have also extended the forecast to give us a seasonally adjusted prediction for the following two years.

The maximum and Minimum Residual Variation in this case is also only around ± 5.5% which in terms of a thirteen-week period is less than a week's difference in the reading of the gas meter – quite a plausible situation in reality. We would conclude that this is a good model. Figure 8.22 shows the closeness of the modelled values to the actual values.

The application of Additive/Subtractive Seasonal Variations to an underlying exponential trend can be achieved using the same procedure.

8.4.2 The Multiplicative Model

The derivation and application of Multiplicative Seasonality Factors around underlying linear or exponential trends follow a similar procedure as Additive/Subtractive Models:

1. Flag each season with a separate identifier
2. If we have calculated the underlying trend by more than one technique, choose one (*We can check our analysis using one or both of the others if we want.*)
3. Determine the value indicated by the underlying trend for each period of time
4. Calculate the differential factor between the Observed Value (Actual) and the Underlying trend (i.e. Actual Value divided by Trend Value)
5. Calculate the Mean of the Factors for each season:

The ups and downs of Time Series Analysis | 441

    a. Use the Arithmetic Mean or simple Average for a linear trend
    b. Use the Geometric Mean for an exponential trend
6.   Apply the relevant Mean Factors from Step 5 to the underlying trend from step 3 for each season as appropriate

Let's bring this to life using our domestic gas consumption model in Table 8.19 with the assumption of an underlying exponential trend.

Here we have expressed the Actual Gas Units Consumed for any quarter as a percentage of the equivalent underlying exponential trend value. (We have used the Offset Moving Geometric Mean Trend technique this time in line with our discussion in Section 8.3.5.)

We can then use the Geometric Mean for each season as our four seasonality factors in our Time Series Model in Table 8.20. To complete our Time Series Model we have simply taken the underlying trend and multiplied by the corresponding average seasonality factor for that quarter. For completeness we have calculated the Residual Variation between the Actual Gas Units Consumed and our modelled data. The spread of Residual Variation around the underlying trend is similar in overall terms to that for the previous Additive/Subtractive Model. As before, we have also projected forward beyond our actuals to get a seasonally adjusted forecast for the next two years, as shown in Figure 8.23

Application of Multiplicative Seasonal Variations to an Underlying Exponential Trend can be achieved using the same procedure, and pragmatically this is often a more natural choice for seasonal variation in many real-life instances with an Exponential Trend.

Table 8.19 Calculation of Multiplicative Seasonal Variations Using Classical Decomposition

Date	Quarter Number (Underlying Trend)	Recurring Quarter (Season)	Actual Gas Units Consumed	Moving Average Offset Qtr	4Q Moving Geometric Mean	Underlying Linear Trend	% Variation from Underlying Trend				Seasonally Adjusted Trend	Residual Variation
							Q1	Q2	Q3	Q4		
Dec-10	1	1	1046			1053	99%				1117	-71
Mar-11	2	2	1549			1042		149%			1590	-41
Jun-11	3	3	1009			1031			98%		989	20
Sep-11	4	4	670	2.5	1023	1020				66%	646	24
Dec-11	5	1	1127	3.5	1042	1010	112%				1071	56
Mar-12	6	2	1464	4.5	1028	999		147%			1525	-61
Jun-12	7	3	961	5.5	1015	989			97%		948	13
Sep-12	8	4	594	6.5	985	978				61%	619	-25
Dec-12	9	1	1035	7.5	964	968	107%				1026	9
Mar-13	10	2	1491	8.5	969	958		156%			1461	30
Jun-13	11	3	912	9.5	956	948			96%		909	3
Sep-13	12	4	592	10.5	955	938				63%	593	-1
Dec-13	13	1	990	11.5	945	928	107%				984	6
Mar-14	14	2	1468	12.5	941	918		160%			1401	67
Jun-14	15	3	840	13.5	922	908			92%		871	-31
Sep-14	16	4	572	14.5	914	899				64%	569	3

Average	1020					1020	0
GeoMean	969			973	106%	153%	96%	63%	969
Exponent		0.99			Seasonal Adjustments
Intercept		1064.61			

Table 8.20 Completed Time Series Model with Multiplicative Seasonal Variations

Date	Quarter Number (Underlying Trend)	Recurring Quarter (Season)	Actual Gas Units Consumed	Underlying Linear Trend	Q1	Q2	Q3	Q4	Seasonally Adjusted Trend	Residual Variation	% Residual Variation
Dec-10	1	1	1046	1053	106%				1117	-71	-6.4%
Mar-11	2	2	1549	1042		153%			1590	-41	-2.6%
Jun-11	3	3	1009	1031			96%		989	20	2.0%
Sep-11	4	4	670	1020				63%	646	24	3.8%
Dec-11	5	1	1127	1010	106%				1071	56	5.3%
Mar-12	6	2	1464	999		153%			1525	-61	-4.0%
Jun-12	7	3	961	989			96%		948	13	1.3%
Sep-12	8	4	594	978				63%	619	-25	-4.0%
Dec-12	9	1	1035	968	106%				1026	9	0.8%
Mar-13	10	2	1491	958		153%			1461	30	2.0%
Jun-13	11	3	912	948			96%		909	3	0.3%
Sep-13	12	4	592	938				63%	593	-1	-0.2%
Dec-13	13	1	990	928	106%				984	6	0.6%
Mar-14	14	2	1468	918		153%			1401	67	4.8%
Jun-14	15	3	840	908			96%		871	-31	-3.6%
Sep-14	16	4	572	899				63%	569	3	0.6%
Dec-14	17	1		889	106%				943		
Mar-15	18	2		880		153%			1343		
Jun-15	19	3		871			96%		835		
Sep-15	20	4		862				63%	545		
Dec-15	21	1		853	106%				904		
Mar-16	22	2		844		153%			1287		
Jun-16	23	3		835			96%		801		
Sep-16	24	4		826				63%	522		

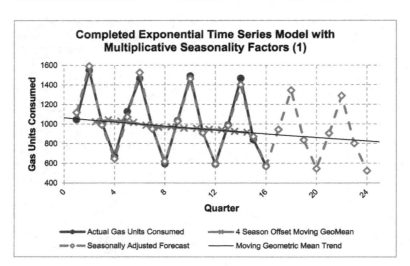

Figure 8.23 Completed Exponential Time Series Model with Multiplicative Seasonality Factors (1)

The ups and downs of Time Series Analysis | 443

In the case of underlying linear trends as we have discussed here, whilst the results of the two analyses are not wildly different, there is a case to be made that in the longer term the multiplicative seasonal adjustment may be the more sensible choice with a decreasing trend (to avoid negative seasonal values), we have a choice over techniques for short-term forecasting.

8.5 Multi-Linear Regression: A holistic approach to Time Series?

Given that the Classical Decomposition Technique has no single way of determining the appropriate underlying trend, wouldn't it be great if we had a technique that obviated the need for that choice and just gave us the results which included the underlying trend and all the seasonal variations in a one-shot technique?

Enter Multi-Linear Regression — the Time Series equivalent of the knight on a white charger . . . or is it? We shall see.

Not only will Regression calculate the numerical values for us, it will also give us that comfort blanket of the statistical significance (or otherwise) of the results.

The procedure to follow is also simpler:

1. Determine the number of seasons involved
2. Identify the functional type of relationship that will be used, i.e. linear or exponential
3. Determine whether the model should use Additive/Subtractive Seasonal Variations or Multiplicative Seasonality Factors
4. Prepare the input data table: Flag each season with a separate identifier and create an on/off categorical variable (Dummy variable) for each season. We will also need a separate variable to indicate the overall time period as a predictor of the Underlying Trend
5. Run the Multi-Linear Regression Model (see Chapters 4 and 6)
6. Check the significance of the results

If the number of seasons is not intuitive from the context of the data, we can determine the most appropriate using a progressive sequence of Moving Averages (see Chapter 3). We can also use the Moving Average (with the selected number of seasons) to determine whether we think that the Underlying Trend is linear or exponential in nature.

In terms of whether we should be using Additive/Subtractive Seasonal Variations or Multiplicative Seasonality Factors we can use the same rationale as we did in Section 8.4. If the data is funnelling, towards or away from the underlying trendline then this is indicative of a Multiplicative Model. Parallel seasonal data suggests that an Additive/Subtractive Model is more appropriate.

Let's use our domestic gas consumption model again to illustrate the next steps.

8.5.1 The Additive/Subtractive Linear Model

Table 8.21 illustrates how we might arrange our input data. Here, we have identified a counter for the quarters, taking values 1 through to 16 and have nominally referred to this as variable x0. For each of the quarters autumn, winter, spring and summer (notionally referred to as x1 to x4), we have assigned values of zero or one. One signifies that the season is 'active' or 'on' and zero that it is 'inactive' or 'off' (i.e. Binary On/Off Switches that we discussed in Section 6.2.2). Clearly, only one season can be 'active' at any one time. (As TRUE and FALSE in Microsoft Excel have the notional values of 1 and 0, it may have seemed reasonable, or more intuitive to use TRUE or FALSE flags but unfortunately these cause Excel to return an error when we run the Data Analysis Regression utility.)

We can now run a Multi-Linear Regression through the Origin, i.e. with the intercept set to zero. *Sorry, did I spring that one on you unexpectedly? Not convinced that this will necessarily give us the optimum result?* OK, let's do it one step at a time first, and run the Multi-Linear Regression without any constraints.

Table 8.22 gives us the results of the regression based on Table 8.21 input data (*and our immediate reaction may be 'Oh dear!' or something along those lines.*)

Table 8.21 Multi-Linear Regression Input Data – Additive/Subtractive Time Series Model

Date	Recurring Quarter (Season)	Actual Gas Units Consumed	Sequential Quarter No	Autumn	Winter	Spring	Summer
		y	x0	x1	x2	x3	x4
Dec-10	1	1046	1	1	0	0	0
Mar-11	2	1549	2	0	1	0	0
Jun-11	3	1009	3	0	0	1	0
Sep-11	4	670	4	0	0	0	1
Dec-11	1	1127	5	1	0	0	0
Mar-12	2	1464	6	0	1	0	0
Jun-12	3	961	7	0	0	1	0
Sep-12	4	594	8	0	0	0	1
Dec-12	1	1035	9	1	0	0	0
Mar-13	2	1491	10	0	1	0	0
Jun-13	3	912	11	0	0	1	0
Sep-13	4	592	12	0	0	0	1
Dec-13	1	990	13	1	0	0	0
Mar-14	2	1468	14	0	1	0	0
Jun-14	3	840	15	0	0	1	0
Sep-14	4	572	16	0	0	0	1
Average		1020					

For the Formula-phobes: Time Series Regression through the Origin

When we have a Time Series with Additive/Subtractive Seasonal Variations either side of an underlying trend. We have in effect a set of parallel lines – one more than the number of seasons.

In reality the underlying trendline is superfluous to our needs from an output perspective. We could just as easily define any one of the seasons as the reference line and make the Additive/Subtractive adjustments to that to get the parallel lines for the other seasons. Similarly, we could take a reference line through the origin and construct our parallel Seasonal predictions in relation to that.

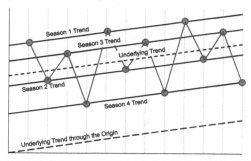

For the Formula-philes: Additive/Subtractive seasonal variation with linear trend

Consider a Time Series variable y_t over consecutive time periods t with an underlying linear trend of slope m, and that there are k seasonal variations with values v_1 to v_k

Let S_{1t} to S_{kt} be categorical binary variables representing each season with values 0 or 1 at time t such that

$$\sum_{i=1}^{k} S_{it} = 1$$

A Linear Time Series with Additive/Subtractive Seasonal Variations can be expressed as:

$$y_t = mt + \sum_{i=1}^{k} v_i S_{it}$$

However, let's think this through logically. The results are telling us that variable x1 (autumn) is superfluous to our needs as it returns a Coefficient of zero with a massively large t-Statistic and an associated incalculably small p-value. (*More worryingly perhaps, the significance of the coefficient of x2 is also looking somewhat suspect!*) Table 8.23 shows the

446 | The ups and downs of Time Series Analysis

Table 8.22 Multi-Linear Regression Output Report – All Parameters Selected

SUMMARY OUTPUT

Regression Statistics	
Multiple R	0.995883299
R Square	0.991783545
Adjusted R Square	0.897886652
Standard Error	35.08845964
Observations	16

Despite an excellent R-Square, Adjusted R-Square and F-Statistic, the extraordinarily high t-Stat for x1 and incalculable P-value suggests that it is a redundant term, and there is something strange happening with x2

ANOVA

	df	SS	MS	F	Significance F
Regression	5	1634758.8	326951.76	331.9442008	8.97091E-11
Residual	11	13543.2	1231.2		
Total	16	1648302			

	Coefficients	Standard Error	t Stat	P-value	Lower 95%	Upper 95%
Intercept	1107.6	22.27840883	49.71629745	2.67774E-14	1058.565553	1156.634447
x0	-8.3	1.961504525	-4.231445758	0.001408988	-12.61724235	-3.98275765
x1	0	0	65535	#NUM!	0	0
x2	451.8	24.88870226	18.15281469	#NUM!	397.0203357	506.5796643
x3	-102.4	25.11951433	-4.076511937	0.00183103	-157.6876783	-47.11232174
x4	-417.6	25.49955882	-16.37675393	4.50708E-09	-473.7241506	-361.4758494

Table 8.23 Multi-Linear (Stepwise) Regression Output Report – Excluding Variable x1

SUMMARY OUTPUT

Regression Statistics	
Multiple R	0.995883299
R Square	0.991783545
Adjusted R Square	0.988795743
Standard Error	35.08845964
Observations	16

The excellent R-Square, Adjusted R-Square and F-Statistic, and the P-values of all our coefficients suggest that this is a statistically viable model

ANOVA

	df	SS	MS	F	Significance F
Regression	4	1634758.8	408689.7	331.9442008	2.19103E-11
Residual	11	13543.2	1231.2		
Total	15	1648302			

	Coefficients	Standard Error	t Stat	P-value	Lower 95%	Upper 95%
Intercept	1107.6	22.27840883	49.71629745	2.67774E-14	1058.565553	1156.634447
x0	-8.3	1.961504525	-4.231445758	0.001408988	-12.61724235	-3.98275765
x2	451.8	24.88870226	18.15281469	1.50748E-09	397.0203357	506.5796643
x3	-102.4	25.11951433	-4.076511937	0.00183103	-157.6876783	-47.11232174
x4	-417.6	25.49955882	-16.37675393	4.50708E-09	-473.7241506	-361.4758494

results of following a Stepwise Regression without variable x1. This now confirms that all remaining candidates give us a statistically significant result. (The potential issue with x2 has gone away and was probably just a quirky knock on effect of x1.)

We do not have to worry about the removal of x1. All this implies is that autumn has been selected as the default season and can be calculated as the linear combination of

The ups and downs of Time Series Analysis | 447

the intercept and the slope of x0 multiplied by the value of x0. All other seasons add or subtract a constant to the default autumn condition. This is equivalent to changing the Intercept but keeping the same slope of line i.e. parallel.

If instead we now were to run the alternative Multi-Linear Regression through the Origin we would get the results in Table 8.24, a statistically significant set of results first time! Worryingly perhaps it gives us an apparently different set of values – but not so, they just camouflaged; Table 8.25 should soothe our worried brows, demonstrating that the two are in fact equivalent.

In short, we can perform the Multi-Linear Regression of our Time Series data by either route and get the corresponding answer. Whichever we choose there are advantages and disadvantages:

- Regression through the origin will give us a result quicker than the Stepwise approach but will not give us a truly comparable F-Statistic to determine the overall validity of the model (see Chapter 4 on the issues of forcing a regression through the origin and the implications and consequences of so doing.) In this particular case, there are no particular concerns about the 'inflated F-Statistic, but if the true underlying model was more 'suspect', then the inflated F-Statistic may misguide us into a Type 1 Error (i.e. accepting a null hypothesis that we should be rejecting.)
- The Stepwise Regression approach will take a little longer and loses a little of the transparency in the model relationships but will more often than not, give us a valid F-Statistic.

Table 8.24 Multi-Linear Regression Output Report – Enforcing an Intercept of Zero

SUMMARY OUTPUT

Regression Statistics	
Multiple R	0.999629792
R Square	0.99925972
Adjusted R Square	0.908081437
Standard Error	35.08845964
Observations	16

The excellent R-Square, Adjusted R-Square and F-Statistic, and the P-values of all our coefficients suggest that this is a statistically viable model

ANOVA

	df	SS	MS	F	Significance F
Regression	5	18281158.8	3656231.76	2969.648928	1.6185E-15
Residual	11	13543.2	1231.2		
Total	16	18294702			

	Coefficients	Standard Error	t Stat	P-value	Lower 95%	Upper 95%
Intercept	0	#N/A	#N/A	#N/A	#N/A	#N/A
x0	-8.3	1.961504525	-4.231445758	0.001408988	-12.61724235	-3.98275765
x1	1107.6	22.27840883	49.71629745	2.67774E-14	1058.565553	1156.634447
x2	1559.4	23.5380543	66.25016581	1.1494E-15	1507.593092	1611.206908
x3	1005.2	24.88870226	40.38780285	2.603E-13	950.4203357	1059.979664
x4	690	26.31634473	26.21944678	2.87862E-11	632.0781158	747.9218842

448 | The ups and downs of Time Series Analysis

Table 8.25 Multi-Linear Regression Output Report Comparison

Regression Parameter	Stepwise Regression Output Coefficients	Numerical Mapping	Regression through the Origin Coefficients	Regression through Origin of Stepwise Regression
Intercept	1107.6	Moves to x1	0	
x0	-8.3	No change	-8.3	Underlying Trend Matched
x1	0	0 + 1107.6 =	1107.6	Intercept
x2	451.8	451.8 + 1107.6 =	1559.4	Intercept + x2
x3	-102.4	-102.4 + 1107.6 =	1005.2	Intercept + x3
x4	-417.6	-417.6 + 1107.6 =	690	Intercept + x4

In Table 8.26 we again compare the Residual Variation between our Observed data and our seasonally adjusted modelled data based on our Regression Output results. We can also produce our two-year forward projection to compare with the equivalent one created using Classical Decomposition. (To save us all looking back, the forecast generated using that technique was slightly less than this, but were we to calculate the Sum of Squares of the two models, this result (being a Least Squares technique) gives us the smaller squared error. Figure 8.24 shows the closeness of fit of the Seasonally Adjusted Time Series to our observed values. (*Don't expect all Time Series Models to be this good – I run a tight ship on the gas bills!*)

Table 8.26 Modelled Data and Two-Year Forecast of Gas Consumption Using Multi-Linear Regression

			Coefficients (Regression Through the Origin)						
			-8.3	1107.6	1559.4	1005.2	690		
Date	Recurring Quarter (Season)	Actual Gas Units Consumed	Sequential Quarter No	Autumn	Winter	Spring	Summer	Seasonally Adjusted Trend	Residual Variation
		y	x0	x1	x2	x3	x4		
Dec-10	1	1046	1	1	0	0	0	1099	-53
Mar-11	2	1549	2	0	1	0	0	1543	6
Jun-11	3	1009	3	0	0	1	0	980	29
Sep-11	4	670	4	0	0	0	1	657	13
Dec-11	1	1127	5	1	0	0	0	1066	61
Mar-12	2	1464	6	0	1	0	0	1510	-46
Jun-12	3	961	7	0	0	1	0	947	14
Sep-12	4	594	8	0	0	0	1	624	-30
Dec-12	1	1035	9	1	0	0	0	1033	2
Mar-13	2	1491	10	0	1	0	0	1476	15
Jun-13	3	912	11	0	0	1	0	914	-2
Sep-13	4	592	12	0	0	0	1	590	2
Dec-13	1	990	13	1	0	0	0	1000	-10
Mar-14	2	1468	14	0	1	0	0	1443	25
Jun-14	3	840	15	0	0	1	0	881	-41
Sep-14	4	572	16	0	0	0	1	557	15
Dec-14	1		17	1	0	0	0	967	
Mar-15	2		18	0	1	0	0	1410	
Jun-15	3		19	0	0	1	0	848	
Sep-15	4		20	0	0	0	1	524	
Dec-15	1		21	1	0	0	0	933	
Mar-16	2		22	0	1	0	0	1377	
Jun-16	3		23	0	0	1	0	814	
Sep-16	4		24	0	0	0	1	491	

The ups and downs of Time Series Analysis

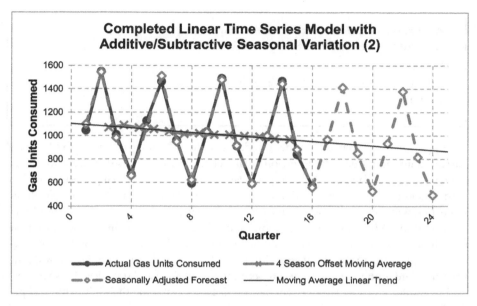

Figure 8.24 Completed Linear Time Series Model with Additive/Subtractive Seasonal Variation (2)

8.5.2 The Additive/Subtractive Exponential Model

As discussed previously, especially with decreasing trends, a linear model is often not sustainable indefinitely and an exponential model may be more appropriate in the longer term. However, from Chapters 5 and 6 we saw that we can transform an Exponential Function into a linear format suitable for Linear Regression by taking the logarithm of the dependent y-variable. The same is true for a Time Series with an exponential trend; we just need to work with the logarithmic value of the output data, or in the case of our example, the Log of the Gas Units Consumed.

For the Formula-philes: Additive/Subtractive seasonal variation with exponential trend

Consider a Time Series variable y_t over consecutive time periods t with an underlying exponential trend of power m, and that there are k seasonal variations with values v_1 to v_k.

Let S_{1t} to S_{kt} be categorical binary variables, one for each season with values 0 or 1 at time t such that

$$\sum_{i=1}^{k} S_{it} = 1 \qquad (1)$$

(Continued)

450 | The ups and downs of Time Series Analysis

An Exponential Time Series with a base b, and with Additive/Subtractive Seasonal Variations can be expressed as

$$\log_b y_t = mt + \sum_{i=1}^{k} v_i S_{it} \quad (2)$$

Taking the antilog of (2) by raising the base b to the power of both sides:

$$y_t = b^{mt + \sum_{i=1}^{k} v_i S_{it}}$$

Suppose now we were to consider our data to exhibit Additive or Subtractive Seasonal Variation around an underlying exponential trend. Table 8.27 illustrates how we might arrange our input data. It will be very similar to that for an underlying linear trend but with an additional column to transform the Units of Gas Consumed into a logarithmic value. This will then enable us to run a Multi-Linear Regression on the transformed data.

As with the Multi-Linear Regression model for an underlying linear trend, we have two choices on whether to go directly to a Regression Result passing through the origin, or to perform a Stepwise Regression Procedure. We will get an equivalent result (as we did before with the same trade-off between expediency and statistical transparency.)

The results of our endeavours can be seen in Table 8.28 using the slightly more expedient Regression through the Origin. As with all linear transformations of exponential data, the implied Regression Best Fit Line will be expressed in logarithmic terms, so to get a meaningful result we have had to take the anti-log of Linear Regression Output. To do this we raise ten (or whatever base we used) to the power of the output. Figure 8.25 illustrates the model fit to the actual data based on the two-year forward forecast of seasonally adjusted data in Table 8.29.

Table 8.27 Transformed Exponential Regression Input Data – Additive/Subtractive Time Series Model

Date	Recurring Quarter (Season)	Actual Gas Units Consumed	Log of Gas Units Consumed	Sequential Quarter Number	Autumn	Winter	Spring	Summer
			y	x0	x1	x2	x3	x4
Dec-10	1	1046	3.0195	1	1	0	0	0
Mar-11	2	1549	3.1901	2	0	1	0	0
Jun-11	3	1009	3.0039	3	0	0	1	0
Sep-11	4	670	2.8261	4	0	0	0	1
Dec-11	1	1127	3.0519	5	1	0	0	0
Mar-12	2	1464	3.1655	6	0	1	0	0
Jun-12	3	961	2.9827	7	0	0	1	0
Sep-12	4	594	2.7738	8	0	0	0	1
Dec-12	1	1035	3.0149	9	1	0	0	0
Mar-13	2	1491	3.1735	10	0	1	0	0
Jun-13	3	912	2.9600	11	0	0	1	0
Sep-13	4	592	2.7723	12	0	0	0	1
Dec-13	1	990	2.9956	13	1	0	0	0
Mar-14	2	1468	3.1667	14	0	1	0	0
Jun-14	3	840	2.9243	15	0	0	1	0
Sep-14	4	572	2.7574	16	0	0	0	1

Average	1020	2.9861	< Arithmetic Mean (Average) of Log Values	
Geometric Mean	968.60	2.9861	< Log of Geometric Mean of the Raw Values	

Table 8.28 Transformed Exponential Regression Output Report – Enforcing an Intercept of Zero

SUMMARY OUTPUT

Regression Statistics	
Multiple R	0.999989188
R Square	0.999978376
Adjusted R Square	0.909061421
Standard Error	0.016766247
Observations	16

The excellent R-Square, Adjusted R-Square and F-Statistic values, and the P-values of all our coefficients suggest that this is a statistically viable model

ANOVA

	df	SS	MS	F	Significance F
Regression	5	142.9911774	28.59823549	101734.3216	3.44409E-23
Residual	11	0.003092178	0.000281107		
Total	16	142.9942696			

	Coefficients	Standard Error	t Stat	P-value	Lower 95%	Upper 95%
Intercept	0	#N/A	#N/A	#N/A	#N/A	#N/A
x0	-0.003998611	0.000937262	-4.266269677	0.001328957	-0.00606151	-0.001935712
x1	3.048498065	0.010645247	286.371758	1.18173E-22	3.025068035	3.071928096
x2	3.205937939	0.011247141	285.0447076	1.24367E-22	3.181183149	3.230692728
x3	3.003709671	0.011892518	252.5713769	4.70372E-22	2.977534415	3.029884927
x4	2.822380859	0.012574686	224.4494174	1.72291E-21	2.794704162	2.850057555

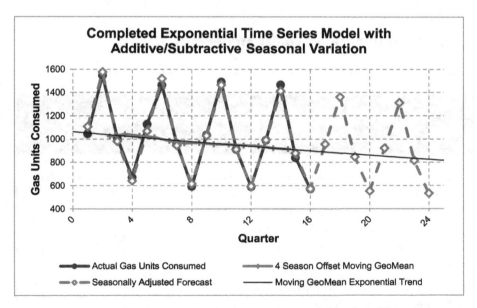

Figure 8.25 Completed Exponential Time Series Model with Additive/Subtractive Seasonal Variation

The ups and downs of Time Series Analysis

Table 8.29 Two-Year Forecast of Gas Consumption Using Seasonally Adjusted Exponential Regression

Date	Recurring Quarter (Season)	Actual Gas Units Consumed	Log of Gas Units Consumed	Sequential Quarter No	Autumn	Winter	Spring	Summer	Seasonally Adjusted Trend	Residual Variation
					Coefficients (Regresssion Through the Origin)					
				-0.004	3.048	3.206	3.004	2.822		
			y	x0	x1	x2	x3	x4		
Dec-10	1	1046	3.0195	1	1	0	0	0	1108	-62
Mar-11	2	1549	3.1901	2	0	1	0	0	1577	-28
Jun-11	3	1009	3.0039	3	0	0	1	0	981	28
Sep-11	4	670	2.8261	4	0	0	0	1	640	30
Dec-11	1	1127	3.0519	5	1	0	0	0	1068	59
Mar-12	2	1464	3.1655	6	0	1	0	0	1520	-56
Jun-12	3	961	2.9827	7	0	0	1	0	946	15
Sep-12	4	594	2.7738	8	0	0	0	1	617	-23
Dec-12	1	1035	3.0149	9	1	0	0	0	1029	6
Mar-13	2	1491	3.1735	10	0	1	0	0	1465	26
Jun-13	3	912	2.9600	11	0	0	1	0	911	1
Sep-13	4	592	2.7723	12	0	0	0	1	595	-3
Dec-13	1	990	2.9956	13	1	0	0	0	992	-2
Mar-14	2	1468	3.1667	14	0	1	0	0	1412	56
Jun-14	3	840	2.9243	15	0	0	1	0	878	-38
Sep-14	4	572	2.7574	16	0	0	0	1	573	-1
Dec-14	1			17	1	0	0	0	956	
Mar-15	2			18	0	1	0	0	1361	
Jun-15	3			19	0	0	1	0	847	
Sep-15	4			20	0	0	0	1	553	
Dec-15	1			21	1	0	0	0	922	
Mar-16	2			22	0	1	0	0	1312	
Jun-16	3			23	0	0	1	0	816	
Sep-16	4			24	0	0	0	1	533	

8.5.3 The Multiplicative Linear Model

This is where it starts to get very messy, but sometimes we just have to put up with a little mess in our lives!

For the Formula-philes: Multiplicative Seasonality Factors with linear trends

Consider a Time Series variable over consecutive time periods t with an underlying linear trend of slope m and intercept c, and that there are k seasonality factors with values $f_1 \ldots f_k$. We have two options we can model

Let S_{1t} to S_{kt} be categorical binary variables, one for each season, with values 0 or 1 at time t such that

$$\sum_{i=1}^{k} S_{it} = 1 \qquad (1)$$

A Linear Time Series with Multiplicative Seasonality Factors can be expressed as:

... a product of the underlying trend and the weighted sum of the seasonality factors and the Binary Seasonal variable

$$y_t = (c + mt) \sum_{i=1}^{k} f_i S_{it} \qquad (2)$$

The ups and downs of Time Series Analysis | 453

...a product of the underlying trend and the product of the seasonality factors raised to the power of the Binary Seasonal variable

$$y_t = (c + mt)\prod_{i=1}^{k} f_i^{S_{it}} \quad (3)$$

In both cases of (2) and (3), (1) imposes that only one value of Seasonality Factor f_i is active for any value of t

For any value of t, a Linear Time Series with Multiplicative Seasonality Factors can be expressed as:

$$y_t = cf_i + mtf_i \quad (4)$$

In Table 8.30 we have compiled the data that reflects this for our domestic gas consumption:

- x0 takes any value of t from 1 to 16 and is included for reference only and plotting the graph of the results
- x1 to x4 represent the seasons with a binary on/off (1 or 0) switch. These correspond to coefficient values of cf_i where $i = 1$ to 4
- x5 to x8 represent the time at which each season occurs, taking the corresponding value of x0 or t for the 'on-season' and zero in every 'off season'. Their coefficients relate to the values mf_i

Table 8.30 Regression Output for Multiplicative Linear Regression Model for Domestic Gas Consumption

Date	Recurring Quarter (Season)	Actual Gas Units Consumed	Sequential Quarter No	Autumn	Winter	Spring	Summer	Autumn	Winter	Spring	Summer
		y		x1	x2	x3	x4	x5	x6	x7	x8
Dec-10	1	1046	1	1	0	0	0	1	0	0	0
Mar-11	2	1549	2	0	1	0	0	0	2	0	0
Jun-11	3	1009	3	0	0	1	0	0	0	3	0
Sep-11	4	670	4	0	0	0	1	0	0	0	4
Dec-11	1	1127	5	1	0	0	0	5	0	0	0
Mar-12	2	1464	6	0	1	0	0	0	6	0	0
Jun-12	3	961	7	0	0	1	0	0	0	7	0
Sep-12	4	594	8	0	0	0	1	0	0	0	8
Dec-12	1	1035	9	1	0	0	0	9	0	0	0
Mar-13	2	1491	10	0	1	0	0	0	10	0	0
Jun-13	3	912	11	0	0	1	0	0	0	11	0
Sep-13	4	592	12	0	0	0	1	0	0	0	12
Dec-13	1	990	13	1	0	0	0	13	0	0	0
Mar-14	2	1468	14	0	1	0	0	0	14	0	0
Jun-14	3	840	15	0	0	1	0	0	0	15	0
Sep-14	4	572	16	0	0	0	1	0	0	0	16

454 | The ups and downs of Time Series Analysis

However, it may or may not have escaped our attention that there is a high degree of multicollinearity between the two groups of variables (all the zeroes line up across the equivalent pairs (x1 with x5, x2 with x6, etc.) and from our formula-phile section (don't look at it if that stuff scares you!) we would expect the difference between the corresponding pairs to be in the ratio of the underlying trend's intercept and slope. Not surprisingly then, we get a highly questionable output from a regression of this type as shown in Table 8.31, which fails our expectation of a constant ratio between coefficient pairs.

Allowing the regression to pass through the origin would not resolve our multicollinearity problem. The thing that really makes this difficult to solve in this way is that we have in effect a Generalised Exponential Function at work here, and we saw how difficult they could be in the last chapter.

Consequently, we have four options we can consider:

i.   Give up! (Hardly the spirit though, is it?)
ii. Revert to using the Classical Decomposition technique (as previously discussed)
iii. Break the Regression down into two steps – a sort of Stepwise Decomposition Regression
iv. Create a pseudo Non-Linear Regression model and use Microsoft Excel's Solver

We will look at option iv in Section 8.6, but before that let's explore option iii.

Table 8.31 Input Data for Multiplicative Linear Regression Model for Domestic Gas Consumption

SUMMARY OUTPUT

Regression Statistics	
Multiple R	0.999725632
R Square	0.999451338
Adjusted R Square	0.87397126
Standard Error	35.42174473
Observations	16

Despite an acceptable Adjusted R-Square and excellent F-Statistic, the p-Values for a number of the dummy variables are highly questionable

ANOVA

	df	SS	MS	F	Significance F
Regression	8	18284664.4	2285583.05	1821.617159	6.49455E-11
Residual	8	10037.6	1254.7		
Total	16	18294702			

	Coefficients	Standard Error	t Stat	P-value	Lower 95%	Upper 95%
Intercept	0	#N/A	#N/A	#N/A	#N/A	#N/A
x1	1095	32.89648537	33.28623066	7.24189E-10	1019.140569	1170.859431
x2	1536.2	36.29648743	42.32365467	1.0705E-10	1452.50015	1619.89985
x3	1055.6	39.80023555	26.52245609	4.39163E-09	963.8204922	1147.379508
x4	681	43.3826002	15.69753765	2.7077E-07	580.9595445	781.0404555
x5	-6.5	3.960271455	-1.641301631	0.139362571	-15.63240235	2.632402353
x6	-5.4	3.960271455	-1.363542894	0.209842113	-14.53240235	3.732402353
x7	-13.9	3.960271455	-3.509860410	0.007963214	-23.03240235	-4.767597647
x8	-7.4	3.960271455	-1.86855878	0.098625211	-16.53240235	1.732402353

The ups and downs of Time Series Analysis | 455

The two stages in our Stepwise Decomposition Model are to perform a suitable regression to determine the underlying trend. We will then create a factor model to measure the percentage relationship between the Observed data and the underlying trend regression output. With this we can then run another regression using dummy Categorical Variables to determine the individual seasonality factors.

In this section, we will assume an underlying linear trend, and in Section 8.5.4 we will consider an underlying exponential trend.

From Section 8.3 we discussed and demonstrated that we cannot rely on a Simple Linear Regression of the raw Time Series Data to determine the underlying trend as it varies depending on our starting quarter. Instead we considered three options that gave more palatable results in terms of their stability and consistency. Rather than repeat the example we will utilise the output we generated previously from Tables 8-7 and 8-17 based on a Simple Linear Regression of the Offset Moving Average as the first step. The example and technique gave us an intercept of 1103.62 and a slope of -9.54.

Using this output, we can develop our input table for the second part of our Stepwise Decomposition regression (Table 8.32) in which we divide the Observed Data by the Trendline Value.

Running the regression through the origin as we are looking at the degree to which the four seasons are above or below the underlying trend in terms of a factor, we get the output in Table 8.33, which now has only four variables, each of which is statistically significant based on their p-values (see Chapter 4.)

Table 8.32 Stepwise Decomposition – Regression Input for Seasonality Factors in a Multiplicative Model

Previous Regression Results based on Offset Moving Average		Slope	Intercept						
		-9.54	1103.62						
Date	Recurring Quarter (Season)	Actual Gas Units Consumed	Sequential Quarter No	Underlying Linear Trend	Actual Gas Units / Trend	Autumn	Winter	Spring	Summer
		y				x1	x2	x3	x4
Dec-10	1	1046	1	1094	0.956	1	0	0	0
Mar-11	2	1549	2	1085	1.428	0	1	0	0
Jun-11	3	1009	3	1075	0.939	0	0	1	0
Sep-11	4	670	4	1065	0.629	0	0	0	1
Dec-11	1	1127	5	1056	1.067	1	0	0	0
Mar-12	2	1464	6	1046	1.399	0	1	0	0
Jun-12	3	961	7	1037	0.927	0	0	1	0
Sep-12	4	594	8	1027	0.578	0	0	0	1
Dec-12	1	1035	9	1018	1.017	1	0	0	0
Mar-13	2	1491	10	1008	1.479	0	1	0	0
Jun-13	3	912	11	999	0.913	0	0	1	0
Sep-13	4	592	12	989	0.599	0	0	0	1
Dec-13	1	990	13	980	1.011	1	0	0	0
Mar-14	2	1468	14	970	1.513	0	1	0	0
Jun-14	3	840	15	961	0.875	0	0	1	0
Sep-14	4	572	16	951	0.601	0	0	0	1

456 | The ups and downs of Time Series Analysis

Using this output, we can present a two-year forward forecast (Figure 8.26) based on the data in Table 8.34. This model does show the funnelling effect of the Multiplicative Seasonality Factors, and appears to be a reasonably good representation of our historical gas consumption . . . with the possible exception of the winter Season which is showing signs of diverging from the regression result. In fact we could make the observation that the underlying trend through the Offset Moving Average Data is probably a shallow curve rather than a straight line, (*but we could equally be accused of splitting hairs.*) However, that thought leads us nicely to considering seasonality factors around an exponential trend.

8.5.4 The Multiplicative Exponential Model

At the risk of over-emphasising the linkage between an Exponential Function and a Linear one, we can transform the dependent y-variable of an Exponential Function into a Linear Function by taking its logarithmic value; this will form a linear relationship with its independent x-variables. We can apply the same rule in relation to an Exponential Time Series as discussed in Section 8.5.2.

For the Formula-philes: Multiplicative Seasonality Factors with linear trends

Consider a Time Series variable over consecutive time periods t with an underlying exponential trend of exponent m and intercept c, and that there are k seasonality factors with values . . . . $f_1 \cdots f_k$

For any value of t, an Exponential Time Series with Multiplicative Seasonality Factors can be expressed as:

$$\log_b y_t = cf_i + mtf_i$$

Table 8.35 considers our domestic gas problem from the perspective of it having an underlying exponential trend. Based our previous example in Table 8.16, back in Section 8.3.5 using the Best Fit Exponential Trendline through the Offset Moving Geometric Mean, we determined that there was an underlying exponential trend with an intercept of 1064.6 and an exponent of 0.9895. The model then allows a dummy categorical variable as a binary switch for each season.

Based on the Regression Output in Table 8.36, we would accept this as a statistically significant model with a high R-Square, Adjusted R-Square and very large t-statistics with correspondingly low p-values for each seasonality factor.

The ups and downs of Time Series Analysis | 457

Table 8.33 Stepwise Decomposition – Regression Output for Seasonality Factors in a Multiplicative Model

SUMMARY OUTPUT

Regression Statistics	
Multiple R	0.999491896
R Square	0.998984051
Adjusted R Square	0.91539673
Standard Error	0.038350551
Observations	16

ANOVA

	df	SS	MS	F	Significance F
Regression	4	17.35446006	4.338615015	2949.90412	1.38909E-16
Residual	12	0.017649177	0.001470765		
Total	16	17.37210924			

	Coefficients	Standard Error	t Stat	P-value	Lower 95%	Upper 95%
Intercept	0	#N/A	#N/A	#N/A	#N/A	#N/A
x1	1.012740508	0.019175275	52.81491312	1.39615E-15	0.970961171	1.054519844
x2	1.454894197	0.019175275	75.87344441	1.82929E-17	1.413114861	1.496673533
x3	0.913308057	0.019175275	47.62946216	4.79872E-15	0.871528721	0.955087393
x4	0.601766522	0.019175275	31.38241864	6.90013E-13	0.559987186	0.643545858

Table 8.34 Stepwise Decomposition – Two-Year Forward Forecast

Previous Regression Results based on Offset Moving Average		Slope	Intercept		Regression Seasonality Factors					
		-9.54	1103.62		1.013	1.455	0.913	0.602		
Date	Recurring Quarter (Season)	Actual Gas Units Consumed	Sequential Quarter No	Underlying Linear Trend	Actual Gas Units / Trend	Autumn	Winter	Spring	Summer	Seasonally Adjusted Trend	Residual Variation
		y				x1	x2	x3	x4		
Dec-10	1	1046	1	1094	0.956	1	0	0	0	1108	-62
Mar-11	2	1549	2	1085	1.428	0	1	0	0	1578	-29
Jun-11	3	1009	3	1075	0.939	0	0	1	0	982	27
Sep-11	4	670	4	1065	0.629	0	0	0	1	641	29
Dec-11	1	1127	5	1056	1.067	1	0	0	0	1069	58
Mar-12	2	1464	6	1046	1.399	0	1	0	0	1522	-58
Jun-12	3	961	7	1037	0.927	0	0	1	0	947	14
Sep-12	4	594	8	1027	0.578	0	0	0	1	618	-24
Dec-12	1	1035	9	1018	1.017	1	0	0	0	1031	4
Mar-13	2	1491	10	1008	1.479	0	1	0	0	1467	24
Jun-13	3	912	11	999	0.913	0	0	1	0	912	0
Sep-13	4	592	12	989	0.599	0	0	0	1	595	-3
Dec-13	1	990	13	980	1.011	1	0	0	0	992	-2
Mar-14	2	1468	14	970	1.513	0	1	0	0	1411	57
Jun-14	3	840	15	961	0.875	0	0	1	0	877	-37
Sep-14	4	572	16	951	0.601	0	0	0	1	572	0
Dec-14	1		17	941		1	0	0	0	953	
Mar-15	2		18	932		0	1	0	0	1356	
Jun-15	3		19	922		0	0	1	0	842	
Sep-15	4		20	913		0	0	0	1	549	
Dec-15	1		21	903		1	0	0	0	915	
Mar-16	2		22	894		0	1	0	0	1300	
Jun-16	3		23	884		0	0	1	0	808	
Sep-16	4		24	875		0	0	0	1	526	

Figure 8.27 summarises the completed Time Series Analysis based on the analysis depicted in Table 8.37. It appears to be another good fit in general . . . until we look at the winter season in 2014, and again it seems to be a little understated.

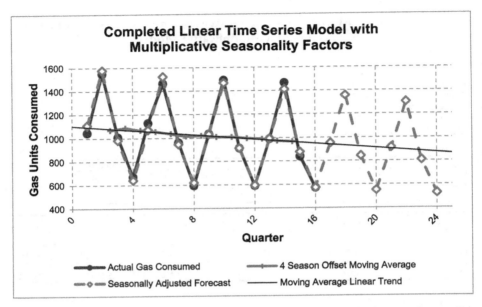

Figure 8.26 Completed Linear Time Series Model with Multiplicative Seasonality Factors

Table 8.35 Stepwise Decomposition – Regression Input for Seasonality Factors in a Multiplicative Model

Previous Regression Results based on Offset Moving Average		Exponent	Intercept					
		0.99	1064.61					
Date	Recurring Quarter (Season)	Actual Gas Units Consumed y	Sequential Quarter No	Underlying Exponential Trend	Actual Gas Units / Trend	Autumn x1	Winter x2	Spring x3	Summer x4
Dec-10	1	1046	1	1053	0.993	1	0	0	0
Mar-11	2	1549	2	1042	1.486	0	1	0	0
Jun-11	3	1009	3	1031	0.978	0	0	1	0
Sep-11	4	670	4	1020	0.657	0	0	0	1
Dec-11	1	1127	5	1010	1.116	1	0	0	0
Mar-12	2	1464	6	999	1.465	0	1	0	0
Jun-12	3	961	7	989	0.972	0	0	1	0
Sep-12	4	594	8	978	0.607	0	0	0	1
Dec-12	1	1035	9	968	1.069	1	0	0	0
Mar-13	2	1491	10	958	1.557	0	1	0	0
Jun-13	3	912	11	948	0.962	0	0	1	0
Sep-13	4	592	12	938	0.631	0	0	0	1
Dec-13	1	990	13	928	1.067	1	0	0	0
Mar-14	2	1468	14	918	1.599	0	1	0	0
Jun-14	3	840	15	908	0.925	0	0	1	0
Sep-14	4	572	16	899	0.636	0	0	0	1

Table 8.36 Stepwise Decomposition – Regression Output for Seasonality Factors in a Multiplicative Model

SUMMARY OUTPUT

Regression Statistics	
Multiple R	0.999417361
R Square	0.998835062
Adjusted R Square	0.915210494
Standard Error	0.043101305
Observations	16

Based on R-Square, Adjusted R-Square the Significance of the F-Statistic and the Significance of the F-Statistic and the coefficient p-values, we have a statistically viable result

ANOVA

	df	SS	MS	F	Significance F
Regression	4	19.11406464	4.77851616	2572.244348	2.94836E-16
Residual	12	0.02229267	0.001857722		
Total	16	19.13635731			

	Coefficients	Standard Error	t Stat	P-value	Lower 95%	Upper 95%
Intercept	0	#N/A	#N/A	#N/A	#N/A	#N/A
x1	1.06135099	0.021550652	49.24913493	3.21865E-15	1.014396152	1.108305829
x2	1.52680406	0.021550652	70.84723134	4.15668E-17	1.479849222	1.573758898
x3	0.959369672	0.021550652	44.51696643	1.07519E-14	0.912414834	1.00632451
x4	0.632873945	0.021550652	29.36681136	1.51593E-12	0.585919107	0.679828783

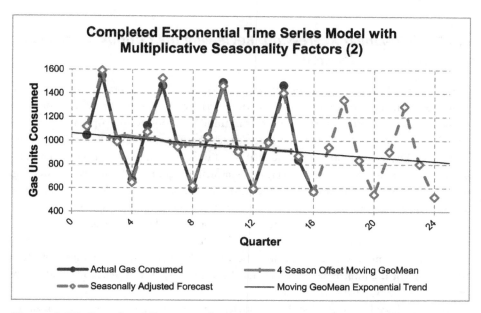

Figure 8.27 Completed Exponential Time Series Model with Multiplicative Seasonality Factors (2)

460 | The ups and downs of Time Series Analysis

Table 8.37 Stepwise Decomposition – Two-Year Forward Forecast

Previous Regression Results based on Offset Moving Average			Exponent 0.99	Intercept 1064.61		Regression Seasonality Factors					
						1.061	1.527	0.959	0.633		
Date	Recurring Quarter (Season)	Actual Gas Units Consumed y	Sequential Quarter Number	Underlying Exponential Trend	Actual Gas Units / Trend	Autumn	Winter	Spring	Summer	Seasonally Adjusted Trend	Residual Variation
						x1	x2	x3	x4		
Dec-10	1	1046	1	1053	0.993	1	0	0	0	1118	-72
Mar-11	2	1549	2	1042	1.486	0	1	0	0	1591	-42
Jun-11	3	1009	3	1031	0.978	0	0	1	0	989	20
Sep-11	4	670	4	1020	0.657	0	0	0	1	646	24
Dec-11	1	1127	5	1010	1.116	1	0	0	0	1072	55
Mar-12	2	1464	6	999	1.465	0	1	0	0	1525	-61
Jun-12	3	961	7	989	0.972	0	0	1	0	948	13
Sep-12	4	594	8	978	0.607	0	0	0	1	619	-25
Dec-12	1	1035	9	968	1.069	1	0	0	0	1027	8
Mar-13	2	1491	10	958	1.557	0	1	0	0	1462	29
Jun-13	3	912	11	948	0.962	0	0	1	0	909	3
Sep-13	4	592	12	938	0.631	0	0	0	1	593	-1
Dec-13	1	990	13	928	1.067	1	0	0	0	985	5
Mar-14	2	1468	14	918	1.599	0	1	0	0	1402	66
Jun-14	3	840	15	908	0.925	0	0	1	0	871	-31
Sep-14	4	572	16	899	0.636	0	0	0	1	569	3
Dec-14	1		17	889		1	0	0	0	944	
Mar-15	2		18	880		0	1	0	0	1344	
Jun-15	3		19	871		0	0	1	0	835	
Sep-15	4		20	862		0	0	0	1	545	
Dec-15	1		21	853		1	0	0	0	905	
Mar-16	2		22	844		0	1	0	0	1288	
Jun-16	3		23	835		0	0	1	0	801	
Sep-16	4		24	826		0	0	0	1	523	

8.5.5 Multi-Linear Regression: Reviewing the options to make an informed decision

We should not be too surprised or concerned as yet because good practice in estimating should encourage us to consider different views of modelling reality to inform our view of the most appropriate estimate. With that in mind let's consider our options.

Whilst in some cases, it may be fairly clear whether we should be looking at an underlying linear or exponential trend, that is not the case here. It may also be evident in some cases from a converging or diverging (seasonal funnelling) whether we are dealing with a multiplicative adjustment to the underlying trend or an Additive/Subtractive adjustment. (Again, in this particular case, it is not absolutely clear, but the recurring concerns we have had so far about the more recent winter levels, indicates that perhaps an Additive/Subtractive variation is the model with the 'better' all round fit.

Table 8.38 summarises our results from Sections 8.5.1 to 8.5.4, along with some additional (but not unfamiliar statistics) . . . and the winner is, on this occasion . . . the first one (whichever way we measure it).

In this case, we have measured the Coefficient of Determination, R-Square in relation to the model's values as a predictor of the Observed or Actual Values. We would expect an excellent value in each case. (Note that these values are different to those generated during either stage of the two-step Regression Analysis as they are measuring different things.)

If we consider the degree of variation in the Residuals, we see that the absolute range between the Maximum and Minimum Variations, the standard deviation of the

The ups and downs of Time Series Analysis | 461

Table 8.38 Comparison of Stepwise Decomposition Regression Results

				Linear	Linear	Exponential	Exponential
		Underlying Trend					
		Adjustment Type		Add/Subtract	Multiplicative	Add/Subtract	Multiplicative
Date	Sequential Quarter No x0	Recurring Quarter (Season)	Actual Gas Units Consumed	LA Model Value	LM Model Value	EA Model Value	EM Model Value
Dec-10	1	1	1046	1099	1108	1108	1118
Mar-11	2	2	1549	1543	1578	1577	1591
Jun-11	3	3	1009	980	982	981	989
Sep-11	4	4	670	657	641	640	646
Dec-11	5	1	1127	1066	1069	1068	1072
Mar-12	6	2	1464	1510	1522	1520	1525
Jun-12	7	3	961	947	947	946	948
Sep-12	8	4	594	624	618	617	619
Dec-12	9	1	1035	1033	1031	1029	1027
Mar-13	10	2	1491	1476	1467	1465	1462
Jun-13	11	3	912	914	912	911	909
Sep-13	12	4	592	590	595	595	593
Dec-13	13	1	990	1000	992	992	985
Mar-14	14	2	1468	1443	1411	1412	1402
Jun-14	15	3	840	881	877	878	871
Sep-14	16	4	572	557	572	573	569

R-Square (Actual, Model)	0.992	0.989	0.988	0.987
Sum of Residuals	0	3	-5	8
Sum of Squares Residuals	13543	18980	19022	21898
Min Residual (Max Negative)	-61	-58	-59	-66
Max Residual (Max Positive)	53	62	62	72
Residual Range (Max - Min)	114	120	121	138
Standard Deviation of Residuals	30.0	35.6	35.6	38.2
Coefficient of Variation of Residuals	2.9%	3.5%	3.5%	3.7%

Residuals, and the Coefficient of Variation, all point to the first model being more closely aligned with the Observed Values.

Every case will be different, and we cannot generalise that this will be so in all other cases. Furthermore, we cannot imply that the model will continue indefinitely; we should review our modelling at the end of each new seasonal cycle.

8.6 Excel Solver technique for Time Series Analysis

The principles of this technique apply to all candidate Time Series, whether they exhibit Additive/Subtractive or Multiplicative Seasonal Variation around an underlying linear or exponential trend ... or even a weird and whacky untransformable nonlinear trend. As a consequence, we will only do two examples here, the first being that Perfect World scenario we started with in Section 8.3, just to prove the concept gives us what we expected, and the second being the actual gas consumption data.

For this we will need to build a Time Series Model in Microsoft Excel from first principles.

462 | The ups and downs of Time Series Analysis

8.6.1 The Perfect World scenario

In Table 8.39 we have taken our Perfect World scenario from Section 8.3 which had an underlying linear trend with an intercept of 710, a slope of -10, and seasonal variations of 100, 300, -100 and -300. Now let's pretend that we don't know these values and want to use Microsoft Excel's **Solver** to determine what they are for us.

The procedure is similar to other Solver examples we have used in previous chapters:

1. Decide which type of Time Series Model we are assuming. In this case our Perfect World example exhibits an underlying linear trend with Additive/Subtractive Seasonal Variations

2. Create a range of input variables for our Time Series model so that we have a single variable for each consecutive time period (often integer, but at least in a manner that follows a regular pattern) and a number of Categorical Binary Variables, one for each season. Here we have defined four discrete Seasonal variables so that one and only one is set to the value 'one' at any one time (*that's a lot of ones at once*). We need to ensure that the 'switching' sequence between seasons follows a regular pattern, i.e. winter is always preceded immediately by autumn and followed by spring.

3. Identify the six variables with unknown values we want Solver to be able to change to determine our best solution: Trend Intercept, Trend Slope, and four Seasonal Variations to be added or subtracted from the underlying trend

4. Choose some starting values (either at random or better still by an informed guess) for the Solver facility. Here we have chosen 685, -15, and four zeroes, but more or

Table 8.39 Solver Set-Up for Perfect Additive/Subtractive Seasonal Variation Around a Linear Trend

				Intercept	Slope	Season 1	Season 2	Season 3	Season 4		
	For Reference Only			685.0	-15.00	0.0	0.0	0.0	0.0		
Time Period	Recurring Season	True Underlying Trend	True Seasonal Variation	Observed Value (Perfect Model)	Sequential Time Period	Season 1	Season 2	Season 3	Season 4	Seasonally Adjusted Trend	Residual Variation
1	1	700	100	800	1	1	0	0	0	670	130
2	2	690	300	990	2	0	1	0	0	655	335
3	3	680	-100	580	3	0	0	1	0	640	-60
4	4	670	-300	370	4	0	0	0	1	625	-255
5	1	660	100	760	5	1	0	0	0	610	150
6	2	650	300	950	6	0	1	0	0	595	355
7	3	640	-100	540	7	0	0	1	0	580	-40
8	4	630	-300	330	8	0	0	0	1	565	-235
9	1	620	100	720	9	1	0	0	0	550	170
10	2	610	300	910	10	0	1	0	0	535	375
11	3	600	-100	500	11	0	0	1	0	520	-20
12	4	590	-300	290	12	0	0	0	1	505	-215
13	1	580	100	680	13	1	0	0	0	490	190
14	2	570	300	870	14	0	1	0	0	475	395
15	3	560	-100	460	15	0	0	1	0	460	0
16	4	550	-300	250	16	0	0	0	1	445	-195
		Average	625.0						Average	557.5	67.5
								Sum of Squares Error			849400

less anything will normally do. As Figure 8.28 shows, they do not have to be a particularly good guess.

5. Create a Time Series Model (labelled seasonally adjusted trend here) based on our initial parameters from step 4 that generates an output value by adding the intercept to the Time Period multiplied by the slope plus each seasonal variation multiplied by its binary on/off switch variable. The easiest way of doing this is to use the Microsoft Excel function **SUMPRODUCT** with the variable values and the variable parameter guesses as the two ranges plus the Intercept
6. Calculate the difference between the Modelled and Observed values (it doesn't matter which way round). These are our Residual Variations or Error terms
7. Calculate the Average of the Residual Variations and the Sum of the Squares of those terms. We can use Microsoft Excel's **AVERAGE** and **SUMSQ** functions for this.
8. Open the Solver dialog box and set the Sum of Squares from step 7 to be the objective or target that we want to minimise (*yes, we are using the principles of Least Squares again*)
9. Add the list of parameters from step 3 that Solver can change
10. Add a constraint that the Average of the Residual Variations from step 7 should be zero (unbiased)
11. Make sure that the tick box labelled 'Make Unconstrained Variables Non-negative' is **not** ticked or checked
12. Click Solve to get the results in Table 8.40 and Figure 8.29

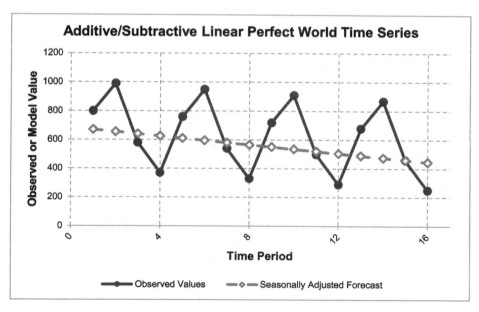

Figure 8.28 Solver Time Series Set-Up with Initial Parameter Starting Values

Table 8.40 Solver Result for Perfect Additive/Subtractive Seasonal Variation Around a Linear Trend

				Linear Trend		Additive/Subtractive Variations					
				Intercept	Slope	Season 1	Season 2	Season 3	Season 4		
	For Reference Only			710.0	-10.00	100.0	300.0	-100.0	-300.0		
Time Period	Recurring Season	True Underlying Trend	True Seasonal Variation	Observed Value (Perfect Model)	Sequential Time Period	Season 1	Season 2	Season 3	Season 4	Seasonally Adjusted Trend	Residual Variation
1	1	700	100	800	1	1	0	0	0	800	0
2	2	690	300	990	2	0	1	0	0	990	0
3	3	680	-100	580	3	0	0	1	0	580	0
4	4	670	-300	370	4	0	0	0	1	370	0
5	1	660	100	760	5	1	0	0	0	760	0
6	2	650	300	950	6	0	1	0	0	950	0
7	3	640	-100	540	7	0	0	1	0	540	0
8	4	630	-300	330	8	0	0	0	1	330	0
9	1	620	100	720	9	1	0	0	0	720	0
10	2	610	300	910	10	0	1	0	0	910	0
11	3	600	-100	500	11	0	0	1	0	500	0
12	4	590	-300	290	12	0	0	0	1	290	0
13	1	580	100	680	13	1	0	0	0	680	0
14	2	570	300	870	14	0	1	0	0	870	0
15	3	560	-100	460	15	0	0	1	0	460	0
16	4	550	-300	250	16	0	0	0	1	250	0
			Average	625.0					Average	625.0	0.0
									Sum of Squares Error		0

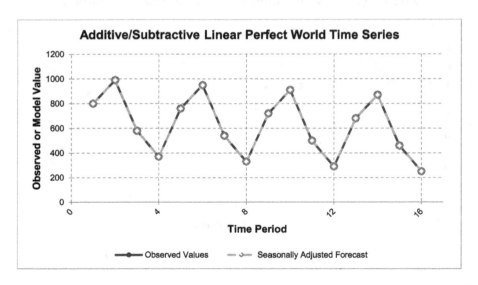

Figure 8.29 Solver Results for Our Perfect World Scenario

The ups and downs of Time Series Analysis | 465

This nice and conveniently is the hypothetical model with which we started. (*Sorry, was my sigh of relief that loud?*) So, we can see that in a hypothetically perfect world the technique works, so now let's see if it works back in reality . . .

8.6.2 The Real World scenario

We can repeat our procedure for our Domestic Gas Consumption in Table 8.41. Again, we are assuming an underlying linear trend with Additive/Subtractive Seasonal Variations. Figure 8.30 illustrates our initial starting parameters with an intercept of 1000, flat (zero) slope and no seasonal variations, i.e. a flat line steady state trend – about as unbiased as we can get. (In this way we are giving Solver absolute freedom to iterate values for each season in any direction, *although looking at the actual data, it is pretty obvious in which direction two of the seasonal variations will be.*)

Running Solver with the same objective and constraint as we did for the Perfect World scenario gives us the results in Table 8.42 and Figure 8.31.

If this output looks somewhat familiar to you, then you really do need to get out more. Otherwise look back at Table 8.26 in Section 8.5.1. It is equivalent to the Multi-Linear Regression Model we did. The only thing that is different is that here we have calculated the nominal underlying linear trend's intercept, whereas before it has embedded in the seasonal adjustments as shown in Table 8.43.

Table 8.41 Solver Set-Up for Domestic Gas Consumption Time Series

	Linear Trend		Additive/Subtractive Variations						
	Intercept	Slope	Season 1	Season 2	Season 3	Season 4			
	1020.0	0.00	0.0	0.0	0.0	0.0			
Time Period	Recurring Quarter (Season)	Actual Gas Units Consumed	Sequential Season No	Season 1	Season 2	Season 3	Season 4	Seasonally Adjusted Trend	Residual Variation
1	1	1046	1	1	0	0	0	1020	26
2	2	1549	2	0	1	0	0	1020	529
3	3	1009	3	0	0	1	0	1020	-11
4	4	670	4	0	0	0	1	1020	-350
5	1	1127	5	1	0	0	0	1020	107
6	2	1464	6	0	1	0	0	1020	444
7	3	961	7	0	0	1	0	1020	-59
8	4	594	8	0	0	0	1	1020	-426
9	1	1035	9	1	0	0	0	1020	15
10	2	1491	10	0	1	0	0	1020	471
11	3	912	11	0	0	1	0	1020	-108
12	4	592	12	0	0	0	1	1020	-428
13	1	990	13	1	0	0	0	1020	-30
14	2	1468	14	0	1	0	0	1020	448
15	3	840	15	0	0	1	0	1020	-180
16	4	572	16	0	0	0	1	1020	-448
	Average	1020.0				Average		1020.0	0.0
						Sum of Squares Error			1648302

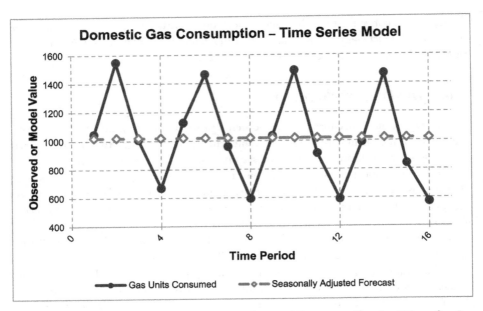

Figure 8.30 Solver Time Series Set-Up with Initial Parameter Starting Values for Gas Consumption

Table 8.42 Solver Result for Domestic Gas Consumption Time Series

Time Period	Recurring Quarter (Season)	Actual Gas Units Consumed	Sequential Season Number	Season 1	Season 2	Season 3	Season 4	Seasonally Adjusted Trend	Residual Variation
		Linear Trend		Additive/Subtractive Variations					
		Intercept	Slope	Season 1	Season 2	Season 3	Season 4		
		1090.5	-8.30	17.1	468.9	-85.3	-400.5		
1	1	1046	1	1	0	0	0	1099	-53
2	2	1549	2	0	1	0	0	1543	6
3	3	1009	3	0	0	1	0	980	29
4	4	670	4	0	0	0	1	657	13
5	1	1127	5	1	0	0	0	1066	61
6	2	1464	6	0	1	0	0	1510	-46
7	3	961	7	0	0	1	0	947	14
8	4	594	8	0	0	0	1	624	-30
9	1	1035	9	1	0	0	0	1033	2
10	2	1491	10	0	1	0	0	1476	15
11	3	912	11	0	0	1	0	914	-2
12	4	592	12	0	0	0	1	590	2
13	1	990	13	1	0	0	0	1000	-10
14	2	1468	14	0	1	0	0	1443	25
15	3	840	15	0	0	1	0	881	-41
16	4	572	16	0	0	0	1	557	15
	Average	1020.0			Average			1020.0	0.0
					Sum of Squares Error				13543

The ups and downs of Time Series Analysis | 467

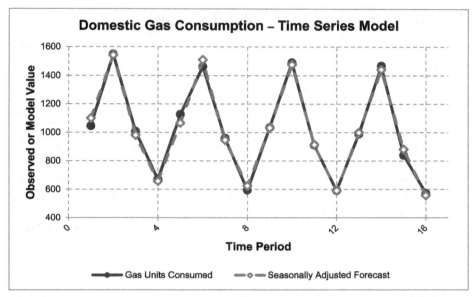

Figure 8.31 Solver Time Series Output with Optimised Parameter Values for Gas Consumption

Table 8.43 Equivalence Between Solver Technique and Stepwise Decomposition Regression

Technique	Intercept	Slope	Season 1	Season 2	Season 3	Season 4
Multi-linear Regression	0.00	-8.30	1107.60	1559.40	1005.20	690.00
Least Squares Residual Error Solver	1090.54	-8.30	17.05	468.85	-85.35	-400.55
Difference	-1090.54	0.00	1090.55	1090.55	1090.55	1090.55

We may find that Solver is the easier of the two techniques to use insomuch that there is only one step to perform, but the Multi-Linear Regression technique does have the advantage of calibrating whether our model is statistically significant or not. In this genuine example, it is pretty much a 'no brainer' but in other situations the results may be more marginal and the statistical summary provided by Excel's Regression utility will help to inform our decision appropriately.

468 | The ups and downs of Time Series Analysis

8.6.3 Wider examples of the Solver technique

The Excel Solver technique can be used in a wider number of circumstances where the other techniques are inappropriate:

- Time Series Analysis with independent seasonal data trends, i.e. diverging or converging trends in the seasonal data (e.g. summers getting hotter, winters getting colder)
- Untransformable non-linear trends (so long as we can articulate the nature of the relationship by means of a formula)
- Discontinuous trends in which, for example, there has been one underlying trend previously and more recently, a different underlying trend. This does imply of course that we need more data
- Step changes in seasonal variations due to some external event (e.g. increased sales following a major TV Advertising campaign)

8.7 Chapter review

In reviewing this chapter, we will have experienced some highs and some lows in respect of Time Series data ... and that is not just a play on words reflecting that Seasonal Variations will account for different levels in the data. It also relates to the fact that there are some techniques, for which we might have had high hopes of success, that have let us down unexpectedly.

During this 'journey through time' we recognised that the most common forms of Time Series models are those with an underlying linear or exponential trend against which we could make Additive/Subtractive or multiplicative seasonal variations. Where data spans a longer period we may also not that there is the possibility of expressing cyclical variations in much the same way. Any difference between reality and our data Model is referred to as a residual variation or error (*for those who like to call a spade a shovel.*)

The main techniques that we have discussed have centred on:

- Classical Decomposition
- Multi-Linear Regression
- Microsoft Solver

Classical Decomposition method requires us to break down the Time Series into its component parts of trend, seasonal variation, cyclical variation and residual variation, and determining their values in discrete steps. Based on previous chapters we may have expected that the recommended technique for determining the underlying trend would be a Simple Linear Regression of the raw data, a technique recommended by a number of authors. However, this faith is misplaced and dramatically different trends can be

The ups and downs of Time Series Analysis | 469

Table 8.44 Times Series Trend Analysis – Combination of Regression and Various Moving Measures

Linear Trend	Exponential Trend	Offset Back to
Offset Moving Average	Offset Moving Geometric Mean	Mid-point of the Moving Series
Offset Cumulative Average	Offset Cumulative Geometric Mean	Mid-point between first and latest data points
Disaggregated Cumulative Quadratic	N/A	Mid-point between first and latest data points
Average or Geometric Mean of individual Seasonal Trends depending on the type of Seasonal Variation to be used		N/A

reported based on different start points. We showed that rather than a Linear Regression of the raw seasonally affected data, a Linear Regression or Linear Transformation Regression, of the following would be more reliable:

In terms of a one-shot Multi-Linear Regression we found that this works well for Additive/Subtractive models but can be unreliable and downright temperamental for Multiplicative Models. For the latter, we introduced a hybrid concept that we called the Stepwise Decomposition Regression, taking the concept of Decomposition from the Classical techniques in order to identify the underlying trend first using one of the techniques above, and then a second regression to determine the values of the seasonality factors.

Finally, we discovered that Microsoft Solver may be the easiest technique to use, building a Time Series Model from scratch and populating the parameters using Solver's in-built algorithm to find the Least Squares Best Fit. However, this does not allow us to consider the statistical significance of the model that Regression techniques provide.

I can see from the look on your faces that it is time to move on!

Reference

Office of National Statistics (2009) 'AEI Average Earnings Index', August 2011, London, ONS.

Glossary of estimating and forecasting terms

This Glossary reflects those Estimating Terms that are either in common usage or have been defined for the purposes of this series of guides. Not all the terms are used in every volume, but where they do occur, their meaning is intended to be consistent.

3-Point Estimate   A 3-Point Estimate is an expression of uncertainty around an Estimate Value. It usually expresses Optimistic, Most Likely and Pessimistic Values.

Accuracy   Accuracy is an expression of how close a measurement, statistic or estimate is to the true value, or to a defined standard.

Actual Cost (AC)   See Earned Value Management Abbreviations and Terminology

ACWP (Actual Cost of Work Performed) or Actual Cost (AC)   See Earned Value Management Terminology

Additive/Subtractive Time Series Model   See Time Series Analysis

Adjusted R-Square   Adjusted R-Square is a measure of the "Goodness of Fit" of a Multi-Linear Regression model to a set of data points, which reduces the Coefficient of Determination by a proportion of the Unexplained Variance relative to the Degrees of Freedom in the model, divided by the Degrees of Freedom in the Sum of Squares Error.

ADORE (Assumptions, Dependencies, Opportunities, Risks, Exclusions)   See Individual Terms.

Alternative Hypothesis   An Alternative Hypothesis is that supposition that the difference between an observed value and another observed or assumed value or effect, cannot be legitimately attributable to random sampling or experimental error. It is usually denoted as H_1.

Analogous Estimating Method or Analogy   See Analogical Estimating Method.

Analogical Estimating Method   The method of estimating by Analogy is a means of creating an estimate by comparing the similarities and/or differences between two things, one of which is used as the reference point against which rational adjustments for differences between the two things are made in order establish an estimate for the other.

Approach   See Estimating Approach.

Glossary | 471

Arithmetic Mean or Average The Arithmetic Mean or Average of a set of numerical data values is a statistic calculated by summating the values of the individual terms and dividing by the number of terms in the set.

Assumption An Assumption is something that we take to be broadly true or expect to come to fruition in the context of the Estimate.

Asymptote An Asymptote to a given curve is a straight line that tends continually closer in value to that of the curve as they tend towards infinity (positive or negative). The difference between the asymptote and its curve reduces towards but never reaches zero at any finite value.

AT (Actual Time) See Earned Value Management Abbreviations and Terminology.

Average See Arithmetic Mean.

Average (Mean) Absolute Deviation (AAD) The Mean or Average Absolute Deviation of a range of data is the average 'absolute' distance of each data point from the Arithmetic Mean of all the data points, ignoring the sign depicting whether each point is less than or greater than the Arithmetic Mean.

Axiom An Axiom is a statement or proposition that requires no proof, being generally accepted as being self-evidently true at all times.

BAC (Budget At Completion) See Earned Value Management Abbreviations and Terminology.

Base Year Values 'Base Year Values' are values that have been adjusted to be expressed relative to a fixed year as a point of reference e.g., for contractual price agreement.

Basis of Estimate (BoE) A Basis of Estimate is a series of statements that define the assumptions, dependencies and exclusions that bound the scope and validity of an estimate. A good BoE also defines the approach, method and potentially techniques used, as well as the source and value of key input variables, and as such supports Estimate TRACEability.

BCWP (Budgeted Cost of Work Performed) See Earned Value Management Abbreviations and Terminology.

BCWS (Budgeted Cost of Work Scheduled) See Earned Value Management Abbreviations and Terminology.

Benford's Law Benford's Law is an empirical observation that in many situations the first or leading digit in a set of apparently random measurements follows a repeating pattern that can be predicted as the Logarithm of one plus the reciprocal of the leading digit. It is used predominately in the detection of fraud.

Bessel's Correction Factor In general, the variance (and standard deviation) of a data sample will understate the variance (and standard deviation) of the underlying data population. Bessel's Correction Factor allows for an adjustment to be made so that the sample variance can be used as an unbiased estimator of the population variance. The adjustment requires that the Sum of Squares of the Deviations from the Sample Mean be divided one less than the number of observations or data points i.e. n-1 rather than the more intuitive the number of observations. Microsoft Excel takes this adjustment into account.

Bottom-up Approach In a Bottom-up Approach to estimating, the estimator identifies the lowest level at which it is appropriate to create a range of estimates based on the task definition available, or that can be inferred. The overall estimate, or higher level summaries, typically through a Work Breakdown Structure, can be produced through incremental aggregation of the lower level estimates. A Bottom-up Approach requires a good definition of the task to be estimated, and is frequently referred to as detailed estimating or as engineering build-up.

Chauvenet's Criterion A test for a single Outlier based on the deviation Z-Score of the suspect data point.

472 | Glossary

Chi-Squared Test or χ^2-Test The Chi-Squared Test is a "goodness of fit" test that compares the variance of a sample against the variance of a theoretical or assumed distribution.

Classical Decomposition Method (Time Series) Classical Decomposition Method is a means of analysing data for which there is a seasonal and/or cyclical pattern of variation. Typically, the underlying trend is identified, from which the average deviation or variation by season can be determined. The method can be used for multiplicative and additive/subtractive Time Series Models.

Closed Interval A Closed Continuous Interval is one which includes its endpoints, and is usually depicted with square brackets: [Minimum, Maximum].

Coefficient of Determination The Coefficient of Determination is a statistical index which measures how much of the total variance in one variable can be explained by the variance in the other variable. It provides a measure of how well the relationship between two variables can be represented by a straight line.

Coefficient of Variation (CV) The Coefficient of Variation of a set of sample data values is a dimensionless statistic which expresses the ratio of the sample's Standard Deviation to its Arithmetic Mean. In the rare cases where the set of data is the entire population, then the Coefficient of Variation is expressed as the ratio of the population's Standard Deviation to its Arithmetic Mean. It can be expressed as either a decimal or percentage.

Collaborative Working Collaborative Working is a term that refers to the management strategy of dividing a task between multiple partners working towards a common goal where there a project may be unviable for a single organisation. There is usually a cost penalty of such collaboration as it tends to create duplication in management and in integration activities.

Collinearity & Multicollinearity Collinearity is an expression of the degree to which two supposedly independent predicator variables are correlated in the context of the observed values being used to model their relationship with the dependent variable that we wish to estimate. Multicollinearity is an expression to which collinearity can be observed across several predicator variables.

Complementary Cumulative Distribution Function (CCDF) The Complementary Cumulative Distribution Function is the theoretical or observed probability of that variable being greater than a given value. It is calculated as the difference between 1 (or 100%) and the Cumulative Distribution Function, 1-CDF.

Composite Index A Composite Index is one that has been created as the weighted average of a number of other distinct Indices for different commodities.

Concave Curve A curve in which the direction of curvature appears to bend towards a viewpoint on the x-axis, similar to one that would be observed when viewing the inside of a circle or sphere.

Cone of Uncertainty A generic term that refers to the empirical observation that the range of estimate uncertainty or accuracy improves through the life of a project. It is typified by its cone or funnel shape appearance.

Confidence Interval A Confidence Interval is an expression of the percentage probability that data will lie between two distinct Confidence Levels, known as the Lower and Upper Confidence Limits, based on a known or assumed distribution of data from either a sample or an entire population.
See also Prediction Interval.

Confidence Level A Confidence Level is an expression of the percentage probability that data selected at random from a known or assumed distribution of data (either a sample or an entire population), will be less than or equal to a particular value.

Glossary | 473

Confidence Limits The Lower and Upper Confidence Limits are the respective Confidence Levels that bound a Confidence Interval, and are expressions of the two percentage probabilities that data will be less or equal to the values specified based on the known or assumed distribution of data in question from either a sample or an entire population.
See also Confidence Interval.

Constant Year Values 'Constant Year Values' are values that have been adjusted to take account of historical or future inflationary effects or other changes, and are expressed in relation to the Current Year Values for any defined year. They are often referred to as 'Real Year Values'.

Continuous Probability Distribution A mathematical expression of the relative theoretical probability of a random variable which can take on any value from a real number range. The range may be bounded or unbounded in either direction.

Convex Curve A curve in which the direction of curvature appears to bend away from a viewpoint on the x-axis, similar to one that would be observed when viewing the outside of a circle or sphere.

Copula A Copula is a Multivariate Probability Distribution based exclusively on a number Uniform Marginal Probability Distributions (one for each variable).

Correlation Correlation is a statistical relationship in which the values of two or more variables exhibit a tendency to change in relationship with one other. These variables are said to be positively (or directly) correlated if the values tend to move in the same direction, and negatively (or inversely) correlated if they tend to move in opposite directions.

Cost Driver See Estimate Drivers.

Covariance The Covariance between a set of paired values is a measure of the extent to which the paired data values are scattered around the paired Arithmetic Means. It is the average of the product of each paired variable from its Arithmetic Mean.

CPI (Cost Performance Index) See Earned Value Management Abbreviations and Terminology.

Crawford's Unit Learning Curve A Crawford Unit Learning Curve is an empirical relationship that expresses the reduction in time or cost of each unit produced as a power function of the cumulative number units produced.

Critical Path The Critical Path at a point in time depicts the string of dependent activities or tasks in a schedule for which there is no float or queuing time. As such the length of the Critical Path represents the quickest time that the schedule can be currently completed based on the current assumed activity durations.

Cross-Impact Analysis A Cross-Impact Analysis is a qualitative technique used to identify the most significant variables in a system by considering the impact of each variable on the other variables.

Cumulative Average A Point Cumulative Average is a single term value calculated as the average of the current and all previous consecutive recorded input values that have occurred in a natural sequence.
A Moving Cumulative Average, sometimes referred to as a Cumulative Moving Average, is an array (a series or range of ordered values) of successive Point Cumulative Average terms calculated from all previous consecutive recorded input values that have occurred in a natural sequence.

Cumulative Distribution Function (CDF) The Cumulative Distribution Function of a Discrete Random Variable expresses the theoretical or observed probability of that

474 | Glossary

variable being less than or equal to any given value. It equates to the sum of the probabilities of achieving that value and each successive lower value.

The Cumulative Distribution Function of a Continuous Random Variable expresses the theoretical or observed probability of that variable being less than or equal to any given value. It equates to the area under the Probability Density Function curve to the left of the value in question.

See also the Complementary Cumulative Distribution Function.

Current Year (or Nominal Year) Values 'Current Year Values' are historical values expressed in terms of those that were current at the historical time at which they were incurred. In some cases, these may be referred to as 'Nominal Year Values'.

CV (Cost Variance) See Earned Value Management Abbreviations and Terminology

Data Type Primary Data is that which has been taken directly from its source, either directly or indirectly, without any adjustment to its values or context.

Secondary Data is that which has been taken from a known source, but has been subjected to some form of adjustment to its values or context, the general nature of which is known and has been considered to be appropriate.

Tertiary Data is data of unknown provenance. The specific source of data and its context is unknown, and it is likely that one or more adjustments of an unknown nature have been made, in order to make it suitable for public distribution.

Data Normalisation Data Normalisation is the act of making adjustments to, or categorisations of, data to achieve a state where data the can be used for comparative purposes in estimating.

Decile A Decile is one of ten subsets from a set of ordered values which nominally contain a tenth of the total number of values in each subset. The term can also be used to express the values that divide the ordered values into the ten ordered subsets.

Degrees of Freedom Degrees of Freedom are the number of different factors in a system or calculation of a statistic that can vary independently.

DeJong Unit Learning Curve A DeJong Unit Learning Curve is a variation of the Crawford Unit Learning Curve that allows for an incompressible or 'unlearnable' element of the task, expressed as a fixed cost or time.

Delphi Technique The Delphi Technique is a qualitative technique that promotes consensus or convergence of opinions to be achieved between diverse subject matter experts in the absence of a clear definition of a task or a lack of tangible evidence.

Dependency A Dependency is something to which an estimate is tied, usually an uncertain event outside of our control or influence, which if it were not to occur, would potentially render the estimated value invalid. If it is an internal dependency, the estimate and schedule should reflect this relationship

Descriptive Statistic A Descriptive Statistic is one which reports an indisputable and repeatable fact, based on the population or sample in question, and the nature of which is described in the name of the Statistic.

Discount Rate The Discount Rate is the percentage reduction used to calculate the present-day values of future cash flows. The discount rate often either reflects the comparable market return on investment of opportunities with similar levels of risk, or reflects an organisation's Weighted Average Cost of Capital (WACC), which is based on the weighted average of interest rates paid on debt (loans) and shareholders' return on equity investment.

Discounted Cash Flow (DCF) Discounted Cash Flow (DCF) is a technique for converting estimated or actual expenditures and revenues to economically comparable values at a common point in time by discounting future cash flows by an agreed percentage

Glossary | 475

discount rate per time period, based on the cost to the organisation of borrowing money, or the average return on comparable investments.

Discrete Probability Distribution A mathematical expression of the theoretical or empirical probability of a random variable which can only take on predefined values from a finite range.

Dixon's Q-Test A test for a single Outlier based on the distance between the suspect data point and its nearest neighbour in comparison with the overall range of the data.

Driver See Estimate Drivers.

Earned Value (EV) See Earned Value Management Terminology.

Earned Value Management (EVM) Earned Value Management is a collective term for the management and control of project scope, schedule and cost.

Earned Value Analysis Earned Value Analysis is a collective term used to refer to the analysis of data gathered and used in an Earned Value Management environment.

Earned Value Management Abbreviations and Terminology (Selected terms only)

ACWP (Actual Cost of Work Performed) sometimes referred to as Actual Cost (AC) Each point represents the cumulative actual cost of the work completed or in progress at that point in time. The curve represents the profile by which the actual cost has been expended for the value achieved over time.

AT (Actual Time) AT measures the time from start to time now.

BAC (Budget At Completion) The BAC refers to the agreed target value for the current scope of work, against which overall performance will be assessed.

BCWP (Budget Cost of Work Performed) sometimes referred to as Earned Value (EV) Each point represents the cumulative budgeted cost of the work completed or in progress to that point in time. The curve represents the profile by which the budgeted cost has been expended over time. The BCWP is expressed in relation to the BAC (Budget At Completion).

BCWS sometimes referred to as Planned Value (PV) Each point represents the cumulative budgeted cost of the work planned to be completed or to be in progress to that point in time. The curve represents the profile by which the budgeted cost was planned to be expended over time. The BCWS is expressed in relation to the BAC (Budget At Completion).

CPI (Cost Performance Index) The CPI is an expression of the relative performance from a cost perspective and is the ratio of Earned Value to Actual Cost (EV/AC) or (BCWP/ACWP).

CV (Cost Variance) CV is a measure of the cumulative Cost Variance as the difference between the Earned Value and the Actual Cost (EV − AC) or (BCWP − ACWP).

ES (Earned Schedule) ES measures the planned time allowed to reach the point that we have currently achieved.

EAC (Estimate At Completion) sometimes referred to as FAC (Forecast At Completion) The EAC or FAC is the sum of the actual cost to date for the work achieved, plus an estimate of the cost to complete any outstanding or incomplete activity or task in the defined scope of work

ETC (Estimate To Completion) The ETC is an estimate of the cost that is likely to be expended on the remaining tasks to complete the current scope of agreed work. It is the difference between the Estimate At Completion and the current Actual Cost (EAC − ACWP or AC).

476 | Glossary

SPI (Schedule Performance Index) The SPI is an expression of the relative schedule performance expressed from a cost perspective and is the ratio of Earned Value to Planned Value (EV/PV) or (BCWP/BCWS). It is now considered to be an inferior measure of true schedule variance in comparison with SPI(t).

SPI(t) The SPI(t) is an expression of the relative schedule performance and is the ratio of Earned Schedule to Actual Time (ES/AT).

SV (Schedule Variance) SV is a measure of the cumulative Schedule Variance measured from a Cost Variance perspective, and is the difference between the Earned Value and the Planned Value (EV − PV) or (BCWP − BCWS). It is now considered to be an inferior measure of true schedule variance in comparison with SV(t).

SV(t) SV(t) is a measure of the cumulative Schedule Variance and is the difference between the Earned Schedule and the Actual Time (ES − AT).

Equivalent Unit Learning Equivalent Unit Learning is a technique that can be applied to complex programmes of recurring activities to take account of Work-in-Progress and can be used to give an early warning indicator of potential learning curve breakpoints. It can be used to supplement traditional completed Unit Learning Curve monitoring.

ES (Earned Schedule) See Earned Value Management Abbreviations and Terminology

Estimate An Estimate for 'something' is a numerical expression of the approximate value that might reasonably be expected to occur based on a given context, which is described and is bounded by a number of parameters and assumptions, all of which are pertinent to and necessarily accompany the numerical value provided.

Estimate At Completion (EAC) and **Estimate To Completion (ETC)** See Earned Value Management Abbreviations and Terminology.

Estimate Drivers A Primary Driver is a technical, physical, programmatic or transactional characteristic that either causes a major change in the value being estimated or in a major constituent element of it, or whose value itself changes correspondingly with the value being estimated, and therefore, can be used as an indicator of a change in that value.

A Secondary Driver is a technical, physical, programmatic or transactional characteristic that either causes a minor change in the value being estimated or in a constituent element of it, or whose value itself changes correspondingly with the value being estimated and can be used as an indicator of a subtle change in that value.

Cost Drivers are specific Estimate Drivers that relate to an indication of Cost behaviour.

Estimate Maturity Assessment (EMA) An Estimate Maturity Assessment provides a 'health warning' on the maturity of an estimate based on its Basis of Estimate, and takes account of the level of task definition available and historical evidence used.

Estimating Approach An Estimating Approach describes the direction by which the lowest level of detail to be estimated is determined.

See also Bottom-up Approach, Top-down Approach and Ethereal Approach.

Estimating Method An Estimating Method is a systematic means of creating an estimate, or an element of an estimate. An Estimating Methodology is a set or system of Estimating Methods.

See also Analogous Method, Parametric Method and Trusted Source Method.

Estimating Metric An Estimating Metric is a value or statistic that expresses a numerical relationship between a value for which an estimate is required, and a Primary or Secondary Driver (or parameter) of that value, or in relation to some fixed reference point.

See also Factor, Rate and Ratio.

Glossary | 477

Estimating Procedure An Estimating Procedure is a series of steps conducted in a certain manner and sequence to optimise the output of an Estimating Approach, Method and/ or Technique.

Estimating Process An Estimating Process is a series of mandatory or possibly optional actions or steps taken within an organisation, usually in a defined sequence or order, in order to plan, generate and approve an estimate for a specific business purpose.

Estimating Technique An Estimating Technique is a series of actions or steps conducted in an efficient manner to achieve a specific purpose as part of a wider Estimating Method. Techniques can be qualitative as well as quantitative.

Ethereal Approach An Ethereal Approach to Estimating is one in which values are accepted into the estimating process, the provenance of which is unknown and at best may be assumed. These are values often created by an external source for low value elements of work, or by other organisations with acknowledged expertise. Other values may be generated by Subject Matter Experts internal to the organisation where there is insufficient definition or data to produce an estimate by a more analytical approach.
The Ethereal Approach should be considered the approach of last resort where low maturity is considered acceptable. The approach should be reserved for low value elements or work, and situations where a robust estimate is not considered critical.

Excess Kurtosis The Excess Kurtosis is an expression of the relative degree of Peakedness or flatness of a set of data values, relative to a Normal Distribution. Flatter distributions with a negative Excess Kurtosis are referred to as Platykurtic; Peakier distributions with a positive Excess Kurtosis are termed Leptokurtic; whereas those similar to a Normal Distribution are said to be Mesokurtic. The measure is based on the fourth power of the deviation around the Arithmetic Mean.

Exclusion An Exclusion is condition or set of circumstances that have been designated to be out of scope of the current estimating activities and their output.

Exponential Function An Exponential Function of two variables is one in which the Logarithm of the dependent variable on the vertical axis produces a monotonic increasing or decreasing Straight Line when plotted against the independent variable on the horizontal axis.

Exponential Smoothing Exponential Smoothing is a 'single-point' predictive technique which generates a forecast for any period based on the forecast made for the prior period, adjusted for the error in that prior period's forecast.

Extrapolation The act of estimating a value extrinsic to or outside the range of the data being used to determine that value. See also Interpolation.

Factored or Expected Value Technique A technique that expresses an estimate based on the weighted sum of all possible values multiplied by the probability of arising.

Factors, Rates and Ratios See individual terms: Factor Metric, Rate Metric and Ratio Metric

Factor Metric A Factor is an Estimating Metric used to express one variable's value as a percentage of another variable's value.

F-Test The F-Test is a "goodness of fit" test that returns the cumulative probability of getting an F-Statistic less than or equal to the ratio inferred by the variances in two samples.

Generalised Exponential Function A variation to the standard Exponential Function which allows for a constant value to exist in the dependent or predicted variable's value. It effectively creates a vertical shift in comparison with a standard Exponential Function.

Generalised Extreme Studentised Deviate A test for multiple Outliers based on the deviation Z-Score of the suspect data point.

478 | Glossary

Generalised Logarithmic Function A variation to the standard Logarithmic Function which allows for a constant value to exist in the independent or predictor variable's value. It effectively creates a horizontal shift in comparison with a standard Logarithmic Function.

Generalised Power Function A variation to the standard Power Function which allows for a constant value to exist in either or both the independent and dependent variables' value. It effectively creates a horizontal and/or vertical shift in comparison with a standard Power Function.

Geometric Mean The Geometric Mean of a set of n numerical data values is a statistic calculated by taking the nth root of the product of the n terms in the set.

Good Practice Spreadsheet Modelling (GPSM) Good Practice Spreadsheet Modelling Principles relate to those recommended practices that should be considered when developing a Spreadsheet in order to help maintain its integrity and reduce the risk of current and future errors.

Grubbs' Test A test for a single Outlier based on the deviation Z-Score of the suspect data point.

Harmonic Mean The Harmonic Mean of a set of n numerical data values is a statistic calculated by taking the reciprocal of the Arithmetic Mean of the reciprocals of the n terms in the set.

Heteroscedasticity Data is said to exhibit Heteroscedasticity if data variances are not equal for all data values.

Homoscedasticity Data is said to exhibit Homoscedasticity if data variances are equal for all data values.

Iglewicz and Hoaglin's M-Score (Modified Z-Score) A test for a single Outlier based on the Median Absolute Deviation of the suspect data point.

Index An index is an empirical average factor used to increase or decrease a known reference value to take account of cumulative changes in the environment, or observed circumstances, over a period of time. Indices are often used as to normalise data.

Inferential Statistic An Inferential Statistic is one which infers something, often about the wider data population, based on one or more Descriptive Statistics for a sample, and as such, it is open to interpretation ... and disagreement.

Inherent Risk in Spreadsheets (IRiS) IRiS is a qualitative assessment tool that can be used to assess the inherent risk in spreadsheets by not following Good Practice Spreadsheets Principles.

Interdecile Range The Interdecile Range comprises the middle eight Decile ranges and represents the 80% Confidence Interval between the 10% and 90% Confidence Levels for the data.

Internal Rate of Return The Internal Rate of Return (IRR) of an investment is that Discount Rate which returns a Net Present Value (NPV) of zero, i.e. the investment breaks even over its life with no over or under recovery.

Interpolation The act of estimating an intermediary or intrinsic value within the range of the data being used to determine that value. See also Extrapolation.

Interquantile Range An Interquantile Range is a generic term for the group of Quantiles that form a symmetrical Confidence Interval around the Median by excluding the first and last Quantile ranges.

Interquartile Range The Interquartile Range comprises the middle two Quartile ranges and represents the 50% Confidence Interval between the 25% and 75% Confidence Levels for the data.

Glossary | 479

Interquintile Range   The Interquintile Range comprises the middle three Quintile ranges and represents the 60% Confidence Interval between the 20% and 80% Confidence Levels for the data.

Jarque-Bera Test   The Jarque-Bera Test is a statistical test for whether data can be assumed to follow a Normal Distribution. It exploits the properties of a Normal Distribution's Skewness and Excess Kurtosis being zero.

Kendall's Tau Rank Correlation Coefficient   Kendall's Tau Rank Correlation Coefficient for two variables is a statistic that measures the difference between the number of Concordant and Discordant data pairs as a proportion of the total number of possible unique pairings, where two pairs are said to be concordant if the ranks of the two variables move in the same direction, or are said to be discordant if the ranks of the two variables move in opposite directions.

Laspeyres Index   Laspeyres Indices are time-based indices which compare the prices of commodities at a point in time with the equivalent prices for the Index Base Period, based on the original quantities consumed at the Index Base Year.

Learning Curve   A Learning Curve is a mathematical representation of the degree at which the cost, time or effort to perform one or more activities reduces through the acquisition and application of knowledge and experience through repetition and practice.

Learning Curve Breakpoint   A Learning Curve Breakpoint is the position in the build or repetition sequence at which the empirical or theoretical rate of learning changes.

Learning Curve Cost Driver   A Learning Curve Cost Driver is an independent variable which affects or indicates the rate or amount of learning observed.

Learning Curve Segmentation   Learning Curve Segmentation refers to a technique which models the impact of discrete Learning Curve Cost Drivers as a product of multiple unit-based learning curves.

Learning Curve Step-point   A Learning Curve Step-point is the position in the build or repetition sequence at which there is a step function increase or decrease in the level of values evident on the empirical or theoretical Learning Curve.

Learning Exponent   A Learning Exponent is the power function exponent of a Learning Curve reduction and is calculated as the Logarithmic value of the Learning Rate using a Logarithmic Base equivalent to the Learning Rate Multiplier.

Learning Rate and Learning Rate Multiplier   The Learning Rate expresses the complement of the percentage reduction over a given Learning Rate Multiplier (usually 2). For example, an 80% Learning Rate with a Learning Multiplier of 2 implies a 20% reduction every time the quantity doubles.

Least Squares Regression   Least Squares Regression is a Regression procedure which identifies the 'Best Fit' of a pre-defined functional form by minimising the Sum of the Squares of the vertical difference between each data observation and the assumed functional form through the Arithmetic Mean of the data.

Leptokurtotic or Leptokurtic   An expression that the degree of Excess Kurtosis in a probability distribution is peakier than a Normal Distribution.

Linear Function   A Linear Function of two variables is one which can be represented as a monotonic increasing or decreasing Straight Line without any need for Mathematical Transformation.

Logarithm   The Logarithm of any positive value for a given positive Base Number not equal to one is that power to which the Base Number must be raised to get the value in question.

480 | Glossary

Logarithmic Function A Logarithmic Function of two variables is one in which the dependent variable on the vertical axis produces a monotonic increasing or decreasing Straight Line, when plotted against the Logarithm of the independent variable on the horizontal axis.

Mann-Whitney U-Test sometimes known as Mann-Whitney-Wilcoxon U-Test A U-Test is used to test whether two samples could be drawn from the same population by comparing the distribution of the joint ranks across the two samples.

Marching Army Technique sometimes referred to as Standing Army Technique The Marching Army Technique refers to a technique that assumes that costs vary directly in proportion with a schedule.

Mathematical Transformation A Mathematical Transformation is a numerical process in which the form, nature or appearance of a numerical expression is converted into an equivalent but non-identical numerical expression with a different form, nature or appearance.

Maximum The Maximum is the largest observed value in a sample of data, or the largest potential value in a known or assumed statistical distribution. In some circumstances, the term may be used to imply a pessimistic value at the upper end of potential values rather than an absolute value.

Mean Absolute Deviation See Average Absolute Deviation (AAD).

Measures of Central Tendency Measures of Central Tendency is a collective term that refers to those descriptive statistics that measure key attributes of a data sample (Means, Modes and Median).

Measures of Dispersion and Shape Measures of Dispersion and Shape is a collective term that refers to those descriptive statistics that measure the degree and/or pattern of scatter in the data in relation to the Measures of Central Tendency.

Median The Median of a set of data is that value which occurs in the middle of the sequence when its values have been arranged in ascending or descending order. There are an equal number of data points less than and greater than the Median.

Median Absolute Deviation (MAD) The Median Absolute Deviation of a range of data is the Median of the 'absolute' distance of each data point from the Median of those data points, ignoring the "sign" depicting whether each point is less than or greater than the Median.

Memoryless Probability Distribution In relation to Queueing Theory, a Memoryless Probability Distribution is one in which the probability of waiting a set period of time is independent of how long we have been waiting already. The probability of waiting longer than the sum of two values is the product of the probabilities of waiting longer than each value in turn. An Exponential Distribution is the only Continuous Probability Distribution that exhibits this property, and a Geometric Distribution is the only discrete form.

Mesokurtotic or Mesokurtic An expression that the degree of Excess Kurtosis in a probability distribution is comparable with a Normal Distribution.

Method See Estimating Method.

Metric A Metric is a statistic that measures an output of a process or a relationship between a variable and another variable or some reference point.
See also Estimating Metric.

Minimum The Minimum is the smallest observed value in a sample of data, or the smallest potential value in a known or assumed statistical distribution. In some circumstances, the

Glossary 481

term may be used to imply an optimistic value at the lower end of potential values rather than an absolute value.

Mode   The Mode of a set of data is that value which has occurred most frequently, or that which has the greatest probability of occurring.

Model Validation and Verification   See individual terms: Validation and Verification.

Monotonic Function   A Monotonic Function of two paired variables is one that when values are arranged in ascending numerical order of one variable, the value of the other variable either perpetually increases or perpetually decreases.

Monte Carlo Simulation   Monte Carlo Simulation is a technique that models the range and relative probabilities of occurrence, of the potential outcomes of a number of input variables whose values are uncertain but can be defined as probability distributions.

Moving Average   A Moving Average is a series or sequence of successive averages calculated from a fixed number of consecutive input values that have occurred in a natural sequence. The fixed number of consecutive input terms used to calculate each average term is referred to as the Moving Average Interval or Base.

Moving Geometric Mean   A Moving Geometric Mean is a series or sequence of successive geometric means calculated from a fixed number of consecutive input values that have occurred in a natural sequence. The fixed number of consecutive input terms used to calculate each geometric mean term is referred to as the Moving Geometric Mean Interval or Base.

Moving Harmonic Mean   A Moving Harmonic Mean is a series or sequence of successive harmonic means calculated from a fixed number of consecutive input values that have occurred in a natural sequence. The fixed number of consecutive input terms used to calculate each harmonic mean term is referred to as the Moving Harmonic Mean Interval or Base.

Moving Maximum   A Moving Maximum is a series or sequence of successive maxima calculated from a fixed number of consecutive input values that have occurred in a natural sequence. The fixed number of consecutive input terms used to calculate each maximum term is referred to as the Moving Maximum Interval or Base.

Moving Median   A Moving Median is a series or sequence of successive medians calculated from a fixed number of consecutive input values that have occurred in a natural sequence. The fixed number of consecutive input terms used to calculate each median term is referred to as the Moving Median Interval or Base.

Moving Minimum   A Moving Minimum is a series or sequence of successive minima calculated from a fixed number of consecutive input values that have occurred in a natural sequence. The fixed number of consecutive input terms used to calculate each minimum term is referred to as the Moving Minimum Interval or Base.

Moving Standard Deviation   A Moving Standard Deviation is a series or sequence of successive standard deviations calculated from a fixed number of consecutive input values that have occurred in a natural sequence. The fixed number of consecutive input terms used to calculate each standard deviation term is referred to as the Moving Standard Deviation Interval or Base.

Multicollinearity   See Collinearity.

Multiplicative Time Series Model   See Time Series Analysis.

Multi-Variant Unit Learning   Multi-Variant Unit Learning is a technique that considers shared and unique learning across multiple variants of the same or similar recurring products.

482 | Glossary

Net Present Value   The Net Present Value (NPV) of an investment is the sum of all positive and negative cash flows through time, each of which have been discounted based on the time value of money relative to a Base Year (usually the present year).

Nominal Year Values   'Nominal Year Values' are historical values expressed in terms of those that were current at the historical time at which they were incurred. In some cases, these may be referred to as 'Current Year Values'.

Norden-Rayleigh Curve   A Norden-Rayleigh is an empirical relationship that models the distribution of resource required in the non-recurring concept demonstration or design and development phases.

Null Hypothesis   A Null Hypothesis is that supposition that the difference between an observed value or effect and another observed or assumed value or effect, can be legitimately attributable to random sampling or experimental error. It is usually denoted as H_0.

Open Interval   An Open Continuous Interval is one which excludes its endpoints, and is usually depicted with rounded brackets: (Minimum, Maximum).

Opportunity   An Opportunity is an event or set of circumstances that may or may not occur, but if it does occur an Opportunity will have a beneficial effect on our plans, impacting positively on the cost, quality, schedule, scope compliance and/or reputation of our project or organisation.

Optimism Bias   Optimism Bias is an expression of the inherent bias (often unintended) in an estimate output based on either incomplete or misunderstood input assumptions.

Outlier   An outlier is a value that falls substantially outside the pattern of other data. The outlier may be representative of unintended atypical factors or may simply be a value which has a very low probability of occurrence.

Outturn Year Values   'Outturn Year Values' are values that have been adjusted to express an expectation of what might be incurred in the future due to escalation or other predicted changes. In some cases, these may be referred to as 'Then Year Values'.

Paasche Index   Paasche Indices are time-based indices which compare prices of commodities at a point in time with the equivalent prices for the Index Base Period, based on the quantities consumed at the current point in time in question

Parametric Estimating Method   A Parametric Estimating Method is a systematic means of establishing and exploiting a pattern of behaviour between the variable that we want to estimate, and some other independent variable or set of variables or characteristics that have an influence on its value.

Payback Period   The Payback Period is an expression of how long it takes for an investment opportunity to break even, i.e. to pay back the investment.

Pearson's Linear Correlation Coefficient   Pearson's Linear Correlation Coefficient for two variables is a measure of the extent to which a change in the value of one variable can be associated with a change in the value of the other variable through a linear relationship. As such it is a measure of linear dependence or linearity between the two variables, and can be calculated by dividing the Covariance of the two variables by the Standard Deviation of each variable.

Peirce's Criterion   A test for multiple Outliers based on the deviation Z-Score of the suspect data point.

Percentile   A Percentile is one of a hundred subsets from a set of ordered values which each nominally contain a hundredth of the total number of values in each subset. The term can also be used to express the values that divide the ordered values into the hundred ordered subsets.

Planned Value (PV)   See Earned Value Management Abbreviations and Terminology.

Glossary | 483

Platykurtotic or Platykurtic An expression that the degree of Excess Kurtosis in a probability distribution is shallower than a Normal Distribution.

Power Function A Power Function of two variables is one in which the Logarithm of the dependent variable on the vertical axis produces a monotonic increasing or decreasing Straight Line when plotted against the Logarithm of the independent variable on the horizontal axis.

Precision

(1) Precision is an expression of how close repeated trials or measurements are to each other.

(2) Precision is an expression of the level of exactness reported in a measurement, statistic or estimate.

Primary Data See Data Type.

Primary Driver See Estimate Drivers.

Probability Density Function (PDF) The Probability Density Function of a Continuous Random Variable expresses the rate of change in the probability distribution over the range of potential continuous values defined, and expresses the relative likelihood of getting one value in comparison with another.

Probability Mass Function (PMF) The Probability Mass Function of a Discrete Random Variable expresses the probability of the variable being equal to each specific value in the range of all potential discrete values defined. The sum of these probabilities over all possible values equals 100%.

Probability of Occurrence A Probability of Occurrence is a quantification of the likelihood that an associated Risk or Opportunity will occur with its consequential effects.

Quadratic Mean or Root Mean Square The Quadratic Mean of a set of n numerical data values is a statistic calculated by taking the square root of the Arithmetic Mean of the squares of the n values. As a consequence, it is often referred to as the Root Mean Square.

Quantile A Quantile is the generic term for a number of specific measures that divide a set of ordered values into a quantity of ranges with an equal proportion of the total number of values in each range. The term can also be used to express the values that divide the ordered values into such ranges.

Quantity-based Learning Curve A Quantity-based Learning Curve is an empirical relationship which reflects that the time, effort or cost to perform an activity reduces as the number of repetitions of that activity increases.

Quartile A Quartile is one of four subsets from a set of ordered values which nominally contain a quarter of the total number of values in each subset. The term can also be used to express the values that divide the ordered values into the four ordered subsets.

Queueing Theory Queueing Theory is that branch of Operation Research that studies the formation and management of queuing systems and waiting times.

Quintile A Quintile is one of five subsets from a set of ordered values which nominally contain a fifth of the total number of values in each subset. The term can also be used to express the values that divide the ordered values into the five ordered subsets.

Range The Range is the difference between the Maximum and Minimum observed values in a dataset, or the Maximum and Minimum theoretical values in a statistical distribution. In some circumstances, the term may be used to imply the difference between pessimistic and optimistic values from the range of potential values rather than an absolute range value.

484 | Glossary

Rate Metric A Rate is an Estimating Metric used to quantify how one variable's value changes in relation to some measurable driver, attribute or parameter, and would be expressed in the form of a [Value] of one attribute per [Unit] of another attribute.

Ratio Metric A Ratio is an Estimating Metric used to quantify the relative size proportions between two different instances of the same driver, attribute or characteristic such as weight. It is typically used as an element of Estimating by Analogy or in the Normalisation of data.

Real Year Values 'Real Year Values' are values that have been adjusted to take account of historical or future inflationary effects or other changes, and are expressed in relation to the Current Year Values for any defined year. They are often referred to as 'Constant Year Values'.

Regression Analysis Regression Analysis is a systematic procedure for establishing the Best Fit relationship of a predefined form between two or more variables, according to a set of Best Fit criteria.

Regression Confidence Interval The Regression Confidence Interval of a given probability is an expression of the Uncertainty Range around the Regression Line. For a known value of a single independent variable, or a known combination of values from multiple independent variables, the mean of all future values of the dependent variable will occur within the Confidence Interval with the probability specified.

Regression Prediction Interval A Regression Prediction Interval of a given probability is an expression of the Uncertainty Range around future values of the dependent variable based on the regression data available. For a known value of a single independent variable, or a known combination of values from multiple independent variables, the future value of the dependent variable will occur within the Prediction Interval with the probability specified.

Residual Risk Exposure The Residual Risk Exposure is the weighted value of the Risk, calculated by multiplying its Most Likely Value by the complement of its Probability of Occurrence (100% – Probability of Occurrence). It is used to highlight the relative value of the risk that is not covered by Risk Exposure calculation.

Risk A Risk is an event or set of circumstances that may or may not occur, but if it does occur a Risk will have a detrimental effect on our plans, impacting negatively on the cost, quality, schedule, scope compliance and/or reputation of our project or organisation.

Risk Exposure A Risk Exposure is the weighted value of the Risk, calculated by multiplying its Most Likely Value by its Probability of Occurrence.
See also Residual Risk Exposure.

Risk & Opportunity Ranking Factor A Risk & Opportunity Ranking Factor is the relative absolute exposure of a Risk or Opportunity in relation to all others, calculated by dividing the absolute value of the Risk Exposure by the sum of the absolute values of all such Risk Exposures.

Risk Uplift Factors A Top-down Approach to Risk Analysis may utilise Risk Uplift Factors to quantify the potential level of risk based on either known risk exposure for the type of work being undertaken based on historical records of similar projects, or based on a Subject Matter Expert's Judgement.

R-Square (Regression) R-Square is a measure of the "Goodness of Fit" of a simple linear regression model to a set of data points. It is directly equivalent to the Coefficient of Determination that shows how much of the total variance in one variable can be explained by the variance in the other variable.
See also Adjusted R-Square.

Glossary | 485

Schedule Maturity Assessment (SMA) A Schedule Maturity Assessment provides a 'health warning' on the maturity of a schedule based on its underpinning assumptions and interdependencies, and takes account of the level of task definition available and historical evidence used.

Secondary Data See Data Type.

Secondary Driver See Estimate Drivers.

Skewness Coefficient The Fisher-Pearson Skewness Coefficient is an expression of the degree of asymmetry of a set of values around their Arithmetic Mean. A positive Skewness Coefficient indicates that the data has a longer tail on the right-hand side, in the direction of the positive axis; such data is said to be Right or Positively Skewed. A negative Skewness Coefficient indicates that the data has a longer tail on the left-hand side, in the direction of the negative axis; such data is said to be Left or Negatively Skewed. Data that is distributed symmetrically returns a Skewness Coefficient of zero.

Slipping and Sliding Technique A technique that compares and contrasts a Bottom-up Monte Carlo Simulation Cost evaluation of Risk, Opportunity and Uncertainty with a holistic Top-down Approach based on Schedule Risk Analysis and Uplift Factors.

Spearman's Rank Correlation Coefficient Spearman's Rank Correlation Coefficient for two variables is a measure of monotonicity of the ranks of the two variables, i.e. the degree to which the ranks move in the same or opposite directions consistently. As such it is a measure of linear or non-linear interdependence.

SPI (Schedule Performance Index – Cost Impact) See Earned Value Management Abbreviations and Terminology.

SPI(t) (Schedule Performance Index – Time Impact) See Earned Value Management Abbreviations and Terminology.

Spreadsheet Validation and Verification See individual terms: Validation and Verification.

Standard Deviation of a Population The Standard Deviation of an entire set (population) of data values is a measure of the extent to which the data is dispersed around its Arithmetic Mean. It is calculated as the square root of the Variance, which is the average of the squares of the deviations of each individual value from the Arithmetic Mean of all the values.

Standard Deviation of a Sample The Standard Deviation of a sample of data taken from the entire population is a measure of the extent to which the sample data is dispersed around its Arithmetic Mean. It is calculated as the square root of the Sample Variance, which is the sum of squares of the deviations of each individual value from the Arithmetic Mean of all the values divided by the degrees of freedom, which is one less than the number of data points in the sample.

Standard Error The Standard Error of a sample's statistic is the Standard Deviation of the sample values of that statistic around the true population value of that statistic. It can be approximated by the dividing the Sample Standard Deviation by the square root of the sample size.

Stanford-B Unit Learning Curve A Stanford-B Unit Learning Curve is a variation of the Crawford Unit Learning Curve that allows for the benefits of prior learning to be expressed in terms of an adjustment to the effective number of cumulative units produced.

Statistics

(1) The science or practice relating to the collection and interpretation of numerical and categorical data for the purposes of describing or inferring representative values of the whole data population from incomplete samples.

486 | Glossary

(2) The numerical values, measures and context that have been generated as outputs from the above practice.

Stepwise Regression Stepwise Regression by Forward Selection is a procedure by which a Multi-Linear Regression is compiled from a list of independent candidate variables, commencing with the most statistically significant individual variable (from a Simple Linear Regression perspective) and progressively adding the next most significant independent variable, until such time that the addition of further candidate variables does not improve the fit of the model to the data in accordance with the accepted Measures of Goodness of Fit for the Regression.

Stepwise Regression by Backward Elimination is a procedure by which a Multi-Linear Regression is compiled commencing with all potential independent candidate variables and eliminating the least statistically significant variable progressively (one at a time) until such time that all remaining candidate variables are deemed to be statistically significant in accordance with the accepted Measures of Goodness of Fit.

Subject Matter Expert's Opinion (Expert Judgement) Expert Judgement is a recognised term expressing the opinion of a Subject Matter Expert (SME).

SV (Schedule Variance – Cost Impact) See Earned Value Management Abbreviations and Terminology.

SV(t) (Schedule Variance – Time Impact) See Earned Value Management Abbreviations and Terminology.

Tertiary Data See Data Type.

Then Year Values 'Then Year Values' are values that have been adjusted to express an expectation of what might be incurred in the future due to escalation or other predicted changes. In some cases, these may be referred to as 'Outturn Year Values'.

Three-Point Estimate See 3-Point Estimate.

Time Series Analysis Time Series Analysis is the procedure whereby a series of values obtained at successive time intervals is separated into its constituent elements that describe and calibrate a repeating pattern of behaviour over time in relation to an underlying trend.

An Additive/Subtractive Time Series Model is one in which the Predicted Value is a function of the forecast value attributable to the underlying Trend plus or minus adjustments for its relative Seasonal and Cyclical positions in time.

A Multiplicative Time Series Model is one in which the Predicted Value is a function of the forecast value attributable to the underlying Trend multiplied by appropriate Seasonal and Cyclical Factors.

Time-Based Learning Curve A Time-based Learning Curve is an empirical relationship which reflects that the time, effort or cost to produce an output from an activity decreases as the elapsed time since commencement of that activity increases.

Time-Constant Learning Curve A Time-Constant Learning Curve considers the output or yield per time period from an activity rather than the time or cost to produce a unit. The model assumes that the output increases due to learning, from an initial starting level, before flattening out asymptotically to a steady state level.

Time-Performance Learning Curve A Time-Performance Learning Curve is an empirical relationship that expresses the reduction in the average time or cost per unit produced per period as a power function of the cumulative number periods since production commenced.

Glossary | 487

Top-down Approach In a top-down approach to estimating, the estimator reviews the overall scope of work in order to identify the major elements of work and characteristics (drivers) that could be estimated separately from other elements. Typically, the estimator might consider a natural flow down through the Work Breakdown Structure (WBS), Product Breakdown Structure (PBS) or Service Breakdown Structure (SBS). The estimate scope may be broken down to different levels of WBS etc as required; it is not necessary to cover all elements of the task at the same level, but the overall project scope must be covered. The overall project estimate would be created by aggregating these high-level estimates. Lower level estimates can be created by subsequent iterations of the estimating process when more definition becomes available, and bridging back to the original estimate.

TRACEability A Basis of Estimate should satisfy the principles of TRACEability:

Transparent – clear and unambiguous with nothing hidden

Repeatable – allowing another estimator to reproduce the same results with the same information

Appropriate – it is justifiable and relevant in the context it is to be used

Credible – it is based on reality or a pragmatic reasoned argument that can be understood and is believable

Experientially-based – it can be underpinned by reference to recorded data (evidence) or prior confirmed experience

Transformation See Mathematical Transformation.

Trusted Source Estimating Method The Trusted Source Method of Estimating is one in which the Estimate Value is provided by a reputable, reliable or undisputed source. Typically, this might be used for low value cost elements. Where the cost element is for a more significant cost value, it would not be unreasonable to request the supporting Basis of Estimate, but this may not be forthcoming if the supporting technical information is considered to be proprietary in nature.

t-Test A t-Test is used for small sample sizes (< 30) to test probability of getting a sample's test statistic (often the Mean), if the equivalent population statistic has an assumed different value. It is also used to test whether two samples could be drawn from the same population.

Tukey's Fences A test for a single Outlier based on the Inter-Quartile Range of the data sample.

Type I Error A Type I Error is one in which we accept a hypothesis we should have rejected.

Type II Error A Type II Error is one in which we reject a hypothesis we should have accepted.

U-Test See Mann-Whitney U-Test.

Uncertainty Uncertainty is an expression of the lack of exactness around a variable, and is frequently quantified in terms of a range of potential values with an optimistic or lower end bound and a pessimistic or upper end bound.

Validation (Spreadsheet or Model) Validation is the process by which the assumptions and data used in a spreadsheet or model are checked for accuracy and appropriateness for their intended purpose.

See also Verification.

Variance of a Population The Variance of an entire set (population) of data values is a measure of the extent to which the data is dispersed around its Arithmetic Mean. It is calculated as the average of the squares of the deviations of each individual value from the Arithmetic Mean of all the values.

488 | Glossary

Variance of a Sample   The Variance of a Sample of data taken from the entire population is a measure of the extent to which the sample data is dispersed around its Arithmetic Mean. It is calculated as the sum of squares of the deviations of each individual value from the Arithmetic Mean of all the values divided by the degrees of freedom, which is one less than the number of data points in the sample.

Verification (Spreadsheet or Model)   Verification is the process by which the calculations and logic of a spreadsheet or model are checked for accuracy and appropriateness for their intended purpose.
See also Validation.

Wilcoxon-Mann-Whitney U-Test   See Mann-Whitney U-Test.

Wright's Cumulative Average Learning Curve   Wright's Cumulative Average Learning Curve is an empirical relationship that expresses the reduction in the cumulative average time or cost of each unit produced as a power function of the cumulative number units produced.

Z-Score   A Z-Score is a statistic which standardises the measurement of the distance of a data point from the Population Mean by dividing by the Population Standard Deviation.

Z-Test   A Z-Test is used for large sample sizes (< 30) to test probability of getting a sample's test statistic (often the Mean), if the equivalent population statistic has an assumed different value.

Legend for Microsoft Excel Worked Example Tables in Greyscale

Cell type	Potential Good Practice Spreadsheet Modelling Colour	Greyscale used in Book	Example of Greyscale Used in Book
Header or Label	Light Grey	Text on grey	Text
Constant	Deep blue	Bold white numeric on black	1
Input	Pale Yellow	Normal black numeric on pale grey	23
Calculation	Pale Green	Normal black numeric on mid grey	45
Solver variable	Lavender	Bold white numeric on mid grey	67
Array formula	Bright Green	Bold white numeric on dark grey	89
Random Number	Pink	Bold black numeric on dark grey	0.0902
Comment	White	Text on white	Text

Index

3-Point Estimate 470; *see also* Accuracy *and* Precision

Accuracy 470; *see also* Precision *and* 3-Point Estimate
Actual Cost (AC) *see* EVM Abbreviations and Terminology
Actual Cost of Work Performed (ACWP) *see* EVM Abbreviations and Terminology
Actual Time (AT) *see* EVM Abbreviations and Terminology
Adjusted R-Square 165, 470; *see also* R-Square
ADORE 470; Assumption 471; Dependency 474; Exclusion 477; Opportunity 482; Risk 484
Alternative Hypothesis *see* Hypothesis Testing
Analogical Estimating Method 10, 470
Analogous Method *see* Analogical Estimating Method
Analogy 113–210, 470 *see* Analogical Estimating Method
ANOVA (Analysis of Variance) 144, 152, 165, 335–6
Approach *see* Estimating Approach
Arithmetic Mean 115–16, 120, 181, 471
Array Formula 185–92
Assumption 471; *see also* ADORE
Asymptote 181, 471
Auxiliary Regression 175–9
Average *see* Arithmetic Mean
Average Absolute Deviation (AAD) 471; *see also* Measures of Dispersion and Shape
Axiom 471

Backward Elimination 193, 197–201, 206–8, 310, 323–30; Stepping Back 325; Stepping Sideways 328; Stepping Up 323; *see also* Stepwise Regression
Base Year Values 471
Basis of Estimate 9, 471
Benford's Law 471; fraud detection 10
Bessel's Correction Factor 471
Beta Distribution 395–7
Binary Switch 313; *see also* Categorical variable
Bisection Technique 260–3
Budget At Completion (BAC) *see* EVM Abbreviations and Terminology
Budget Profiling 30–3
Budgeted Cost of Work Performed (BCWP) *see* EVM Abbreviations and Terminology
Budgeted Cost of Work Scheduled (BCWS) *see* EVM Abbreviations and Terminology

Categorical Variable 133, 302, 313–16, 455–6
cause and effect 114
chance occurrence *see* F-Statistic
Chauvenet's Criterion 139–40, 471; *see also* Outlier Test
Chi-Squared Distribution 321
Chi-Squared Test 472
choice of regression model 208–9; Adjusted R-Square 208–9; R-Square 208–9
Classical Decomposition Method 438–9; Additive/Subtractive 438–9; Convergence Funnels 430–1; Cumulative Average 422–9; Exponential trend 430–6; Linear Regression 416–20; Moving Average 420, 424–9; Multiplicative 440–3 procedure

492 | Index

440–2; seasonal smoothing 420–4; Trend 420–37
Closed Interval 472; *see also* Open Interval
Coefficient of Determination 201, 472; Goodness of Fit 141–9; *see also* R-Square
Coefficient of Variation (CV) 472
Collaborative Working 12; penalty 472
Collinearity *see* Multicollinearity
Comparative Estimating Method *see* Analogical Estimating Method
concave curve *or* function 227, 339, 472; *see also* convex curve
Confidence Interval 288–90, 472, 484; *see also* Prediction Interval
Confidence Level 472
Confidence Limit 473
Continuous Linear Function 26–33
convex curve *or* function 227, 339, 473; *see also* concave curve
Correlation 11, 17–19, 473; Copula 473; Pearson's Linear Correlation Coefficient 482; *see also* Measures of Linearity, Dependence and Correlation; *see also* Linear Correlation *and* Rank Correlation
Correlation Matrix 328–9
Cost Driver *see* Estimate Drivers
Cost Performance Index (CPI) *see* EVM Abbreviations and Terminology
Cost Variance (CV) *see* EVM Abbreviations and Terminology
Covariance 473; *see also* Measures of Dispersion and Shape
Crawford's Unit Learning Curve 473; *see also* Quantity-based Learning Curve
Critical Path 473
Cross-Impact Analysis 473
Cubic regression 368–78
Cumulative Average 21–5, 473; Lag 98; Slope 98–100
Cumulative Average Smoothing 96–111; Cumulative Batch Data 103; Equivalent Units 103–5; Missing values 101–3; Moving Cumulative Average 97; Point Cumulative Average 96; where and when to use it 97–101
Cumulative Distribution Function (CDF) 472; Complementary Cumulative Distribution Function (CCDF) 472; Continuous 474; Discrete 473
Cumulative Smoothing 105–11
Cumulative value 21–6, 26–33, 361–9
Current Year Value 474
Curve Fitting 47

CV: *see* Coefficient of Variation *and* Cost Variance

Data Normalisation 10, 474
data scatter 295; Lognormal Distribution 295; Normal Distribution 289, 295; Student's t-Distribution 289, 295
Data Type 474
deadweight 308
Deciles 474; *see also* Percentiles, Quantiles, Quartiles *and* Quintiles; *see also* Measures of Dispersion and Shape
Degrees of Freedom 474
DeJong Unit Learning Curve 474; *see also* Quantity-based Learning Curve
Delphi Technique 474
Dependency 474; *see also* ADORE
dependent and independent variables 114, 130–1
Descriptive Statistic 474; *see also* Inferential Statistic
Dimmer Switch 313–16, *see also* Categorical variable
Discounted Cash Flow (DCF) 474; Discount Rate 474
Discrete Linear Function 21–6
Dixon's Q-Test 475; *see also* Outlier Test
Drivers *see* Estimate Drivers
Dummy Variable *see* Categorical Variable

Earned Schedule (ES) *see* EVM Abbreviations and Terminology
Earned Value (EV) *see* BCWP (Budget Cost of Work Performed)
Earned Value Analysis 475 *see* EVM Abbreviations and Terminology
Earned Value Management (EVM) 475; *see* EVM Abbreviations and Terminology 475
Equivalent Units 27–33
Estimate 476
Estimate At Completion (EAC) *see* EVM Abbreviations and Terminology
Estimate Drivers 476
Estimate Maturity Assessment (EMA) 9, 476
Estimate To Completion (ETC) *see* EVM Abbreviations and Terminology
Estimating Approach 9, 476; Bottom-up Approach 471; Ethereal Approach 477; Top-down Approach 487
Estimating Method 9, 476
Estimating Metric 476, 477; Factors 477; Rates 484; Ratios 484
Estimating Procedure 4777; Classical Decomposition Method 438–41; Multiple

Index   493

Linear Regression 443; Time Series Analysis 411, 438–41, 462–5
Estimating Process 9, 477
Estimating Technique 9, 477
EVM Abbreviations and Terminology 271–4; AC (Actual Cost) 272–3; ACWP (Actual Cost of Work Performed) 272–3, 475; AT (Actual Time) 273, 475; BAC (Budget At Completion) 272, 475; BCWP (Budget Cost of Work Performed) 272–3, 475; BCWS (Budget Cost of Work Scheduled) 272–3, 475; CPI (Cost Performance Index) 273, 475; CV (Cost Variance) 273, 475; EAC (Estimate At Completion) 272–3, 475; ES (Earned Schedule) 273, 475; ETC (Estimate To Completion) 475; EV (Earned Value) 272–3; PV (Planned Value) 272–3; SPI (Schedule Performance Index - Cost Impact) 273, 476; SPI(t) 273–4; SPI(t) (Schedule Performance Index - Time Impact) 476; SV (Schedule Variance - Cost Impact) 273, 476; SV(t) (Schedule Variance - Time Impact) 273–4, 476
Excel function see Microsoft Excel functions
Excess Kurtosis 477; see also Measures of Dispersion and Shape
Exclusion 477; see also ADORE
Expected Value Technique see Factored Value Technique
Exponential Function 456, 477; combining with Power Function 299; Generalised 249–50, 348–50, 454; Linear Transformation 231–2; offset constant 249–50; reversing the Linear Transformation 233; Standard 230–3, 249–50
Exponential Smoothing 89–96, 477; choice of Damping Factor 92–4; choice of Smoothing Constant 92–4; Damping Factor 89–92; Double Exponential Smoothing 95–6; Smoothing Constant 89–92; Triple Exponential Smoothing 96; where and when to use it 94–5
Exponential trend 237–41
Exponentially Weighted Moving Average see Exponential Smoothing
Extrapolation 47, 114, 477

Factored Value Technique 477
Factors, Rates and Ratios 10
Formulaic Estimating Method see Parametric Estimating Method
F-Statistic 160–71
F-Test 477

Function Type 244–57; Generalised 256–7; logarithm 222–44; Standard 256–7
Function Type Identification 260–3; Bisection Technique 260–3; Generalised 259–60; Microsoft Excel Goal Seek 263–70; Microsoft Solver 263–70; Standard 259–60

Gamma Distribution 321, 399–406
Generalised Exponential Function see Exponential Function
Generalised Extreme Studentised Deviate (GESD) 477; see also Outlier Test
Generalised Logarithmic Function see Logarithmic Function
Generalised Power Function see Power Function
Geometric Mean 441, 478; see also Measures of Central Tendency
Good Practice Spreadsheet Modelling (GPSM) 9, 478
Grubbs' Test 139–40, 478; see also Outlier Test

Halving Rule
Harmonic Mean 478; see also Measures of Central Tendency
Heteroscedasticity 174–9, 478
Homoscedasticity 174–9, 478
Hypothesis Testing 11; Alternative Hypothesis 470; Null Hypothesis 156, 159–60, 162, 482; P-value 163–5, 169

Iglewicz and Hoaglin's Modified Z-Score 478; see also Outlier Test
Indices 66–7, 478; Composite 472; Constant Year Value 473; Current Year Value 474; Laspeyres 479; Nominal Year Value 474; Paasche 482; Then Year Value 486; Outturn Year Value 486
Inferential Statistic 478; see also Descriptive Statistic
Inherent Risk in Spreadsheets (IRiS) 9, 478
Intercept 160
Interdecile Range 478; see also Measures of Dispersion and Shape
Internal Rate of Return (IRR) 478; Discounted Cash Flow (DCF) 478
Interpolation 47, 114, 478
Interquantile Range 478; see also Measures of Dispersion and Shape
Interquartile Range 478; see also Measures of Dispersion and Shape
Interquintile Range 479; see also Measures of Dispersion and Shape

494 | Index

Laspeyres Index 479
Learning Curve 298, 319–21, 351–2, 479; Breakpoint 479; Cost Drivers 479; Segmentation 479; Step-point 479; Learning Exponent 479; Learning Rate and Learning Rate Multiplier 479; *see also* Quantity-based Learning Curves *and* Time-based Learning Curves
Least Squares: Error 94, 115–20, 381–406; Regression 479; Sum-to-Zero Properties 120–1; *see also* Residuals
Least Squares Regression 479
Leptokurtic or Leptokurtotic 479; *see also* Measures of Dispersion and Shape
Line of Best Fit 34–6, 39–41
Linear Correlation 122–3, 473; Actual Cost (AC) 274–5; AT (Actual Time) 275; Earned Value (EV) 274–5; ES (Earned Schedule) 275; Pearson's Linear Correlation Coefficient 482; *see also* Measures of Linearity, Dependence and Correlation; *see also* Rank Correlation
Linear Function 223–5, 479
Linear Regression 154–5; Adjusted R-Square 154–5; Coefficient of Determination 141–9; Coefficient of Variation (CV) 172–4; Common Sense 171–2; Confidence Interval 179–92; degrees of freedom 149–55; F-Statistic 149–56; Goodness of Fit 140–79, 198, 203; Least Squares Error 115–20; Prediction Interval 179–92; Regression through the Origin 157, 162–71; R-Square 154; t-Statistic 156–62; Uncertainty 179–92; White's Test 174–9
Linear Transformation 211–83, 288, 291–5, 299; choosing an option 254–7; Cumulative Value Disaggregation 279–81; EVM Disintegration 271–8; Microsoft Excel 237–42; reversing the Linear Transformation 242–4, 293; *see also* Mathematical Transformation
Linear trend 237–41
Logarithm 212–22, 216–22, 479; Additive property 217–18; Common logs 216–18; Multiplicative property 217–19; Natural logs 216; Reciprocal property 219–20
Logarithmic Function 480; Generalised 245–8, 342–8; Linear Transformation 227–8; offset constant 245–8; reversing the Linear Transformation 230; Standard 225–30, 245–8

Logarithmic trend 237–41
Lower and Upper Data Half Means 117

MAID Technique 58–66, 83
Mann-Whitney U-Test 480
Marching Army Technique 480
Mathematical Transformation: definition 4, 480; *see also* Logarithm
Maximum 480
MDAL (Master Data and Assumptions List) *see* ADORE
Mean Absolute Deviation *see* Average Absolute Deviation (AAD)
Measures of Central Tendency 10, 480
Measures of Dispersion and Shape 10, 480
Measures of Linearity, Dependence and Correlation 11
Median 181, 480; *see also* Measures of Central Tendency
Median Absolute Deviation (MAD) 480; *see also* Measures of Central Tendency
Memoryless Probability Distribution 480
Mesokurtic or Mesokurtotic 480; *see also* Measures of Dispersion and Shape
Method *see* Estimating Method
Microsoft Excel 125–9, 133–6, 157; Data Analysis Add-in 125–9, 133–6, 157; Goal Seek 263–70; *see also* Microsoft Excel Functions *and* Microsoft Excel Solver
Microsoft Excel Functions AVERAGE 50, 170, 420, 463; BETA.DIST 395; CHISQ.DIST.RT 178; CHISQ.INV.RT 178; CORREL 130, 201; EXP 222; F.DIST. RT 152, 167; FORECAST 124, 128; GEOMEAN 432; GROWTH 295; INDEX 128, 136, 155, 201, 259, 296, 324, 362, 430, 433; INTERCEPT 124, 127, 424; LEN 190; LINEST 128, 131, 136–7, 154–5, 201, 259, 296–7, 324, 362; LN 222, 240; LOG 220, 222, 240; LOG10 220, 240; LOGEST 296–7, 430, 433; MAX 76; MEDIAN 420, 424; MIN 76; MINVERSE 188–9; MMULT 188–9; NORM.DIST 388, 395; PEARSON 131, 201; PERCENTILE.EXC 84; PERCENTILE.INC 84; QUARTILE. INC 381; RSQ 201, 262, 267, 323; SLOPE 124, 127, 422; STDEV.S 60, 76; SUM 170; SUMPRODUCT 55, 69, 187, 190, 192, 362, 400, 463; SUMSQ 170, 336, 463; T.DIST.2T 157; T.INV.2T 183–4, 186, 190; TRANSPOSE 188–9, 192; TREND 123–4, 128, 295

Microsoft Excel Solver 263–70, 338–57, 461–7; R-Square 267–9; Sum of Squares Error 267; TRACEability 263, 269
Minimum 480
Minimum Absolute Deviation 116–20
Missing values 47
Mode 181, 481; *see also* Measures of Central Tendency
Model Validation and Verification *see* individual terms: Validation *and* Verification
Monotonic function *and* monotonicity 15, 481
Monte Carlo Simulation 481
Moving Average 48–80, 481; comparison between simple and weighted 58–8; Missing values 49–50, 70–1; Moving Average of Moving Average 66–8; Moving Interval 48, 50–4, 58–66; Interval Difference Technique 58–66, 83; Moving Interval Lag 52–4, 56–8; Profiling Recurring Tasks 68–9; Simple 49, 50–4; Uncertainty Range 71–80; Weighted 49, 54–8; where and when to use them 49–50; *see also* Arithmetic Mean
Moving Geometric Mean 87, 481; *see also* Geometric Mean
Moving Harmonic Mean 87–8, 481; *see also* Harmonic Mean
Moving Interval Lag 79–80
Moving Maximum 71–4, 481; *see also* Maximum
Moving Measures 45–112
Moving Median 81–5, 481; Missing values 84; Moving Interval 83; Uncertainty Range 84–5; *see also* Median
Moving Minimum 71–4, 481; *see also* Minimum
Moving Mode 88
Moving Percentiles 84–5
Moving Standard Deviation 74–80, 481; *see also* Standard Deviation
Multicollinearity 160, 195–8, 303–5, 321, 454, 472, 481
Multi-linear *or* Multiple Linear Regression 129–37; Adjusted R-Square 146–9; Categorical Variable 131–3; degrees of freedom 146; Microsoft Excel 133–7; procedure 443
Multi-Variant Learning 481
Naperian logs *see* Natural logs
Natural logs 216; Euler's Number 216
Net Present Value (NPV) 482; *see also* Discounted Cash Flow

Nominal Year Value *see* Current Year Value
Nonlinear Curve Fitting 399–406; Continuous Probability Distribution 391–9; Discrete Probability Distribution 381–92; Least Squares 381–406
Nonlinear Function combinations 300–22; combined Exponential and Power Transformations 320–2; combined Linear and Logarithmic Transformations 306–7; Multiple Exponential Transformations 312–16; Multiple Logarithmic Transformations 302; Multiple Power Transformations 316–22
Nonlinear Regression 284–379; Exponential Function 291–7; Generalised Nonlinear Function 337–59; Geometric Mean 285–7; Goodness of Fit 287, 333–7; Logarithm 284–300; Logarithmic Function 288–91; Nonlinear Function combinations 300–22; Polynomial Function 359–78; Power Function 298–9; Quadratic Curve 361–8; reversing the Linear Transformation 299–300; Stepwise Regression 322–33
Norden-Rayleigh Curve 482
Normal Distribution 75–6, 79, 386–92, 395–6
Normality Test 479; Jarque-Bera Test 479
Null Hypothesis *see* Hypothesis Testing

offset constant 257–70; finding the best fit 257–70
Open Interval 482; *see also* Closed Interval
Opportunity 482; *see also* Risk *and* Uncertainty
Optimism Bias 482
Ordinary Least Squares Regression *see* Simple Linear Regression
Outlier 482; Regression 138–41
Outlier Tests 11
Outturn Year Value *see* Then Year Value

Paasche Index 482
parameter relevance *see* t-Statistic
Parametric Estimating Method 482
Payback Period 482; *see also* Discounted Cash Flow
Pearson's Linear Correlation Coefficient 482
pedagogical features 2–7; Caveat Augur 5–6; definition 3–4; Excel functions and facilities 6–7; Formula-philes 4–5; Formula-phobes 5; references 7; Worked Examples 6
Peirce's Criterion 482; *see also* Outlier Test

496 | Index

Percentiles 482; *see also* Deciles, Quantiles, Quartiles *and* Quintiles; *see also* Measures of Dispersion and Shape
Planned Value (PV) *see* EVM Abbreviations and Terminology
Platykurtic or Platykurtotic 483; *see also* Measures of Dispersion and Shape
Polynomial trend 242
powers 213–22; additive property 213–16; multiplicative property 213–16
Power Function 483; combined Exponential and Power Transformations 299; Generalised 250–4, 351–7; Linear Transformation 233–6; offset constant 250–3; Reciprocal 253–4; reversing the Linear Transformation 236–7; Standard 233–7, 250–3; *see also* Logarithm
Power trend 237–41
Precision 483; *see also* Accuracy *and* 3-Point Estimate
Prediction Interval 288–90, 295, 299, 319–20, 374–8, 484
predictive power *see* Adjusted R-Square
Primary Data 483; *see also* Data Type
Primary Drivers 483; *see also* Secondary Drivers
Probability Density Function (PDF) 483
Probability Distribution 10–11, 473; Continuous 473; Discrete 475
Probability Mass Function (PMF) 483
Probability of Occurrence 483; *see also* Risk and Opportunity
P-value *see* t-Statistic

Quadratic Curve *and* Equation 21–32, 34–43, 279–81; cumulative difference property 39–43; General Solution 36–7, 361–9; missing values 42–3
Quadratic Mean 136, 483 *see also* Root Mean Square; *see also* Measures of Dispersion and Shape
Quadratic Smoothing *see* Cumulative Smoothing
Qualitative Techniques 10
Quantiles 382–3, 483; *see also* Deciles, Percentiles, Quartiles *and* Quintiles; *see also* Measures of Dispersion and Shape
Quantity-based Learning Curve 11–12, 483; Equivalent Unit Learning 12, 476; Learning Curve Cost Driver Segmentation 11–12; Multi-Variant Learning 12; Unlearning and Re-learning 12
Quartiles 483; *see also* Median; *see also* Deciles, Percentiles, Quantiles, Quartiles *and*

Quintiles; *see also* Measures of Dispersion and Shape
Queueing Theory 483
Quintiles 483; *see also* Deciles, Percentiles, Quantiles *and* Quartiles; *see also* Measures of Dispersion and Shape
Quotations *see* a word (or two) from the wise

Range 483
Rank Correlation 483; Kendall's Tau Rank Correlation Coefficient 479
Real Year Value 484
References 14, 44, 112, 210, 283, 379, 406, 469
Regression Analysis 113–21, 484; Outliers 138–40; *see also* Linear Regression, Multiple Linear Regression, Nonlinear Regression *and* Simple Linear Regression
Regression Confidence Interval *see* Confidence Interval
Regression parameters 157–9; Confidence Interval 157–9
Regression Prediction Interval *see* Prediction Interval
Residual Risk Exposure 484; *see also* Risk Exposure
Residuals 137–40
Risk 484
Risk and Opportunity Ranking Factor 484
Risk Exposure 484; *see also* Residual Risk Exposure
Risk Uplift Factors 484
Rolling Average *see* Moving Average
Rolling Mean *see* Moving Average
Root Mean Square *see also* Quadratic Mean
R-Square (Regression) 165, 167–70, 237–41, 257–9, 279–82, 337–60, 456, 460, 484 *see also* Coefficient of Determination *and* Adjusted R-Square
Schedule Maturity Assessment (SMA) 485
Schedule Performance Index (SPI(t)): Time Impact 276–8; *see also* EVM Abbreviations and Terminology
Schedule Performance Index (SPI): Cost Impact *see* EVM Abbreviations and Terminology
Schedule Variance (SV(t)): Time Impact *see* EVM Abbreviations and Terminology
Schedule Variance (SV): Cost Impact *see* EVM Abbreviations and Terminology
Seasonal variation: Binary Switch 444–5, 450–1, 453, 456
Secondary Data 485; *see also* Data Type *and* Primary Driver

Index | 497

Series overview 1–14; *see also* Volume Overviews

Simple Linear Regression 122–9; Microsoft Excel 123–9

Skewness Coefficient 485

Slipping and Sliding Technique 485

slope 160

Solver *see* Microsoft Excel Solver

Spearman's Rank Correlation Coefficient 485

Spreadsheet Validation and Verification 485; *see also* individual terms: Validation and Verification

Standard Deviation 485; Population 485; Sample 58–66, 485

Standard Error 485

Standard Error of the Estimate (*SEE*) 333–6

Standing Army Technique *see* Marching Army Technique

Stanford-B Unit Learning Curve 485; *see also* Quantity-based Learning Curve

Statistics 485

Stepwise Regression 193–209, 446–8, 449, 454, 486; Backward Elimination 193, 197–201, 206–8, 310, 323–30; Choice of regression model 208–9; Choice of technique 206–8; Forward Selection 193, 201–5, 206–8, 330–3; *see also* Regression Analysis

straight line 19–20; difference property 19–20; linear properties 15–20, 18–19, 19–20; nonlinear properties 21–6, 21–43, 26–33, 34–43, 361–7; parameters 15–17

Student's t-Distribution 180–5

Sum of Squares Error 257

Tertiary Data 486; *see also* Data Type

Then Year Value 486; *see also* Indices

Three-Point Estimate *see* 3-Point Estimate

Tightness of Fit *see* Coefficient of Variation (CV)

Time Series Analysis 407–69, 486; Additive/Subtractive 412–13, 444–52; Classical Decomposition 415–43; Classical Decomposition Method 472; Cyclical variation 408–10; Exponential trend 449–52; Linear trend 444–5, 449–56; Microsoft Excel 461–8; Multi-linear Regression 443–61; Multiplicative 413–15, 449–60; procedure 411, 462–5; Residual variation 409–10; Seasonal Data 81; Seasonal variation 408–11; Trend 408–11, 415–20

Time-Based Learning Curve 11, 486; Time-Constant Learning Curve 486; Time-Performance Learning Curve 486; *see also* Quantity-based Learning Curve

TRACEability 9, 66, 138, 175, 256, 342, 344, 359, 487

Transformation *see* Mathematical Transformation

Trend Analysis 45–112; batch size effects 46, 47–8; characteristics 45–8; Missing values 47–8; non-time-based 47; where and when to use it 45–8

Trend Smoothing 45–112; with Cumulative and Cumulative Average 96–111; with Exponential Smoothing 89–96; with Moving Averages 48–80; with Moving Medians 81–5; with other Moving Measures 85–8; *see also* Trend Analysis

Trendlines: Exponential 237–41; Linear 237–41; Logarithmic 237–41; Polynomial 242; Power 237–41; *see also* Time Series Analysis

Triangular Distribution 395–9

Trusted Source Estimating Method 487

t-Statistic; Adjusted R-Square 160; F-Statistic 156, 160; P-value 157; R-Square 156, 160

t-Test 487

Tukey Fences 139, 487; *see also* Outlier Test

Type I Error (False Positive) 487

Type II Error (False Negative) 487

Uncertainty 487; Cone of Uncertainty 472

Uniform Distribution 75–6, 79, 393–5

Validation 487; *see also* Verification

Variance 487; Population 487; Sample 488

Verification 488; *see also* Validation

Volume I overview 9–10

Volume II overview 10–11

Volume III overview 7–8

Volume IV overview 11–12

Volume V overview 12–13

White's Test 174–9; Chi-Squared Distribution 176, 178; F-Statistic 178; P-value 177–8

Wilcoxon-Mann-Whitney U-Test *see* Mann-Whitney U-Test

word (or two) from the wise 3, 13, 27, 38, 43, 53, 97, 172, 211, 284, 380, 407

Wright's Cumulative Average Learning Curve 488; *see also* Quantity-based Learning Curve

Z-Score 488

Z-Test 488